Strain Gage
Voltage vs. Current
Def. charge

Electricity
and
Electronics

Electricity
and
Electronics

DALE R. PATRICK
STEPHEN W. FARDO

Department of Industrial Education and Technology
Eastern Kentucky University

Prentice-Hall, Inc., Englewood Cliffs, N.J. 07632

Library of Congress Cataloging in Publication Data

Patrick, Dale R.
  Electricity and electronics.

  Includes index.
  1 Electric engineering.   2. Electronics.   I. Fardo,
Stephen W.   II. Title.
TK146.P34 1984   621.3   83-3423
ISBN 0-13-248344-0

Editorial/production supervision and interior design: BARBARA BERNSTEIN
Manufacturing buyer: TONY CARUSO
Cover design: Photo Plus Art, CELINE BRANDES

Printed in the United States of America

10 9 8 7 6

ISBN 0-13-248344-0

PRENTICE-HALL INTERNATIONAL, INC., *London*
PRENTICE-HALL OF AUSTRALIA PTY. LIMITED, *Sydney*
EDITORA PRENTICE-HALL DO BRASIL, LTDA., *Rio de Janeiro*
PRENTICE-HALL CANADA INC., *Toronto*
PRENTICE-HALL OF INDIA PRIVATE LIMITED, *New Delhi*
PRENTICE-HALL OF JAPAN, INC., *Tokyo*
PRENTICE-HALL OF SOUTHEAST ASIA PTE. LTD., *Singapore*
WHITEHALL BOOKS LIMITED, *Wellington, New Zealand*

To
"Our Three Sons"

STEVEN,

BRIAN,

and

DAVID

Contents

ELECTRICAL CIRCUITS 54

MAGNETISM AND ELECTROMAGNETISM 80

ELECTRICAL ENERGY SOURCES 95

ALTERNATING-CURRENT ELECTRICITY 123

ELECTRICAL ENERGY CONVERSION 171

ELECTRICAL INSTRUMENTS 202

ELECTRONIC BASICS 223

14

AMPLIFYING SYSTEMS 353

15

OSCILLATORS 382

16

COMMUNICATION SYSTEMS 409

17

DIGITAL ELECTRONIC SYSTEMS 457

18

ELECTRONIC POWER CONTROL 486

Preface

Electricity and Electronics is an introductory text which explores many aspects of electricity and electronics in a very basic and easy-to-understand way. The key concepts presented in this book are discussed using a "big picture" or "systems" approach which greatly enhances student learning. A great many applications, testing procedures, and operational aspects of equipment and devices are discussed in this comprehensive textbook. The mathematics used is kept to the very minimum and discussed clearly through applications and illustrations.

The book is written primarily for high school students and vocational-technical school students. It would also make an excellent textbook for industrial training programs, home-study courses, or as a reference for college introductory courses in electricity and electronics.

The book is divided into two sections, one dealing with the basics of electricity and the other providing an overview of electronics. The chapters are organized in the following way to aid student understanding:

Introduction
Important terms defined
Major content
Review
Student activities

Definitions of important terms are listed in the beginning of each chapter. The review section at the end of each chapter is used to help the student review the more important topics of each chapter. Electrical safety is stressed in the book. A Teacher's Guide which provides answers to the reviews at the end of each chapter and other helpful aids in teaching is available. The student activities at the end of each chapter stress the practical applications and problem solving involved in electricity and electronics. The expense of the equipment required is kept to a minimum. The student activities are suggested low-cost or no-cost activities which can be done in a school lab or shop. They are very simple and easy to understand.

The authors would like to thank the many companies that provided photographs and technical information for the preparation of the manuscript. The authors also wish to express their sincere thanks to their wives, Kay and Helen, and to their families for their help and understanding during the period of manuscript preparation.

Dale R. Patrick
Stephen W. Fardo

Electricity
and
Electronics

1

Basics of Electricity

Electricity is a very fascinating science that we use in many different ways. It would be difficult to think of the many ways that we use electricity each day. It is very important to have an understanding of electricity.

This chapter deals with the most basic topics in the study of electricity. These include the discovery of electricity, basic electrical systems, energy and power, the structure of matter, electrical charges, static electricity, electrical current, voltage, resistance, and electrical safety. This chapter, as well as all others, has definitions of important terms in the beginning. Preview these terms to gain a better understanding of what is discussed in the chapter. As the chapter is studied, go back to the definitions whenever the need arises. There is also a review at the end of this chapter and all other chapters. These will aid in understanding the material in the chapter. Several student activities are suggested at the end of this and other chapters. They may be completed in the lab or shop.

IMPORTANT TERMS

Before reading the rest of this chapter, review the following terms. These terms provide a basic understanding of some of the concepts that are discussed in the chapter. Also, as other chapters are read, it may be necessary to look back at the terms as the need arises.

Ampere (symbol: I).   The electrical charge movement, which is the basic unit of measurement for current flow in an electrical circuit.

Atom.   The smallest particle that an element can be reduced to and still retain its characteristics.

Atomic number.   The number of particles called protons in the nucleus (center) of an atom.

Closed circuit.   A circuit which forms a complete path so that electrical current can flow through it.

Compound.   The chemical combination of two or more elements to make an entirely different material.

Conductor.   A material that allows electrical current to flow through it easily.

Control.   The part of an electrical system that effects what the system does; a switch to turn a light on and off is a type of control.

Conventional current flow.   Current flow that is assumed to be in a direction from high charge concentration (+) to low charge concentration (−).

Coulomb.   A unit of electrical charge that represents a large number of electrons.

Current.   The movement of electrical charge; the flow of electrons through an electrical circuit.

Electromotive force (EMF).   The "pressure" or "force" which causes electrical current to flow.

Electron.   A small particle which is part of an atom that is said to have a negative (−) electrical charge; electrons are the means by which the transfer of electrical energy can take place.

Electron current flow.   Current flow that is assumed to be in the direction of electron movement from a negative (−) potential to a positive (+) potential.

Electrostatic field.   The space around a charged material in which the influence of the electrical charge is experienced.

Element.   The basic materials that make up all other materials; they exist by themselves (such as copper, hydrogen, carbon) or in combination with other elements (water is a combination of the elements hydrogen and oxygen).

Energy.   The capacity to do work.

Free electrons.   Electrons located in the outer orbit of an atom which are easily removed and result in electrical current flow.

Indicator.   The part of an electrical system that shows if it is on or off or indicates a specific quantity.

Insulator.   A material that offers a high resistance to electrical current flow.

Kinetic energy.   Energy that exists because of movement.

Load.   The part of an electrical system that con-verts electrical energy into another form of energy, such as an electric motor which converts electrical energy into mechanical energy.

Matter.   Any material that makes up the world; anything that occupies space and has weight; can be a solid, a liquid, or a gas.

Metallic bonding.   The method by which loosely held atoms are bound together in metals.

Molecule.   The smallest particle that a compound can be reduced to before being broken down into its basic elements.

Neutron.   A particle in the nucleus (center) of an atom which has no electrical charge or is neutral.

Nucleus.   The core or center part of an atom, which contains protons having a positive charge and neutrons having no electrical charge.

Ohm (Ω).   The unit of measurement of electrical resistance.

Open circuit.   A circuit which has a broken path so that no electrical current can flow through it.

Orbit.   The path along which electrons travel around the nucleus of an atom.

Path.   The part of an electrical system through which electrons travel from a source to a load, such as the electrical wiring used in a building.

Potential energy.   Energy that exists due to position.

Power.   The rate of which work is done.

Proton.   A particle in the center of an atom which has a positive (+) electrical charge.

Resistance (R).   The opposition to the flow of electrical current in a circuit; its unit of measurement is the ohm (Ω).

Semiconductor.   A material that has a value of electrical resistance between that of a conductor and an insulator, and is used to manufacture solid-state devices such as diodes and transistors.

Short circuit.   A circuit that forms a direct path across a voltage source so that a very high and possibly unsafe electrical current flows.

Source.   The part of an electrical system that supplies energy to other parts of the system, such as a battery which supplies energy for a flashlight.

Stable atom. An atom that will not release electrons under normal conditions.

Static charge. A charge on a material which is said to be either positive or negative.

Static electricity. Electricity "at rest" which is due to positive negative electrical charges.

Valence electrons. Electrons in the outer orbit of an atom.

Volt (V). The unit of measurement of electrical potential.

Voltage. Electrical force or "pressure" which causes current to flow in a circuit.

Watt (W). The unit of measurement of electrical power.

Work. The transforming or transferring of energy.

DISCOVERY OF ELECTRICITY

Many years ago, the ancient Greeks discovered that pieces of amber would attract small objects when rubbed with a wool cloth. They called this unknown force *electricity*. The Greeks noticed that these pieces of amber would attract dry leaves or wood shavings. The amber had actually been *charged* due to what we now call static electricity.

Several people have made important discoveries about electricity. Benjamin Franklin is given much credit for the discovery of how electricity can be used. Franklin and other scientists in his time thought that electricity was a force that could have "positive" or "negative" charges.

Today, we know that electricity is produced by the effect of tiny particles called *electrons*. Electrons are too small to be seen with the human eye. They exist in all materials. The production of electricity has become a very important part of our society. Electricity is one of the most important forms of energy in our world. Think of some of the things we could not have without electricity—lights, radios, TV sets, computers, telephones, home appliances—to name only a few. These things, which we take for granted, were not developed until relatively recently. The use of large quantities of electricity began only in the twentieth century. Most electrical items that we use were invented less than 50 years ago. Now electricity is used practically everywhere.

ELECTRICAL SYSTEMS

A simple electrical *system* block diagram and pictorial diagram are shown in Fig. 1-1. Using a block diagram allows a better understanding of electrical

Figure 1-1. Electrical system: (a) block diagram; (b) pictorial diagram.

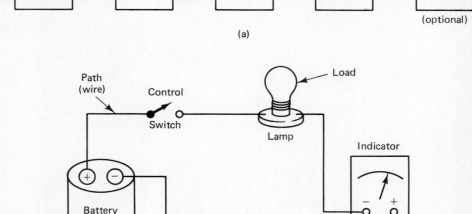

equipment and provides a simple way to "fit pieces together." The system block diagram can be used to simplify many types of electrical circuits and equipment.

The parts of an electrical system are the source, path, control, load, and indicator. The concept of "electrical systems" allows discussion of some complex things in a simplified manner. This method is used to present much of the material in this book—to make it easier to understand.

The *systems* concept serves as a "big picture" in the study of electricity and electronics. In this way a system can be divided into a number of parts. The role played by each part then becomes clearer. It is easy to understand the operation of a complete electrical system. When the function of the system is known, it is easier to understand the operation of each part. In this way, it is possible to see how the pieces of any electrical system fit together to make something that operates.

Each block of an electrical system has an important role to play in the operation of the system. Hundreds and even thousands of components are sometimes needed to form an electrical system. Regardless of the complexity of the system, each block must achieve its function when the system operates.

The *source* of an electrical system provides electrical energy for the system. Heat, light, chemical, and mechanical energy may be used as sources of electrical energy. Figure 1-2 shows some sources of electrical energy.

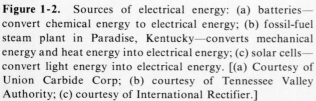

Size "C"   Size "D"   Size "AA"   Size "AAA"   (b)   (c)

(a)

Figure 1-2. Sources of electrical energy: (a) batteries—convert chemical energy to electrical energy; (b) fossil-fuel steam plant in Paradise, Kentucky—converts mechanical energy and heat energy into electrical energy; (c) solar cells—convert light energy into electrical energy. [(a) Courtesy of Union Carbide Corp; (b) courtesy of Tennessee Valley Authority; (c) courtesy of International Rectifier.]

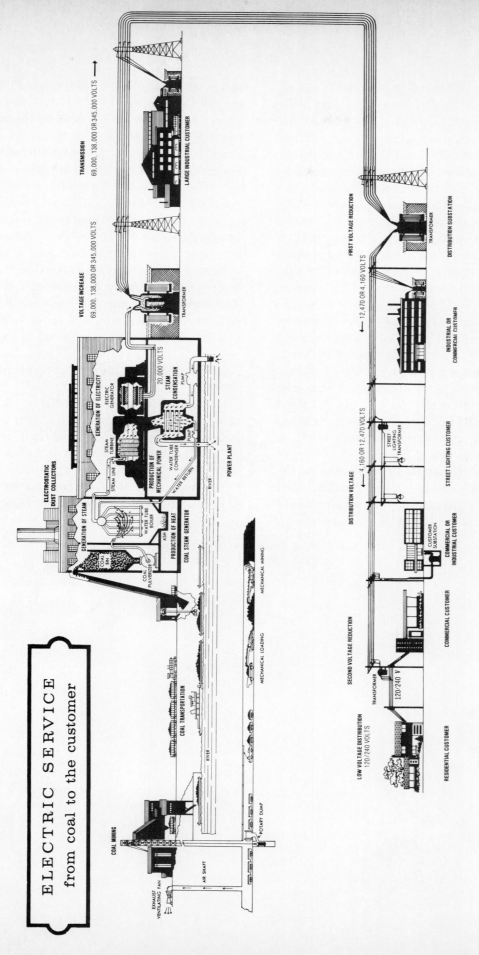

Figure 1-3. Distribution path for electrical power from its source to where it is used. (Courtesy of Kentucky Utilities Co.)

The *path* of an electrical system is simple compared with other system parts. This part of the system provides a path for the transfer of electrical energy. It starts with the energy source and continues through the load. In some cases this path is an electrical wire. In other systems a complex supply line is placed between the source and the load and a return line from the load to the source is used. There are usually many paths within a complete electrical system. Figure 1-3 shows the distribution path of electrical power from its source to where it is used.

The *control* section of an electrical system is the most complex part of the system. In its simplest form, control is achieved when a system is turned on or off. Control of this type takes place anywhere between the source and the load. The term *full control* is used to describe this operation. A system may also use some type of partial control. *Partial control* causes some type of operational change in the system other than turning it on or off. A change in the amount of electrical current is a type of change that is achieved by partial control. Some common control devices are shown in Fig. 1-4.

The *load* of an electrical system is the part or group of parts which do some type of work. Work occurs when energy goes through a transformation or change. Heat, light, and mechanical motion are forms of work produced by loads. Much of the energy produced by the source is changed to another type by the load. The load is usually the most obvious part of the system because of the work it does. An example is a light bulb which produces light. Some common loads are shown in Fig. 1-5.

The *indicator* of an electrical system displays a particular operating condition. In some systems the indicator is an optional part that is not really needed. In other systems it is necessary for proper operation. In some cases adjustments are made by using indicators. In other cases an indicator is attached temporarily to the system to make measurements. Test lights, panel meters, oscilloscopes, and chart recorders are common indicators used in electrical systems. Electrical indicators are shown in Fig. 1-6.

Examples of Electrical Systems

Nearly everyone has used a flashlight. This device is designed to serve as a light source. Flashlights are a simple type of electrical system. Figure 1-7 is a cutaway drawing of a flashlight with each part shown.

The battery of a flashlight serves as the energy *source* of the system. Chemical energy in the battery is changed into electrical energy to cause the system

Figure 1-4. Common control devices: (a) potentiometers—partial control; (b) switches—full control. [(a) Courtesy of Allen-Bradley Co.]

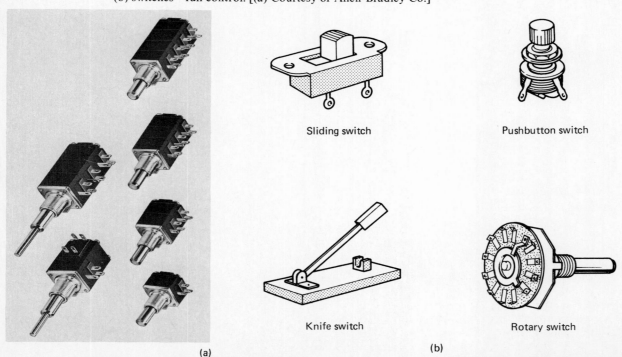

Sliding switch

Pushbutton switch

Knife switch

Rotary switch

(a)

(b)

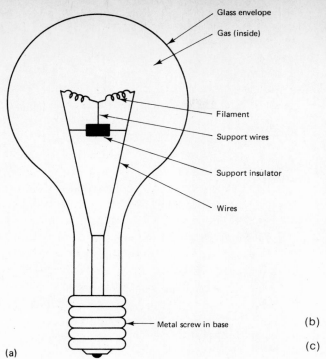

Glass envelope

Gas (inside)

Filament

Support wires

Support insulator

Wires

Metal screw in base

(a)

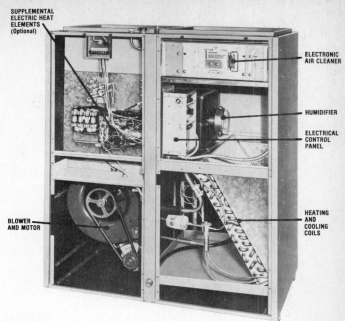

SUPPLEMENTAL
ELECTRIC HEAT
ELEMENTS
(Optional)

ELECTRONIC
AIR CLEANER

HUMIDIFIER

ELECTRICAL
CONTROL
PANEL

HEATING
AND
COOLING
COILS

BLOWER
AND MOTOR

(b)

(c)

Figure 1-5. Common electrical loads: (a) incandescent light bulb—converts electrical energy to light energy; (b) heat pump home heating system—converts electrical energy to heat energy; (c) electric motor—converts electrical energy into mechanical energy. [(b) Courtesy of Williamson Co.; (c) courtesy of Delco Products Division—General Motors Corp.]

(a)

(b)

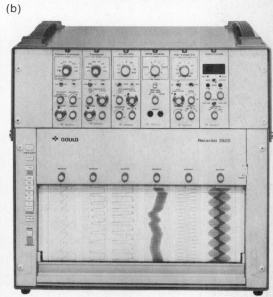

Figure 1-6. Electrical indicators: (a) meter which measures electrical quantities; (b) chart recorder which makes a permanent record of some quantity. [(a) Courtesy of Triplett Corp.; (b) courtesy of Gould, Inc., Instruments Division, Cleveland, Ohio.]

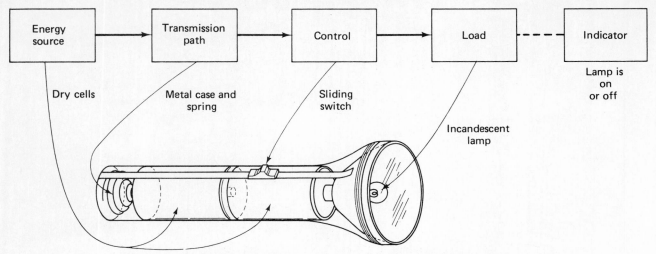

Figure 1-7. Cutaway drawing of a flashlight. (From Patrick/Fardo, *Energy Management and Conservation*, Prentice-Hall, Englewood Cliffs, N.J., 1982.)

to operate. The energy source of a flashlight—the battery—may be thrown away. Batteries are replaced periodically when they lose their ability to produce energy.

The *path* of a flashlight is a metal case or a small metal strip. Copper, brass, or plated steel are used as paths.

The *control* of electrical energy in a flashlight is achieved by a slide switch or a pushbutton switch. This type of control closes or opens the path between the source and the load device. Flashlights have only a means of full control, which is operated manually by a person.

The *load* of a flashlight is a small lamp bulb. When electrical energy from the source passes through the lamp, the lamp produces a bright glow. Electrical energy is then changed into light energy. A certain amount of work is done by the lamp when this energy change takes place.

Flashlights do not use an *indicator* as part of the system. Operation is indicated, however, when the lamp produces light. The load of this system also acts as an indicator. In many electrical systems, the indicator is an optional part.

A more complex example of a system is the electrical power system that supplies energy to buildings such as our homes. Figure 1-8 shows a sketch of a simple electrical power system.

The energy *source* of an electrical power system is much more complex than that of a flashlight. The source of energy may be coal, oil, natural gas, atomic fuel, or moving water. This type of energy is needed to produce mechanical energy. The mechanical energy develops the motion needed to turn a turbine. Large generators are then rotated by the turbine to produce electrical energy. The energy conversion of this system is very complex from start to finish. The functions of the parts of the system remain the same regardless of complexity.

In an electrical power system, the *path* is composed of many electrical conductors. Copper wire and aluminum wire are ordinarily used as conductors. Metal, water, the earth, and the human body can all be made paths for electrical energy transfer.

The *control* function of an electrical power system is performed in many different ways. *Full-control* devices include switches, circuit breakers, and fuses. *Partial control* of an electrical power system is achieved in many ways. For example, transformers are used throughout the system. These partial-control devices are designed to change the amount of voltage of the system.

The *load* of an electrical power system includes everything that uses electrical energy from the source. The total load of an electrical power system changes continually. The load is the part of the system that actually does work. Motors, lamps, electrical ovens, welders, and power tools are some common load devices. Loads are classified according to the type of work they produce. These include light, heat, and mechanical loads.

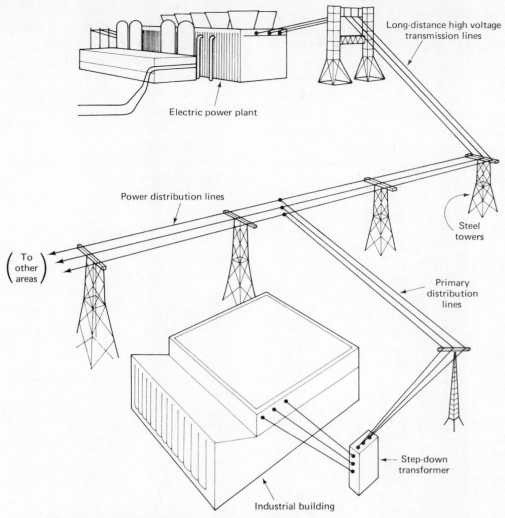

Figure 1-8. Electrical power system. (From Patrick/Fardo, *Energy Management and Conservation*, Prentice-Hall, Englewood Cliffs, N.J., 1982.)

The *indicator* of an electrical power system is designed to show the presence of electrical energy. It may also be used to measure electrical quantities. Panel-mounted meters, oscilloscopes, and chart recording instruments are some of the indicators used in this type of system. Indicators of this type provide information about the operation of the system.

The systems concept is a method that may be used to study electricity and electronics. This method provides a common organizational plan that applies to most electrical systems. The energy source, path, control, load, and indicator are basic to all systems. An understanding of a basic electrical system helps to overcome some of the problems involved in understanding complex systems.

ENERGY, WORK, AND POWER

An understanding of the terms, "energy," "work," and "power" is necessary in the study of electricity and electronics. The first term, *energy*, means the capacity to do work. For example, the capacity to light a light bulb, to heat a home, or to move something requires energy. Energy exists in many forms, such as electrical, mechanical, chemical, and heat. If energy exists because of the movement of some item, such as a ball rolling down a hill, it is called *kinetic energy*. If it exists because of the position of something, such as a ball at the top of the hill but not yet rolling, it is called *potential energy*. Energy has become one of the most important factors in our society.

A second important term is *work*. Work is the transferring or transforming of energy. Work is done when a force is exerted to move something over a certain distance against opposition. Work is done when a chair is moved from one side of a room to the other. An electrical motor used to drive a machine performs work. When force is applied to open a door, work is performed. Work is performed when motion is accomplished against the action of a force that tends to oppose the motion. Work is also done each time energy changes from one form into another.

A third important term is *power*. Power is *the rate at which work is done*. It considers not only the work that is performed, but the amount of time in which the work is done. For instance, electrical power is the rate at which work is done as electrical current flows through a wire. Mechanical power is the rate at which work is done as an object is moved against opposition over a certain distance. Power is either the rate of production or the rate of use of energy. The *watt* is the unit of measurement of electrical power.

STRUCTURE OF MATTER

In the study of electricity and electronics, it is necessary to understand why electrical energy exists. To gain this understanding, let's look first at how certain natural materials are made. It is then easier to see why electrical energy exists.

Let's begin with some basic scientific terms. These terms are often used in the study of chemistry. They are also very important in the study of electricity. First, we say that *matter* is anything that occupies space and has weight. Matter can be either a *solid*, a *liquid*, or a *gaseous* material. Solid matter includes such things as metal and wood; liquid matter is exemplified by water or gasoline; and gaseous matter includes such things as oxygen and hydrogen. Solids can be converted into liquids. Liquids can be made into gases. For example, water can be a solid in the form of ice. Water can also be a gas in the form of steam. The difference is that the particles of which they are made move when heated. As they move, they strike one another, causing them to move farther apart. Ice is converted into a liquid by adding heat. If heated to a high temperature, water becomes a gas (steam). All forms of matter

exist in their most familiar form because of the amount of heat they contain. Some materials require more heat than others to become a liquid or a gas. However, all materials can be made to change from a solid to a liquid or from a liquid to a gas if enough heat is added. Also, these materials can change into liquids or solids if heat is taken from them.

The next important term in the study of the structure of matter is the *element*. An element is considered to be the basic material that makes up all matter. Materials such as hydrogen, aluminum, copper, iron, and iodine are a few of the over 100 elements known to exist. A table of elements is shown in Appendix 1. Some elements exist in nature and some are human-made. Everything around us is made of elements.

There are many more materials in our world than there are elements. Other materials are made by combining elements. A combination of two or more elements is called a *compound*. For example, water is a compound made from the elements hydrogen and oxygen. Salt is made from sodium and chlorine.

Another important term is *molecule*. A molecule is said to be the smallest particle that a compound can be reduced to before breaking down into its basic elements. For example, one molecule of water has two hydrogen atoms and one oxygen atom.

In an even deeper look into the structure of matter, it is believed that there are particles called atoms. Within these atoms are the forces that cause electricity to exist. An atom is considered to be the smallest particle that an element can be reduced to and still have the properties of that element. If an atom were broken down any further, the element would no longer exist. The smallest particles that are found in all atoms are called *electrons*, *protons*, and *neutrons*. Elements differ from one another on the basis of the amounts of these particles found in their atoms. The relationship of matter, elements, compounds, molecules, atoms, electrons, protons, and neutrons is shown in Fig. 1-9.

The simplest atom, hydrogen, is shown in Fig. 1-10. The hydrogen atom has a center part called a *nucleus*, which has one *proton*. A proton is a particle which is said to have a positive (+) charge. The hydrogen atom has one *electron*, which orbits around the nucleus of the atom. The electron is said to have a negative (−) charge. Most atoms also have neutrons in the nucleus. A *neutron* has neither a

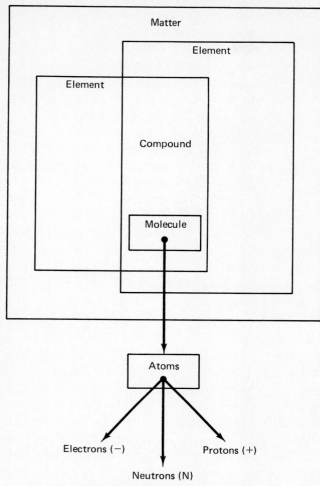

Figure 1-9. Structure of matter.

positive nor a negative charge and is considered neutral. A carbon atom is shown in Fig. 1-11. A carbon atom has six protons (+), six neutrons (N), and six electrons (−). The protons and the neutrons are in the nucleus and the electrons *orbit* around the nucleus. The carbon atom has two orbits or circular paths. In the first orbit, there are two electrons. The other four electrons are in the second orbit.

Turn to the table of elements in Appendix 1. Notice there are a different number of protons in the nucleus of each atom. This causes each element to be different. For example, hydrogen has 1 proton, carbon has 6, oxygen has 8, and lead has 82. The number of protons that each atom has is called its *atomic number.*

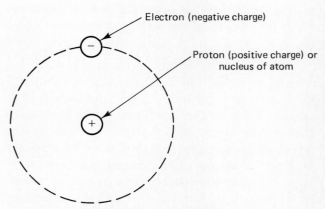

Figure 1-10. Hydrogen atom.

Figure 1-11. Carbon atom.

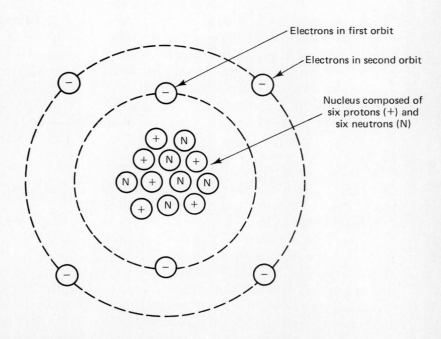

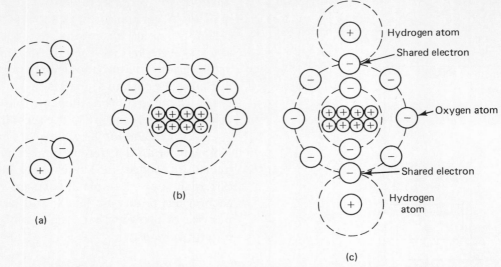

Figure 1-12.   Water formed by combining hydrogen and oxygen: (a) hydrogen atoms; (b) oxygen atom; (c) water molecule.

The nucleus of an atom contains protons (+) and neutrons (N). Since neutrons have no charge and protons have positive charges, the nucleus of an atom has a positive charge. Protons are believed to be about one-third the diameter of electrons. The *mass* or weight of a proton is thought to be over 1800 times more than that of an electron. Electrons move easily in their orbits around the nucleus of an atom. It is the movement of electrons that causes electrical energy to exist.

Atoms are similar in many ways to a miniature solar system. An atom has a nucleus in the center with electrons revolving around it in orbits. In a solar system, planets orbit around the sun. The electrons of an atom are negative (−) charges which are held in orbit by the positive (+) charges of the nucleus. Atoms have the same number of protons (+) and electrons (−). The atom itself has no charge. An atom, unlike the solar system, can have more than one electron in an orbit. Each planet in the solar system has its own orbit. Atoms that have more electrons have more orbits.

An exact pattern is thought to be followed in the placement of electrons of an atom. The first orbit, or *shell*, contains up to *2* electrons. The next shell contains up to *8* electrons. The third contains up to *18* electrons. Eighteen is the largest quantity any shell can contain. New shells are started as soon as shells nearer the nucleus have been filled with the maximum number of electrons. An atom with an incomplete outer shell wants to fill that shell. Atoms with incomplete outer shells are very *active*. When two unlike atoms with incomplete outer shells come together, they try to *share* their outer electrons. When their combined outer electrons are enough to make up one complete shell, *stable* atoms are formed. For example, oxygen has eight electrons, two in the first shell and six in its outer shell. There is room for eight electrons in the outer shell. Hydrogen has one electron in its outer shell. When two hydrogen atoms come near, oxygen combines with the hydrogen atoms by sharing the electrons of the two hydrogen atoms. Water is formed as shown in Fig. 1-12. All the electrons are then bound tightly together and a very stable water molecule is formed. The electrons in the incomplete outer shell of an atom are known as *valence electrons*. They are the only electrons that will combine with other atoms to form compounds. They are also the only electrons used to cause electric current to flow. It is for this reason that it is necessary to understand the structure of matter.

ELECTRICAL CHARGES

In the preceding section, the positive and negative charges of particles called protons and electrons were studied. Protons and electrons are parts of atoms that make up all things in our world. The positive charge of a proton is similar to the negative charge of an electron. However, a positive charge is the *opposite* of a negative charge. These charges are called *electrostatic charges*. Figure 1-13 shows how

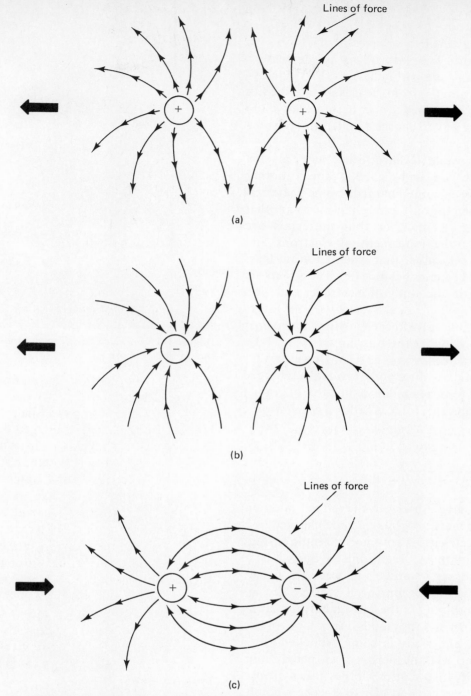

Figure 1-13. Electrostatic charges: (a) positive charges repel; (b) negative charges repel; (c) positive and negative charges attract.

electrostatic charges affect each other. Each charged particle is surrounded by an *electrostatic field*.

The effect that electrostatic charges have on each other is very important. They either repel (move away) or attract (come together) each other. This action is as follows:

1. Positive charges repel each other [see Fig. 1-13(a)].

2. Negative charges repel each other [see Fig. 1-13(b)].

3. Positive and negative charges attract each other [see Fig. 1-13(c)].

Therefore, it is said that *like charges repel* and *unlike charges attract*.

The atoms of some materials can be made to gain or lose electrons. The material then becomes charged. One way to do this is to rub a glass rod with a piece of silk cloth. The glass rod loses electrons (−), so it now has a positive (+) charge. The silk cloth pulls away electrons (−) from the glass. Since the silk cloth gains new electrons, it now has a negative (−) charge. Another way to charge a material is to rub a rubber rod with fur.

It is also possible to charge other materials. If a charged rubber rod is touched against another material, it may become charged. Some materials are charged when they are brought close to another charged object. Remember that materials are charged due to the movement of electrons and protons. Also, remember that when an atom loses electrons (−), it becomes positive (+). These facts are very important in the study of electricity and electronics.

Charged materials affect each other due to lines of force. Try to visualize these as shown in Fig. 1-13. These imaginary lines cannot be seen. However, they exert a force in all directions around a charged material. Their force is similar to the force of gravity around the earth. This force is called a *gravitational field*.

STATIC ELECTRICITY

Most people have observed the effect of *static electricity*. Whenever objects become charged, it is due to static electricity. A common example of static electricity is lightning. Lightning is caused by a difference in charge (+ and −) between the earth's surface and the clouds during a storm. The arc produced by lightning is the movement of charges between the earth and the clouds. Another common effect of static electricity is being "shocked" by a doorknob after walking across a carpeted floor. Static electricity also causes clothes taken from a dryer to cling together and hair to stick to a comb.

There are two types of electricity: (1) static electricity, and (2) current electricity. The use of electricity today depends mostly upon *current electricity*. Current electricity is the controlled movement of electrical charges. Current electricity is discussed in detail in this book. Static electricity has some practical uses, such as the Van de Graff generator shown in Fig. 1-14 used in nuclear research. A Van de Graff generator is a piece of equipment which produces electrical charges. A

Figure 1-14. Van de Graaff generator. (Courtesy of Westinghouse Electric Corp.)

Figure 1-15. Electrostatic generator—Van de Graaff type. (Courtesy of Sargent-Welch Scientific Co.)

small electrostatic generator is shown in Figure 1-15. Electrical charges are used to filter dust and soot in devices called electrostatic filters. Electrostatic precipitators are used in power plants to filter the exhaust gas that goes into the air. Static electricity is

Conductors

A material through which current flows is called a *conductor*. A conductor passes electrical current very easily. Copper and aluminum wire are commonly used as conductors. Conductors are said to have low *resistance* to electrical current flow. Conductors usually have three or fewer electrons in the outer orbit of their atoms. Remember that the electrons of an atom orbit around the nucleus. Many metals are electrical conductors. Each metal has a different ability to conduct electrical current. For example, silver is a better conductor than copper. Silver is too expensive to use in large amounts. Aluminum does not conduct electrical current as well as copper. The use of aluminum is common, since it is cheaper and lighter than other conductors. Copper is used more than any other conductor. Materials with only one outer orbit or valence electron (gold, silver, copper) are the best conductors.

Insulators

There are some materials that do not allow electrical current to flow easily. The electrons of materials that are insulators are difficult to release. Some insulators have their outer orbits filled with eight electrons. Others have outer orbits that are over half-filled with electrons. The atoms of materials that are insulators are said to be stable. Insulators have high resistance to the movement of electrical current. Some examples of insulators are plastic and rubber. Figure 1-17 shows some types of insulators. A test facility for large electrical power line insulators is shown in Fig. 1-18.

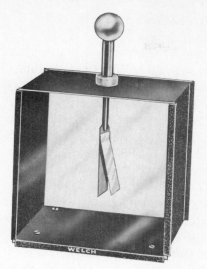

Figure 1-16. Electroscope. (Courtesy of Sargent-Welch Scientific Co.)

also used in the manufacture of sandpaper and in the spray painting of automobiles. A device called an electroscope, shown in Fig. 1-16, is used to detect a negative or positive charge.

ELECTRICAL CURRENT

Static electricity is caused by *stationary* charges. However, electrical current is the *motion* of electrical charges from one point to another. Electrical current is produced when electrons (−) are removed from their atoms. Some electrons in the outer orbits of the atoms or certain elements are easy to remove. A force or pressure applied to a material causes electrons to be removed. The movement of electrons from one atom to another is called *electrical current flow*.

Figure 1-17.   Some types of insulators. (Courtesy of The Foxboro Co.)

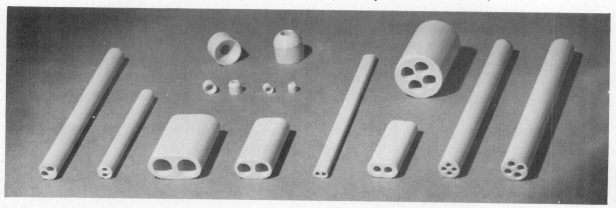

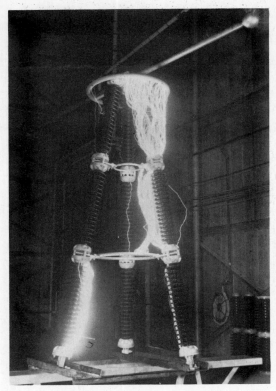

Figure 1-18. Testing electrical power line insulators. (Courtesy of Lapp Insulator, Leroy, N.Y.)

Semiconductors

Materials called semiconductors have become very important in electronics. Semiconductor materials are not conductors and not insulators. They will not conduct electrical current easily and are not good insulators. Their classification also depends on the number of electrons their atoms have in the *outer* orbit. Semiconductors have *four* electrons in their outer orbits. Remember that conductors have outer orbits *less* than half-filled and insulators ordinarily have outer orbits *more* than half-filled. Figure 1-19 compares conductors, insulators, and semiconductors. Some common types of semiconductor materials are silicon, germanium, and selenium. Semiconductors are discussed in detail in Chapter 9.

Current Flow

The usefulness of electricity is due to what is called electrical current flow. Current flow is the movement of electrical charges along a conductor. Static electricity or electricity at rest has some practical uses due to electrical charges. Electrical current flow allows us to use electrical energy to do many types of work.

The movement of outer orbit electrons of conductors produces electrical current. The electrons on the outer orbit of the atoms of a conductor are called *free electrons*. Energy released by these electrons as they move allows work to be done. As more electrons move along a conductor, more energy is released. This is called an increased electrical current flow. The movement of electrons along a conductor is shown in Fig. 1-20.

To understand how current flow takes place, it is necessary to know about the atoms of conductors. Conductors, such as copper, have atoms which are loosely held together. Copper is said to have atoms connected together by *metallic bonding*. A copper atom has one outer orbit electron which is loosely held to the atom. These atoms are so close together that their outer orbits overlap each other. Electrons can easily move from one atom to another. In any conductor, the outer orbit electrons continually move in a random manner from atom to atom.

The random movement of electrons does *not* result in current flow. Electrons must move in the same direction to cause current flow. If electrical charges are placed on each end of a conductor, the free electrons move in one direction. Figure 1-20 shows current flow through a conductor caused by negative (−) and positive (+) electrical charges. Current flow takes place because there is a *difference* in the charges at each end of the conductor. Remember that like charges *repel* and unlike charges *attract*.

When an electrical charge is placed on each end of the conductor, the free electrons move. Free electrons have a negative (−) charge, so they are repelled by the negative (−) charge on the left of Fig. 1-20. The free electrons are attracted to the positive (+) charge on the right. The free electrons move to the right from one atom to another. If the charges on each end of the conductor are increased, more free electrons will move. This increased movement will cause more electrical current flow.

Current flow is the result of electrical energy caused as electrons change orbits. This impulse moves from one electron to another. When one electron (−) moves out of its orbit, it enters another atom's orbit. An electron (−) is then *repelled* from that atom. This action goes on in all parts of a conductor. Remember that electrical current flow is a *transfer* of energy.

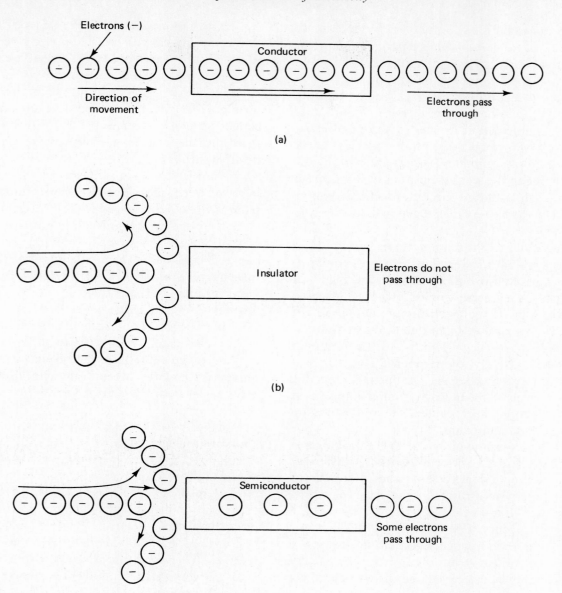

Figure 1-19. Comparison of (a) conductors, (b) insulators, and (c) semiconductors.

Figure 1-20. Current flow through a conductor.

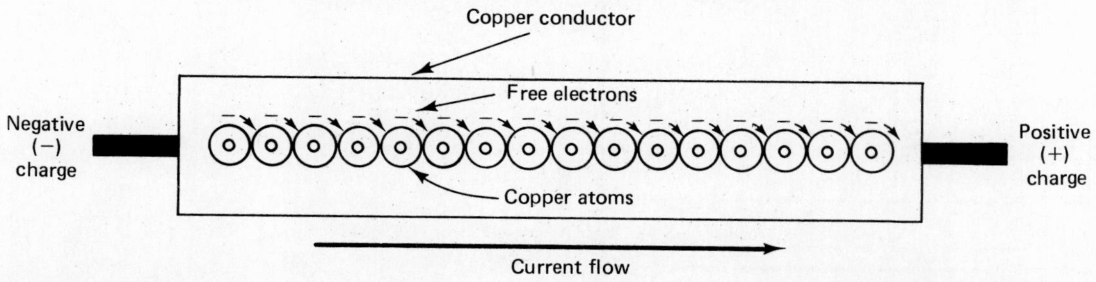

ELECTRICAL CIRCUITS. Current flow takes place in electrical circuits. A *circuit* is a path or conductor for electrical current flow. Electrical current flows only when it has a complete or *closed-circuit* path. There must be a *source* of electrical energy to cause current to flow along a closed path. Figure 1-21 shows a battery used as an energy source to cause current to flow through a light bulb. Notice that the path or circuit is *complete*. Light is given off by the light bulb due to the work done as electrical current flows through a closed circuit. Electrical energy produced by the battery is changed to light energy in this circuit.

Electrical current cannot flow if a circuit is open. An *open circuit* does not provide a complete path for current flow. If the circuit of Fig. 1-21 became open, no current would flow. The light bulb would not glow. Free electrons of the conductor would no longer move from one atom to another. An example of an open circuit is when a light bulb "burns out." Actually, the filament (part that produces light) has become open. The open filament of a light bulb stops current flow from the source of electrical energy. This causes the bulb to stop burning or producing light.

Another common circuit term is called a *short circuit*. In electrical work, a short circuit can be very harmful. A short circuit occurs when a conductor connects directly across the terminals of an electrical energy source. If a wire is placed across a battery, a short circuit occurs. For safety purposes, a short circuit should never happen. Short circuits cause *too much* current to flow from the source. The battery would probably be destroyed and the wire could get hot or possibly melt due to a short circuit.

DIRECTION OF CURRENT FLOW. Electrical current flow is the movement of electrons along a conductor. Electrons are negative charges. Negative charges are *attracted* to positive charges and *repelled* by other negative charges. Electrons move from the negative terminal of a battery to the positive terminal. This is called *electron current flow*. Electron current flow is in a direction of electron movement from negative to positive through a circuit.

Another way to look at electrical current flow is in terms of charges. Electrical charge movements is from an area of high charge to an area of low charge. A high charge can be considered positive and a low charge, negative. Using this method, an electrical charge is considered to move from a high charge to a low charge. This is called *conventional current flow*. Conventional current flow is the movement of charges from positive to negative.

Electron and conventional current flow should not be confusing. They are two different ways of looking at current flow. One deals with electron movement and the other deals with charge movement. In this book, *electron current flow* is used.

AMOUNT OF CURRENT FLOW (THE AMPERE). The amount of electrical current that flows through a circuit depends on the number of electrons that pass a point in a certain time. The *coulomb* is a unit of measurement of electrical current. In electricity, many units of measurement are used. A coulomb is a large quantity of electrons. It is estimated that one coulomb is 6,280,000,000,000,000,000 electrons (6.28×10^{18} in scientific notation form). Since electrons are very small, it takes many to make one unit of measurement. When one coulomb passes a point

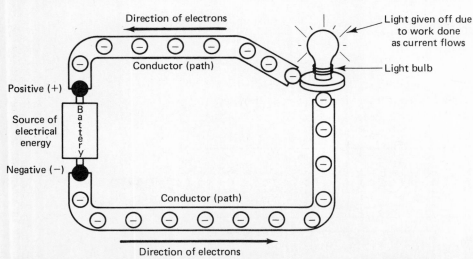

Figure 1-21. Current flow in a closed circuit.

on a conductor in one second, one *ampere* of current flows in the circuit. The unit is named for A. M. Ampere, an eighteenth-century scientist who studied electricity. Current is commonly measured in units called *milli*amperes and *micro*amperes. These are smaller units of current. A milliampere is one thousandth (1/1000) of an ampere and a micro-ampere is one millionth (1/1,000,000) of an ampere. Electrical units of measurement such as these are discussed in detail in Chapter 2.

CURRENT FLOW COMPARED TO WATER FLOW. An electrical circuit is a path in which an electric current flows. Current flow is similar to the flow of water through a pipe. Electric current and water flow can be compared in some ways. Water flow is used to show how current flows in an electrical circuit. When water flows in a pipe, something causes it to move. The pipe offers opposition or resistance to the flow of water. If the pipe is small, it is more difficult for the water to flow.

In an electric circuit, current flows through wires (conductors). The wires of an electric circuit are similar to the pipes through which water flows. If the wires are made of a material that has high resistance, it is difficult for current to flow. The result is the same as water flow through a pipe which has a rough surface. If the wires are large, it is easier for current to flow in an electrical circuit. In the same way, it is easier for water to flow through a large pipe. Electrical current and water flow are compared in Fig. 1-22. Current flows from one place to another

in an electrical circuit. Similarly, water that leaves a pump moves from one place to another. The rate of water flow through a pipe is measured in gallons per minute. In an electric circuit, the current is measured in *amperes*. The flow of electrical current is measured by the number of *coulombs* which pass a point on a conductor each second. A gallon of water is a certain number of atoms of water. A coulomb is a certain number of electrons. A current flow of one coulomb per second makes one ampere of current flow.

ELECTRICAL FORCE (VOLTAGE)

Water pressure is needed to force water along a pipe. Similarly, electrical pressure is needed to force current along a conductor. Water pressure is usually measured in pounds per square inch (*psi*). Electrical pressure is measured in *volts*. If a motor is rated at 120 volts, it requires 120 volts of electrical pressure applied to the motor to force the proper amount of current through it. More pressure would increase the current flow and less pressure would not force enough current to flow. The motor would not operate properly with too much or too little voltage. Water pressure produced by a pump causes water to flow through pipes. Pumps produce pressure that causes water to flow. The same is true of an electrical energy source. A source, such as a battery or generator, produces current flow through a circuit.

Figure 1-22. Comparison of electrical current and water flow: (a) water pipes; (b) electrical conductors.

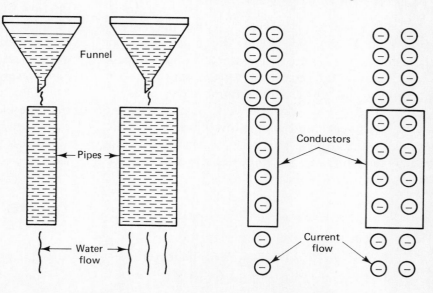

(a)   (b)

As voltage is increased, the amount of current in a circuit is also increased. Voltage is also called *electromotive force* (EMF).

RESISTANCE

The opposition to current flow in electrical circuits is called *resistance*. Resistance is not the same for all materials. The number of free electrons in a material determines the amount of opposition to current flow. Atoms of some materials give up their free electrons easily. These materials offer low opposition to current flow. Other materials hold their outer electrons and offer high opposition to current flow.

Electric current is the movement of free electrons in a material. Electric current needs a source of electrical pressure to move the free electrons through a material. An electric current will not flow if the source of electrical pressure is removed. A material will not release electrons until enough force is applied. With a constant amount of electrical force (voltage) and more opposition (resistance) to current flow, the number of electrons flowing (current) through the material is smaller. With constant voltage, current flow is increased by decreasing resistance. Decreased current results from more resistance. By increasing or decreasing the amount of resistance in a circuit, the amount of current flow can be changed.

Materials that are good conductors have many free electrons. Insulating materials do not easily give up the electrons in the outer orbits of their atoms. Metals are the best conductors, with copper, aluminum, and iron wire being the most common. Carbon and water are two nonmetals which are also conductors. Materials such as glass, paper, rubber, ceramics, and plastics are common types of insulators.

Even very good conductors have some resistance which limits the flow of electric current through them. The resistance of any material depends on four factors:

1. The material of which it is made
2. The length of the material
3. The cross-sectional area of the material
4. The temperature of the material

The material of which an object is made affects its resistance. The ease with which different materials give up their outer electrons is very important in determining resistance. Silver is an excellent conductor of electricity. Copper, aluminum, and iron have more resistance, but are more commonly used since they are less expensive. All materials conduct an electric current to some extent, even though some (insulators) have very, very high resistance. Length also affects the resistance of a conductor. The longer a conductor, the *greater* the resistance. The shorter a conductor, the *lower* its resistance. A material resists the flow of electrons because of the way in which each atom holds onto its outer electrons. The more material that is in the path of an electric current, the less current flow the circuit will have. If the length of a conductor is doubled, there would be *twice* as much resistance in the circuit. Another factor that affects resistance is the cross-sectional area of a material. The greater the cross-sectional area of a material, the *lower* the resistance. The smaller this area, the *higher* the resistance of the material. If two conductors have the same length but twice the cross-sectional area, the current flow would be twice as much through the wire with the larger cross-sectional area. This happens because there is a wider path for electric current to flow through. Twice as many free electrons are available to allow current flow. Temperature also affects resistance. For *most* materials, the higher the temperature, the more resistance it offers to the flow of electric current. The colder the temperature, the less resistance a material offers to the flow of electric current. This effect is produced because a change in the temperature of a material changes the ease with which a material releases its outer electrons. A few materials, such as carbon, have lower resistance as the temperature increases. The effect of temperature on resistance varies with the type of material. The effect of temperature on resistance is the least important of the factors that affect resistance. A device called a resistor which is used in electrical circuits is shown in Fig. 1-23.

VOLTAGE, CURRENT, AND RESISTANCE

We depend on electricity to do many things which are sometimes taken for granted. It is very important

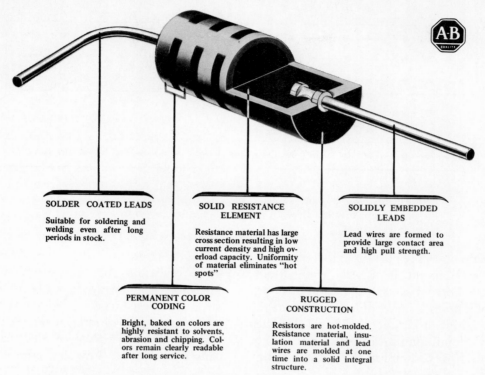

SOLDER COATED LEADS

Suitable for soldering and welding even after long periods in stock.

SOLID RESISTANCE ELEMENT

Resistance material has large cross section resulting in low current density and high overload capacity. Uniformity of material eliminates "hot spots"

SOLIDLY EMBEDDED LEADS

Lead wires are formed to provide large contact area and high pull strength.

PERMANENT COLOR CODING

Bright, baked on colors are highly resistant to solvents, abrasion and chipping. Colors remain clearly readable after long service.

RUGGED CONSTRUCTION

Resistors are hot-molded. Resistance material, insulation material and lead wires are molded at one time into a solid integral structure.

Figure 1-23. Resistor used in electrical circuits. (Courtesy of Allen-Bradley Co.)

to learn some of the basic electrical terms which are commonly used in the study of electricity and electronics. There are three basic electrical terms which must be understood. These terms are "voltage," "current," and "resistance." These terms are very commonly used. The term "voltage" is best understood by looking at a flashlight battery. The battery is a source of voltage. It is capable of supplying electrical energy to a light connected to it. The voltage the battery supplies should be thought of as electrical "pressure." The battery has positive (+) and negative (−) terminals.

For a battery to supply electrical pressure, an *electrical circuit* must be made. An electrical circuit in its simplest form has a source, a conductor, and a load. An electrical circuit is shown in Fig. 1-21. The battery is a source of electrical pressure or *voltage*. The conductor is a path to allow the electrical current to pass the load. The lamp is called a load because it changes electrical energy to light energy. When the source, conductors, and load are connected together, a complete circuit is made. When the battery or voltage source is connected to the light bulb by conductors, a current will flow. *Current*

flows because of the electrical pressure produced by the battery. The battery is similar to a water pump. A water pump also supplies pressure. Water in pipes is somewhat similar to the flow of current through a conductor. When the conductor is connected to the lamp, a current flows. The current flow causes the lamp to light. Electrical current is the flow of electrons through the conductor. Electrons move because of the pressure produced by the battery. Remember that electrons have a negative charge (−).

The movement of electrons through a conductor takes place at a rate based on the *resistance* of a circuit. A lamp offers resistance to the flow of electrical current. Resistance is opposition to the flow of electrical current. More resistance in a circuit causes *less* current to flow. Resistance can be explained by using the example of two water pipes shown in Fig. 1-24. If a water pump is connected to a large pipe such as pipe 1, water flows easily. The pipe offers a small amount of resistance to the flow of water. However, if the same water pump is connected to a small pipe, such as pipe 2, there is more opposition to the flow of water. The water flow through pipe 2 is less.

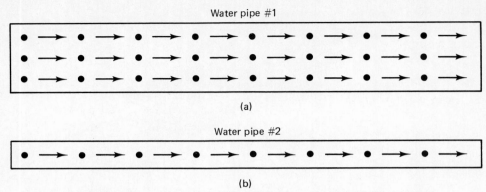

Figure 1-24. Water pipes showing the effect of resistance: (a) water pipe 1—many drops of water flow through; (b) water pipe 2—only a few drops of water flow through.

Inside a lamp bulb, the part that glows is called a filament. The filament is a wire that offers resistance to the flow of electrical current. If the filament wire of a lamp is made of large wire, much current flows as shown in circuit (a) of Fig. 1-25. The filament offers a small amount of resistance to the flow of current. However, circuit (b) shows a lamp with a filament of small wire. The small wire has more resistance or opposition to current flow. Therefore, less current flows in circuit (b) because it has higher resistance.

The terms *voltage*, *current*, and *resistance* are very important. *Voltage* is electrical pressure that causes current to flow in a circuit. *Current* is the

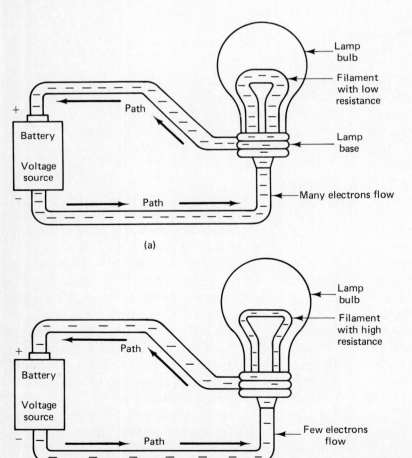

Figure 1-25. How lamp filament size affects current flow.

movement of particles called electrons through a conductor in a circuit. *Resistance* is opposition to the flow of current in a circuit.

VOLTS, OHMS, AND AMPERES

There are many similarities between water systems and electrical systems. These similarities help in understanding basic electrical quantities. The volt (unit of electrical pressure) is compared to the pressure that causes water to flow in pipes. Since the volt is a unit of electrical pressure, it is always measured *across* two points. An electrical pressure of 120 volts exist across the terminals of electrical outlets in the home. This value is measured with an electrical instrument called a *voltmeter*.

The ampere, often called simply "amp," is a measure of the rate of flow of electrical current or electron movement. Electrical current is similar to the rate of flow of water in a pipe. Water flow is measured in gallons per minute. An ampere is a number of electrons per unit of time flowing in an electrical circuit. An ampere is a measure of the *rate* of flow. An *ammeter* is used to measure the number of electrons that flow in a circuit.

When pressure is applied to a water pipe, water flows. The rate of flow is limited by friction in the pipe. When an electrical pressure (voltage) is applied to an electrical circuit, the number of electrons (current) that flows is limited by the resistance of the path. Resistance is measured with a meter called an *ohmmeter*, since the basic unit of resistance is the *ohm*.

ELECTRICAL SAFETY

Electrical safety is very important. Many dangers are not easy to see. For this reason, safety should be based on understanding basic electrical principles. Common sense is also important. The physical arrangement of equipment in the electrical lab or work area should be done in a safe manner. Well-designed electrical equipment should always be used. For economic reasons, electrical equipment is often improvised. It is important that all equipment be made as safe as possible. This is especially true for equipment and circuits that are designed and built in the lab or shop.

Work surfaces in the shop or lab should be covered with a material that is nonconducting and the floor of the lab or shop should also be nonconducting. Concrete floors should be covered with rubber tile or linoleum. A fire extinguisher that has a nonconducting agent should be placed in a convenient location. Extinguishers should be used with caution. Their use should be explained by the teacher.

Electrical circuits and equipment in the lab or shop should be plainly marked. Voltages at outlets require special plugs for each voltage. Several voltage values are ordinarily used with electrical lab work. Storage facilities for electrical supplies and equipment should be neatly kept. Neatness encourages safety and helps keep equipment in good condition. Tools and small equipment should be maintained in good condition and stored in a tool panel or marked storage area. Tools that have insulated handles should be used. Tools and equipment plugged into convenience outlets should be wired with three-wire cords and plugs. The purpose of the third wire is to prevent electrical shocks by grounding all metal parts connected to the outlet.

Soldering irons are often used in the electrical shop or lab. They can be a fire hazard. They should have a metal storage rack. Irons should be unplugged while not in use. Soldering irons can also cause burns if not used properly. *Rosin-core* solder should always be used in the electrical lab or shop.

Adequate laboratory space is needed to reduce the possibility of accidents. Proper ventilation, heat, and light also provides a safe working environment. Wiring in the electrical lab or shop should conform to specifications of the *National Electrical Code®* (NEC). The NEC governs all electrical wiring in buildings.

Lab or Shop Practices

All student activities should be done with low voltages whenever possible. Instructions should be written, with clear directions for performing lab activities. All lab or shop work should emphasize safety. Students should always have the teacher check experimental circuits before they are plugged into a power source. Electrical lab projects should be constructed to provide maximum safety when used. "Horseplay" of any type in the lab or shop should not be allowed.

Electrical equipment should be disconnected from the source of power before working on it. When testing electronic equipment, such as TV sets or other 120-volt devices, use an *isolation transformer*. This isolates the chassis ground from the ground of the equipment. This eliminates the shock hazard when working with 120-volt equipment.

Electrical Hazards

A good first-aid kit should be in every electrical shop or lab. The phone number of an ambulance service or other medical services available should be in the lab or work area in case of emergency. Any accident should be reported immediately to the proper school officials. Teachers should be proficient in the treatment of minor cuts and bruises. They should also be able to apply artificial respiration. In case of electrical shock, when breathing stops, artificial respiration must be immediately started. Extreme care should be used in moving a shock victim from the circuit that caused the shock. An insulated material should be used so that someone else does not come in contact with the same voltage. It is not likely that a high-voltage shock will occur. However, students should know what to do in case of emergency.

Normally, the human body is not a good conductor of electricity. When wet skin comes in contact with an electric conductor, the body is a better conductor. A slight shock from an electrical circuit should be a warning that something is wrong. Equipment that causes a shock should be immediately checked and repaired or replaced. Proper grounding is important in preventing shock.

Safety devices called ground-fault circuit interrupters (GFIs) are now used for bathroom and outdoor power receptacles. They have the potential of saving many lives by preventing shock. GFIs immediately cut off power if a shock occurs. The National Electric Code® specifies where GFIs should be used.

Electricity causes many fires each year. Electrical wiring with too many appliances connected to a circuit overheats wires. Overheating may set fire to nearby combustible materials. Defective and worn equipment can allow electrical conductors to touch one another to cause a "short circuit." A short circuit causes a blown fuse. It could also cause a spark or arc which might ignite insulation or other combustible materials or burn electrical wires.

Fuses and Circuit Breakers

Fuses and circuit breakers are important safety devices. When a fuse "blows," it means that something is wrong in the circuit. Causes of blown fuses could be:

1. A short circuit caused by two wires touching
2. Too much equipment on the same circuit
3. Worn insulation allowing bare wires to touch grounded metal objects such as heat radiators or water pipes

After correcting the problem, a new fuse of *proper size* should be installed. Power should be turned off to replace a fuse. Never use a makeshift device in place of a new fuse of the correct size. This destroys the purpose of the fuse. Fuses are used to cut off the power and prevent overheated wires.

Circuit breakers are now very common. Circuit breakers operate on spring tension. They can be turned on or off like wall switches. If a circuit breaker "opens," something is wrong in the circuit. Locate and correct the cause and *then* reset the breaker.

Always remember to use "common sense" whenever working with electrical equipment or circuits. Safe practices should be followed in the electrical lab or shop as well as in the home. Detailed safety information is available from the National Safety Council and other organizations. *It is always wise to be safe.*

Review

1. How was electricity discovered?
2. What are the five parts of any electrical system?
3. Discuss each of the parts of an electrical system.

4. What are some types of energy?
5. What is work?
6. What is power?
7. What are the three types of matter?
8. What are elements, compounds, and molecules?
9. Discuss the three types of particles that make up atoms.
10. Discuss the terms "orbit" and "valence electrons."
11. Discuss electrostatic charges.
12. What are some applications of static electricity?
13. What are conductors, insulators, and semiconductors?
14. What is (a) an open circuit, (b) a closed circuit, and (c) a short circuit?
15. Discuss conventional current flow and electron current flow. Which type is used in this book?
16. How does electrical current flow compare to water flow in a pipe?
17. What is voltage?
18. What is resistance?
19. What factors determine the resistance of a material?
20. List at least five important things to remember about electrical safety.

Student Activities

Electrical Systems (see Fig. 1-7)
1. Obtain one or more flashlights.
2. Examine the construction of the flashlight.
3. On a sheet of paper, discuss the following parts of the flashlight:
   a. Source of energy
   b. Path of current flow
   c. Control method
   d. Load

Structure of Matter (see Fig. 1-10)
1. Obtain some golf balls and rigid wire.
2. Make a model that shows the arrangement of electrons, protons, and neutrons in a simple atom, such as hydrogen.
3. Paint the golf balls different colors to represent electrons, protons, and neutrons.
4. Drill holes in the golf balls or attach the wire in some other way to make the model sturdy.
5. Discuss the arrangement of electrons, protons, and neutrons in an atom on a sheet of paper.

Static Electricity (see Fig. 1-13)
1. Obtain a small piece of plastic and some fur or wool cloth.

2. Tear off six to eight small peices of paper.

3. Rub the plastic with the wool or cloth.

4. Hold the plastic near the pieces of paper.

5. On a sheet of paper, explain what happens and why it happens.

Electrical Safety

1. Use the inspection checklist shown in Fig. 1-26.

2. Go through your classroom or shop area in teams of two or three to look for safety hazards.

3. Make notes on the checklist on each part of the inspection.

4. On the sheet of paper, discuss the inspection and the importance of electrical safety.

Figure 1-26. Electrical safety inspection checklist.

Use this checklist for a laboratory or shop area. A safe and healthy environment is always necessary. Inspections that are similar to this should be done periodically. This is not a detailed inspection.

For the items on the checklist, place the number to the left of the item that describes its condition as follows:

0 = Does not apply to this area
1 = Very good (needs no attention)
2 = Good (needs little attention)
3 = Acceptable (needs some attention)
4 = Bad (needs attention soon)
5 = Very bad (needs immediate attention)

Place comments to the right of each item that needs attention.

A. General room environment
_____ 1. Temperature
_____ 2. Ventilation
_____ 3. Lighting
_____ 4. Condition of equipment
_____ 5. Condition of floors
_____ 6. Condition of walls
         and windows
_____ 7. Availability of fire
         extinguishers
_____ 8. Condition of fire
         extinguishers
_____ 9. Fire escape plan
_____10. Hanging objects on walls
         or ceilings
_____11. Electrical power outlets
_____12. General appearance (neatness)
_____13. Condition of benches
_____14. Condition of supply rooms
_____15. Storage of materials
         (especially hazardous ones)
_____16. Other

B. Equipment
_____ 1. High-voltage and danger
         areas identified
_____ 2. Machines and equipment well
         guarded and protected
_____ 3. All moving gears, belts, etc.,
         protected properly
_____ 4. Tools in good condition
_____ 5. Machines in good condition
_____ 6. Equipment controls easy to
         reach
_____ 7. Main switch or circuit breaker
         to turn off all equipment easy
         to reach
_____ 8. Other

C. Electrical distribution
_____ 1. No temporary wiring used
_____ 2. Controls are enclosed
         properly
_____ 3. Power outlet voltages (other
         than 120 V) are identified
_____ 4. Equipment is wired properly
_____ 5. Overload protection is used
         on equipment
_____ 6. Master switch available to
         turn off equipment
_____ 7. Other

D. Miscellaneous
_____ 1. First-aid kit available
_____ 2. Accident reports available
_____ 3. Evidence of safety
         awareness
_____ 4. Safety glasses and protective
         equipment available
         where needed
_____ 5. Other

E. General safety rating of the area checked:
    Comments:

Soldering (study Appendix 2 before proceeding)

1. Obtain a soldering iron and some rosin-core solder.
2. Complete any of the following soldering activities assigned by your teacher.
    a. Component mounting
        (1) Obtain a component-mounting strip and a resistor or some other component from your teacher.
        (2) Allow the soldering iron to heat to its operating temperature.
        (3) Apply a small amount of solder to the tip of the iron to check its temperature.
        (4) Solder the resistor onto the component mounting strip.
    b. Printed circuit board soldering
        (1) Obtain a used printed circuit (PC) board and a 3-in. length of insulation wire (No. 22 AWG size will work) to practice soldering.
        (2) Place the wire near the holes in the PC board where you intend to solder it. Do this to find the length of wire required.
        (3) Strip $\frac{3}{8}$ to $\frac{1}{2}$ in. of insulation from one end of the wire with a wire stripper.
        (4) Cut off any excess wire and strip the insulation off the other end of the wire. The wire should lie flat on the board when soldered.
        (5) Solder both ends of the wire to the PC board.
    c. Printed circuit board desoldering
        (1) Obtain a desoldering tool or use a soldering iron.
        (2) Using the proper desoldering techniques, remove a component from a PC board without damaging the PC board.
    d. Installing a terminal connector
        (1) Obtain a "solderless" terminal connector, crimping tool, and a piece of insulated wire from your instructor.
        (2) Strip enough insulation from the wire so that the terminal connector fits properly.
        (3) Use the crimping tool to fasten the terminal connector to the wire.

Electricity in the Home

1. Complete this activity at home.
    a. List several electrical devices and equipment in your home. See if the completed list is longer than those of others in the class.
2. Turn in the list to your teacher when finished.

Electrical Conductors

1. Obtain an American Wire Gage (see Fig. 1-27) and several conductors from your teacher.
2. Measure the diameter of each conductor with the gage.
3. Record the values on a sheet of paper.
4. Also record whether the conductor is copper or aluminum.

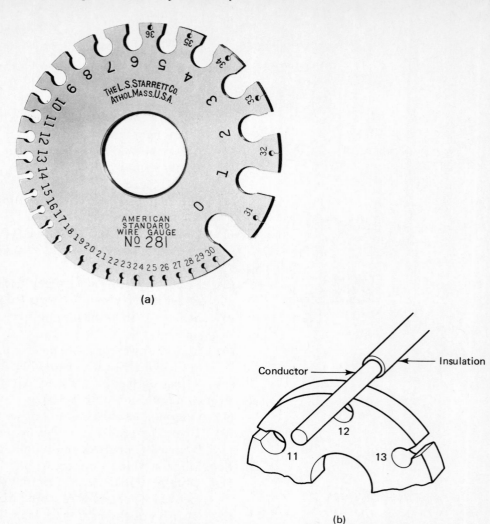

(a)

(b)

Figure 1-27.   (a) American Wire Gage (AWG) used to measure conductor size; (b) using the AWG. (Courtesy of L. S. Starrett Co.)

Electrical Tool Identification

1. Study the illustrations in Appendix 3.
2. Learn the names of the electrical tools shown.
3. Discuss the uses of at least five tools assigned by your teacher.
4. Your teacher can now give you a test to see how many electrical tools you can identify properly.

Electrical Components and Measurements

Most electrical equipment is made of several parts or components that work together. It would be almost impossible to explain how electrical equipment operates without using symbols and diagrams. Electrical diagrams show how the component parts of equipment fit together. Common electrical components are easy to identify. It is also easy to learn the symbols used to represent electrical components. The components of electrical equipment work together to form an electrical system.

Another important activity in the study of electricity and electronics is measurement. Measurements are made in many types of electrical circuits. The proper ways of measuring resistance, voltage, and current should be learned. These are the three most common electrical measurements.

IMPORTANT TERMS

Before reading this chapter, review the following terms. These terms provide a basic understanding of electrical components and measurements.

Ammeter. A meter used to measure current (amperes).

Battery. An electrical energy source consisting of two or more cells connected together.

Block diagram. A diagram used to show how the parts of a system fit together.

Cell. An electrical energy source that converts chemical energy into electrical energy.

Component. An electrical device that is used in a circuit.

Conductor. A material that allows electrical current to flow through it easily.

Continuity check. A test to see if a circuit is an open or closed path.

Lamp. An electrical load device that converts electrical energy into light energy.

Multifunction meter. A meter that measures two or more electrical quantities, such as a volt-ohm-milliammeter (VOM), which measures voltage, resistance, and current.

Multirange meter. A meter that has two or more ranges to measure an electrical quantity.

Ohmmeter. A meter used to measure resistance (ohms).

Polarity. The direction of an electrical potential (− or +) or a magnetic charge (north or south).

Potentiometer. A variable-resistance component used as a control device in electrical circuits.

Precision resistor. A resistor used when a high degree of accuracy is needed.

Resistor. A component used to control either the amount of current flow or the voltage distribution in a circuit.

Schematic diagram. A diagram used to show how the components of electrical circuits are wired together.

Scientific notation. The use of "powers of 10" to simplify large and small numbers.

Switch. A control device used to turn a circuit on or off.

Symbol. Used as a simple way to represent a component.

Voltage drop. The electrical potential (voltage) that exists across two points of an electrical circuit.

Voltmeter. A meter used to measure voltage.

Volt-ohm-milliammeter (VOM). A multifunction, multirange meter which usually is designed to measure voltage, current, and resistance.

Wiring diagram. A diagram which shows how wires are connected by showing the point-to-point wiring and the path followed by each wire.

COMPONENTS, SYMBOLS, AND DIAGRAMS

Anyone who studies electricity and electronics should be able to identify the components used in simple electrical circuits. *Components* are represented by symbols. *Symbols* are used to make diagrams. A *diagram* shows how the components are connected together in a circuit. For example, it is easier to show symbols for a battery connected to a lamp than to draw a pictorial diagram of the battery and the lamp connected together. There are several

symbols which it will be necessary for you to learn to recognize. These symbols are used in many electrical diagrams. Diagrams are used for installing, troubleshooting, and repairing electrical equipment. The use of symbols makes it easy to draw diagrams and to understand the purpose of each circuit. Common electrical symbols are listed in Appendix 4.

Most electrical equipment uses wires (conductors) to connect its components or parts together. The symbol for a conductor is a narrow line. If two conductors cross one another on a diagram, they may be shown by using symbols. Figure 2-1(a) shows two conductors crossing one another. If two conductors are connected together, they may also be identified by symbols, as shown in Fig. 2-1(b).

The symbols for two lamps connected across a battery are shown in Fig. 2-2 using symbols for the battery and lamps. Notice the part of the diagram where the conductors are connected together.

A common electrical component is a switch such as the toggle switch shown in Fig. 2-3. The simplest switch is a single-pole, single-throw (SPST) switch. This switch turns a circuit on or off. In Fig. 2-4(a), the symbol for a switch in the "OFF" or open position is shown. There is no path for current to flow from the battery to the lamp. The lamp will be off when the switch is open. Fig. 2-4(b) shows a

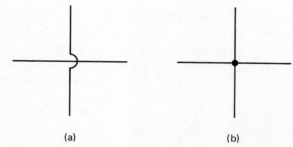

(a)   (b)

Figure 2-1. Symbols for electrical conductors: (a) conductors crossing; (b) conductors connected.

Figure 2-2. Symbols for a battery connected across two lamps.

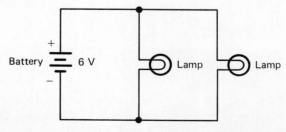

Figure 2-3. Electrical switch. (Courtesy of Cutler-Hammer.)

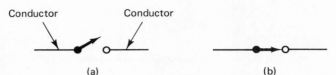

Conductor   Conductor

(a)   (b)

Figure 2-4. Symbol for a single-pole, single-throw (SPST) switch: (a) open or off condition; (b) closed or on condition.

switch in the "ON" or closed position. This switch position completes the circuit and allows current to flow.

In many electrical circuits, a component called a resistor is used. Resistors are usually small cylinder-shaped components, such as those shown in Fig. 2-5. They are used to control the flow of electrical current. A typical color-coated resistor and its symbol are shown in Fig. 2-6. The most common type of resistor uses color coding to mark its value. Resistor value is always in *ohms*. For instance, a resistor might have a value of 100 ohms. The symbol for ohms is the Greek capital letter omega (Ω). Each color on the resistor represents a specific number. Resistor color-code values are easy to learn.

Another type of resistor is called a potentiometer or "pot." A pot is a variable resistor whose value can be changed by adjusting a rotary shaft. For example, a 1000-Ω pot can be adjusted to any resistance value from zero to 1000 Ω by rotating the shaft. The pictorial and symbol of this component are shown in Fig. 2-7(a) and (b). In the example shown in Fig. 2-7(c), potentiometer 1 is adjusted so that the resistance between points A and B is zero.

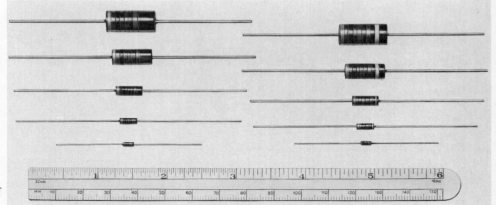

Figure 2-5. Resistors. (Courtesy of Allen-Bradley Co.)

Figure 2-6. (a) Color-coded resistor; (b) symbol for a resistor. [(a) Courtesy of Allen-Bradley Co.]

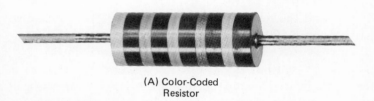

(A) Color-Coded Resistor

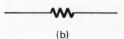

(b)

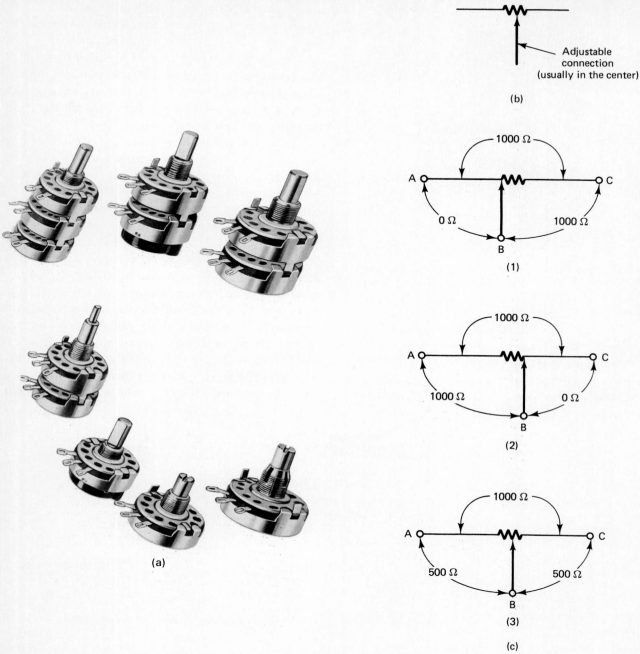

Figure 2-7. Potentiometers: (a) pictorials; (b) symbol; (c) examples. [(a) Courtesy of Allen-Bradley Co.]

The resistance between points *B* and *C* is 1000 Ω. By turning the shaft as far in the opposite direction as it will go, the resistance between points *B* and *C* becomes zero (see potentiometer 2). Between points *A* and *B*, the resistance is now 1000 Ω. By rotating the shaft to the center of its movement, as shown by potentiometer 3, the resistance is split in half. Now the resistance from point *A* to point *B* is about 500 Ω

and the resistance from point *B* to point *C* is also about 500 Ω.

The symbol for a battery is shown in Fig. 2-2. The symbol for any battery over 1.5 volts (V) is indicated by two sets of lines. A 1.5-V battery or cell is shown with one set of lines. The voltage of a battery is marked near its symbol. The long line in the symbol is always the positive (+) side and the

Figure 2-8. Common types of batteries. (Courtesy of Union Carbide Corp.)

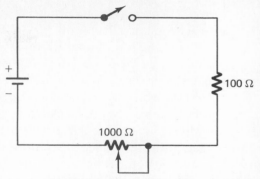

Figure 2-9. Simple circuit diagram.

short line is the negative (−) side of the battery. Some common types of batteries are shown in Fig. 2-8.

A simple circuit diagram using symbols is shown in Fig. 2-9. This diagram shows a 1.5-V battery connected to an SPST switch, a 100-Ω resistor, and a 1000-Ω potentiometer. Since symbols are used, no words have to be written beside them. Anyone using this diagram should recognize the components represented and how they fit together to form a circuit.

RESISTORS

There is some resistance in all electrical circuits. Resistance is added to a circuit to control current flow. Devices that are used to cause proper resistance in a circuit are called *resistors*. A wide variety of resistors are used. Some have a fixed value, and others are variable. Resistors are made of either special carbon material or of metal film. Wire-wound resistors are ordinarily used to control large currents and carbon resistors control currents which are smaller. Some types of resistors are shown in Fig. 2-10.

Figure 2-10. (a) Metal film resistor; (b) molded wire-wound resistor; (c) high-wattage wire-wound resistor; (d) 16-pin dual-in-line package (DIP) resistor network used in PC boards. (Courtesy of TRW/IRC Resistors.)

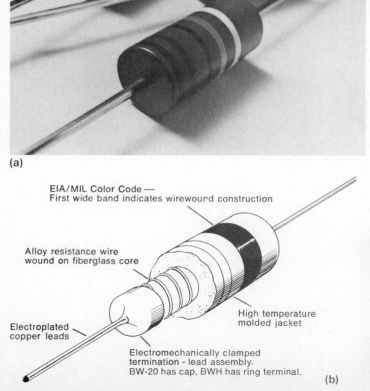

(a)

(b)

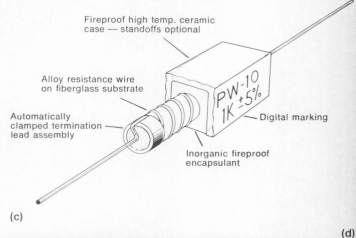

(c)

(d)

Wire-wound resistors are constructed by winding resistance wire on an insulating material. The wire ends are attached to metal terminals. An enamel coating is used to protect the wire and to conduct heat away from it. Wire-wound resistors may have fixed taps which can be used to change the resistance value in steps. They may also have sliders which can be adjusted to change the resistance to any fraction of their total resistance. *Precision-wound resistors* are used where the resistance value must be very accurate, such as in measuring instruments.

Carbon resistors are constructed of a small cylinder of compressed material. Wires are attached to each end of the cylinder. The cylinder is then covered with an insulating coating.

Variable resistors are used to change resistance while equipment is in operation. They are called potentiomenters or rheostats. Both carbon and wire-wound *variable* resistors are made by winding on a circular form [see Fig. 2-11(a)]. A contact arm is attached to the form to make contact with the wire.

The contact arm can be adjusted to any position on the circular form by a rotating shaft. A wire connected to the movable contact is used to vary the resistance from the contact arm to either of the two outer wires of the variable resistor. For controlling smaller currents, carbon variable resistors are made by using a carbon compound mounted on a fiber disk [see Fig. 2-11(b)]. A contact on a movable arm varies the resistance as the arm is turned by rotating a metal shaft.

Resistor Color Codes

It is usually easy to find the value of a resistor by its color code or marked value. Most wire-wound resistors have resistance values (in ohms) *printed* on the resistor. If they are not marked in this way, an ohmmeter must be used to measure the value. Most carbon resistors use colored bands to identify their value. Carbon resistors are of two types: radial and axial, as shown in Fig. 2-12. They differ by the way

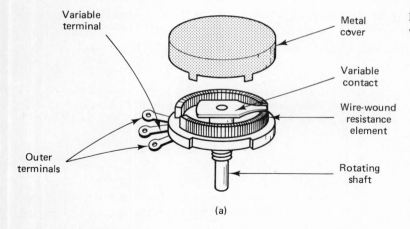

(a)

Figure 2-11. Variable resistor construction: (a) wire-wound variable resistor; (b) carbon variable resistor.

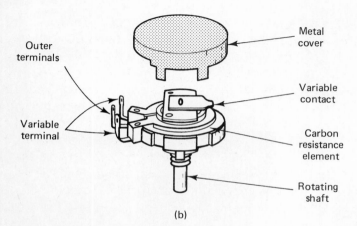

(b)

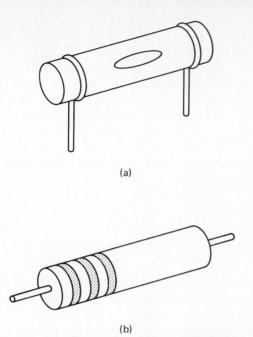

(a)

(b)

Figure 2-12. Carbon resistors: (a) radial; (b) axial.

Color	1st digit	2nd digit	Number of zeros or multiplier	Tolerance (%)
Black	0	0	1	
Brown	1	1	10	
Red	2	2	100	
Orange	3	3	1,000	
Yellow	4	4	10,000	
Green	5	5	100,000	
Blue	6	6	1,000,000	
Violet	7	7		
Gray	8	8		
White	9	9		
Gold*	–	–	0.1	5
Silver*	–	–	0.01	10
No color	–	–		20

1st digit
2nd digit
Multiplier (number of zeros)
Tolerance

*When resistors have a value of less than 10 Ω, the third color band is a decimal multiplier. The two colors used are: gold = × 0.1 and silver = × 0.01.

Figure 2-13. Resistor color code.

in which their wires are connected to the body of the resistor. Both types use the same color code. However, the colors are put on them in a different way. Radial resistors are made with the wire leads wound around the ends of a carbon cylinder. The wire leads come off at right angles. The resistor body is painted *but not insulated.* This type of resistor must be mounted where it will not come into contact with other parts of a circuit. Radial lead resistors are seldom found in modern equipment. Axial resistors are made with the body of the resistor. The carbon cylinder is coated with an insulating material.

Radial resistors are color coded with the *body-end-dot* system. Most axial resistors are color coded by an *end-to-center color band* system of marking. In each color-coding system, three colors are used to indicate the *resistance value* in ohms. A fourth color is used to indicate the *tolerance* of the resistor. The colors are read in the correct order from the end of a resistor. Numbers from the resistor color code, shown in Fig. 2-13, are substituted for the colors. Through practice using the resistor color code, the value of a resistor may be determined at a glance.

It is difficult to manufacture a resistor to the exact value required. For many uses, the actual resistance value can be as much as 20% higher or lower than the value marked on the resistor without causing any problem. In most uses, the actual

resistance does not need to be any closer than 10% higher or lower than the marked value. This percentage of variation between the marked color code value and the actual of a resistor is called *tolerance*. A resistor with a 5% tolerance should be no more than 5% higher or lower than the marked value. Resistors with tolerances of lower than 5% are called *precision* resistors. A few resistors use the body-end-dot system of marking. They are coded with the body of the resistor a solid color. One end has another color, and a dot of a third color is placed near the middle of the resistor. For example, a resistor might have a red body, brown end, and a yellow dot. The body color indicates the first digit, the end color is the second digit, and the dot is the number of zeros to be added to the digits. The value of the resistor is 210,000 Ω, as shown in Fig. 2-14.

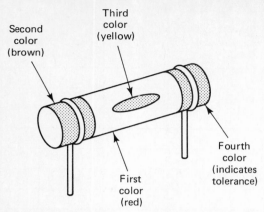

Figure 2-14. Body-end-dot system for radial resistors, resistor value = 210,000 Ω.

Axial resistors are the most common type of resistor. They are marked with color bands at one end of the resistor. The body color is not used as part of the resistor value. For example, a resistor that has three color bands (yellow, violet, and brown) at one end has the color bands read from the end toward the center, as shown in Fig. 2-15. The resistance value is 470 Ω. Remember that black as the center dot or third color means that *no* zeros are to be added to the digits. A resistor with a green band, a red band, and a black band, has a value of 52 Ω.

An example of a resistor with the band color-code system is shown in Fig. 2-16. Read the colors from left to right from the end of the resistor where the bands begin. Use the resistor color chart to determine the value of the resistor in ohms, its tolerance, and its failure rate.

The first color is orange, so the first digit in the value of the resistor is 3. The second color is black, so the second digit in the value of the resistor is 0. The

Figure 2-15. End-to-center system for axial resistors, resistor value = 470 Ω.

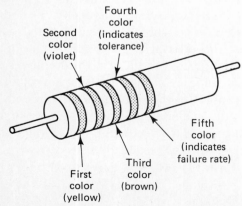

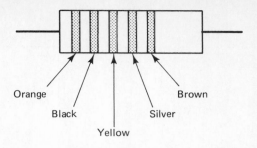

Resistor value = 300,000 Ω; 10% tolerance; failure rate = 1%

Figure 2-16. Resistor color code example.

first two digits of the resistor value is 20. The third color is yellow, which indicates the number by which the first two digits are to be multiplied. Sometimes it is easier to think of the third color as the number of zeros to add to the first two digits. The color yellow is a multiplier of 10,000.

Multiply 30 × 10,000 to obtain the value of the resistor (30 × 10,000 = 300,000 Ω). A yellow color means that four zeros are added to the first two digits to find the value of the resistor [30 with four zeros (0000) added equals 300,000 Ω].

The fourth color is silver. The tolerance of the resistor is given with this band. The tolerance shows how close the actual value of the resistor should be to the color-code value. Tolerance is a percentage of the actual value. In this example, silver shows a tolerance of ∓10%. This means that the resistor should be within 10% of 300,000 Ω in either direction. The value of the resistor could be as low as 270,000 Ω (300,000 × 0.10 = 30,000) and 300,000 − 30,000 = 270,000). The value could be as high as 330,000 (300,000 × 0.10 = 30,000) and (300,000 + 30,000 = 330,000).

The fifth color is brown, which shows the percentage of failure per 1000 hours of use. The color brown shows a failure rate of 1% per 1000 hours of resistor use.

Sometimes there is no fourth color on the body of the resistor. Then the tolerance is 20%. Frequently, no fifth color appears on the resistor. Using the color code, it is easy to list the ohms value, tolerance, and failure rate of resistors. Standard values of 5% and 10% tolerance color-coded resistors are listed in Table 2-1.

Another way to determine the value of color-coded resistors is to remember the following:

Big Brown Rabbits Often Yield Great Big Vocal Groans When Gingerly Slapped.

Table 2-1
STANDARD VALUES OF COLOR-CODED RESISTORS (Ω)

10% tolerance

0.27	1.2	5.6	27	120	560	2700	12,000	56,000	0.27M	1.2M	5.6M
0.33	1.5	6.8	33	150	680	3300	15,000	68,000	0.33M	1.5M	6.8M
0.39	1.8	8.2	39	180	820	3900	18,000	82,000	0.39M	1.8M	8.2M
0.47	2.2	10	47	220	1000	4700	22,000	0.1M	0.47M	2.2M	10M
0.56	2.7	12	56	270	1200	5600	27,000	0.12M	0.56M	2.7M	12M
0.68	3.3	15	68	330	1500	6800	33,000	0.15M	0.68M	3.3M	15M
0.82	3.9	18	82	390	1800	8200	39,000	0.18M	0.82M	3.9M	18M
1.0	4.7	22	100	470	2200	10,000	47,000	0.22M	1.0M	4.7M	22M

5% tolerance

0.24	1.1	5.1	24	110	510	2400	11,000	51,000	0.24M	1.1M	5.1M
0.27	1.2	5.6	27	120	560	2700	12,000	56,000	0.27M	1.2M	5.6M
0.30	1.3	6.2	30	130	620	3000	13,000	62,000	0.30M	1.3M	6.2M
0.33	1.5	6.8	33	150	680	3300	15,000	68,000	0.33M	1.5M	6.8M
0.36	1.6	7.5	36	160	750	3600	16,000	75,000	0.36M	1.6M	7.5M
0.39	1.8	8.2	39	180	820	3900	18,000	82,000	0.39M	1.8M	8.2M
0.43	2.0	9.1	43	200	910	4300	20,000	91,000	0.43M	2.0M	9.1M
0.47	2.2	10	47	220	1000	4700	22,000	0.1M	0.47M	2.2M	10M
0.51	2.4	11	51	240	1100	5100	24,000	0.11M	0.51M	2.4M	11M
0.56	2.7	12	56	270	1200	5600	27,000	0.12M	0.56M	2.7M	12M
0.62	3.0	13	62	300	1300	6200	30,000	0.13M	0.62M	3.0M	13M
0.68	3.3	15	68	330	1500	6800	33,000	0.15M	0.68M	3.3M	15M
0.75	3.6	16	75	360	1600	7500	36,000	0.16M	0.75M	3.6M	16M
0.82	3.9	18	82	390	1800	8200	39,000	0.18M	0.82M	3.9M	18M
0.91	4.3	20	91	430	2000	9100	43,000	0.20M	0.91M	4.3M	20M
1.0	4.7	22	100	470	2200	10,000	47,000	0.22M	1.0M	4.7M	22M

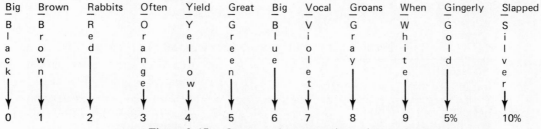

Figure 2-17. Quote used to memorize resistor color code.

The first letter of each word in the quote is the same as the first letter in each of the colors used in the color code. The words of the quote are counted (beginning with *zero*) to find the word corresponding to each digit or the number of zeros to be added (refer to Fig. 2-17). Some method such as this quote should be used to remember the color code.

Power Rating of Resistors

The size of a resistor helps to determine its power rating. Larger resistors are used in circuits that have high power ratings. Small resistors will become damaged if they are put in high-power circuits. The power rating of a resistor indicates its ability to give off or dissipate heat. Common power (wattage) ratings of color-coded resistors are $\frac{1}{8}, \frac{1}{4}, \frac{1}{2}$, 1, and 2 watts (W). Notice the difference in the sizes of the resistors in Fig. 2-5. Resistors that are larger in physical size will give off more heat and have higher power ratings.

ELECTRICAL UNITS

All quantities can be measured. The distance between two points may be measured in meters, kilometers, inches, feet, or miles. The weight of an object may be measured in ounces, pounds, grams, or kilograms. Electrical quantities may also be measured. The more common electrical units of measurement are discussed below.

There are four common units of electrical measurement. Voltage is used to indicate the force that causes electron movement. Current is a measure of the amount of electron movement. Resistance is the opposition to electron movement. The amount of work done or energy used in the movement of electrons in a given period of time is called power. Table 2-2 shows the four basic units of electrical measurement.

Small Units

The electrical unit used to measure a certain value is often less than a whole unit (less than 1). Examples of this are 0.6 V, 0.025 A, 0.0550 W. When this occurs, *prefixes* are used. Some prefixes are shown in Table 2-3.

For example, a millivolt is 1/1000 of a volt and a microampere is 1/1,000,000 of an ampere. The prefixes of Table 2-3 may be used with any electrical unit of measurement. The unit is divided by the fractional part of the unit. For example, if 0.6 V is to be changed to millivolts, 0.6 V is divided by the fractional part of the unit. So 0.6 V equals 600 millivolts (mV) or 0.6 divided by 0.001 = 600 mV. If 0.0005 A is changed to microamperes, 0.0005 A is equal to 500 microamperes (μA) or 0.0001 divided by 0.000005 = 500 μA. When changing a basic electrical unit to a prefix unit, move the decimal point of the unit to the right by the same number of places in the fractional prefix. To change 0.8 V to millivolts, the decimal point of 0.8 V is moved three places to the right (8.0.0.) since the prefix "milli" has three decimal places. So 0.8 V equals 800 mV. The same method is used for converting any electrical unit to a unit with a smaller prefix.

Often an electrical unit with a prefix is converted back to the basic unit. For example, milliamperes may be converted back to amperes. Microvolts sometimes are converted back to volts. When a unit with a prefix is converted back to a basic unit, the prefix must be multiplied by the fractional part of the whole unit of the prefix. For example, 68 mV converted to volts is equal to 0.068 V. When 68 mV is multiplied by the fractional part of the whole unit (0.001 for the prefix milli) this equals 0.0068 V (68 mV × 0.001 = 0.0068 V).

When changing a fractional prefix unit into a basic electrical unit, move the decimal in the prefix unit to the left by the same number of places of the prefix. To change 225 millivolts to volts, move the

Table 2-2
BASIC UNITS OF ELECTRICAL MEASUREMENT

Electrical quantity	Unit of measurement	Symbol	Description
Voltage	Volt (V)	*E*	Electrical "pressure" that causes current flow
Current	Ampere (A)	*I*	Amount of electron movement through a circuit
Resistance	Ohm (Ω)	*R*	Opposition to current flow
Power	Watt (W)	*P*	Amount of work done as current flows through a circuit

Table 2-3
PREFIXES OF UNITS SMALLER THAN 1

Prefix	Abbreviation	Fractional part of a whole unit
milli	m	1/1000 or 0.001 (3 decimal places)
micro	μ	1/1,000,000 or 0.000001 (6 decimal places)
nano	n	1/1,000,000,000 or 0.000000001 (9 decimal places)
pico	p	1/1,000,000,000,000 or 0.000000000001 (12 decimal places)

decimal point in 225 three places to the *left* (.225) since the prefix milli has three decimal places. So 225 mV equals 0.225 V. This same method is used when changing any fractional prefix unit back to the basic electrical unit.

Large Units

Sometimes electrical units of measurement are very large. Units such as 20,000,000 W, 50,000 Ω, or 38,000 V are very large. When this occurs, prefixes are needed to make these large numbers easier to use. Some prefixes used for large electrical values are shown in Table 2-4. To change a large value to a smaller unit, divide the large value by the number of the prefix. For example, 48,000,000 Ω is changed to 48 megohms (MΩ) by dividing (48,000,000 divided by 1,000,000 = 48 MΩ). To convert 7000 V to 7 kilovolts (kV), divide 7000 by 1000 (7000 divided by 1000 = 7 kV). To change a large value to a prefix unit, move the decimal point in the large value to the *left* by the number of zeros of the prefix. Thus 3600 V equals 3.6 kV (3.600). To convert a prefix unit back to a large number, the decimal point is moved to the *right* by the same number of places in the unit. Also, the number may be multiplied by the number of the prefix. If 90 MΩ is converted to ohms, the decimal point is moved six places to the right (90,000,000).

Table 2-4
PREFIXES OF LARGE UNITS

Prefix	Abbreviation	Number of times larger than 1
Kilo	k	1000
Mega	M	1,000,000
Giga	G	1,000,000,000

The 90-MΩ value may also be multiplied by the number of the prefix, which is 1,000,000. Thus 90 MΩ × 1,000,000 = 90,000,000 Ω.

The simple conversion scale shown in Fig. 2-18 is useful when converting large and small units to units of measurement with prefixes. This scale uses either powers of 10 or decimals to express the units.

SCIENTIFIC NOTATION

Using powers of 10 or scientific notation greatly simplifies math operations. A number that has many zeros to the right or to the left of the decimal point is made by simpler by putting it in the form of scientific notation (powers of 10). For example, 0.0000035 × 0.000025 is difficult to multiply. It can be put in the form $(3.5 \times 10^{-6}) \times (2.5 \times 10^{-5})$. Notice the number of places that the decimal point is moved in each number.

Table 2-5 lists some of the powers of 10. In a whole number, the power to which the number is raised is *positive*. It equals the number of zeros following the 1. In decimals, the power is *negative* and equals the number of places the decimal point is moved to the left of the 1. Easy powers of 10 to remember are $10^2 = 100$ (10 × 10) and $10^3 = 1000$ (10 × 10 × 10).

Any number written as a multiple of a power of 10 and a number between 1 and 10 is said to be expressed in *scientific notation*. For example:

$$81,000,000 = 8.1 \times 10,000,000 \text{ or } 8.1 \times 10^7$$
$$500,000,000 = 5 \times 100,000,000 \text{ or } 5 \times 10^8$$
$$0.0000000004 = 4 \times 0.0000000001 \text{ or } 4 \times 10^{-10}$$

CONVERSION SCALE FOR LARGE OR SMALL NUMBERS

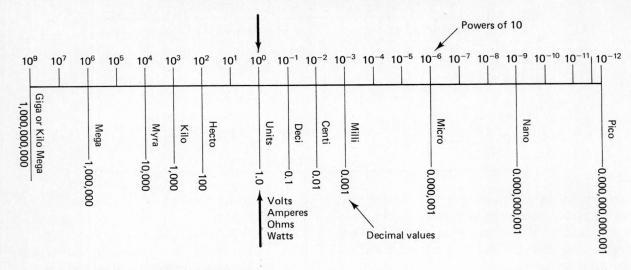

Directions for using the conversion scale:

1. Find the position of the term as expressed in its original form. ⟶ 20 μA
2. Select the position of the conversion unit on the scale. ⟶ Amperes
3. Write the original number as a whole number or in powers of 10. ⟶ 20 μA or 20 × 10⁻⁶ A
4. Shift the decimal point *in the direction of* the desired unit. ⟶ 0.000020 A or 20 × 10⁻⁶ A

Decimal point moved six places
to the left

Figure 2-18. Simple conversion scale for large or small numbers.

Table 2-5
POWERS OF 10

	Number	Power of 10
Whole numbers	1,000,000	10^6
	100,000	10^5
	10,000	10^4
	1000	10^3
	100	10^2
	10	10^1
	1.0	10^0
Decimals	0.1	10^{-1}
	0.01	10^{-2}
	0.001	10^{-3}
	0.0001	10^{-4}
	0.00001	10^{-5}
	0.000001	10^{-6}

Scientific notation simplifies multiplying and dividing large numbers of small decimals. For example:

$$4800 \times 0.000045 \times 800 \times 0.0058$$
$$= (4.8 \times 10^3) \times (4.5 \times 10^{-5}) \times (8 \times 10^2) \times (5.8 \times 10^{-3})$$
$$= (4.8 \times 4.5 \times 8 \times 5.8) \times (10^{3-5+2-3})$$

$$= 1002.24 \times 10^{-3}$$
$$= 1.00224$$

$$95,000 \div 0.0008$$
$$= \frac{9.5 \times 10^4}{8 \times 10^4}$$
$$= \frac{9.5 \times 10^{4-(-4)}}{8}$$
$$= \frac{9.5 \times 10^8}{8} = 1.1875 \times 10^8$$
$$= 118,750,000$$

With some practice the use of scientific notation becomes easy.

SCHEMATIC DIAGRAMS

Schematic diagrams are used to represent the parts of electrical equipment or circuits. They show how the components or parts of each circuit fit together. Schematic diagrams are used to show the *details* of the electrical connections of any type of circuit or system. Schematics are used by manufacturers of electrical equipment showing operation and as an

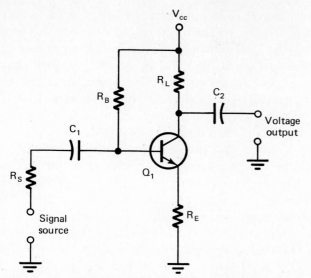

Figure 2-19. Schematic diagram of a transistor amplifier circuit.

aid in servicing the equipment. A typical schematic diagram is shown in Fig. 2-19. Notice the symbols that are used. Symbols are used to represent electrical components in schematic diagrams. Standard electrical symbols are used by all equipment manufacturers. Some common basic electrical symbols are shown in Appendix 4. These symbols should be memorized.

BLOCK DIAGRAMS

Another way to show how electrical equipment operates is to use block diagrams. Block diagrams show the functions of the subparts of any electrical system. A block diagram of an electrical system was shown in Fig. 1-1(a). The same type of diagram is used to show the parts of a radio in Fig. 2-20. Inside the blocks, symbols or words are used to describe the function of the block. Block diagrams usually show the operation of the whole system. They provide an idea of how a system operates. However, they do not show detail like a schematic diagram. It is easy to see the major subparts of a system by looking at a block diagram.

WIRING DIAGRAMS

Another type of electrical diagram is called a wiring diagram (sometimes called a cabling diagram). Wiring diagrams show the actual location of parts and wires on equipment. The details of each *connection* are shown on a wiring diagram. Schematic and block diagrams show only how parts fit together *electrically*. Wiring diagrams show the details of actual connections. A simple wiring diagram is shown in Fig. 2-21.

MEASURING RESISTANCE

Many important electrical tests may be made by measuring resistance. Resistance is opposition to the flow of current in an electrical circuit. The current that flows in a circuit depends upon the amount of resistance in that circuit. You should learn to measure resistance in an electrical circuit by using a meter.

Figure 2-20. Block diagram of the parts of a radio.

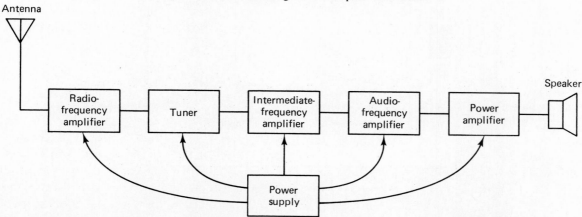

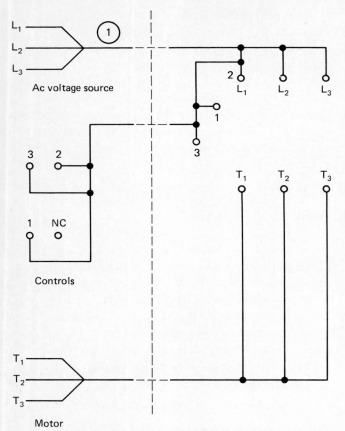

Figure 2-21.   Simple wiring diagram.

A volt-ohm-millimeter (VOM) is used to measure resistance. The VOM, such as the one shown in Fig. 1-6, is one of the most used meters for doing electrical work. A VOM is used to measure resistance, voltage, or current. The type of measurement is changed by adjusting the "function select switch" to the desired measurement. Figure 2-22 shows the controls of a common type of VOM.

Notice that the function select switch is in the center of the meter. Also notice that part of the ranges are for measuring "ohms" or resistance. This is called a multirange, multifunction meter. The ohms measurement ranges are divided into four portions: ×1, ×10, ×1,000, and ×100,000. Most VOMs are very similar to the example shown. The VOM is adjusted to any of the four positions for measuring resistance. The test leads used with the VOM are ordinarily black and red. These colors are used to help identify which lead is the positive or negative side of the meter. This is important when measuring direct-current (dc) values. Red indicates positive polarity (+) and black indicates the negative (−) polarity.

Refer again to the diagram of the meter controls shown in Fig. 2-22. The red test lead is put in the

Figure 2-22.   Controls of a typical VOM.

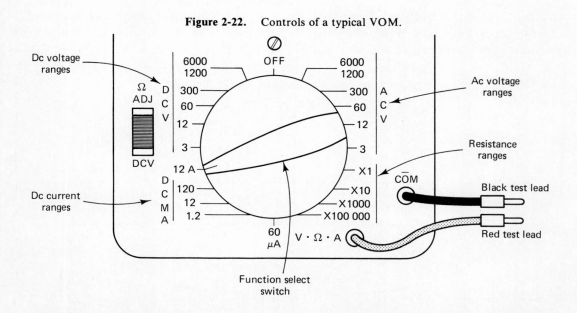

hold or "jack" marked with "V-Ω-A" or volts-ohms-amperes. The black test lead is put in the hold or jack labeled "—COM" or negative common. The function select switch should be placed on one of the resistance ranges. When the test leads are touched together or "shorted," the meter needle moves from the left side of the meter scale to the right side. This test shows that the meter is operational.

The meter's *scale* should now be studied. Figure 2-23 shows the scale of a very common type of VOM. Notice that the top scale, from 0 to infinity (∞), is labeled "OHMS." This scale is used for measuring ohms *only*. On most VOMs, the top scale is the resistance or ohms scale. To measure any resistance, first select the proper meter range. On the meter range shown in Fig. 2-22, there are four ranges: ×1, ×10, ×1,000, and ×100,000. These values are called multipliers. The ohmmeter must be properly "zeroed" before attempting to measure resistance accurately. To "zero" the ohmmeter properly, touch the two test leads together. This should cause the needle to move from infinity (∞) on the left to zero (0) on the right. Infinity represents a very high resistance. Zero represents a very low resistance. If

the needle does not reach zero or goes past zero when the test leads are touched or shorted, the control marked "Ohms Adjust" is used. The needle is adjusted to zero when the test leads are touched together. The ohms adjust control is indicated by "Ω ADJ" on the diagram. The ohmmeter should always be zeroed prior to each resistance measurement and after changing ranges. If the meter is not zeroed, measurements will be incorrect.

A more accurate measurement of resistance is made when the meter's needle stops somewhere between the center of the ohms scale and zero. Choosing the proper range adjustments controls how far the needle moves. If the same range selected is ×1, this means that the number the needle points to is multiplied by 1. If the function select switch is adjusted to the ×100K range, this means that the number the needle points to is multiplied by 100,000. The meter needle should always move to somewhere between the center of the scale and zero. Always zero the meter when changing ranges, and always multiply the number indicated on the scale by the multiplier of the range. Never measure the resistance of a component until it has been *disconnected* or the

Figure 2-23. VOM scale.

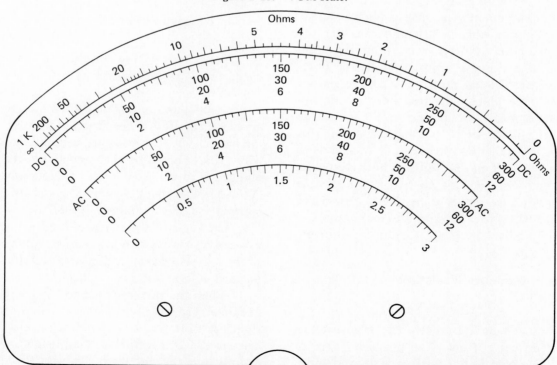

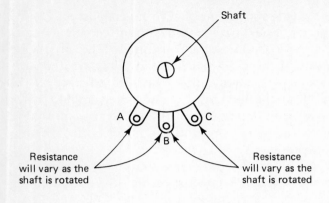

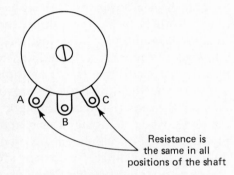

Figure 2-24.   Measuring the resistance of a potentiometer.

reading may be wrong. Voltage should *never* be applied to a component when measuring resistance.

A VOM may also be used to measure the resistance of a potentiometer, as shown in Fig. 2-24. If the shaft of the pot is adjusted while the ohmmeter is connected to points *A* and *C*, no resistance change will take place. The resistance of the potentiometer is measured in this way. Connecting to points *B* and *C*, or to points *B* and *A* allow changes in resistance as the shaft is turned. The potentiometer shaft may be adjusted both clockwise and counterclockwise. This adjustment affects the measured resistance across points *B* and *C* or *B* and *A*. The resistance varies from zero to maximum and from maximum back to zero as the shaft is adjusted.

How to Measure Resistance with a VOM

Remember that resistance is the opposition to flow of electrical current. For example, a lamp connected to a battery has *resistance*. Its resistance value is determined by the size of the filament wire. The filament wire opposes the flow of electrical

current from the battery. The battery causes current to flow through the lamp's filament. The amount of current through the lamp depends on the filament resistance. If it offers little opposition to current flow from the battery, a *large* current flows in the circuit. If the lamp filament has high resistance, it offers much opposition to current flow from the battery. Then a *small* current flows in the circuit.

Resistance tests are sometimes called *continuity checks*. A continuity check is made to see if a circuit is open or closed (a "continuous" path). An ohmmeter is also used to measure exact values of resistance. Resistance must *always* be measured with *no voltage* applied to the component being measured. The ohmmeter ranges of a volt-ohm-milliammeter (VOM) are used to measure resistance. This type of meter is often used by electrical technicians since it measures resistance, voltage, or current. By adjusting the rotary function select switch, the meter can be set to measure either resistance, voltage, or current. The meter switch shown in Fig. 2-22 has settings for:

1. Direct current (dc) voltage
2. Direct current (dc) amps and milliamps
3. Alternating current (ac) voltage
4. Resistance (ohms)

Notice that the lower right part of the function select switch is for measuring resistance or ohms. The ohms measurement settings are marked as ×1, ×10, ×1,000, and ×100,000. When measuring resistance with an ohmmeter, first put the test leads into the meter. The test leads are usually black and red wires that plug into the meter. The red wire is plugged into the hole marked "volts-ohms-amps" (V-Ω-A). The black wire is plugged into the hole marked "negative common" (—COM).

The scale of the meter is used to indicate the value of resistance in ohms. Notice that the right side of the scale is marked with a zero and the left side is marked with an infinity (∞) sign.

When the test leads are touched together, the needle on the scale of the meter should move to the right side of the scale. This indicates *zero* resistance. The needle of the meter is adjusted so that it is exactly over the zero mark. This is called "*zeroing*" the meter. This *must* be done to measure any resistance accurately. The "ohms adjust" (Ω ADJ) control is used to zero the needle of the meter. The meter should be zeroed before each resistance measurement is made.

It is important to be able to read the scale of the meter. Notice that the top scale shown in Fig. 2-23 is labeled with a *zero* on the right side and *infinity* (∞) sign on the left side. This scale is used for measuring resistance only. Most accurate readings are made when the meter's needle moves to somewhere between the center of the scale and zero. Notice that this is due to the *greater distance* between the numbers on the right of the scale.

To measure any resistance accurately, first select the proper range. The ranges for measuring resistance are on the lower right part of the meter's function select switch. This switch has resistance ranges marked as ×1, ×10, ×1,000, and ×100,000. If the meter's range setting is on the ×1, this means that the reading on the meter's scale must be multiplied by 1.

Some examples with the meter range set on different multipliers are done as follows. The meter must be zeroed whenever a range is changed. The test leads are then placed across a resistance. Assume that the needle of the meter moves to point A on the scale of Fig. 2-25. The resistance equals 7.5 × 1 = 7.5 Ω. Now change the meter range to ×1,000. The reading at point B equals 5.5 × 1000 = 5500 Ω. At point C, the reading is 0.3 × 1000 = 300 Ω. The same procedure is used for the ×100,000 range. If the needle moves to 2.2 (point D) on the scale, the reading would be equal to 2.2 × 100,000, or 220,000 Ω. If the meter range is set on ×100,000 and the needle moves to 3.9 (point E) on the scale, the reading is 3.9 × 100,000, or 390,000 Ω.

Remember to zero the meter by touching the test leads together and using the "ohms adjust" control before making a resistance measurement. Each time the meter range is changed, the meter needle must be zeroed on the scale. If this procedure

is not followed, the meter reading will not be accurate.

To learn to measure resistance, it is easy to use color-coded resistors. These resistors are small and easy to handle. Practice in the use of the meter to measure several values of resistors makes reading the meter much easier.

MEASURING VOLTAGE

Voltage is applied to electrical equipment to cause it to operate. It is important to be able to measure voltage to check the operation of equipment. Many electrical problems develop due to either too much or too little voltage being applied to the equipment. A voltmeter is used to measure voltage in an electrical circuit. A VOM is also used to measure voltage. Refer to the controls of a VOM shown in Fig. 2-22. The voltage ranges shown are 3, 12, 60, 300, 1200, and 6000 V. When the function select switch is adjusted to 3 V on the *dc volts range*, the meter measures *up to* 3 V. The same is true for the other ranges of dc voltage. The voltage values of each range is the *maximum* value of voltage that may be measured with the VOM set on that range.

When making voltage measurements, adjust the function select switch to the highest range of dc voltage. Connect the red and black test leads to the meter by putting them into the proper jacks. The red test lead should be put into the jack labeled "V-Ω-A." The black test lead should be put into the jack labeled "—COM."

It is easy to become familiar with the part of the meter's scale that is used to measure dc voltage. Refer to the VOM scale of Fig. 2-23. Notice that the part of the scale below the ohms scale is the dc voltage scale. This scale is usually black. Notice that there are three dc voltage scales: 0 to 12 V, 0 to 60 V,

Figure 2-25. Ohm's scale of a VOM.

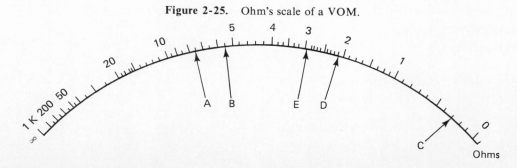

0 to 300 V. All dc voltages are measured using one of these scales. Notice that each of the dc voltage ranges on the function select switch corresponds to a number on the right side of the meter scale *or a* number that can be easily multiplied or divided (by 10) to equal the number on the function switch.

Notice that when the 12, 60, or 300 V range is used, the scale is read directly. On these ranges, the number to which the needle points is the actual value of the voltage being measured. When the 3, 1200, or 6000 V range is used, the number to which the needle points must be *multiplied or divided*. If the meter's needle points to the number 50 while the meter is adjusted to the 60 V range, the measured voltage is 50 V. If the meter's needle points to the number 250 while the meter is adjusted to the 3 V range, the measured voltage is 2.5 V (250 divided by 100 = 2.5). When the 1200 V range is used, the numbers on the 0 to 12 V scale are read and then multiplied by 100. Most VOMs have several scales. Some of these scales are read directly, while others require multiplication or division.

Before making any measurements, the proper dc voltage range is chosen. The value of the range being used is the *maximum* value of voltage that can be measured on that range. For example, when the range selected is 12 V, the maximum voltage the meter can measure is 12 V. Any voltage above 12 V could damage the meter. To measure a voltage that is unknown (no indication of its value), start by using the *highest* range on the meter. Then slowly adjust the range downward until a voltage reading is indicated on the right side of the meter scale.

Matching the meter *polarity* to the voltage polarity is very important when measuring dc voltage. The meter needle moves backward, possibly causing damage to the meter if polarities are not connected properly. Meter polarity is simple to determine. The positive (+) red test lead is connected to the positive side of the dc voltage being measured. The negative (−) black test lead is connected to the negative side of the dc voltage being measured. The meter is always connected *across* (in *parallel* with) the dc voltage being measured.

How to Measure DC Voltage with a VOM

Voltage is the electrical pressure that causes current to flow in a circuit. A common voltage source is a battery. Batteries come in many sizes and voltage values. The voltage applied to a component determines how much current will flow through it.

A dc voltmeter or the dc voltage ranges of a VOM are used to measure dc voltage. The upper left part of the VOM function select switch of Fig. 2-22 is used for measuring dc voltage. The dc voltage ranges are 3, 12, 60, 300, 1200, and 6000 V.

When measuring voltage with a VOM, first put the test leads into the meter. The red test lead is plugged into the hole marked "volts-ohms-amps" (V-Ω-A). The black test lead is plugged into the hole marked "negative common" (−COM). The scale of the meter is used to indicate voltage (in volts). Notice that the left side of the dc voltage ranges is marked zero and the right side is marked 300, 60, and 12. The meter needle rests on the zero until some voltage is measured. These three scales are used to measure dc voltages on the sample meter scale.

To measure a dc voltage, first select the proper range. The ranges for measuring dc voltage are on the upper left part of the meter's function select switch. If the meter's range setting is on the 3 V range, this means that the voltage being measured cannot be larger than 3 V. If the voltage is greater than 3 V, the meter would probably be damaged. You must be careful to use a meter range that is larger than the voltage being measured. Notice that each of the dc voltage ranges on the function select switch corresponds to a number on the right side of the meter scale or a number that can be easily multiplied or divided to equal the number on the function select switch. When the 12, 60, or 300 V range is used, the dc voltage scale is read directly. On these ranges, the number that the needle points to is the actual value of the voltage being measured. When the 3, 1200, or 6000 V range is used, the number that the needle points to is multiplied or divided by 100. If the meter's needle points to the number 850 while the 3 V range is being used, the measured voltage is 8.5, since 850 divided by 100 equals 8.5.

Some examples of dc voltage measurements with the meter set on the 3 V range are given below. If the test leads of the meter are placed across a voltage source and the meter's needle moves to point *A* on the scale of Fig. 2-26, the dc voltage is equal to 100 divided by 100, or 1 V. The reading at point *B* is 165 divided by 100, or 1.65 V. At point *C*, the reading is 280 divided by 100, or 2.8 V. There is some difficulty in reading the voltage divisions on the scales. Look at the division marks from 200 to 250.

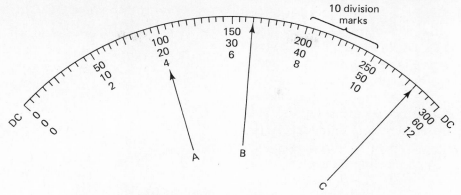

Figure 2-26. Dc voltage scale of a VOM.

The difference between 200 and 250 is 50 units (250 minus 200 equals 50). There are 10 division marks between 200 and 250. The voltage per division mark is 50 divided by 10, or 5 V per division. So each division mark between 200 and 250 is equal to 5 V. This procedure is like reading a ruler or other types of scales.

If the range switch is changed to the 12 V position, the voltage is read directly from the meter scale. For example, if the range is set on 12 V and the meter needle moved on point *A* on Fig. 2-26, the voltage equals 4 V. The reading at point *B* equals 6.6 V. At point *C*, the reading is 11.2 V. The same procedure is used for all other ranges.

When measuring voltage, always be sure to select the proper range. The range used is the *maximum* value of voltage that can be measured on that range. For example, when the range selected is 12 V, the maximum voltage that the meter can measure is 12 V. Any voltage above 12 V could damage the meter. When measuring an unknown voltage, start with the highest range setting on the meter. Then slowly adjust the range setting to lower values until the meter needle moves to somewhere between the center and right side of the meter scale.

When measuring direct current voltage, polarity is very important. The proper matching of meter polarity and voltage source polarity must be assured. The negative (black) test lead of the meter is connected to the negative polarity of the voltage being measured. The positive (red) test lead is connected to the positive polarity of the voltage. If the polarities are reversed, the meter needle will move backward and possibly damage the meter.

A certain amount of voltage is needed to cause electrical current to flow through a resistance in a circuit. This voltage is called *voltage drop*. Voltage

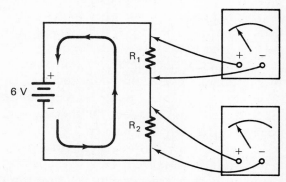

Figure 2-27. Measuring voltage drop in a dc circuit.

drop is measured across any component through which current flows. The polarity of a voltage drop depends on the direction of current flow. Current flows from the negative polarity of a battery to the positive polarity. In Fig. 2-27, the bottom of each resistor is negative. The top of each resistor is positive. The negative test lead of the meter is connected to the bottom of the resistor. The positive test lead is connected to the top. The meters are connected as shown to measure each of the voltage drops in the circuit. If the meter polarity is reversed, the meter needle would move in the wrong direction.

MEASURING CURRENT

Current flows through a complete electrical circuit when voltage is applied. Many important tests are made by measuring current flow in electrical circuits. The current values in an electrical circuit depend on the amount of resistance in the circuit. Learning to use an *ammeter* to measure current in an electrical circuit is very important.

Most VOMs will also measure dc current. Refer to the controls of the VOM shown in Fig. 2-22. The function select switch may be adjusted to any of five ranges of direct current. These are: 12 A, 120 mA, 12 mA, 1.2 mA, and 60 μA. For example, when the function select switch is placed in the 120 mA range, the meter is capable of measuring up to 120 mA of current. The value of the current set on the range is the maximum value that can be measured on that range. The function select switch should first be adjusted to the highest range of direct current. Current is mesured by connecting the meter into a circuit as shown in Fig. 2-28. This is referred to as connecting the meter in *series* with the circuit. Series circuits are discussed in detail in Chapter 3.

Current flows from a voltage source when some device that has resistance is connected to the source. When a lamp is connected to a battery, a current flows from the battery through the lamp. In the circuit of Fig. 2-28, electrons flow from the negative battery terminal, through the lamp, and back to the positive battery terminal. Electrons are so small that the human eye cannot see them, but their movement can be measured with an ammeter.

As the voltage applied to a circuit increases, the current also increases. So, if 12 V is applied to the lamp in Fig. 2-28, a larger current will flow through the lamp. If 24 V is applied to the same lamp, an even larger current will flow. As resistance gets smaller, current increases. Resistance is the opposition to current flow. When a circuit has more resistance, it has less current flow.

Figure 2-28.   Meter connection for measuring dc current.

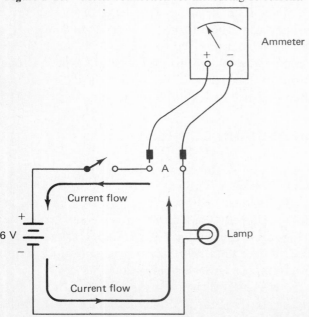

How to Measure DC Current with a VOM

Refer to the direct current ranges of the VOM shown in Fig. 2-22. Notice that the ranges begin with 12 A. The next ranges are for measuring 120 mA, 12 mA, 1.2 mA, and 60 μA. There are a total of five current ranges. The function select switch is adjusted to any of these five ranges for measuring direct current. When measuring current, always start with the meter set on its *highest* range. If this procedure is practiced, it will become a habit. This habit helps in using a meter properly. Always start on the highest range. Then move the range setting to a lower value if the meter needle only moves a small amount. The most accurate reading is when the meter needle is between the center of the scale and the right side. The same scales on the VOM are often used for measuring direct current and dc voltage.

If the meter range is set on the 12 A range, the scale is read directly. The bottom dc scale, which has the number "12" on the right side, is used. Some examples are shown in Fig. 2-29 with the meter set on the 12 A range. At point *A* on the scale, the reading is 4.6 A. The reading at point *B* is 8.8 A.

The 60 μA range on the meter is for measuring very small currents. This range is also read directly on the meter scale. Notice that the number "60" is the middle number on the right of the dc scale.

When the meter is set on the 120 mA range, the meter will measure up to 120 mA of direct current. The readings on the scale are multiplied by 10 on this range setting. The readings at the points shown in Fig. 2-29 for the 120 mA range are:

Point *A* = 46 (4.6 times 10) mA
Point *B* = 88 (8.8 times 10) mA
Point *C* = 100 (10 times 10) mA
Point *D* = 113 (11.3 times 10) mA

Notice that the reading at point *D* is between two of the scale divisions. Point *D* is halfway between the 11.2 and 11.4 divisions on the scale. So the reading is 11.3 times 10, or 113 mA.

The test lead polarity of the VOM is also important when measuring direct current. The VOM is connected to allow current to flow through the meter in the right direction. The negative test lead is connected nearest to the negative side of the voltage source. The meter is then connected *into* the circuit.

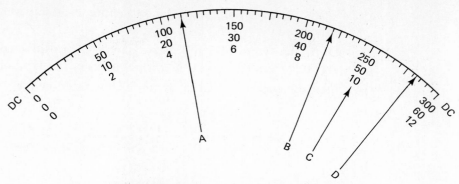

Figure 2-29. Dc current scale of a VOM.

To measure current, a wire is removed from the circuit to place the meter into the circuit. No voltage should be applied to the circuit when connecting the current meter. The meter is placed *in series* with the circuit. Series circuits have *one path* for current flow.

The proper procedure for measuring current through point *A* in the circuit of Fig. 2-28 is:

1. Turn off the circuit's voltage source by opening the switch.
2. Set the meter to the highest current range (12 A).
3. Remove the wire at point *A*.
4. Connect the negative test lead of the meter to the negative side of the voltage source.
5. Connect the positive test lead to the end of the wire which was removed from point *A*.
6. Turn on the switch to apply voltage to the circuit.
7. Look at the meter needle to see how far it has moved up the scale.
8. Adjust the meter range until the needle moves to between the center of the scale and the right side.

Always remember the following safety tips when measuring current:

1. Turn off the voltage before connecting the meter, so as not to get an electrical shock. This is an important habit to develop. Always remember to turn off the voltage before connecting the meter.
2. Set the meter to its highest current range. This assures that the meter needle will not move too far to the right of the scale and possibly cause damage to the meter.
3. A wire is disconnected from the circuit and the meter is put in series with the circuit. Always remember to disconnect a wire and reconnect the wire to one of the meter test leads. If a wire is not removed to put the meter into the circuit, the meter will not be connected properly.
4. Use the proper meter polarity. The negative test lead is connected so that it is nearest the negative side of the voltage source. Similarly, the positive test lead is connected so that it is nearest the positive side of the voltage source.

Review

1. Why are symbols used to represent electrical components?
2. Draw the symbols for the components listed below.
   - **a.** Resistor
   - **b.** Ground connection
   - **c.** SPST switch
   - **d.** Potentiometer
   - **e.** Lamp
   - **f.** Fuse

3. Draw the symbols for (a) conductors crossing and (b) conductors connected.

4. Draw the symbol for a battery placing the positive (plus) sign and the negative (minus) sign at the proper sides of the symbol.

5. Identify each of the meter symbols illustrated below.

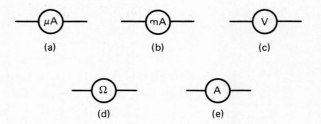

6. Draw a diagram of a circuit that has one battery, two resistors, a single-throw switch, and a current meter connected to form one path.

7. Why is scientific notation used?

8. Why is it necessary to convert large or small electrical units by using prefixes?

9. On which part of the ohms scale of a meter are the most accurate resistance measurements made?

10. What is the ×1000 range of the ohmmeter?

11. What is meant by "zeroing" the meter?

12. If the range of the ohmeter is adjusted to ×100,000 and the needle points to 10.6 on the ohms scale, what is the value of the resistance being measured?

13. What values are shown by the first and second color bands of a color-coded resistor?

14. What value is shown by the third color band of a color-coded resistor?

15. What value is shown by the fourth color band of a color-coded resistor?

16. What unit of electrical measurement is used for resistance?

17. If a fourth color band is not indicated on a resistor, what is its tolerance value?

18. When measuring an unknown voltage, what meter range should be used?

19. What determines how the meter is connected to measure the voltage drop across a resistor?

20. When is it necessary to multiply or divide a scale reading on a VOM?

21. How is the proper voltage range of a VOM selected?

22. Could 4 V be measured on the 3 V range of a VOM? Why?

23. How is the VOM properly connected into a circuit to measure current flow?

24. Why is proper polarity important when measuring direct current?

25. What is the proper procedure to be followed when attempting to measure an unknown current?

Student Activities

Lab Meter Use (see Figs. 2-21 and 2-23)

1. Obtain a meter from the lab.
2. Study the meter and try to answer correctly each of the following questions. Use a sheet of paper to record the answers.
   a. What company manufactured the meter?
   b. What is the model number of the meter?
   c. The meter will measure up to _____ amperes of direct current.
   d. Alternating current (ac) is read on the _____ colored scales.
   e. The "Ohms Adjust" control must be checked each time the resistance range is changed. (True or False.)
   f. To measure a direct current greater than 1 A, the range switch is placed in the _____ position.
   g. For measuring current in a circuit, the meter should be connected in series or parallel?
   h. For measuring voltage, the meter should be connected in series or parallel?
   i. To measure 18 mA of current, the range switch should be placed in the _____ position.
   j. To measure 10 μA of current, the range switch should be placed in the _____ range.
   k. To measure resistance, the red test lead must be placed in the jack marked _____ and the black test lead in the jack marked _____.
   l. The most accurate resistance reading is on the *right* or *left* side of the meter scale?
   m. Polarity is important when measuring ac voltage. (True or False.)
   n. Polarity is *not* important when measuring resistance. (True or False.)
   o. Up to _____ volts can be measured with the meter.
   p. For measuring a resistor valued at 10 Ω, the _____ range should be used.
   q. Polarities must be observed when measuring dc. (True or False.)
   r. It is correct to measure the resistance of a circuit with voltage applied. (True or False.)
   s. When measuring an unknown value of voltage, one should start at the highest scale and work down to the correct scale. (True or False.)
   t. Meters should be handled with care. (True or False.)
3. Turn in the answers to the teacher.

Electrical Symbols

1. Memorize the electrical symbols in Appendix 4.
2. The teacher will give a test on these after they have been memorized.

Electrical Unit Conversions (see Tables 2-2 to 2-4)

1. Answer each of the following electrical unit problems on a sheet of paper.
   a. 0.65 ampere = _____ milliamperes.
   b. 0.12 microfarad = _____ picofarads.

    c. 0.215 millivolt = _____ volts.
    d. 0.0000005 farad = _____ microfarads.
    e. 255 microamperes = _____ amperes.
    f. 45,000 ohms = _____ megaohms.
    g. 0.85 megaohm = _____ ohms.
    h. 6500 watts = _____ kilowatts.
    i. 68,000 volts = _____ kilovolts.
    j. 9200 watts = _____ megawatts.

2. Turn in the answers to the teacher.

Scientific Notation (see Table 2-5)

1. Write the following numbers as powers of 10 on a sheet of paper.
    a. 0.00001
    b. 0.00000001
    c. 10,000,000
    d. 1000
    e. 10
    f. 0.01
    g. 10,000
    h. 0.0001
    i. 1.0
    j. 1,000,000

2. Write the following numbers in scientific notation (as a number between 1 and 10 times a power of 10).
    a. 0.00128
    b. 1520
    c. 0.000632
    d. 0.0030
    e. 28.2
    f. 7,300,000,000
    g. 52.30
    h. 8,800,000
    i. 0.051
    j. 0.000006

3. Turn in the answers to the teacher.

Resistor Color Code

1. Using Fig. 2-13, write the resistance value and tolerance of each example below on a sheet of paper, as _____ ohms; _____ tolerance.

a.			d.					
	(1) Violet		**(1)** Red				**(4)** Silver	
	(2) Green		**(2)** Red			**g.**	**(1)** White	
	(3) Orange		**(3)** Blue				**(2)** Brown	
	(4) Gold	e.	**(1)** Black				**(3)** Orange	
b.	**(1)** Yellow		**(2)** Brown				**(4)** Gold	
	(2) Violet		**(3)** Brown			**h.**	**(1)** Yellow	
	(3) Green		**(4)** Gold				**(2)** Violet	
c.	**(1)** Green	f.	**(1)** Gray				**(3)** Orange	
	(2) Blue		**(2)** Red			**i.**	**(1)** Brown	
	(3) Red		**(3)** Black				**(2)** Green	

(3) Orange		**(4)** Gold		**n.**	**(1)**	Red
j. **(1)** Blue		**l.** **(1)** Brown			**(2)**	Violet
(2) Gray		**(2)** Black			**(3)**	Gold
(3) Black		**(3)** Black			**(4)**	Gold
(4) Gold		**m.** **(1)** Orange		**o.**	**(1)**	Brown
k. **(1)** Green		**(2)** White			**(2)**	Brown
(2) Blue		**(3)** Black			**(3)**	Black
(3) Black		**(4)** Gold			**(4)**	Gold

2. Turn in the answers to the teacher.

Circuit Construction

1. Construct the simple electrical circuit shown in Fig. 2-30 using the materials listed below.

1 1.5-V lamp and socket*
1 sheet metal strip $\frac{1}{2}$ in. wide $\times$ 5 in. long
1 sheet metal strip $\frac{1}{2}$ in. wide $\times$ 2 in. long
1 battery holder (can be made with sheet metal)
2 small nails
1 piece of wood $\frac{1}{2}$ in. $\times$ 3 in. $\times$ 5 in.
3 pieces of small wire
1 1.5-V battery

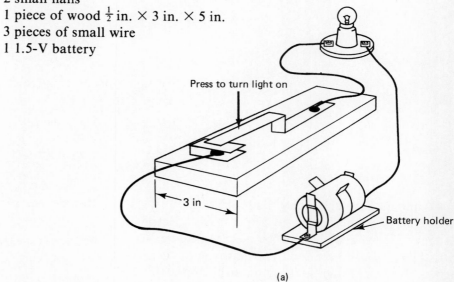

(a)

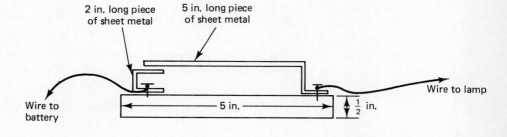

(b)

Figure 2-30. Electrical circuit to build: (a) pictorial view; (b) side view.

2. Turn in the completed project to the teacher for grading.

* Wires can be soldered to the bulb if a socket is not available.

3

Electrical Circuits

To understand electricity and electronics, it is necessary to know how to apply basic electrical theory. Electricity and electronics is a somewhat mathematical subject area. The mathematics is easy to understand, since it has practical applications which are easy to see. The basic theory which is used is called *Ohm's law*. Ohm's law should be learned, as it applies to the basic theory of electrical circuits. All the examples in this chapter are *direct-current* (*dc*) circuits. Alternating-current (ac) circuits are more complex. Ac circuits are studied in Chapter 6.

IMPORTANT TERMS

The following terms should be reviewed before studying Chapter 3.

Branch.   A path of a parallel circuit.

Branch current.   The current through a parallel branch.

Branch resistance.   The total resistance of a parallel branch.

Branch voltage.   The voltage across a parallel branch.

Circuit.   A path through which electrical current flows.

Combination circuit.   A circuit that has one portion connected in series with the voltage source and another part connected in parallel.

Complex circuit.   Another term for a combination circuit.

Current.   The flow of electrical charges through a circuit path or conductor.

Difference in potential.   The voltage across two points of a circuit.

Directly proportional.   When one quantity increases or decreases, causing another quantity to do the same.

Equivalent resistance. A resistance value that would be the same value in a circuit as two or more parallel resistances of a circuit.

Inverse. The value of 1 divided by some quantity, such as $1/R_T$ for finding parallel resistance.

Inversely proportional. When one quantity increases or decreases, causing another quantity to do the opposite.

Kirchhoff's current law. The sum of the current flowing into any point or junction of conductors of a circuit is equal to the sum of the currents flowing away from that point.

Kirchhoff's voltage law. In any current loop of a circuit the sum of the voltage drops is equal to the voltage supplied to that loop; or taken with proper signs (− or +), the algebraic sum of the voltage sources and voltage drops in a circuit is equal to zero.

Ohm's law. The law that explains the relationship of voltage, current, and resistance in electrical circuits.

Parallel circuit. A circuit that has two or more current paths.

Power (P). The rate of doing work in electrical circuits, found by using the equation $P = I \times E$.

Reciprocal. *See* Inverse.

Resistance. Opposition to the flow of current in an electrical circuit.

Series circuit. A circuit that has one path for current flow.

Total current. The current that flows from the voltage source of a circuit.

Total resistance. The total opposition to current flow of a circuit which may be found by removing the voltage source and connecting an ohmmeter across the points where the source was connected.

Total voltage. The voltage supplied by a source.

Voltage. Electrical "pressure" that causes current to flow through a circuit.

Voltage drop. The voltage across two points of a circuit, found by using the equation $E = I \times R$.

Watt (W). The unit of electrical power.

OHM'S LAW

Ohm's law is the most basic and most used of all electrical theories. Ohm's law explains the relationship of voltage (the force that causes current to flow), current (the movement of electrons), and resistance (the opposition to current flow). Ohm's law is stated as follows: An increase in voltage increases current if resistance remains the same. Ohm's law stated in another way is: An increase in resistance causes a decrease in current if voltage remains the same. The electrical values used with Ohm's law are usually represented with capital letters. For example, voltage is represented with the letter E, current with the letter I, and resistance with the letter R. The mathematical relationship of the three electrical units is shown in the following formulas. These should be memorized. The Ohm's law circle of Fig. 3-1 is helpful to remember the formulas.

$$E = I \times R$$
$$I = \frac{E}{R}$$
$$R = \frac{E}{I}$$

Voltage (E) is measured in *volts*. Current (I) is measured in *amperes*. Resistance (R) is measured in *ohms*. If two electrical values are known, the third value can be calculated by using one of the formulas.

Figure 3-1. Ohm's law circle: E, voltage; I, current; R, resistance. To use the circle, cover the value you want to find and read the other values as they appear in the formula: $E = I \times R$, $I = E/R$, $R = E/I$.

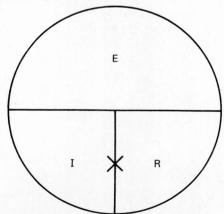

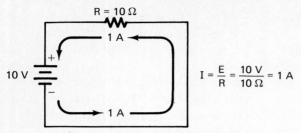

Figure 3-2. Ohm's law example.

Look at Fig. 3-2. Using the Ohm's law current formula, $I = E/R$, the calculated value of I in the circuit is

$$I = \frac{E}{R} = \frac{10\ \text{V}}{10\ \Omega} = 1\ \text{A}$$

If the voltage in the circuit is doubled as shown in Fig. 3-3, the calculated current using the Ohm's law current formula ($I = E/R$) is

$$I = \frac{E}{R} = \frac{20\ \text{V}}{10\ \Omega} = 2\ \text{A}$$

From this example, notice that as voltage is increased, current also increases if resistance remains the same. Also, if voltage is doubled, current is doubled. If voltage is 10 times larger, the current becomes 10 times larger.

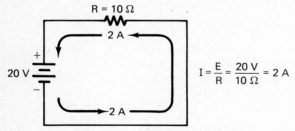

Figure 3-3. Effect of doubling the voltage.

Now look at Fig. 3-4(a). The calculated current flow in the circuit is

$$I = \frac{E}{R} = \frac{10\ \text{V}}{100\ \Omega} = 0.1\ \text{A}$$

In Fig. 3-4(b), the 100-Ω resistor in the circuit is replaced with a 1K-Ω resistor. The calculated value of the current 10 V applied to the 1K-Ω resistance is

$$I = \frac{E}{R} = \frac{10\ \text{V}}{1000\ \Omega} = 0.01\ \text{A}$$

As resistance increases, current decreases.

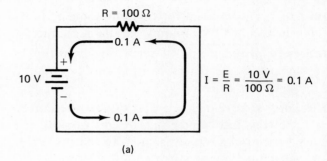

(a)

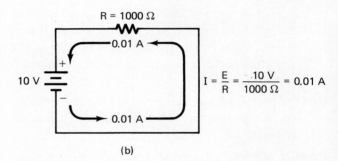

(b)

Figure 3-4. Effect of increasing resistance.

Ohm's law explains the relationship of voltage, current, and resistance in electrical circuits. The circle shown in Fig. 3-1 is used to help remember this relationship. An easy way to remember the Ohm's law formulas used to find voltage, current, and resistance values is to use this circle. To calculate the voltage in a circuit, cover the E on the circle. Note that E is equal to I times R. To find current, cover the I and note that I is equal to E over R. To find resistance, cover the R and note that R is equal to E over I. The circle is easy to remember. It helps in using Ohm's law to solve simple electrical problems.

Another example of using Ohm's law is shown in Fig. 3-5(a). To find the value of current that flows in this circuit, use the Ohm's law circle. Cover the I and find that I equals E over R. In this circuit, $I = 12$ V divided by 2 Ω. So the current is equal to 6 A.

Another example of Ohm's law is shown in Fig. 3-5(b). The voltage of this circuit is equal to 10 V and the current is equal to 2 A. The resistance of the circuit is equal to 10 V divided by 2 A, or 5 Ω. Ohm's law also states that if the resistance of a circuit

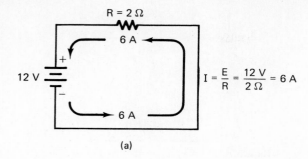

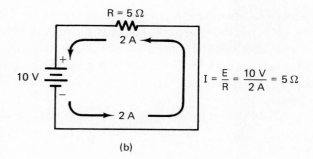

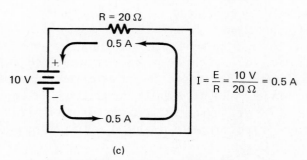

Figure 3-5. Ohm's law examples.

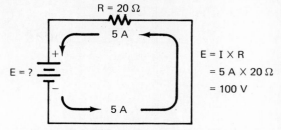

Figure 3-6. Using Ohm's law to find voltage.

Ohm's law is also used to find the value of resistance in a circuit. Assume that a circuit has a known value of voltage and current. The value of resistance required to cause this value of current flow may be found. In the example of Fig. 3-7, the voltage is equal to 70 V and the current equals 10 A. The resistance of the circuit is found by using the Ohm's law circle. Cover the R and find that R is equal to E over I. So the resistance of this circuit is equal to 70 V divided by 10 A (7 Ω).

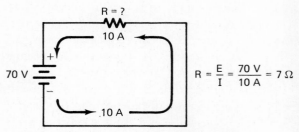

Figure 3-7. Using Ohm's law to find resistance.

increases, the current will decrease if the voltage stays the same. An example of this is shown in Fig. 3-5(c). The resistance of the circuit is increased to 20 Ω. The current of the circuit is now equal to 10 V divided by 20 Ω, or 0.5 A. The current in the previous circuit was 2 A. If the resistance of a circuit is increased four times, the current flow decreases to one-fourth of its original value.

The previous examples dealt mainly with calculating current. Voltage and resistance of a circuit may also be calculated by using Ohm's law. In a circuit in which current and resistance are known, Ohm's law is used to find voltage. In the circuit of Fig. 3-6, assume that resistance is equal to 20 Ω and the current is 5 A. Use the circle and cover the E. This shows that E equals I times R. So the voltage required to cause 5 A of current through a 20-Ω resistance is equal to 5 A times 20 Ω (100 V).

SERIES ELECTRICAL CIRCUITS

One type of electrical circuit is referred to as a series circuit. In a series circuit, there is only one path for current to flow. Since there is only one current path, the current flow is the *same* value in any part of the circuit. The voltages in the circuit depend on the resistance of the components in the circuit. When a series circuit is opened, there is no path for current flow. Thus, the circuit would not operate.

There are three types of electrical circuits. They are called series circuits, parallel circuits, and combination circuits. The easiest type of circuit to understand is the series circuit. Series circuits are different from other types of electrical circuits. It is important to remember the characteristics of a series circuit. In the circuit examples that follow, many *subscripts* (such as R_T, E_T, I_1) are used. Those who are not

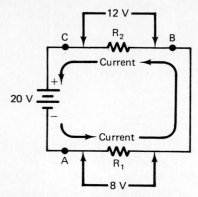

Figure 3-8.   Series electrical circuit.

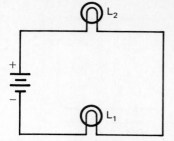

Figure 3-10.   Two lamps connected in series.

familiar with subscripts should review Appendix 5. The main characteristic of a series circuit is that it has only one path for current flow. In the circuit shown in Fig. 3-8, current flows from the negative side of the voltage source, through resistor R_1, through resistor R_2, and then to the positive side of the voltage source. Another characteristic of a series circuit is that the current is the same everywhere in the circuit. A wire at point A could be removed and a VOM placed into the circuit to measure current. The value of current at point A is the same as the current at point B and point C of the circuit.

In any series circuit, the sum of the voltage drops is equal to the voltage applied to the circuit. The circuit is shown in Fig. 3-8 has voltage drops of 12 V plus 8 V, which is equal to 20 V. The voltage across any component in a circuit is called "voltage drop." Another characteristic of a series circuit is that its *total resistance* to current flow is equal to the sum of all resistance in the circuit. In the circuit shown in Fig. 3-9, the total resistance of the circuit is the sum of the two resistances. So the total resistance is equal to 20 Ω plus 10 Ω, or 30 Ω.

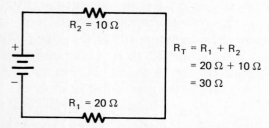

Figure 3-9.   Finding total resistance in a series circuit.

When a series circuit is opened, there is no longer a path for current flow. The circuit will not operate. In the circuit of Fig. 3-10, if lamp 1 is

burned out, its filament is *open*. Since a series circuit has only one current path, that path is broken. No current flows in the circuit. Lamp 2 will not work either. If one light burns out, the others will go out also. This is because the series current path is opened.

Ohm's law is used to explain how a series circuit operates. In the circuit of Fig. 3-11, the total resistance is equal to 2 Ω plus 3 Ω, or 5 Ω. The applied voltage is 10 V. Current is equal to voltage divided by resistance, or $I = E/R$. In the circuit shown, current is equal to 10 V divided by 5 Ω, which is 2 A. If a current meter is connected into this circuit, the current measurement would be 2 A. Voltage drops across each of the resistors may also be found. Voltage is equal to current times resistance ($E = I \times R$). The voltage drop across R_1 (E_1) is equal to the current through R_1 (2 A) times the value of R_1 (2 Ω). The voltage drop across R_1 equals 2 A × 2 Ω, or 4 V. The voltage drop across R_2 (E_2) equals 2 A × 3 Ω, or 6 V. The sum of these voltage drops is equal to the applied voltage. To check these values, add 4 V plus 6 V, which is equal to 10 V.

Figure 3-11.   Using Ohm's law for a series circuit.

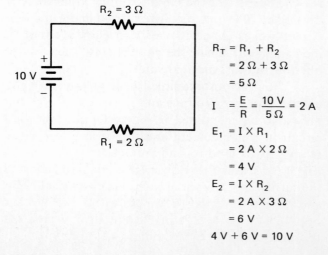

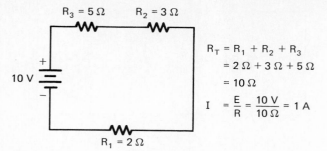

Figure 3-12. Effect of adding resistance to a series circuit.

$$R_T = R_1 + R_2 + R_3$$
$$= 2\,\Omega + 3\,\Omega + 5\,\Omega$$
$$= 10\,\Omega$$
$$I = \frac{E}{R} = \frac{10\text{ V}}{10\,\Omega} = 1\text{ A}$$

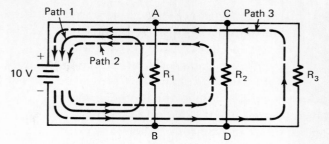

Figure 3-13. Parallel electrical circuit.

If another resistance is added to a series circuit, as shown in Fig. 3-12, resistance increases. Since there is more resistance, the current flow becomes smaller. The circuit now has R_3 (a 5-Ω resistor) added in series to R_1 and R_2. The total resistance is now $2\,\Omega + 3\,\Omega + 5\,\Omega$, or $10\,\Omega$, compared to $5\,\Omega$ in the previous example. The current is now 1 A, compared to 2 A in the other circuit:

$$I = \frac{E}{R} \qquad \text{or} \qquad I = \frac{10\text{ V}}{10\,\Omega} = 1\text{ A}$$

Summary of Series Circuits

There are several important characteristics of series circuits. Remember these basic rules for series circuits:

1. The same current flows through each part of a series circuit.
2. The total resistance of a series circuit is equal to the sum of the individual resistances.
3. The voltage applied to a series circuit is equal to the sum of the individual voltage drops.
4. The voltage drop across a resistor in a series circuit is directly proportional to the size of the resistor.
5. If the circuit is broken at any point, no current will flow.

PARALLEL ELECTRICAL CIRCUITS

It is also important to understand the characteristics of *parallel circuits*. Parallel circuits are different from series circuits in several ways. A parallel circuit

has two or more paths for current to flow from the voltage source. In Fig. 3-13, path 1 is from the negative side of the voltage source, through R_1, and back to the positive side of the voltage source. Path 2 is from the negative side of the voltage source, through R_2, and back to the positive side of the voltage source. Also, path 3 is from the negative side of the power supply, through R_3, and back to the voltage source.

In a parallel circuit, the voltage is the same across every component of the circuit. In Fig. 3-13, the voltage across points A and B is equal to 10 V. This is the value of the voltage applied to the circuit. By following point A to point C, it can be seen that these two points are connected together. Point B and point D are also connected together. So the voltage from point A to point B will be the same as the voltage from points C to D.

Another characteristic of a parallel circuit is that the sum of the currents through each path equals the total current that flows from the voltage source. In Fig. 3-14, the currents through the paths are 1 A, 2 A, and 3 A. One ampere of current could be measured through R_1 at point A or point B. Two amperes of current could be measured through R_2 and 3 A through R_3. The total current is equal to 6A. This value of total current could be measured at point C or point D in the circuit. Remember that the current is the same in every part of a series circuit, and current *divides* through each branch in a parallel

Figure 3-14. Current flow in a parallel circuit.

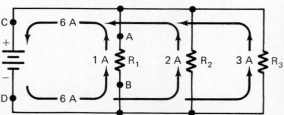

circuit. More resistance causes less current to flow through a parallel *branch*. A branch is a parallel path through a circuit.

The *total resistance* of a parallel circuit is more difficult to calculate than for a series circuit. The formula used is: 1 divided by total resistance ($1/R_T$) is equal to $1/R_1 + 1/R_2 + 1/R_3 + 1$ divided by as many other resistances as there are in the circuit. This is called an inverse or reciprocal formula. Refer to the example of Fig. 3-15. When trying to find the total resistance of a parallel circuit, first write the formula. Notice that 1 is divided by each value . Next, divide each resistance value by 1 and write the values. Then add these values to get a value of 1 divided by total resistance. Do not forget to divide the value obtained by 1 to find the total resistance of the circuit. Also, always remember that the total resistance in a parallel circuit is *less* than any individual resistance in the circuit. In the example shown, 1.33 Ω is less than 2 Ω or 4 Ω.

If there are *only two paths* in a parallel circuit, it is easier to find the total resistance. This formula can be used:

$$R_T = \frac{R_1 \times R_2}{R_1 + R_2}$$

Notice that the two resistances are multiplied and then added. The multiple of the two is put at the top of the formula and the sum is put at the bottom. When using this formula, it is not necessary to divide the resistance values by 1. Remember that this formula may be used *only* when there are two resistances in a parallel circuit. It there are more than two resistances, the reciprocal formula must be used.

Figure 3-15.   Finding total resistance of a parallel circuit.

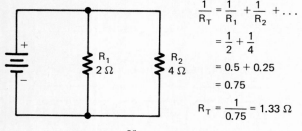

$$\frac{1}{R_T} = \frac{1}{R_1} + \frac{1}{R_2} + \ldots$$
$$= \frac{1}{2} + \frac{1}{4}$$
$$= 0.5 + 0.25$$
$$= 0.75$$
$$R_T = \frac{1}{0.75} = 1.33\ \Omega$$

or
when there are only *two* resistances:

$$R_T = \frac{R_1 \times R_2}{R_1 + R_2} = \frac{2\ \Omega \times 4\ \Omega}{2\ \Omega + 4\ \Omega} = \frac{8}{6} = 1.33\ \Omega$$

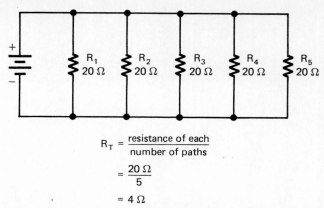

$$R_T = \frac{\text{resistance of each}}{\text{number of paths}}$$
$$= \frac{20\ \Omega}{5}$$
$$= 4\ \Omega$$

Figure 3-16.   Finding total resistance when all resistances are the same.

Another simple method of finding total resistance in a parallel circuit is when *all the resistance values are the same*. An example of a parallel circuit with all resistances the same is a string of Christmas tree lights connected in parallel. Each lamp has the same resistance. When all resistances are equal, to find total resistance, divide the resistance value of each resistor by the number of paths (refer to Fig. 3-16). If five 20-Ω resistors are connected in parallel, the total resistance is equal to 20 divided by 5, or 4 Ω.

When one of the components of a parallel path is opened, the rest of the circuit will continue to operate. Remember that in a series circuit, when one component is opened, no current will flow in the circuit. Since the same voltage is applied to all parts of a parallel circuit, the circuit will operate unless the path from the voltage source is broken (refer to Fig. 3-17). If lamp 3 has a filament burned out, this causes an open circuit. No current flows through lamp 3. However, lamps 1 and 2 will continue to operate. Some types of Christmas tree lights are connected in parallel. If one of these lights burns out, the rest of them will still burn.

A sample problem with a parallel circuit is shown in Fig. 3-18. This circuit has a source voltage

Figure 3-17.   Three lamps connected in parallel.

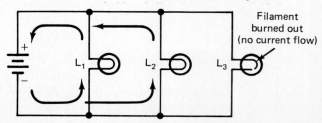

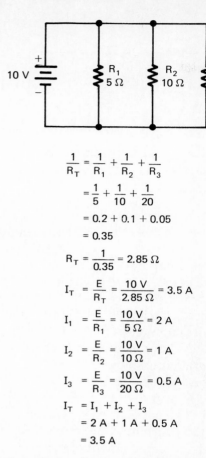

$$\frac{1}{R_T} = \frac{1}{R_1} + \frac{1}{R_2} + \frac{1}{R_3}$$

$$= \frac{1}{5} + \frac{1}{10} + \frac{1}{20}$$

$$= 0.2 + 0.1 + 0.05$$

$$= 0.35$$

$$R_T = \frac{1}{0.35} = 2.85 \ \Omega$$

$$I_T = \frac{E}{R_T} = \frac{10 \ V}{2.85 \ \Omega} = 3.5 \ A$$

$$I_1 = \frac{E}{R_1} = \frac{10 \ V}{5 \ \Omega} = 2 \ A$$

$$I_2 = \frac{E}{R_2} = \frac{10 \ V}{10 \ \Omega} = 1 \ A$$

$$I_3 = \frac{E}{R_3} = \frac{10 \ V}{20 \ \Omega} = 0.5 \ A$$

$$I_T = I_1 + I_2 + I_3$$

$$= 2 \ A + 1 \ A + 0.5 \ A$$

$$= 3.5 \ A$$

Figure 3-18. Sample parallel circuit problem.

of 10 V and resistors of 5, 10, and 20 Ω. First, find the total resistance of the circuit. The total resistance of this circuit is 2.85 Ω. The total opposition to current flow which this circuit has is 2.85 Ω.

Next, find the current that flows from the voltage source. In this circuit the total current is found by dividing the voltage by the total resistance. So 10 V divided by 2.85 Ω = 3.5 A of current. This means that 3.5 A of current flows from the voltage source and divides into the three paths of the circuit.

Now, find the current through each path of the circuit. The current from the voltage source divides into each path of the circuit. A lower resistance in a path causes more current to flow through the path. Current is equal to voltage divided by resistance. The current through each path is found by dividing the voltage (10 V) by the resistance of the path. Notice that the same voltage is across each path. The current values are 2, 1, and 0.5 A. These are added together to equal the total current. The total current of 3.5 A is equal to 2 A + 1 A + 0.5 A.

Parallel Circuit Measurements

To measure current through path 1 in the parallel circuit of Fig. 3-19, use the following procedure:

1. Make sure that no voltage is applied to the circuit by opening the switch.
2. Remove wires 1 and 2 from point *A*.
3. Set the meter to the highest current range.
4. Connect wires 1 and 2 to the positive test lead of the meter.
5. Connect the negative test lead of the meter to point *A*.
6. Turn on the switch to apply voltage to the circuit.

Figure 3-19. Making measurements in a parallel circuit: (a) original circuit; (b) circuit set up to measure current through path 1.

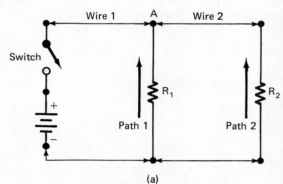

(a)

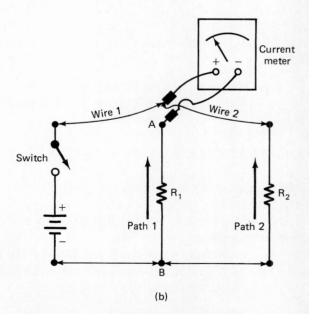

(b)

7. Adjust the meter, if necessary, to a lower range to get an accurate reading.

8. Read the current value on the scale of the meter.

To measure the resistance of a parallel circuit, first remove the voltage source. Prepare the meter to measure resistance. Be sure to zero the meter. Connect the meter across the points where the circuit *was* connected to the voltage source [points *A* and *B* in Fig. 3-19(a)]. Adjust the meter range, if necessary, to get an accurate resistance reading. Be sure to zero the meter each time a change of ranges is made. Once the proper range is selected, accurately read the resistance on the meter scale.

Summary of Parallel Circuits

There are several important characteristics of parallel circuits. Remember these basic rules for parallel circuits:

1. There are two or more paths for current flow.
2. Voltage is the same across each component of the circuit.
3. The sum of the currents through each path is equal to the total current that flows from the source.
4. Total resistance is found by using the formula

$$\frac{1}{R_T} = \frac{1}{R_1} + \frac{1}{R_2} + \frac{1}{R_3} + \cdots$$

5. If one of the parallel paths is broken, current will continue to flow in all the other paths.

COMBINATION ELECTRICAL CIRCUITS

Combination electrical circuits are made up of both series and parallel parts. They are sometimes called series–parallel circuits. Almost all electrical equipment has combination circuits rather than only series circuits or only parallel circuits. However, it is important to understand series *and* parallel circuits in order to work with combination circuits.

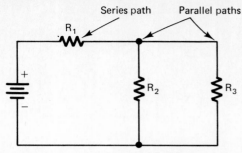

Figure 3-20. Simple combination circuit.

A combination circuit has a series part and a parallel part. Remember that a series circuit has only one path for current flow from the voltage source and a parallel circuit has two or more paths. In the circuit shown in Fig. 3-20, R_1 is in series with voltage source and R_2 and R_3 are in parallel. There are many different types of combination circuits. Some have only one series component and many parallel components. Others have many series components and only a few parallel components. In the circuit shown in Fig. 3-21, R_1 and R_2 are in series with the voltage source. $R_3 - R_4$ and $R_5 - R_6$ are in parallel. Notice that there are two paths in each parallel part of the circuit. The total current of the circuit flows through each series part of the circuit. In this circuit, the current is the same through resistors R_1 and R_2.

Another combination circuit is shown in Fig. 3-22. R_1, R_2, and R_3 are in series with the voltage source. The total current (I_T) flows through R_1, R_2, and R_3. At point A, the current divides through the parallel paths of R_4, R_5, and R_6. The currents I_4, I_5, and I_6 flow through the parallel paths.

Figure 3-21. Combination circuit.

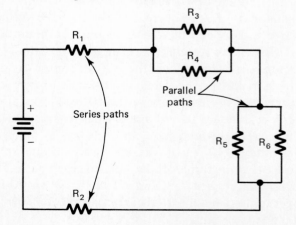

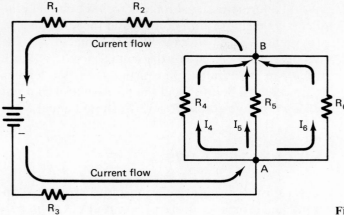

Figure 3-22. Current paths in a combination circuit.

To find the *total resistance* of a combination circuit, the series resistance is added to the parallel resistance. In the circuit shown in Fig. 3-23, R_1 is the only series part of the circuit. R_2 and R_3 are in parallel. First the parallel resistance of R_2 and R_3 is found. Then the series resistance and parallel resistance are added to find the total resistance. Total resistance of this circuit is 4 Ω.

To find the *total current* in the circuit, the same method that was discussed before is used. Current is equal to voltage divided by resistance. The current that flows from the voltage source in this circuit is 2.5 A. The total current also flows through each series part of the circuit. So 2.5 A also flows through

R_1. In this circuit, it is easy to find the current flow through R_2 and R_3. Since R_2 and R_3 are equal resistance values, the same current flows through each of them. There is 2.5 A of current flowing to point *A* of the circuit. This 2.5 A divides into two paths through R_2 and R_3. To find the current through R_2 and R_3, divide the current coming into point *A* (2.5 A) by the number of paths (2). The current through R_2 and R_3 is 1.25 A.

To find voltage across R_1, multiply the current through R_1 by the value of R_1. The voltage across R_1 is equal to 2.5 A times 2 Ω, or 5 V. This is the voltage from point *C* to point *B*. The voltage across R_2 or R_3 is also across points *B* and *A*. The voltage across

Figure 3-23. Combination circuit example.

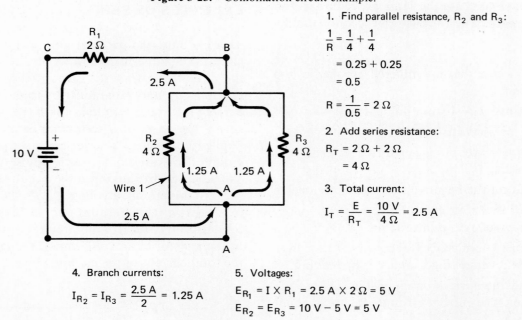

1. Find parallel resistance, R_2 and R_3:

$$\frac{1}{R} = \frac{1}{4} + \frac{1}{4}$$

$$= 0.25 + 0.25$$

$$= 0.5$$

$$R = \frac{1}{0.5} = 2\ \Omega$$

2. Add series resistance:

$$R_T = 2\ \Omega + 2\ \Omega$$

$$= 4\ \Omega$$

3. Total current:

$$I_T = \frac{E}{R_T} = \frac{10\ V}{4\ \Omega} = 2.5\ A$$

4. Branch currents:

$$I_{R_2} = I_{R_3} = \frac{2.5\ A}{2} = 1.25\ A$$

5. Voltages:

$$E_{R_1} = I \times R_1 = 2.5\ A \times 2\ \Omega = 5\ V$$

$$E_{R_2} = E_{R_3} = 10\ V - 5\ V = 5\ V$$

points *B* and *A* is found by subtracting the voltage across R_1 from the applied voltage. The voltage across R_2 and R_3 is equal to 10 V − 5 V, or 5 V.

Combination Circuit Measurements

To measure the *total resistance* of the circuit of Fig. 3-23, first remove the voltage source from the circuit. Prepare the meter to measure resistance. Be sure to zero the meter. Connect the meter across points *A* and *C*. These points are where the voltage source *was* connected into the circuit. Adjust the meter range if necessary to get an accurate resistance reading. Once the proper range is selected, read the measured resistance on the meter scale.

To measure the *current* through R_1, first make sure that no voltage is applied to the circuit. Remove the wire at point *C*. Set the VOM to the highest current range. Connect the negative VOM test lead to point *C* and the positive test lead to the positive power source terminal. Now apply voltage to the circuit. Adjust the meter, if necessary, to a lower range to get an accurate current reading. Then read the current value on the scale of the meter. The current through R_1 should equal 2.5 A. This value is the same as the total current (I_T) of the circuit since R_1 is a series resistance.

To measure the current through R_2 in this combination circuit, use the same procedure as for any parallel path. The following procedure should be followed:

1. Make sure that no voltage is applied to the circuit.
2. Remove wire 1 from point *A*.
3. Set the meter to its highest current range.
4. Connect wire 1 to the positive test lead of the meter.
5. Connect the negative test lead of the meter to point *A*.
6. Apply voltage to the circuit.
7. Adjust the meter, if necessary, to a lower current range to get an accurate reading.
8. Read the current value on the scale of the meter. The current through R_2 should equal 1.25 A.

To measure the voltage across R_1, connect the negative meter lead to point *B* and the positive lead to point *C*. The voltage should equal 5 V. The voltage across R_2 equals the voltage across R_3 since they are in parallel. The negative meter lead is connected to point *A* and the positive lead to point *B*. The voltage across R_2 and R_3 should equal 5 V.

KIRCHHOFF'S LAWS

Ohm's law shows the relationship of voltage, current, and resistance in electrical circuits. *Kirchhoff's voltage law* is also important in solving electrical problems. A scientist named Kirchhoff is given credit for discovering this effect. He found that the sum of voltage drops around any closed-circuit loop must equal the voltage applied to that loop. Figure 3-24(a) shows a simple series circuit to illustrate this law. This law holds true for any series circuit loop.

Kirchhoff's current law is also important in solving electrical problems, especially for parallel circuits. The current law states that at any junction of electrical conductors in a circuit, the total amount of current entering the junction must equal the amount of current leaving the junction. Figure 3-24(b) shows some examples of Kirchhoff's current law.

EXAMPLES OF SERIES CIRCUITS

There is only one path for current flow in a series circuit. The same current flows through each part of the circuit. To find the current through a series circuit, only one value must be found. Kirchhoff's current law states that the sum of the voltage drops across the parts of a series circuit is equal to the applied voltage. In a series circuit, the sum of the voltage drops always equals the source voltage. This can be shown by looking at the circuit shown in Fig. 3-25. In this circuit, a voltage (E_T) of 30 V is applied to a series circuit that has two 10-Ω resistors. The total resistance (R_T) of the circuit is equal to the sum of the resistor values (20 ohms). Using Ohm's law, the total current of the circuit is found:

$$I_T = \frac{E_T}{R_T} = \frac{30 \text{ V}}{20 \text{ }\Omega} = 1.5 \text{ A}$$

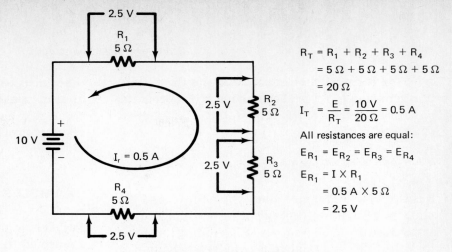

$$R_T = R_1 + R_2 + R_3 + R_4$$
$$= 5\,\Omega + 5\,\Omega + 5\,\Omega + 5\,\Omega$$
$$= 20\,\Omega$$

$$I_T = \frac{E}{R_T} = \frac{10\text{ V}}{20\,\Omega} = 0.5\text{ A}$$

All resistances are equal:

$$E_{R_1} = E_{R_2} = E_{R_3} = E_{R_4}$$

$$E_{R_1} = I \times R_1$$
$$= 0.5\text{ A} \times 5\,\Omega$$
$$= 2.5\text{ V}$$

Source voltage = sum of voltage drops

10 V = 2.5 V + 2.5 V + 2.5 V + 2.5 V

(a)

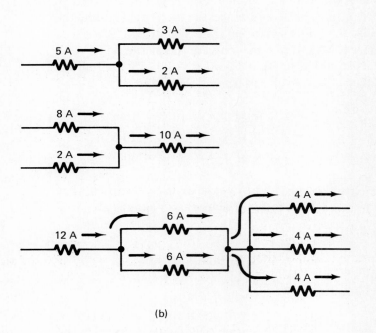

Figure 3-24. Kirchhoff's laws: (a) voltage law example; (b) current law examples.

(b)

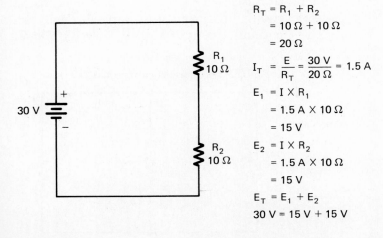

Figure 3-25. Series-circuit example.

$$R_T = R_1 + R_2$$
$$= 10\,\Omega + 10\,\Omega$$
$$= 20\,\Omega$$

$$I_T = \frac{E}{R_T} = \frac{30\text{ V}}{20\,\Omega} = 1.5\text{ A}$$

$$E_1 = I \times R_1$$
$$= 1.5\text{ A} \times 10\,\Omega$$
$$= 15\text{ V}$$

$$E_2 = I \times R_2$$
$$= 1.5\text{ A} \times 10\,\Omega$$
$$= 15\text{ V}$$

$$E_T = E_1 + E_2$$

30 V = 15 V + 15 V

The resistors each equal 10 Ω and the current through them is 1.5 A. The voltage drop across the resistors is then found. The voltage (E_1) across resistor R_1 is

$$E_1 = I \times R_1 = 1.5 \text{ A} \times 10 \text{ }\Omega = 15 \text{ V}$$

Resistor R_2 is the same value as resistor R_1 and the same current flows through it. The voltage drop across resistor R_2 also equals 15 V. Adding these two 15-V drops gives a total voltage of 30 V. The sum of the voltage drops is equal to the applied voltage (15 V + 15 V = 30 V).

The series circuit of Fig. 3-26 has three resistors with values of 10, 20, and 30 Ω. The applied voltage (E_T) may be found since the current through the circuit is known to be 1 A. The circuit is a series circuit, so the same current flows through each resistor. Using Ohm's law ($E = I \times R$), the voltage drop across each resistor can be found:

$$E_1 = I \times R_1 = 1 \text{ A} \times 10 \text{ }\Omega = 10 \text{ V}$$
$$E_2 = I \times R_2 = 1 \text{ A} \times 20 \text{ }\Omega = 20 \text{ V}$$
$$E_3 = I \times R_3 = 1 \text{ A} \times 30 \text{ }\Omega = 30 \text{ V}$$

When the individual voltage drops are known, they may be added to find the applied voltage (E_T):

$$\begin{aligned} E_T &= E_1 + E_2 + E_3 \\ &= 10 \text{ V} + 20 \text{ V} + 30 \text{ V} \\ &= 60 \text{ V} \end{aligned}$$

Notice that the voltage drops in a series circuit are in *direct proportion* (when one increases, the

other does too) to the resistance across which they appear. This is true since the same current flows through each resistor. The larger the value of the resistor, the larger the voltage drop across it will be.

EXAMPLES OF PARALLEL CIRCUITS

Parallel circuits have more than one current path connected to the same voltage source. A parallel-circuit example is shown in Fig. 3-27. Start at the voltage source (E_T) and look at the circuit. Two complete current paths are formed. One path is from the voltage source, through resistor R_1, and back to the voltage source. The other path is from the voltage source, through resistor R_2, and back to the voltage source. Remember that the source voltage divides across the resistors in a *series* circuit. In a parallel circuit, the same voltage is across all the resistors.

Assume that the current through resistor R_1 in Fig. 3-27 is 5 A and the value of the resistor is 10 Ω. The voltage across resistors R_1 and R_2, and the source voltage (E_T) may be found. Using Ohm's law, the voltage across resistor R_1 may be found:

$$E_T = I_1 \times R_1 = 5 \text{ A} \times 10 \text{ }\Omega = 50 \text{ V}$$

Since this is a parallel circuit, the voltages are the same. The source voltage in this circuit is equal to 50 V.

Ohm's law shows that the current in a circuit is *inversely* proportional to the resistance of the circuit. This means that when resistance increases, current decreases. The division of current in parallel circuit paths is based on this fact. Remember that in a series circuit, the current is the same through each part. The total current in a parallel circuit *divides*

Figure 3-26.   Series-circuit example.

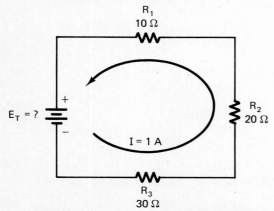

Figure 3-27.   Parallel-circuit example.

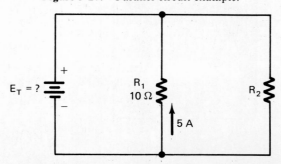

through the paths based on the value of resistance in each path.

The flow of current in a parallel circuit is shown by using the example of Fig. 3-28. Figure 3-28(a) shows a series circuit. The total current passes through one resistor (R_1). The amount of current is:

$$I_1 = \frac{E_T}{R_1} = \frac{10\ \text{V}}{10\ \Omega} = 1.0\ \text{A}$$

Figure 3-28(b) shows the same circuit with another resistor (R_2) connected in parallel across the voltage source. The current through R_2 is the same as through R_1 since their resistances are equal:

$$I_2 = \frac{E_T}{R_2} = \frac{10\ \text{V}}{10\ \Omega} = 1.0\ \text{A}$$

Since 1 A of current flows through each of the two resistors, the total current of two amperes flows from the source. The division of currents through the resistors is shown here. Each current path in the parallel circuit is called a *branch*. Each branch carries part of the total current which flows from the source.

Changing the value of any resistor in a parallel circuit has no effect on the current in the other branches. However, a change in resistance does affect the total current. If R_2 of Fig. 3-28(b) is changed to 5 Ω [Fig. 3-28(c)], the total current would increase:

$$I_1 = \frac{E_T}{R_1} = \frac{10\ \text{V}}{10\ \Omega} = 1.0\ \text{A}$$

$$I_2 = \frac{E_T}{R_2} = \frac{10\ \text{V}}{5\ \Omega} = 2\ \text{A}$$

$$I_T = I_1 + I_2 = 1\ \text{A} + 2\ \text{A} = 3.0\ \text{A}$$

The total resistance of a parallel circuit is *not* equal to the sum of the resistors as in series circuits.

Figure 3-28. Current flow in a parallel circuit: (a) one path; (b) two paths; (c) R_2 changed to 5 Ω.

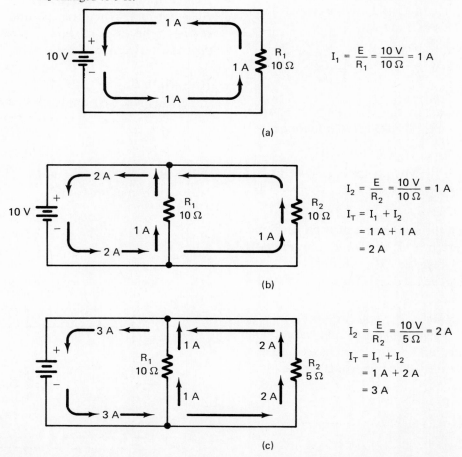

When more branches are added to a parallel circuit, the total current (I_T) increases. In a parallel circuit, the total resistance is *less than any of the branch resistances*. As more parallel resistances are added, the total resistance of the circuit decreases.

There are several ways to find the total resistance of parallel circuits. The method used depends on the type of circuit. Some common methods for finding total parallel resistance are shown below.

1. *Equal resistors*: When two or more equal-value resistors are connected in parallel, their total resistance (R_T) is

$$R_T = \frac{\text{value of one resistance}}{\text{number of paths}}$$

If four 20-Ω resistors are connected in parallel, their total resistance is 20 divided by 4 = 5 Ω. Also, two 50-Ω resistors in parallel − R_T = 50 divided by 2 = 25 Ω.

2. *Product over the sum*: Another shortcut for finding total resistance of *two* parallel resistors is called the product-over-sum method. This method is:

$$R_T = \frac{R_1 \times R_2}{R_1 + R_2}$$

For example, to find the total resistance (R_T) of a 10-Ω and a 20-Ω resistor connected in parallel:

$$R_T = \frac{R_1 \times R_2}{R_1 + R_2} = \frac{10 \times 20}{10 + 20} = \frac{200}{30} = 6.67 \ \Omega$$

Notice that the total resistance of a 10-Ω and a 20-Ω resistor is *less than the smallest* individual resistor. The product-over-sum method can be used only for *two* resistances in parallel.

3. *Reciprocal method*: Most circuits have more than two resistors of unequal value. The reciprocal method must then be used to find total resistance. The reciprocal method is as follows:

$$\frac{1}{R_T} = \frac{1}{R_1} + \frac{1}{R_2} + \frac{1}{R_3} + \cdots$$

(continued for the number of resistances in the circuit)

For example, if a 1-, 2-, 3-, and 4-Ω resistor are connected in parallel:

$$\frac{1}{R_T} = \frac{1}{1} + \frac{1}{2} + \frac{1}{3} + \frac{1}{4} = 1 + 0.5 + 0.33 + 0.25$$

$$= 2.08$$

$$R_T = \frac{1}{2.08}$$

$$= 0.48 \ \Omega$$

EXAMPLES OF COMBINATION CIRCUITS

Most circuits have *both* series and parallel parts. They are called *combination circuits*. The solution of problems for this type of circuit is done by combining series and parallel circuit rules.

Look at the circuit of Fig. 3-29. The value that should first be calculated is the resistance of R_2 and R_3 in parallel. When this quanity is found, it can be added to the value of the series resistor (R_1) to find the total resistance of the circuit:

$$R_T = R_1 + R_2 \| R_3$$

(this means R_2 in parallel with R_3)

$$= 30 \ \Omega + \frac{10 \times 20}{10 + 20} = 30 \ \Omega + 6.67 \ \Omega = 36.67 \ \Omega$$

When the total resistance (R_T) is found, the total current (I_T) may be found:

$$I_T = \frac{E_T}{R_T} = \frac{40 \ \text{V}}{36.67 \ \Omega} = 1.09 \ \text{A}$$

Figure 3-29. Combination-circuit example.

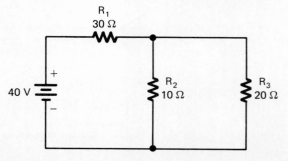

Notice that the total current flows through resistor R_1 since it is in series with the voltage source. The voltage drop across resistor R_1 is

$$E_1 = I_T \times R_1 = 1.09 \text{ A} \times 30 \text{ } \Omega = 32.7 \text{ V}$$

The applied voltage is 40 V and 32.7 V is dropped across resistor R_1. The remaining voltage is dropped across the two parallel resistors (R_2 and R_3). 40 V − 32.7 V = 7.3 V across R_2 and R_3. The currents through R_2 and R_3 are

$$I_2 = \frac{E_2}{R_2} = \frac{7.3 \text{ V}}{10 \text{ } \Omega} = 0.73 \text{ A}$$

$$I_3 = \frac{E_3}{R_3} = \frac{7.3 \text{ V}}{20 \text{ } \Omega} = 0.365 \text{ A}$$

Another type of combination circuit is shown in Fig. 3-30. Resistors R_2 and R_3 are in series. When they are combined by adding their values, the circuit becomes a two-branch parallel circuit. Total resistance (R_T) is found by using the product-over-sum method:

$$R_T = \frac{10 \times 50}{10 + 50} = \frac{500}{60} = 8.33 \text{ } \Omega$$

Total current is

$$I_T = \frac{E_T}{R_T} = \frac{40 \text{ V}}{8.33 \text{ } \Omega} = 4.8 \text{ A}$$

The voltage across R_1 is 40 V since it is in parallel with the voltage source. Voltage drops across R_2 and R_3 are

$$I_2 = \frac{E_T}{R \text{ of branch}} = \frac{40 \text{ V}}{50 \text{ } \Omega} = 0.8 \text{ A}$$

$$E_2 = I \times R_2$$
$$= 0.8 \text{ A} \times 20 \text{ } \Omega$$
$$= 16 \text{ V}$$

$$E_3 = I \times R_3$$
$$= 0.8 \text{ A} \times 30 \text{ } \Omega$$
$$= 24 \text{ V}$$

Notice that E_2 and E_3 (16 V + 24 V) is equal to the source voltage.

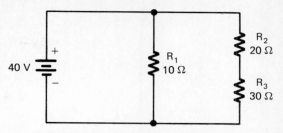

Figure 3-30. Combination-circuit example.

Combination circuit problems may be solved by using the following step-by-step procedure:

1. Combine series and parallel parts to find the total resistance of the circuit.
2. Find the total current that flows through the circuit.
3. Find the voltage across each part of the circuit.
4. Find the current through each resistance of this circuit.

Steps 3 and 4 must often be done in combination with each other, rather than doing one and then the other.

POWER IN ELECTRICAL CIRCUITS

In terms of voltage and current, power (P) is equal to voltage (in volts) multiplied by current (in amperes). The formula is: $P = E \times I$. This formula is an easy way to find electric power. For example, a 120-V electrical outlet with 4 A of current flowing from it has a power value of:

$$P = E \times I \text{ or } 120 \text{ V} \times 4 \text{ A} = 480 \text{ W}$$

The unit of electrical power is the *watt*. In the example, 480 W of power is converted by the load portion of the circuit. Another way to find power is

$$P = \frac{E^2}{R}$$

This formula is used when voltage and resistance are known but current is not known. The formula $P = I^2 \times R$ is used when current and resistance are known.

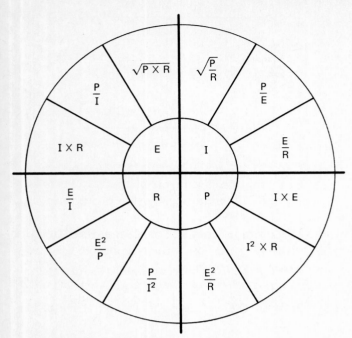

Figure 3-31. Formulas for finding voltage, current, resistance, or power.

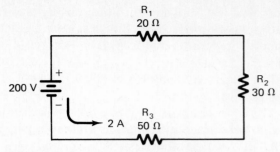

Figure 3-32. Finding power values in a series circuit.

Several formulas are summarized in Fig. 3-31. The quantity in the center of the circle may be found by any of the three formulas along the outer part of the circle in the same part of the circle. This circle is handy to use for making electrical calculations for voltage, current, resistance, or power.

It is easy to find the amount of power converted by each of the resistors in a series circuit, such as the one shown in Fig. 3-32. In the circuit shown, the amount of power converted by each of the resistors and the total power is found as follows:

1. Power converted by resistor R_1:

$$P_1 = I^2 \times R_1 = 2^2 \times 20 \ \Omega = 80 \ W$$

2. Power converted by resistor R_2:

$$P_2 = I^2 \times R_2 = 2^2 \times 30 \ \Omega = 120 \ W$$

3. Power converted by resistor R_3:

$$P_3 = I^2 \times R_3 = 2^2 \times 50 \ \Omega = 200 \ W$$

4. Power converted by the circuit:

$$P_T = P_1 + P_2 + P_3 = 80 \ W + 120 \ W + 200 \ W$$
$$= 400 \ W, \ or$$
$$P_T = E_T \times I = 200 \ V \times 2 \ A = 400 \ W$$

When working with electrical circuits it is possible to check your results by using other formulas.

Power in parallel circuits is found in the same way as for series circuits. In the example of Fig. 3-33, the power converted by each of the resistors and the total power of the parallel circuit is found as follows:

1. Power converted by resistor R_1:

$$P_1 = \frac{E^2}{R_1} = \frac{30^2}{5} = \frac{900}{5} = 180 \ W$$

2. Power converted by resistor R_2:

$$P_2 = \frac{E^2}{R_2} = \frac{30^2}{10} = \frac{900}{10} = 90 \ W$$

3. Power converted by resistor R_3:

$$P_3 = \frac{E^2}{R_3} = \frac{30^2}{20} = \frac{900}{20} = 45 \ W$$

4. Total power converted by the circuit:

$$P_T = P_1 = P_2 + P_3 = 180 \ W + 90 \ W$$
$$+ 45 \ W = 315 \ W$$

The watt is the basic unit of electrical power. To determine an actual quantity of electrical energy, a

Figure 3-33. Finding power values in a parallel circuit.

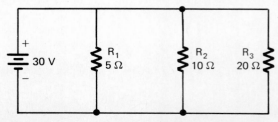

factor that indicates *how long* a power value continued must be used. Such a unit of electrical energy is called a *watt-second*. It is the product of watts (W) and time (in seconds). The watt-second is a very small quantity of energy. It is more common to measure electrical energy in *kilowatt-hours* (kWh). It is the kWh quantity of electrical energy which is used to determine the amount of electric utility bills. A kilowatt-hour is 1000 W in 1 h of time or 3,600,000 W per second.

As an example, if an electrical heater operates on 120 V, and has a resistance of 20 Ω, what is the cost to use the heater for 200 h at a cost of 5 cents per kWh?

1. $P = \dfrac{E^2}{R} = \dfrac{120^2}{20\,\Omega} = \dfrac{14,400}{20\,\Omega}$

$$= 720\ \text{W} = 0.72\ \text{kW}.$$

2. There are 1000 W in a kilowatt (1000 W = 1 kW).

3. Multiply the kW that the heater has used by the hours of use:

$$\text{kW} \times 200\ \text{h} = \text{kilowatt-hours (kWh)}$$
$$0.72 \times 200\ \text{h} = 144\ \text{kWh}$$

4. Multiply the kWh by the cost:

$$\text{kWh} \times \text{cost} = 144\ \text{kWh} \times 0.05 = \$7.20$$

Some simple electrical circuit examples have been discussed in this chapter. They become easy to understand after practice with each type of circuit. It is very important to understand the characteristics of series, parallel, and combination circuits.

Review

1. What is Ohm's law?
2. What are the symbols used for (a) voltage, (b) current, and (c) resistance?
3. What is the relationship of voltage and current in a circuit?
4. What is the relationship of resistance and current in a circuit?
5. What is a series circuit?
6. What is a parallel circuit?
7. What is a combination circuit?
8. What are the voltage, current, and resistance characteristics of (a) series circuits, and (b) parallel circuits?
9. How is total resistance of a series circuit measured?
10. How is total resistance of a parallel circuit measured?
11. How is total current of a series circuit measured?
12. How is total current of a parallel circuit measured?
13. How is voltage drop measured for a series circuit?
14. How is voltage drop measured for a parallel circuit?
15. What are Kirchhoff's laws? Explain them.
16. Explain the three ways used to find total resistance of parallel circuits.
17. What are three ways to find the electrical *power* of a circuit?
18. What is meant by a kilowatt-hour?

Student Activities

Applying Basic Electrical Theory (see Fig. 3-1)

1. Use Ohm's law to solve the following problems. Record your answers on a sheet of paper.
    a. A doorbell requires 0.5 A of current to ring. The voltage applied to the bell is 120 V. What is its resistance? $R =$ _____ Ω.
    b. A relay used to control a motor has a 50-Ω resistance. It draws a current of 0.5 A. What voltage is required to operate the relay? $E =$ _____ V.
    c. An automobile battery supplies a current of 10 A to the starter. It has a resistance of 1.25 Ω. What is the voltage delivered by the battery? $E =$ _____ V.
    d. What voltage is needed to light a lamp if the current required is 3 A and the resistance of the lamp is 80 Ω? $E =$ _____ V.
    e. If the resistance of a radio receiver circuit is 200 Ω and it draws 0.6 A, what voltage is needed? $E =$ _____ V.
    f. A television draws 0.25 A. The operating voltage is 120 V. What is the resistance of the TV circuit? $R =$ _____ Ω.
    g. The resistance of the motor of a vacuum cleaner is 30 Ω. For a voltage of 120 V, find the current. $I =$ _____ A.
    h. The magnet of a speaker carries 0.1 A when connected to a 50-V supply. Find its resistance. $R =$ _____ Ω.
    i. How much current is drawn from a 12-V battery when operating an automobile horn of 16-Ω resistance? $I =$ _____ A.
    j. How much current is drawn by a 2600-Ω clock that plugs into a 120-V outlet? $I =$ _____ A.

2. Turn in your answers to the teacher for grading.

Using Ohm's Law (see Fig. 3-1)

1. Use Ohm's law to solve these problems. Record your answers on a sheet of paper.
    a. $I = 2$ A, $R = 500$ Ω, $E =$ _____ V.
    b. $R = 20$ Ω, $I = 3$ A, $E =$ _____ V.
    c. $I = 1.2$ A, $R = 1000$ Ω, $E =$ _____ V.
    d. $R = 1.5$ kΩ, $I = 130$ mA, $E =$ _____ V.
    e. $R = 120$ kΩ, $I = 20$ mA, $E =$ _____ V.
    f. $E = 2$ V, $R = 1.5$ kΩ, $I =$ _____ A.
    g. $R = 1.8$ kΩ, $E = 10$ V, $I =$ _____ A.
    h. $R = 6.8$ kΩ, $E = 15$ V, $I =$ _____ A.
    i. $R = 5.6$ kΩ, $E = 12$ V, $I =$ _____ A.
    j. $E = 240$ V, $R = 1$ MΩ, $I =$ _____ A.
    k. $E = 20$ V, $I = 5$ A, $R =$ _____ Ω.
    l. $E = 35$ V, $I = 5$ mA, $R =$ _____ Ω.
    m. $I = 10$ μA, $E = 120$ V, $R =$ _____ Ω.

    n. $I = 260$ mA, $E = 120$ V, $R =$ _____ Ω.

    o. $E = 5$ V, $I = 20$ μA, $R =$ _____ Ω.

2. Turn in your answers to the teacher for grading.

Working with Series Circuits (see Fig. 3-11)

1. Solve each of the following series-circuit problems. Record your answers on a sheet of paper.

    a. Find:

        (1) Current (I) = _____ A.

        (2) Voltage across R (E_R) = _____ V.

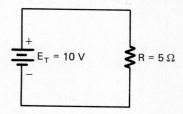

    b. Find:

        (1) Total resistance (R_T) = _____ Ω.

        (2) Current through R_1 (I_1) = _____ A.

        (3) Current through R_2 (I_2) = _____ A.

        (4) Voltage across R_2 (E_2) = _____ V.

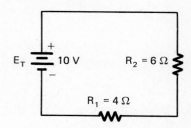

    c. Find:

        (1) Voltage across R_1 (E_1) = _____ V.

        (2) Resistance of R_2 = _____ Ω.

        (3) Voltage across R_2 (E_2) = _____ V.

        (4) Resistance of R_3 = _____ Ω.

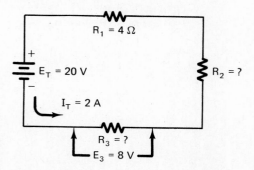

d. Find:

    (1) $E_1 =$ _____ V.

    (2) $R_2 =$ _____ Ω.

    (3) $E_3 =$ _____ V.

    (4) $R_3 =$ _____ Ω.

    (5) $I_3 =$ _____ A.

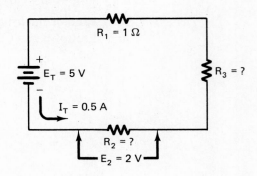

e. Find:

    (1) $I_T =$ _____ A.

    (2) $E_1 =$ _____ V.

    (3) $E_2 =$ _____ V.

    (4) $E_3 =$ _____ V.

    (5)   Power converted by R_1 (P_1) = _____ W.

    (6)   Power converted by R_2 (P_2) = _____ W.

    (7)   Power converted by R_3 (P_3) = _____ W.

    (8)   Total power converted by circuit (P_T) = _____ W.

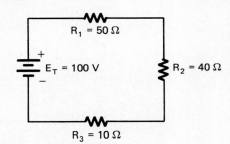

f. Find:

    (1) $E_T =$ _____ V. **(5)**   Total power (P_T) = _____ W.

    (2) $R_T =$ _____ Ω.

    (3) $E_1 =$ _____ V.

    (4) $I_T =$ _____ A.

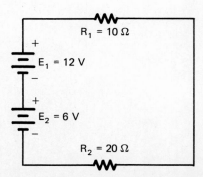

2.   Turn in your answers to the teacher for grading.

Working with Parallel Circuits (see Fig. 3-18)

1. Solve each of the following parallel-circuit problems. Record your answers on a sheet of paper.

   a. Find:
   - **(1)** Total resistance (R_T) = _____ Ω.
   - **(2)** Total current (I_T) = _____ A.
   - **(3)** Current through resistor R_1 (I_1) = _____ A.
   - **(4)** Current through resistor R_2 (I_2) = _____ A.

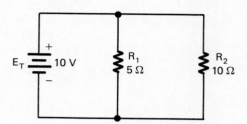

   b. Find:
   - **(1)** Total resistance (R_T) = _____ Ω.
   - **(2)** Total current (I_T) = _____ A.
   - **(3)** Current through resistor R_1 (I_1) = _____ A.
   - **(4)** Current through resistor R_2 (I_2) = _____ A.
   - **(5)** Current through resistor R_3 (I_3) = _____ A.
   - **(6)** Current through resistor R_4 (I_4) = _____ A.
   - **(7)** Total power (P_T) = _____ W.

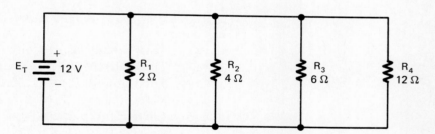

   c. Find:
   - **(1)** Total resistance (R_T) = _____ Ω.
   - **(2)** Total current (I_T) = _____ A.
   - **(3)** Current through resistor R_1 (I_1) = _____ A.
   - **(4)** Current through resistor R_2 (I_2) = _____ A.
   - **(5)** Current through resistor R_3 (I_3) = _____ A.
   - **(6)** Power converted by resistor R_1 (P_1) = _____ W.
   - **(7)** Power converted by resistor R_2 (P_2) = _____ W.
   - **(8)** Power converted by resistor R_3 (P_3) = _____ W.
   - **(9)** Total power (P_T) = _____ W.

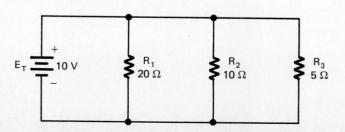

2. Turn in your answers to the teacher for grading.

Working with Combination Circuits (see Fig. 3-23)

1. Solve each of the following combination-circuit problems. Record your answers on a sheet of paper.
   a. Find:
       (1) Total resistance (R_T) = _____ Ω.
       (2) Total current (I_T) = _____ A.
       (3) Voltage across R_1 (E_1) = _____ V.
       (4) Total power (P_T) = _____ W.
       (5) Current through R_2 (I_2) = _____ A.
       (6) Voltage across R_2 (E_2) = _____ V.

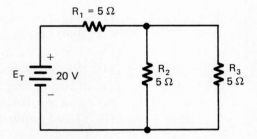

   b. Find:
       (1) Total resistance (R_T) = _____ Ω.
       (2) Total current (I_T) = _____ A.
       (3) Total power (P_T) = _____ W.
       (4) Voltage across R_1 (E_1) = _____ V.
       (5) Current through R_2 (I_2) = _____ A.
       (6) Current through R_4 (I_4) = _____ A.
       (7) Voltage across R_5 (E_5) = _____ V.

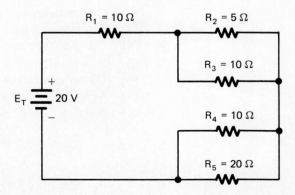

c. Find:

    (1) Total resistance (R_T) = _____ Ω.
    (2) Total current (I_T) = _____ A.
    (3) Voltage across resistor R_1 (E_1) = _____ V.
    (4) Voltage across resistor R_4 (E_4) = _____ V.
    (5) Current through resistor R_2 (I_2) = _____ A.
    (6) Current through resistor R_3 (I_3) = _____ A.

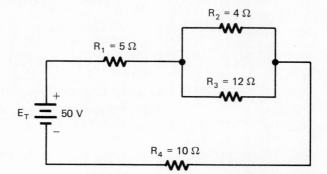

d. Find:

    (1) Total resistance (R_T) = _____ Ω.
    (2) Total current (I_T) = _____ A.
    (3) Voltage across resistor R_1 (E_1) = _____ V.
    (4) Voltage across resistor R_2 (E_2) = _____ V.
    (5) Voltage across resistor R_3 (E_3) = _____ V.
    (6) Current through resistor R_3 (I_3) = _____ A.
    (7) Current through resistor R_2 (I_2) = _____ A.
    (8) Total power (P_T) = _____ W.
    (9) Power converted by R_3 (P_3) = _____ W.

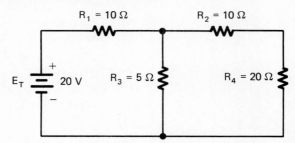

2. Turn in your answers to the teacher for grading.

Series-Circuit Measurement

1. Construct a series circuit similar to Fig. 3-12. Use a 6-V lantern battery or a 1.5-V battery with a holder and any values of resistors R_1, R_2, and R_3 (from 100 Ω to 10 kΩ).

2. Make the following measurements with a VOM and record them on a sheet of paper.

    a. *Before* connecting the circuit, measure:

        (1) $R_1 = $ _____ Ω.

        (2) $R_2 = $ _____ Ω.

        (3) $R_3 = $ _____ Ω.

    b. *After* connecting the battery, measure:

        (1) $E_1 = $ _____ V.

        (2) $E_2 = $ _____ V.

        (3) $E_3 = $ _____ V.

        (4) $I_1 = $ _____ mA.

        (5) $I_2 = $ _____ mA.

        (6) $I_3 = $ _____ mA.

        (7) $I_T = $ _____ mA.

3. Calculate the following power values and record them:

    a. $P_1 = E_1 \times I_1 = $ _____ W.

    b. $P_2 = E_2 \times I_2 = $ _____ W.

    c. $P_3 = E_3 \times I_3 = $ _____ W.

    d. $P_T = E_T \times I_T = P_1 + P_2 + P_3 = $ _____ W.

4. Turn in your answers to the teacher for grading.

Parallel-Circuit Measurement

1. Construct a parallel circuit similar to Fig. 3-13. Use a 6-V lantern battery or a 1.5-V flashlight battery with a holder and any values of resistors R_1, R_2, and R_3 (from 100 Ω to 10 kΩ).

2. Make the following measurements with a VOM and record them on a sheet of paper.

    a. *Before* connecting the circuit, measure:

        (1) $R_1 = $ _____ Ω.

        (2) $R_2 = $ _____ Ω.

        (3) $R_3 = $ _____ Ω.

    b. Connect R_1, R_2, and R_3 as shown in Fig. 3-13. *Do not* connect the battery. Measure:

        (1) $R_T = $ _____ Ω (point *A* to point *B* in Fig. 3-13).

    c. *After* connecting the battery, measure:

        (1) $E_1 = $ _____ V.

        (2) $E_2 = $ _____ V.

        (3) $E_3 = $ _____ V.

        (4) $I_1 = $ _____ A.

        (5) $I_2 = $ _____ A.

        (6) $I_3 = $ _____ A.

        (7) $I_4 = $ _____ A.

3. Calculate the following power values and record them:

    a. $P_1 = E_1 \times I_1 = $ _____ W.

    b. $P_2 = E_2 \times I_2 = $ _____ W.

    c. $P_3 = E_3 \times E_3 = $ _____ W.

    d. $P_T = E_T \times I_T = $ _____ W.

4. Turn in your answers to the teacher for grading.

Combination-Circuit Measurement

1. Construct a combination circuit similar to Fig. 3-20. Use a 6-V lantern battery or a 1.5-V flashlight battery with holder and any values of resistors R_1, R_2, and R_3 (from 100 Ω to 10 kΩ).

2. Make the following measurements with a VOM and record them on a sheet of paper.

    a. *Before* connecting the circuit, measure:

        (1) $R_1 = $ _____ Ω.

        (2) $R_2 = $ _____ Ω.

        (3) $R_3 = $ _____ Ω.

    b. Connect R_1, R_2, and R_3 as shown in Fig. 3-20. Measure:

        (1) $R_1 = $ _____ Ω.

        (2) $R_2 = $ _____ Ω.

        (3) $R_3 = $ _____ Ω.

    c. *After* connecting the battery to the circuit, measure:

        (1) $E_1 = $ _____ V.

        (2) $E_2 = $ _____ V.

        (3) $E_3 = $ _____ V.

        (4) $I_1 = $ _____ mA.

        (5) $I_2 = $ _____ mA.

        (6) $I_3 = $ _____ mA.

        (7) $I_T = $ _____ mA.

3. Turn in your answers to the teacher for grading.

Magnetism and Electromagnetism

Magnetism has been studied for many years. Some metals in their natural state attract small pieces of iron. This property is called magnetism. Materials that have this ability are called natural magnets. The first magnets used were called "lodestones." Now, artificial magnets are made in many different strengths, sizes, and shapes. Magnetism is important since it is used in electric motors, generators, transformers, relays, and many other electrical devices. The earth itself has a magnetic field like a large magnet.

Electromagnetism is magnetism which is brought about due to electrical current flow. There are many electrical machines which operate because of electromagnetism. This chapter deals with magnetism, electromagnetism, and some important applications.

IMPORTANT TERMS

In this chapter, magnetism and electromagnetism are studied. There are many important applications of magnetism and electromagnetism. The terms that follow help to define some of the specifics involved in the study of magnetism and electromagnetism.

Alnico. An alloy of aluminum, nickel, iron, and cobalt used to make permanent magnets.

Ampere-turn. The unit of measurement of magnetic field strength; amperes of current times the number of turns of wire.

Armature. The movable part of a relay.

Coefficient of coupling (*k*). A decimal value that indicates the amount of magnetic coupling between coils.

Core. Iron or steel materials used to make laminated internal sections of electromagnets.

Core saturation. When the atoms of a metal core material are aligned in the same pattern so that no more magnetic lines of force can be developed.

Coupling. The amount of mutual inductance between coils.

Domain theory. A theory of magnetism which assumes that groups of atoms produced by movement of electrons align themselves in groups called "domains" in magnetic materials.

Electromagnet. A coil of wire wound on an iron core so that as current flows through the coil it becomes magnetized.

Flux (Φ). Invisible lines of force that extend around a magnetic material.

Flux density. The number of lines of force per unit area of a magnetic material or circuit.

Gauss. A unit of measurement of magnetic flux density.

Gilbert. A unit of measurement of magnetomotive force (MMF).

Hysteresis. The property of a magnetic material that causes actual magnetizing action to lag behind the force that produces it.

Laws of magnetism. (1) Like magnetic poles repel; (2) unlike magnetic poles attract.

Lines of force. Same as magnetic flux. *See* Flux.

Lodestone. The name used for natural magnets used in early times.

Magnet. A metallic material, usually iron, nickel, or cobalt, which has magnetic properties.

Magnetic circuit. A complete path for magnetic lines of force from a north to a south polarity.

Magnetic field. Magnetic lines of force which extend from a north polarity and enter a south polarity to form a closed loop around the outside of a magnetic material.

Magnetic flux. *See* Flux.

Magnetic materials. Metallic materials such as iron, nickel, and cobalt which exhibit magnetic properties.

Magnetic poles. Areas of concentrated lines of force on a magnet which produce north and south polarities.

Magnetic saturation. A condition that exists in a magnetic material when an increase in magnetizing force does not produce an increase in magnetic flux density around the material.

Magnetomotive force (MMF). A force that produces magnetic flux around a magnetic device.

Magnetostriction. The effect that produces a change in shape of certain materials when they are placed in a magnetic field.

Natural magnet. Metallic materials that have magnetic properties in their natural state.

Permanent magnet. Bars or other shapes of materials that retain their magnetic properties.

Permeability (μ). The ability of a material to conduct magnetic lines of force as compared to air; the ability of a material to magnetize or demagnetize.

Polarities. *See* Magnetic poles.

Relay. An electromagnetic switch that uses a low current to control a higher-current circuit.

Reluctance ($\mathcal{R}$). The opposition of a material to the flow of magnetic flux.

Residual magnetism. The magnetism that remains around a material after the magnetizing force has been removed.

Retentivity. The ability of a material to retain magnetism after a magnetizing force has been removed.

Solenoid. An electromagnetic coil with a metal core which moves when current passes through the coil.

PERMANENT MAGNETS

Magnets are made of iron, cobalt, or nickel materials, usually in an alloy combination. An alloy is a mixture of these materials. Each end of the magnet is called a pole. If a magnet were broken, each part would become a magnet. Each magnet would have two poles. Magnetic poles are always in pairs. When a magnet is suspended in air so that it can turn freely, one pole will point to the North Pole of the earth. The earth is like a large permanent magnet. This is why compasses can be used to determine direction. The north pole of a magnet will *attract* the south pole of another magnet. A north pole *repels* another north pole and a south pole *repels* another south pole. The two laws of magnetism are: (1) like poles repel, and (2) unlike poles attract.

The magnetic field patterns when two permanent magnets are placed end to end are shown in Fig. 4-1. When the magnets are farther apart, a smaller force of attraction or repulsion exists. This type of permanent magnet is called a *bar magnet*.

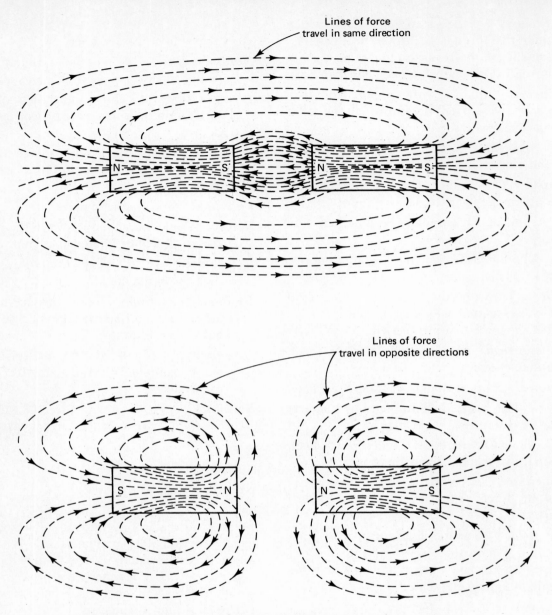

Figure 4-1. Magnetic field patterns when magnets are placed end to end.

Some materials retain magnetism longer than others. Hard steel holds its magnetism much longer than soft steel. A *magnetic field* is set up around any magnetic material. The field is made up of *lines of force* or *magnetic flux*. These magnetic flux lines are invisible. They never cross one another, but they always form individual closed loops around a magnetic material. They have a definite direction from north to south pole along the outside of a magnet. When magnetic flux lines are close together, the magnetic field is strong. When magnetic flux lines are farther apart, the field is weaker. The magnetic

field is strongest near the poles. Line of force pass through all materials. It is easy for lines of force to pass through iron and steel. Magnetic flux passes through a piece of iron as shown in Fig. 4-2.

When magnetic flux passes through a piece of iron, the iron acts like a magnet. Magnetic poles are formed due to the influence of the flux lines. These are called *induced poles*. The induced poles and the magnet's poles attract and repel each other. Magnets attract pieces of soft iron in this way. It is possible to temporarily magnetize pieces of metal by using a bar magnet. If a magnet is passed over the top of a piece

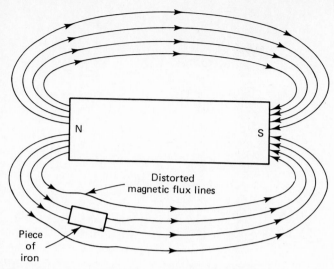

Figure 4-2. Magnetic flux lines distorted while passing through a piece of iron.

of iron several times in the same direction, the soft iron becomes magnetized. It will stay magnetized for a short period of time.

When a compass is brought near the north pole of a magnet, the north-seeking pole of the compass is attracted to it. The polarities of the magnet may be determined by observing a compass brought near each pole. Compasses detect the presence of magnetic fields.

Horseshoe magnets are similar to bar magnets. They are bent in the shape of a horseshoe, as shown in Fig. 4-3. This shape gives more magnetic field strength than a similar bar magnet, since the magnetic poles are closer. The magnetic field strength is more concentrated into one area. Many electrical devices use horseshoe magnets.

A magnetic material can lose some of its magnetism if it is jarred or heated. People must be careful when handling equipment that contains permanent magnets. A magnet also becomes weakened by loss of magnetic flux. Magnets should always be

Figure 4-3. Horseshoe magnet.

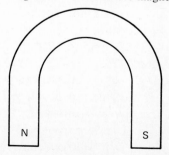

stored with a *keeper*. This is a soft-iron piece used to join magnetic poles. The keeper provides the magnetic flux with an easy path between poles. The magnet will retain its greatest strength for a longer period of time if keepers are used. Bar magnets should always be stored in pairs with a north pole and a south pole placed together. A complete path for magnetic flux is made in this way.

MAGNETIC FIELD AROUND CONDUCTORS

Current-carrying conductors produce a magnetic field. It is possible to show the presence of a magnetic field around a current-carrying conductor. A compass is used to show that the magnetic flux lines are circular in shape. The conductor is in the center of the circular shape. The direction of the current flow and the magnetic flux lines can be shown by using the *left-hand rule* of magnetic flux. A conductor is held in the left hand as shown in Fig. 4-4(a). The thumb points in the direction of current flow from negative to positive. The fingers will then encircle the conductor in the direction of the magnetic flux lines.

A circular magnetic field is produced around a conductor. The field is stronger near the conductor and becomes weaker farther away from the conductor. A cross-sectional end view of a conductor with current flowing toward the observer is shown in Fig. 4-4(b). Current flow toward the observer is shown by a circle with a dot in the center. Notice that the direction of the magnetic flux lines is clockwise. This can be verified by using the left-hand rule.

When the direction of current flow through a conductor is reversed, the direction of the magnetic lines of force is also reversed. The cross-sectional end view of a conductor in Fig. 4-4(c) shows a current flow in a direction away from the observer. Notice that the direction of the magnetic lines of force is now counterclockwise.

The presence of magnetic lines of force around a current-carrying conductor can be observed by using a compass. When a compass is moved around the outside of a conductor, the needle will align itself *tangent* to the lines of force as shown in Fig. 4-4(d). The needle will not point toward the conductor. When current flows in the opposite direction, the compass polarities reverse. The compass needle will align itself tangent to the conductor.

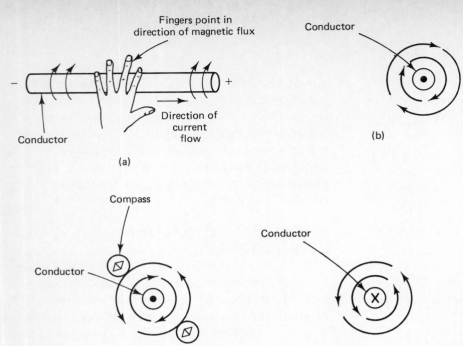

Fingers point in direction of magnetic flux

Direction of current flow

Conductor

Conductor

(a)

Conductor

(b)

Compass

Conductor

Conductor

(c)

Conductor

(d)

Figure 4-4. (a) Left-hand rule of magnetic flux; (b) cross section of conductor with current flow toward observer; (c) compass aligns tangent to magnetic lines of force; (d) cross section of conductor with current flow away from the observer.

MAGNETIC FIELD AROUND A COIL

The magnetic field around one loop of wire is shown in Fig. 4-5. Magnetic flux lines extend around the conductor as shown. Inside the loop, the magnetic flux is in one direction. When many loops are joined together to form a coil, the magnetic flux lines surround the coil as shown in Fig. 4-6. The field around a coil is much stronger than the field of one loop of wire. The field around the coil is the same shape as the field around a bar magnet. A coil that has an iron or steel core inside it is called an *electromagnet*. A core increases the magnetic flux density of a coil.

Figure 4-5. Magnetic field around a loop of wire.

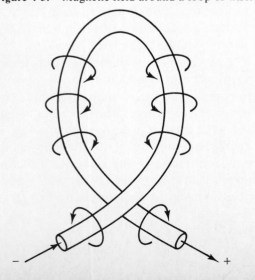

Figure 4-6. Magnetic field around a coil: (a) coil of wire showing current flow; (b) lines of force around two loops that are parallel; (c) cross section of coil showing lines of force.

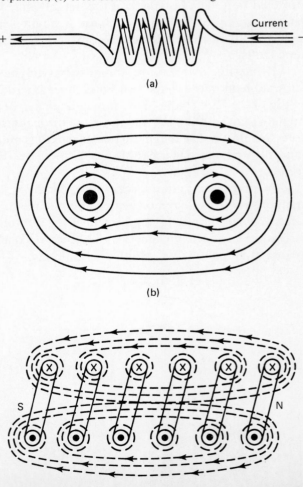

Current

(a)

(b)

(c)

ELECTROMAGNETS

Electromagnets are produced when current flows through a coil of wire as shown in Fig. 4-7. The north pole of a coil of wire is the end where the lines of force come out. The south polarity is the end where the lines of force enter the coil. This is like the field of a bar magnet. To find the north pole of a coil, use the *left-hand rule for polarity*, as shown in Fig. 4-8. Grasp the coil with the left hand. Point the fingers in the direction of current flow through the coil. The thumb points to the north polarity of the coil.

When the polarity of the voltage source is reversed, the magnetic poles of the coil will also reverse. The poles of an electromagnet can be checked with a compass. The compass is placed near a pole of the electromagnet. If the north-seeking pole of the compass points to the coil, that side is the north pole.

Electromagnets have several turns of wire wound around a soft-iron core. An electrical power source is then connected to the ends of the turns of wire. When current flows through the wire, magnetic polarities are produced at the ends of the soft iron core. The three basic parts of an electromagnet are: (1) an iron core, (2) wire windings, and (3) an electrical power source. Electromagnetism is made possible by electrical current flow which produces a magnetic field. When electrical current flows through the coil, the properties of magnetic materials are developed.

(a)

(b)

Figure 4-7. Electromagnets: (a) pictorial; (b) an electromagnet in operation. [(b) Courtesy of O. S. Walker Co.]

Figure 4-8. Left-hand rule for finding the polarities of an electromagnet.

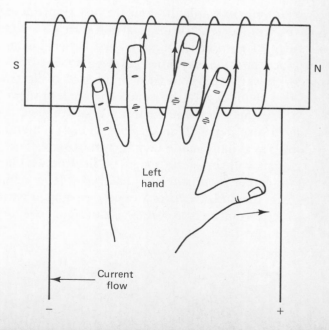

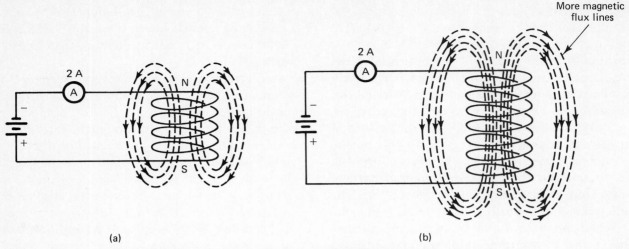

Figure 4-9. Effect of ampere turns on magnetic field strength: (a) five turns, two amperes = 10 ampere-turns; (b) eight turns, two amperes = 16 ampere-turns.

Magnetic Strength of Electromagnets

The magnetic strength of an electromagnet depends on three factors: (1) the amount of current passing through the coil, (2) the number of turns of wire, and (3) the type of core material. The number of magnetic lines of force is increased by increasing the current, by increasing the number of turns of wire, or by using a more desirable type of core material. The magnetic strength of electromagnets is determined by the *ampere-turns* of each coil. The number of ampere-turns is equal to the current in amperes multiplied by the number of turns of wire ($I \times N$). For example, 200 ampere-turns is produced by 2 A of current through a 100-turn coil. One ampere of current through a 200-turn coil would produce the same magnetic field strength. Figure 4-9 shows how the magnetic field strength of an electromagnet changes with the number of ampere-turns.

The magnetic field strength of an electromagnet also depends on the type of core material. Cores are usually made of soft iron or steel. These materials will transfer a magnetic field better than air or other nonmagnetic materials. Iron cores increase the *flux density* of an electromagnet. Figure 4-10 shows that an iron core causes the magnetic flux to be more dense.

Figure 4-10. Effect of an iron core on magnetic strength: (a) coil without core; (b) coil with core.

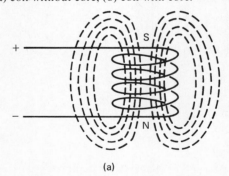

(a)

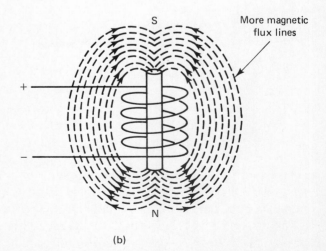

(b)

An electromagnet loses its field strength when the current stops flowing. However, an electromagnet's core retains a small amount of magnetic strength after current stops flowing. This is called *residual magnetism* or "leftover" magnetism. It can be reduced by using soft-iron cores or increased by using hard-steel core material. Residual magnetism is very important in the operation of some types of electrical generators.

In many ways, electromagnetism is similar to magnetism produced by natural magnets such as bar magnets. However, the main advantage of electromagnetism is that it is easily controlled. It is easy to increase the strength of an electromagnet by increasing the current flow through the coil. This is done by increasing the voltage applied to the coil. The second way to increase the strength of an electromagnet is to have more turns of wire around the core. A greater number of turns produces more magnetic lines of force around the electromagnet. The strength of an electromagnet is also affected by the type of core material used. Different alloys of iron are used to make the cores of electromagnets. Some materials aid in the development of magnetic lines of force to a greater extent. Other types of core materials offer greater resistance to the development of magnetic flux around an electromagnet.

OHM'S LAW FOR MAGNETIC CIRCUITS

Ohm's law for electrical circuits was studied in Chapter 3. A similar relationship exists in magnetic circuits. Magnetic circuits have *magnetomotive force* (MMF), *magnetic flux* (Φ), and *reluctance* ($\mathcal{R}$). MMF is the force that causes a magnetic flux to be developed. Magnetic flux is the lines of force around a magnetic material. Reluctance is the opposition to the development of a magnetic flux. These terms may be compared to voltage, current, and resistance in electrical circuits, as shown in Fig. 4-11. When MMF increases, magnetic flux increases. Remember that in an electrical circuit, when voltage increases, current increases. When resistance in an electrical circuit increases, current decreases. When reluctance of a magnetic circuit increases, magnetic flux decreases. The relationship of magnetic and electrical terms in Fig. 4-11 should be studied.

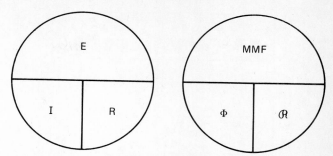

Figure 4-11. Relationship of magnetic and electrical terms.

"DOMAIN" THEORY OF MAGNETISM

A theory of magnetism was presented in the nineteenth century by a German scientist named Wilhelm Weber. Weber's theory of magnetism was called the molecular theory. It dealt with the alignment of molecules in magnetic materials. Weber felt that molecules were aligned in an *orderly* arrangement in magnetic materials. In nonmagnetic materials, he thought that molecules were arranged in a *random* pattern.

Weber's theory has now been modified somewhat to become the "domain" theory of magnetism. This theory deals with the alignment of "domains" in materials rather than molecules. A domain is a group of atoms (about 10^{15} atoms). Each domain acts like a tiny magnet. The rotation of electrons around the nucleus of these atoms is important. Electrons have a negative charge. As they orbit around the nucleus of atoms, their electrical charge moves. This moving electrical field produces a magnetic field. The polarity of the magnetic field is determined by the direction of electron rotation.

The domains of magnetic materials are atoms grouped together. Their electrons are believed to spin in the same directions. This produces a magnetic field due to electrical charge movement. Figure 4-12 shows the arrangement of domains in magnetic, nonmagnetic, and partially magnetized materials. In nonmagnetic materials, half of the electrons spin in one direction and half in the opposite direction. Their charges cancel each other out. There is no magnetic field produced since the charges cancel. Electron rotation in magnetic materials is in the same direction. This causes the domains to act like tiny magnets that align to produce a magnetic field.

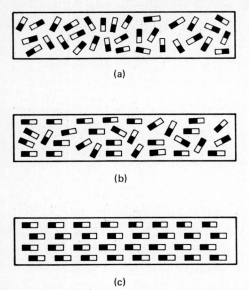

(a)

(b)

(c)

Figure 4-12. Domain theory of magnetism: (a) unmagnetized; (b) slightly magnetized; (c) fully magnetized—saturation.

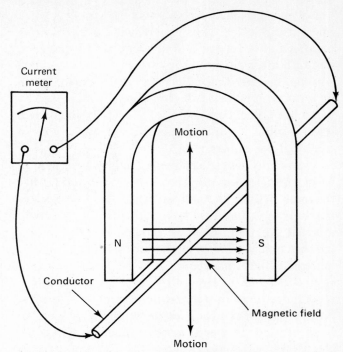

Figure 4-13. Faraday's law—electrical current is produced when there is relative motion between a conductor and a magnetic field.

ELECTRICITY PRODUCED BY MAGNETISM

A scientist named Michael Faraday discovered in the early 1830s that electricity is produced from magnetism. He found that if a magnet is placed inside a coil of wire, electrical current is produced when the magnet is *moved*. Faraday found that electrical current is caused by magnetism and motion.

Faraday's law is stated as follows: When a coil of wire moves across the lines of force of a magnetic field, electrons flow through the wire in one direction. When the coil of wire moves across the magnetic lines of force in the opposite direction, electrons flow through the wire in the opposite direction.

This law is the principle of electrical power generation produced by magnetism. Figure 4-13 shows Faraday's law as it relates to electrical power generation.

Current flows in a conductor placed inside a magnetic field only when there is motion between the conductor and the magnetic field. If a conductor is stopped while moving across the magnetic lines of force, current stops flowing. The operation of electrical generators depends on conductors moving across a magnetic field. This principle, called electromagnetic induction, is discussed in more detail in Chapter 5.

Relays

Relays are electrical devices that rely on magnetism to operate. They control other equipment, such as motors, lights, or heating elements. Relays are very important devices. They are used to start the operation of other equipment. They use a small amount of electrical current to control a larger current, such as the current through a motor.

The basic construction and symbols of a relay is shown in Fig. 4-14. Relays have an electromagnetic coil with electrical power applied to its two external leads. When the power is turned on, the electromagnet is energized. The electromagnet part of the relay controls a set of contacts. The contacts are called *normally open* (NO) or *normally closed* (NC) depending on their condition when the electromagnet is not energized. There is also a *common* contact.

If a lamp and its power source are connected in series with the common and normally open contact as shown in Fig. 4-14(b), the lamp will be off when the relay is not energized. Notice that the lamp or any load connected to the relay contacts requires a separate power source. If the relay is energized by applying power to it, the common contact is attracted to the normally open contact by magnetic energy.

(a)

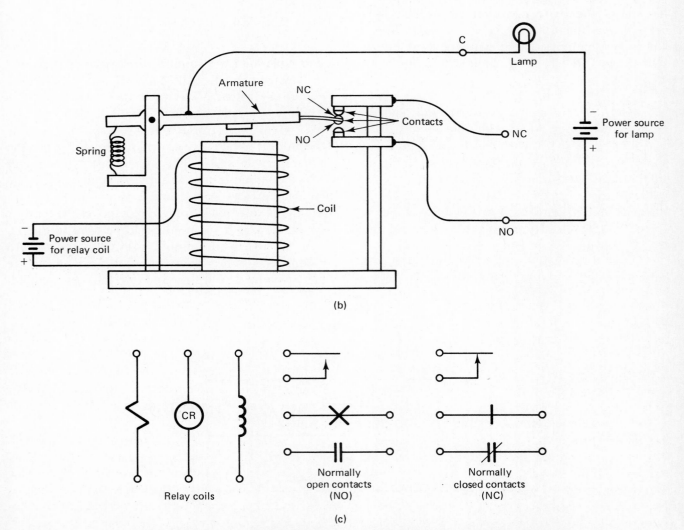

(b)

(c)

Figure 4-14. Construction and symbols of a relay: (a) pictorial; (b) illustration; (c) symbols. [(a) Courtesy of Airpax Corporation, Cambridge Division.]

The common contact is built onto a movable armature that moves when the electromagnet is energized. When the relay is energized, the light connected to the normally open contact will turn on.

In a similar way, the normally closed contacts are used (refer to Fig. 4-15). When the relay is off, the circuit from the common terminal through the power source and lamp 1 is closed. This causes lamp 1, which is in series with the normally closed contacts, to be turned off. Also note that the power source for the lamps is in the common line since it is common to both of the other contacts. When the relay is turned on, the lamps will change states. Lamp 2 connected to the normally open contact will be turned on, while lamp 1 connected to the normally closed contact will be turned off. Notice in Fig. 4-14(b) how the common contact moves from the NC contact so that it touches the NO contact. This shows the basic operation of a relay with normally open and normally closed contacts. Such a relay is very common. It is called a double-pole, single-throw (DPST) relay. There are many other types used.

The coil resistance of a relay is determined by the size of the wire used to wind the coil and the number of windings. A coil with only a few turns of large diameter wire has a low resistance. This causes a high current flow through the relay coil. If a relay coil has many turns of small-diameter wire, it has a high resistance. Remember that small wire has high resistance and large wire offers lower resistance to current flow. Coil resistance is usually marked on a relay. It may also be measured with a meter.

There are some important current ratings for relays. Two of these ratings are called *pickup current*

and *dropout current*. These ratings are usually specified on the relay. They may also be found by using a variable power source and a current meter to measure the actual values. When the voltage applied to the relay coil is increased to a point where the relay turns on, the current indicated on the meter is the *pickup current*. Pickup current is the minimum current required to energize the relay. If the applied voltage is decreased until the relay deenergizes, the meter indicates *dropout current*. Dropout current is the minimum current that will keep the relay energized. The *contact current* rating is also very important. A large current usually flows through the relay contacts and the load connected to the relay. This rating is the maximum current that can safely flow through the contact circuit.

Solenoids

Solenoids are similar to relays since they also use electromagnetic coils. They are sometimes called actuators. Instead of having contacts, a solenoid has a plunger that moves when the coil is energized. The back of the plunger is attached to a spring. It causes the plunger to return to its original position when the solenoid deenergizes. The movement of the plunger of the solenoid is used to activate some type of load connected to it. A solenoid is shown in Fig. 4-16. When the solenoid is energized, it moves the plunger in the center. A solenoid could be used to open a control valve to allow liquid from a tank to flow into a container. When the solenoid is deenergized, the control valve would close and stop the flow of liquid.

Figure 4-15. Schematic diagram of a relay circuit.

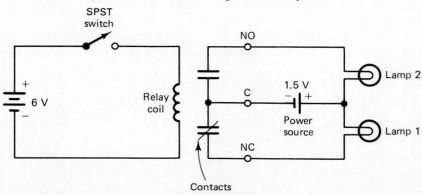

Figure 4-16. Solenoid device. (Courtesy of Heinemann Electric Co.)

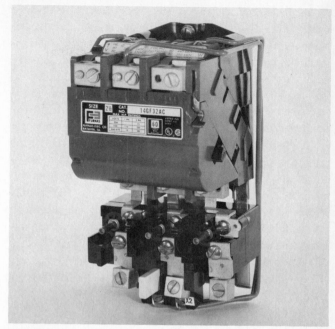

Figure 4-17. Magnetic motor contactor. (Courtesy of Furnas Electric Co.).

Magnetic Motor Contactors

A very important type of relay is an electric motor contactor. A magnetic motor contactor is shown in Fig. 4-17. The motor contactor is a control element that starts and stops motors. It operates due to electromagnetic relay action. A "start" pushbutton switch is pressed to close a contactor. This completes a *low-current* path through the contactor coil. The contactor coil produces a magnetic field that causes a set of contact points to close. The movement of a part called its armature completes an electrical path between the power line and a motor. When this action takes place, the motor will start.

Releasing the "start" button does not deenergize the contactor coil. A path to the coil's voltage source is completed through the "stop" pushbutton switch.

The motor continues to run as long as electric power is applied.

Stopping a contactor-controlled motor is achieved by pushing the "stop" button. This opens the contactor coil's voltage source. The coil will de-energize and causes the armature to move. The contact points then break contact. This removes the power from the motor. The path becomes open and the motor will stop. Motor contactors are designed to latch in place. This holds it in place once the coil is energized. This characteristic is important for motor control.

MAGNETIC EFFECTS

There are several magnetic effects which are important. Among these effects are residual magnetism, permeability, retentivity, and magnetic saturation.

Residual magnetism is an effect that is important in the operation of some types of electrical generators. Residual magnetism is the ability of electromagnets to retain a small magnetic field after electrical current is turned off. A small magnetic field remains around an electromagnet after it is demagnetized. This magnetic field is very weak.

Permeability (μ) is the ability of a magnetic material to transfer magnetic flux. It is the ability of a material to magnetize and demagnetize. Soft-iron material has a high permeability. It transfers magnetic flux very easily. Soft iron magnetizes and demagnetizes rapidly. This makes soft iron a good material to use in the construction of generators, motors, transformers, and other electromagnetic machines.

A related term is *relative permeability* (μ_r). This is a comparison of the permeability of a material to the permeability of air (1.0). Suppose that a material has a relative permeability of 1000. This means that the material will transfer 1000 times more magnetic flux than an equal amount of air. The highest relative permeability of materials is over 5000.

Another magnetic effect is called *retentivity*. The retentivity of a material is its ability to retain a magnetic flux after a magnetizing force is removed. Some materials will retain a magnetic flux for a long period of time. Other materials lose their magnetic flux almost immediately after the magnetizing force is removed.

Magnetic saturation is important in the operation of machines that have electromagnets, especially generators. Saturation is best explained by the curve shown in Fig. 4-18. This is called a magnetization or *B–H* curve. The curve shows the relationship between

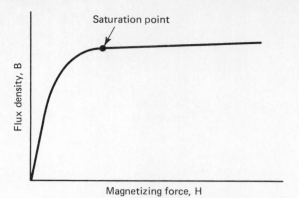

Figure 4-18.   Magnetization or *B–H* curve.

a magnetizing force (H) and flux density (B). Notice that as a magnetizing force increases, so does the flux density. Flux density is the amount of lines of force per unit area of a material. An increase in flux density occurs until *magnetic saturation* is reached. This saturation point depends on the type of material. At the saturation point, the maximum alignment of domains takes place in the material.

Magnetizing force (H) is measured in *oersteds*. The base unit is the number of ampere-turns per meter of length. *Flux density* (B) is the amount of magnetic flux per unit area. The unit of flux density is the *gauss* per square centimeter of area.

Review

1. What were the first magnets called?
2. What is electromagnetism?
3. What three materials are used in the construction of permanent magnets?
4. What are the two laws of magnetism?
5. How can a piece of iron be temporarily magnetized?
6. Why should magnets be stored in a "keeper?"
7. What is the left-hand rule of magnetic flux around a conductor?
8. What is the left-hand rule for determining the polarity of an electromagnet?
9. What are the three basic parts of an electromagnet?
10. What are three ways to increase the strength of an electromagnet?
11. What is residual magnetism?
12. Discuss the relationship of magnetomotive force, magnetic flux, and reluctance in a magnetic circuit.

13. Discuss the "domain" theory of magnetism.
14. What is Faraday's law?
15. How is Faraday's law important in the generation of electrical power?
16. What is a relay?
17. What are NO and NC contacts of relays?
18. What are pickup current and dropout current ratings of relays?
19. What is a solenoid?
20. What is a magnetic contactor?
21. What is permeability? retentivity? saturation? magnetizing force?

Student Activities

Working with Permanent Magnets (see Fig. 4-1)

1. Obtain two permanent magnets, a sheet of paper, and some iron filings.
2. Place one magnet on a flat surface with a sheet of paper placed over it.
3. Carefully pour iron filings onto the paper above the magnet.
4. Carefully lift the paper and place the second magnet under the paper with the two north poles about 1 in. apart. Make a sketch of the magnetic field pattern on a sheet of paper.
5. Place the north and south poles about 1 in. apart. Make a sketch of the magnetic field pattern.
6. Discuss the laws of magnetism on the sheet of paper.
7. Turn in your answers to the teacher for grading.

Magnetizing an Iron Bar

1. Obtain a magnet, an iron bar (about 6 in. long and $\frac{1}{2}$ to 1 in. in diameter), and a compass.
2. Place the iron bar on a flat surface.
3. Gently stroke the iron bar with the magnet, moving in the same direction each time.
4. Place the compass near one end of the iron bar. Describe what happens.

Making an Electromagnet (see Fig. 4-7)

1. Obtain the following materials:

   $\frac{3}{8}$- to $\frac{3}{4}$-in.-diameter bolt, about 4 in. long
   10 to 20 ft of solid copper wire (No. 18 to No. 24 diameter)
   6-V lattern battery
   Compass

2. Use the bolt as an iron core. Start at one end and wind the wire around the core in the *same direction*. Leave 6 to 10 in. of wire on each end, for connecting to the battery. Wrap tape around the wire to secure it in place. The finished coil should have between 100 and 200 turns of wire.

3. Momentarily connect the ends of the coil to a 6-V lantern battery. It may be necessary to remove about $\frac{1}{2}$ in. of insulation from the ends of the wire. This will allow good electrical contact to be made. Don't keep the battery connected for over a few seconds at a time.

4. Test the electromagnet with a compass to see if it has a north polarity at one end and a south polarity at the other end.

5. Try picking up metal objects with the electromagnet.

6. Let the teacher check the electromagnet.

7. On a sheet of paper, discuss the construction and operation of an electromagnet. Be sure to discuss ways to make the electromagnet stronger.

8. Turn in your paper to the teacher.

5

Electrical Energy Sources

Electrical energy sources convert some other form of energy into electrical energy. Batteries and electrical generators are two major sources of electrical energy. Batteries convert chemical energy into electrical energy. There are several types of electrical generators. These types include single-phase ac generators, three-phase ac generators, and dc generators. All electrical generators rely on the principle of electromagnetic induction to covert mechanical energy into electrical energy. Batteries, generators, and other sources of electrical energy are discussed in this chapter. Some sources produce direct-current (dc) energy while others produce alternating-current (ac) energy.

IMPORTANT TERMS

The following terms are used to aid in the understanding of electrical energy sources.

Ac.   An abbreviation used for alternating current.

Alternating current.   The current produced when electrons move first in one direction and then in the opposite direction.

Alternator.   A rotating machine that generates ac voltage.

Ampere-hour.   A capacity rating of storage batteries which refers to the maximum current that can flow from the battery during a specified period of time.

Armature.   The part of a generator into which current is induced.

Armature reaction.   The effect in direct-current generators and motors in which current flow through the armature conductors reacts with the main magnetic field and increases sparking between the brushes and commutator.

Battery.   Several cells connected in series to form a voltage source.

Brush.   A sliding contact made of carbon and graphite which touches the commutator of a generator or motor.

Commutation.   The process of changing ac induced into the rotor of a dc generator into dc applied to the load circuit.

Commutator.   An assembly of copper segments which provide a method of connecting rotating coils to the brushes of a dc generator.

Copper losses.   The heat losses in electrical machines due to the resistance of the copper wire used for windings; also called $I^2 R$ losses.

Core.   Laminated iron or steel used in the internal construction of the magnetic circuit of electrical machines.

Cycle.   The voltage generated by the rotation of a conductor through a magnetic field for one complete (360°) rotation.

Dc.   An abbreviation for direct current.

Dc generator.   A rotating machine that produces a form of direct-current (dc) voltage.

Delta connection.   A method of connecting three-phase alternator stator windings in which the beginning of one-phase winding is connected to the end of the adjacent phase winding; power lines extend from the three beginning–end connections.

Direct current.   The flow of electrons in one direction from negative (−) to positive (+).

Dry cell.   A nonliquid cell that procudes dc voltage by chemical action.

Eddy current.   Induced current in the metal parts of electrical machines which causes heat losses.

Efficiency.   The ratio of output power and input power of a generator:

$$\text{efficiency} = \frac{\text{power output (watts)}}{\text{power input (horsepower)}}$$

Electrode.   The negative or positive element of a cell.

Electrolyte.   The acid solution used in a cell which produces dc.

Field coils.   Electromagnetic coils that develop the magnetic fields of electric machines.

Field pole.   Laminated metal that serves as the core material for field coils.

Frequency.   The number of ac cycles per second, measured in hertz (Hz).

Generator.   A rotating electrical machine that converts mechanical energy into electrical energy.

Hydrometer.   An instrument used to measure the specific gravity or "charge" of the electrolyte of a storage battery.

Induced current.   The current that flows through a conductor due to magnetic transfer of energy.

Induced voltage.   The potential that causes induced current to flow through a conductor which passes through a conductor which passes through a magnetic field.

Ion.   An atom that has lost or gained electrons, making it a negative or positive charge.

Laminations.   Thin sheets of soft iron or steel used in the construction of electrical machines to reduce heat losses due to eddy currents.

Lead–acid cell.   A secondary cell that has positive and negative plates made of lead peroxide and lead and has a liquid electrolyte of sulfuric acid mixed with water.

Left-hand rule.   (1) To determine the *direction of the magnetic field* around a single conductor, point the thumb of the left hand in the direction of current flow (− to +) and the fingers will extend around the conductor in the direction of the magnetic field; (2) to determine the *polarity of an electromagnetic coil*, extend the fingers of the left hand around the coil in the direction of current and the thumb will point to the north polarity; (3) to determine the *direction of induced current flow* in a generator conductor, hold the thumb, forefinger, and middle finger of the left hand at right angles to one another, point the thumb in the direction of motion of the conductor, the forefinger in the direction of the magnetic field (N to S), and the middle finger will point in the direction of induced current.

Lenz's law.   The induced voltage in any circuit is always in such a direction that it will oppose the force that produces it.

Neutral plane.   The theoretical switching position of the commutator and brushes of a dc generator or motor which occurs when no current flows through the armature conductors and the main magnetic field has least distortion.

Nickel–cadmium cell. An alkaline cell which has a longer life than carbon–zinc dry cells.

Photovoltaic cell. A cell that produces dc voltage at the junction of two materials when light shines onto its surface.

Piezoelectric effect. The property of certain crystal materials to produce a voltage when pressure is applied to them.

Primary cell. A cell that cannot be recharged.

Prime mover. A system that supplies the mechanical energy to rotate an electrical generator.

Pulsating dc. A dc voltage that is not in "straight line" or "pure" dc form, such as dc voltage produced by a battery.

Regulation. A measure of the amount of voltage change that occurs in the output of a generator due to changes in load.

Rotating-armature method. The method used when a generator has dc voltage applied to produce a field to the stationary part (stator) of the machine and voltage is induced into the rotating part (rotor).

Rotating-field method. The method used when a generator has dc voltage applied to produce a field to the rotor of the machine and voltage is induced into the stator coils.

Rotor. The rotating part of an electrical generator or motor.

Running neutral plane. The actual switching position of the commutator and brushes of a dc generator or motor that shifts the theoretical neutral plane due to armature reaction.

Secondary cell. A cell that can be recharged by applying dc voltage from a battery charger.

Sine wave. One cycle of single-phase ac voltage.

Single-phase ac generator. A generator that produces single-phase ac voltage in the form of a sine wave.

Slip rings. Copper rings that connect to the rotor coils of an ac generator and connect to a set of brushes for making connection to an external circuit.

Specific gravity. The weight of a liquid as compared to the weight of water, which has a value of 1.0.

Split rings. *See* Commutator.

Stator. The stationary part of an electrical generator or motor.

Storage battery. Another term used for lead–acid secondary cells.

Thermocouple. A device that has two pieces of metal joined together so that when its junction is heated, a voltage is produced.

Three-phase ac generator. A generator that produces three ac sine-wave voltages that are separated in phase by 120 electrical degrees.

Voltaic cell. A cell that produces voltage due to two metal elements which are suspended in an acid solution.

Wye connection. A method of connecting three-phase alternator stator windings in which the beginnings *or* ends of each phase winding are connected to form a common or neutral point; power lines extend from the remaining beginnings or ends.

CHEMICAL SOURCES

Conversion of chemical energy into electrical energy is done by chemical cells. When two or more cells are connected in series or parallel (or a combination of both), they form a battery. A cell is made of two different metals immersed in a liquid or paste called an *electrolyte*. Chemical cells are either primary or secondary cells. *Primary cells* are usable only for a certain time period. *Secondary cells* are renewed after being used to produce electrical energy once again. This is known as *charging*. Both primary and secondary cells have many uses.

Primary Cells

The operation of a primary cell involves the placing of two unlike materials called *electrodes* into a solution called an *electrolyte*. When the materials of the cell are brought together, their molecular structures are changed. During this chemical change, atoms may either gain additional electrons or leave behind some electrons. These atoms then have either a positive or a negative electrical charge. They are called *ions*. Ionization of atoms allows a chemical solution of a cell to produce an electrical current.

A load device such as a lamp may be connected to a cell. Electrons flow from one of the cell's electrodes to the other through the electrolyte material. This creates an electrical current flow through the load. Current leaves the cell through its negative electrode. It passes through the load device and then goes back to the cell through its positive electrode. A complete circuit exists between the cell (source) and the lamp (load).

The voltage output of a primary cell depends on the electrode materials used and the type of electrolyte. The familiar carbon–zinc cell is shown in Fig. 5-1. It produces approximately 1.5 V. The negative electrode of this cell is the zinc container. The positive electrode is a carbon rod. A paste material acts as the electrolyte. It is placed between the two electrodes. This type of cell is called a *dry cell.*

There are many types of primary cells used today. The carbon–zinc cell is the most used type. This cell is low in cost and available in many sizes. Applications are mainly for portable equipment and instruments. For uses that require higher voltage or current than one cell can deliver, several cells are combined in series, parallel, or series–parallel connec-tions. Carbon–zinc batteries are available in many voltage ratings.

Another type of primary cell is the mercury (zinc–mercuric oxide) cell shown in Fig. 5-2. This

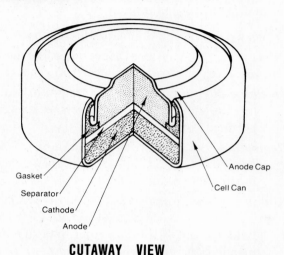

CUTAWAY VIEW

Figure 5-2.   Mercury cell. (Courtesy of Union Carbide Corp.)

Figure 5-3.   Alkaline cell. (Courtesy of Union Carbide Corp.)

Figure 5-1.   Carbon–zinc cell. (Courtesy of Union Carbide Corp.)

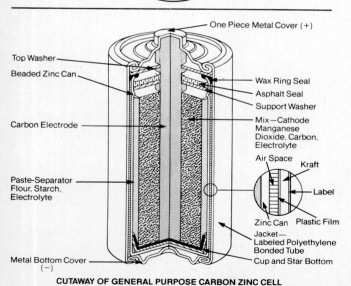

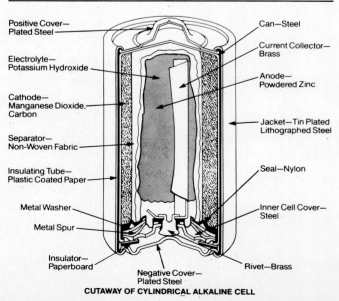

cell is an improvement over the carbon–zinc cell. It has a more constant voltage output, a longer life, and a smaller size. Mercury cells are more expensive. They produce a voltage of about 1.35 V, which is slightly less than carbon–zinc cells.

An alkaline (zinc-manganese dioxide) cell is shown in Fig. 5-3. Alkaline cells have a voltage per cell of 1.5 V. They supply higher-current electrical loads. They have much longer lives than carbon-zinc cells of the same types.

Secondary Cells

Chemical cells that may be reactivated by charging are called *secondary cells* or *storage cells*. Common types are the lead–acid, nickel–iron (Edison), and nickel–cadmium cells. The latter two cells serve as primary electrical sources for industry.

LEAD–ACID CELLS. The lead–acid cell of Fig. 5-4 is a secondary cell. The electrodes of lead–acid

Figure 5-4. Lead–acid cells of a battery. (Courtesy of Exide Corp.)

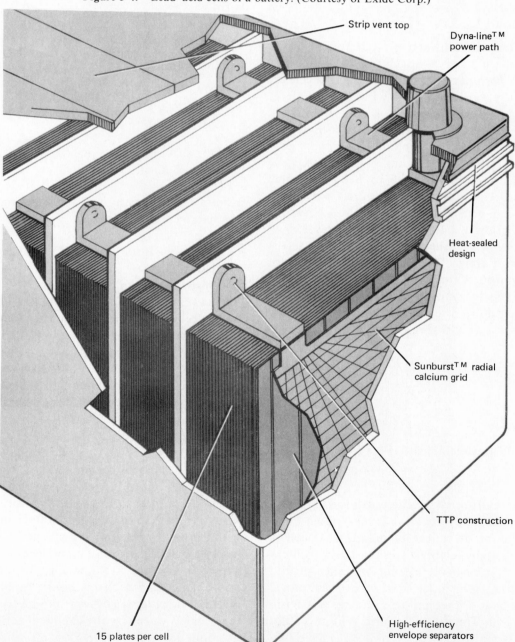

Strip vent top

Dyna-line^{TM} power path

Heat-sealed design

Sunburst^{TM} radial calcium grid

TTP construction

15 plates per cell

High-efficiency envelope separators

cells are made of lead and lead peroxide. The positive plate is lead peroxide (PbO_2). The negative plate is lead (Pb). The electrolyte is sulfuric acid (H_2SO_4). When the lead-acid cell supplies current to a load, the chemical process is written as

$$Pb + PbO_2 + 2H_2SO_4 \rightarrow 2PbSO_4 + 2H_2O$$

or

lead + lead peroxide
+ 2 parts sulfuric acid yields 2 parts lead sulfate
+ 2 parts water

The sulfuric acid ionizes to produce four positive hydrogen ions (H^+) and two negative sulfate (SO_4^-) ions. A negative charge is developed on the lead plate when an SO_4^- ion combines with the lead plate to form lead sulfate ($PbSO_4$). The positive hydrogen ions (H^+) combine with electrons of the lead-peroxide plate. They become neutral hydrogen atoms. The H^+ ions also combine with the oxygen (O) of the lead peroxide plate to become water (H_2O). The lead-peroxide plate then has a positive charge. A lead-acid cell has a voltage between electrodes of about 2.1 V when fully charged.

Cells discharge when supplying current for long periods of time. They are no longer able to develop an output voltage when discharged. Cells may be charged by causing direct current to flow through the cell in the opposite direction. The chemical process of charging is written as

$$2PbSO_4 + 2H_2O \rightarrow Pb + PbO_2 + 2H_2SO_4$$

or

2 parts lead sulfate
+ 2 parts water yields lead + lead peroxide
+ 2 parts sulfuric acid

The original condition of the chemicals is reached by charging. The chemical reaction is reversible.

The amount of charge of a lead-acid cell is measured by a *specific gravity* test. A *hydrometer* is used to test the electrolyte solution. The specific gravity of a liquid is an index of how heavy a liquid is compared to water. Pure sulfuric acid has a specific gravity of 1.840. The dilute sulfuric acid of a fully charged lead-acid cell varies from 1.275 to 1.300.

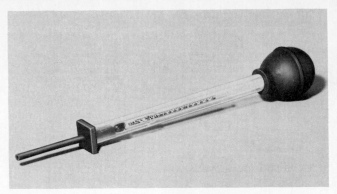

Figure 5-5. Hydrometer used to test batteries. (Courtesy of Exide Corp.)

During the discharge of the cell, water is formed. This reduces the specific gravity of the electrolyte. A specific gravity of between 1.120 and 1.150 indicates a fully charged cell. A hydrometer used to test lead-acid cells is shown in Fig. 5-5.

The *capacity* of a battery made of lead-acid cells is given by an *ampere-hour* rating. A 50-ampere-hour battery is rated to deliver 50 A for 1 h, 25 A for 2 h, or 12.5 A for 4 h. The ampere-hour rating is an approximate value. It depends on the rate of discharge and the operating temperature of the battery.

NICKEL-IRON (EDISON) CELL. The nickel-iron (Edison) cell, shown in Fig. 5-6, is a secondary cell. It is constructed with an unbreakable case of welded, nickel-plated sheet steel. The positive plate is made of nickel tubes. Nickel oxide is contained in the tubes. The tubes are mounted into steel grids. They are forced into position under high pressure and grouped together. The negative plates are made of flat-nickel-plated steel pockets made of iron oxide. They are built into groups of steel grids similar to the positive plates. The electrolyte is a solution of potassium hydroxide.

The chemical action of an Edison cell is very complex. The cells are similar to lead-acid cells in many ways. The electrolyte, potassium hydroxide, breaks up into negative and positive ions. The negative ions move toward the iron plate. They combine with the iron, and give up excess electrons to the plate. The plate then becomes negatively charged. The positive ions move to the nickel oxide plate. They take electrons from the plate, causing the nickel oxide plate to become positively charged. The voltage produced by the chemical action of this cell is about 1.4 V.

Figure 5-6. Edison cell. (Courtesy of McGraw-Edison.)

Edison cells are also charged by applying a direct-current voltage to the positive and negative terminals. This causes electron flow within the cells to reverse. The chemicals are reactivated by charging so that the battery may be used again.

NICKEL-CADMIUM CELLS. Another type of secondary cell is nickel–cadmium cell. These cells are available in many sizes. Figure 5-7 shows nickel–cadmium cells. They are often used in portable equipment. The positive plate of this cell is nickel hydroxide. The negative plate is cadmium hydroxide. The electrolyte is made of potassium hydroxide. These cells have a long life. A fully charged nickel–cadmium cell has a voltage of approximately 1.25 V.

OTHER SECONDARY CELLS. There are other types of secondary cells used today. Many have specialized applications. Among these are silver oxide–zinc cells and silver–cadmium cells. These cells have a high output and long life. They are more expensive than other types of secondary cells of the same size.

Secondary cells have many uses. Storage batteries are used in some buildings to provide emergency

(a)

SEALED NICKEL CADMIUM RECHARGEABLE CYLINDRICAL TYPE CELL EXPLODED VIEW

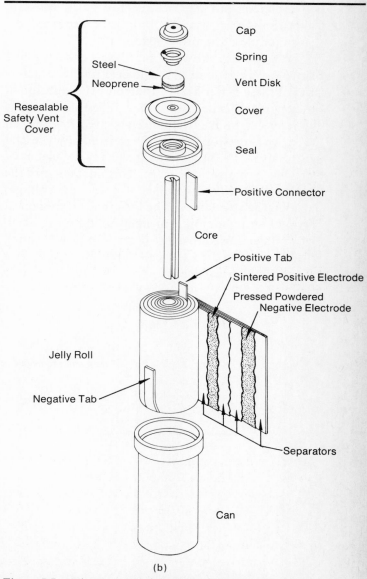

(b)

Figure 5-7. Nickel–cadmium cells: (a) pictorial; (b) cutaway. (Courtesy of Union Carbide Corp.)

power when a power failure occurs. Standby systems are needed, especially for lighting when power is off. Automobiles use storage batteries for their everyday operation. Many types of instruments and portable equipment use batteries for power. Some instruments use rechargable secondary cells and others use primary cells.

LIGHT SOURCES

Light is a form of energy that is easily converted to electrical energy. The device used to convert light energy into electrical energy is called a photovoltaic cell or a *solar cell*. Solar cells are used in our space program. They collect the rays of the sun and convert them into electrical energy. Solar cells are used to power circuits which control space satellites, lunar modules, and other spacecraft.

A solar cell, shown in Fig. 5-8, is usually made of two layers of material. The electrical characteristics of these materials are altered by the addition of other elements called *impurities*. When a solar cell is exposed to light, the two materials interact with each other. An excessive number of electrons is developed on one layer. The other layer then has a deficiency of electrons or a positive charge. This imbalance in the electrons causes a difference of potential (voltage) between the two layers. The difference in potential depends on the amount of light falling on the cell. The voltage is used to cause current to flow through a load connected to the cell. Thus light acts as a source of electrical energy.

HEAT SOURCES

Electrical energy is converted into heat energy when food is cooked and homes are heated. Similarly, heat can be converted into electrical energy. Thomas J. Seebeck, a German scientist, discovered that heat could be converted into electrical energy in the early 1800s. He found that when the ends of two different types of metals are connected together and heated, a small dc voltage is created at the open ends (refer to Fig. 5-9). The amount of dc voltage depends on the amount of heat being applied at the connected ends of the two pieces of metal. Converting heat into electrical energy by this method is known as the *Seebeck effect*. This effect is used to measure temperature and for thermostats.

When metals are heated, their electrons tend to move away from the areas being heated. This causes electrons to be more concentrated in a cool area than in a heated area. When two different types of metal are connected together and heated at their junction, electrons in both metals tend to move away from the heat. Since the two metals are different, there are more electrons at the cool end of one metal than at the other. This causes the metal with the most electrons to be a negative (−) charge. Compared to the other metal, the one with the least electrons is positive (+). The difference in charge between the two cooler ends of the metals develops a voltage. A small voltage, usually in millivolts, is produced. Devices used to convert heat energy into electrical energy are called *thermocouples*.

Figure 5-8. Solar cell.

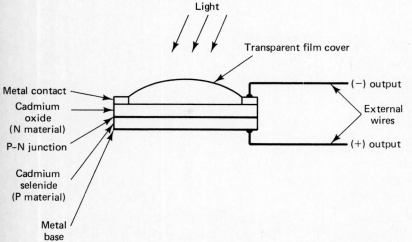

Figure 5-9. Operating principle of a thermocouple.

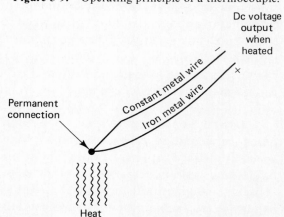

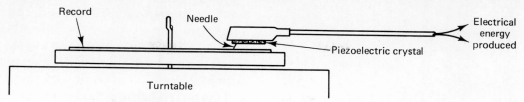

Figure 5-10. Piezoelectric principle of a phonograph cartridge.

PRESSURE SOURCES

Electrical energy is produced by mechanical energy in electrical generators. Mechanical energy is used to rotate prime movers which drive electrical generators. Mechanical energy in the form of *pressure* is also used as a source of electrical energy.

The change of mechanical pressure into electrical energy is called the *piezoelectric effect*. Certain crystal materials may be compressed as pressure is applied to the surfaces. A voltage is created between their top and bottom surfaces. The amount of voltage is determined by the amount of pressure. The greater the pressure, the greater the voltage. The less the pressure, the less the voltage will be for any piezoelectric crystal. These pressure-sensitive crystals are useful as phonograph cartridges to change the pressure applied by the grooves of the record into a voltage. They are also used as pressure sensors to sense and measure pressure in security and industrial systems.

When these crystalline materials are subjected to a mechanical pressure, electrical energy is developed across the material. Crystals such as quartz and Rochelle salt have this characteristic. An application of the piezoelectric principal is the cartridge and needle assembly of a phonograph, shown in Fig. 5-10. The cartridge contains a crystalline material. The crystal vibrates according to the size of the grooves of a phonograph record. The needle is attached to the cartridge to connect the record grooves to the crystal. The crystalline material produces a voltage due to the mechanical vibrations or pressure changes. These small voltage changes are then amplified by a sound system.

It is also possible to convert pressure in the form of sound into electrical energy with piezoelectric crystals. This is done with crystal microphones, as shown in Fig. 5-11. Sound waves are used

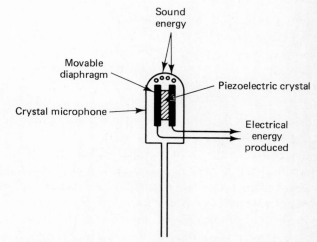

Figure 5-11. Piezoelectric principle of a microphone.

to cause vibration of a piezoelectric crystal. A voltage is developed across the crystal and amplified by the sound system. Higher sound pressure causes more voltage output to be produced.

ELECTROMAGNETIC INDUCTION

Electrical energy can be produced by placing a conductor inside a magnetic field. An experiment by a scientist named Michael Faraday was very important. It showed the following principle: When a conductor moves across the lines of force of a magnetic field, electrons in the conductor tend to flow in a certain direction. When the conductor moves across the lines of force in the opposite direction, electrons in the conductor tend to flow in the opposite direction. This is the principle of electrical power generation. Most of the electrical energy used today is produced by using magnetic energy.

Figure 5-12 shows the principle of electromagnetic induction. Electrical current is produced

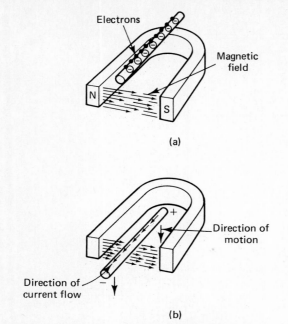

Figure 5-12. Principle of electromagnetic induction: (a) conductor placed inside a magnetic field; (b) conductor moved downward through the magnetic field.

only when there is motion. When the conductor is brought to a stop while crossing lines of force, electrical current stops.

If a conductor or a group of conductors is moved through a strong magnetic field, induced current will flow and a voltage will be produced. Figure 5-13 shows a loop of wire rotated through a magnetic field. The position of the loop inside the magnetic field determines the amount of induced current and voltage. The opposite sides of the loop move across the magnetic lines of force in opposite directions. This movement causes an equal amount of electrical current to flow in *opposite directions* through the two sides of the loop. Notice each position of the loop and the resulting output voltage in Fig. 5-13. The electrical current flows in one direction and then in the opposite direction with every complete revolution of the conductor. This method produces alternating current (ac). One complete rotation is called a *cycle*. The number of cycles per second is known as the *frequency*. Most ac generators produce 60 cycles per second.

The ends of the conductor which move across the magnetic field of the generator shown in Fig. 5-13 are connected to slip rings and brushes. The slip rings are mounted on the same shaft as the conductor.

Carbon brushes are used to make contact with the slip rings. The electrical current induced into the conductor flows through the slip rings to the brushes. When the conductor turns half a revolution, electrical current flows in one direction through the slip rings and the meter. During the next half revolution of the coil, the positions of the two sides of the conductor are opposite. The direction of the induced current is reversed. Current now flows through the meter in the opposite direction.

The conductors that make up the rotor of a generator have many turns. The generated voltage is determined by the number of turns of wire used, the strength of the magnetic field, and by the speed of the prime mover used to rotate the machine.

Direct current can also be produced by electromagnetic induction. A simple dc generator has a *split-ring* commutator instead of two slip rings. The split rings resemble one full ring, except that they are separated by small openings. Induced electrical current still flows in opposite directions to each half of the split ring. However, current flows in the same direction in the load circuit due to the action of the split rings. Dc generator operation is discussed later in this chapter.

GENERATING A VOLTAGE

Electromagnetic induction takes place when a conductor passes through a magnetic field and cuts across lines of force. As a conductor is passed through a magnetic field, it cuts across the magnetic flux lines. As the conductor cuts across the flux lines, the magnetic field develops a force on the electrons of the conductor. The direction of the electron movement determines the polarity of the induced voltage. The left-hand rule is used to determine the direction of electron flow. This rule for generators is stated as follows: Hold the thumb, forefinger, and middle finger of the right hand perpendicular to each other. Point the forefinger in the direction of the magnetic field from north to south. Point the thumb in the direction of the motion of the conductor. The middle finger will then point in the direction of electron current flow.

The amount of voltage induced into a conductor cutting across a magnetic field depends on the number of lines of force cut in a given time. This is

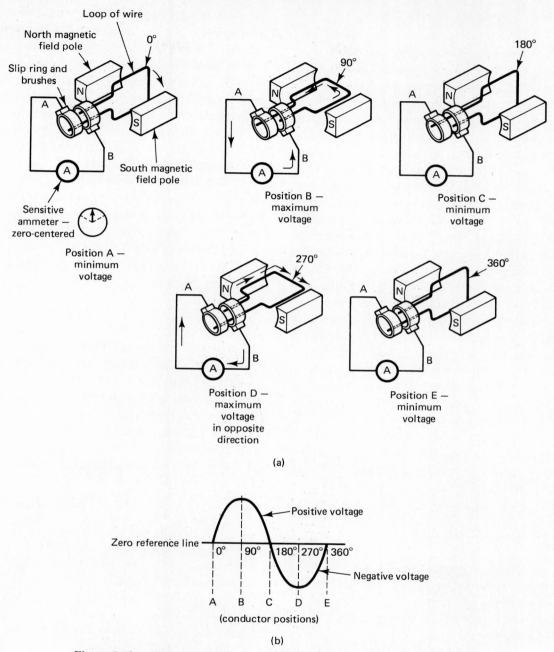

Figure 5-13. (a) Loop of wire rotated through a magnetic field; (b) voltage produced.

determined by these three factors:

1. The *speed* of the relative motion between the magnetic field and the conductors
2. The *strength* of the magnetic field
3. The *length* of the conductor passed through the magnetic field

If the speed of the conductor cutting the magnetic lines of force is increased, the generated voltage increases. If the strength of the magnetic field is increased, the induced voltage is also increased. A longer conductor allows the magnetic field to induce more voltage into the conductor. The induced voltage increases when each of the three quantities listed above is increased.

ELECTRICAL GENERATOR BASICS

Electrical generators are used to produce electrical energy. They require some form of mechanical energy input. The mechanical energy is used to move electrical conductors (turns of wire) through a magnetic field inside the generator. All generators operate due to electromagnetic induction. A generator has a stationary part and a rotating part housed inside a machine assembly. The stationary part is called the *stator* and the rotating part is called the *rotor*. The generator has *magnetic field poles* of north and south polarities. Generators must have a method of producing rotary motion (mechanical energy). This system is called a *prime mover* and is connected to the generator shaft. There must also be a method of electrically connecting the rotating conductors to an external circuit. This is done by a *slip ring* or *split ring* and brush assembly. Slip rings are used with ac generators and split rings with dc generators. Slip rings are made of solid sections of copper and split rings are made of several copper sections which are separated from each other. The rings are permanently mounted on the shaft of a generator. The rings connect to the ends of the conductors of the rotor. When a load is connected to a generator, a complete circuit is made. With all of the generator parts working together, electrical power is produced.

SINGLE-PHASE AC GENERATORS

Single-phase electrical power is often used, particularly in homes. However, very little electrical power is produced by single-phase generators. Alternating-current generators are usually referred to as "alternators." Single-phase electrical power used in homes

is usually produced by three-phase generators at power plants. It is then converted to single-phase electrical energy before it is distributed to homes. There are several uses for single-phase generators.

The current produced by single-phase generators is in the form of a *sine wave*. This waveform is called a sine wave due to its mathematical origin. It is based on the trigonometric sine function used in mathematics. Two sine waves or cycles of single-phase ac voltage are shown in Fig. 5-14.

The voltage induced into the conductors of the armature varies as the sine of the angle of rotation between the conductors and the magnetic field (see Fig. 5-15). The voltage induced at a specific time is called *instantaneous voltage* (E_i). Voltage induced into an armature conductor at a specific time is found by using the formula

$$E_i = E_{max} \times \sin \theta$$

E_{max} is the maximum voltage induced into the conductor. The symbol theta (θ) is the angle of conductor rotation.

For example, at the 60° position, assume that the maximum voltage output is 100 V. The instantaneous voltage induced at 60° is: $E_i = 100 \times \sin \theta$. It may be necessary to study Appendix 6 at this time. In Table A6-1 it is found that the sine of 60° = 0.866. The induced voltage at 60° is 86.6 V ($E_i = 100 \times 0.866 = 86.6$ V). A calculator with trigonometric functions can also be used.

The frequency of the voltage produced by alternators is usually 60 hertz (Hz). A *cycle* of ac is generated when the rotor moves one complete revolution (360°). Cycles per second or hertz refers to the number of revolutions per second. For example, a speed of 60 revolutions per second (3600 rpm) produces a frequency of 60 Hz. The frequency

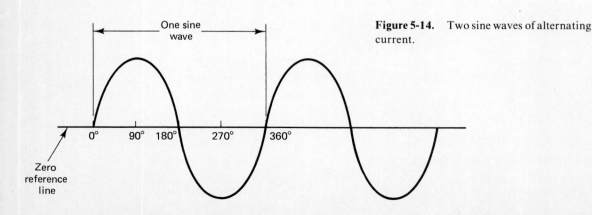

Figure 5-14.   Two sine waves of alternating current.

Degrees	Sine
0/360	0.000
30	0.500
60	0.866
90	1.000
120	0.866
150	0.500
180	0.000
210	−0.500
240	−0.866
270	−1.000
300	−0.866
330	−0.500

(a)

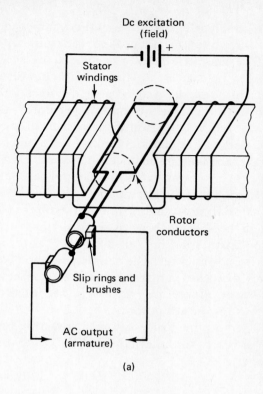

Dc excitation (field)

(a)

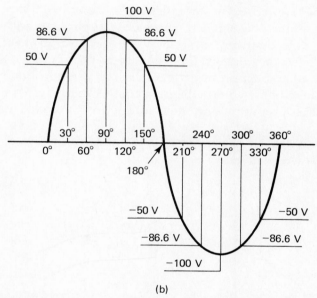

(b)

Figure 5-15. Generation of an ac sine wave: (a) sine values of angles from 0 to 360°; (b) sine wave produced.

of an alternator is found by using the formula

$$f(\text{Hz}) = \frac{\text{number of magnetic poles} \times \text{speed of rotation (rpm)}}{120}$$

The frequency is measured in hertz. If the number of poles (field coils) is increased, the speed of rotation can be reduced and still produce a 60-Hz frequency.

For a generator to convert mechanical energy into electrical energy, three conditions must exist: (1) there must be a magnetic field, (2) there must be conductors cutting the magnetic field, and (3) there must be relative motion between the magnetic field and the conductors. Two methods are used to accomplish these conditions. They are the *rotating armature* method and the *rotating field* method shown in Fig. 5-16.

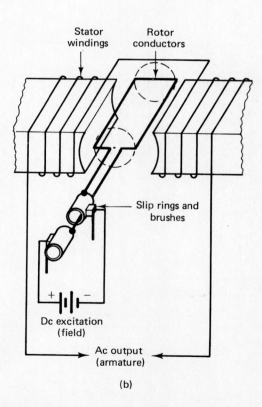

(b)

Figure 5-16. Generating voltage: (a) rotating armature method; (b) rotating field method.

In the rotating armature method, shown in Fig. 5-16(a), ac voltage is induced into the rotor conductors. The magnetic field is developed by a set of stationary field poles. Relative motion between the conductors and the magnetic field comes from a prime mover connected to the generator shaft. Prime movers can be gasoline engines, diesel engines, steam turbines, or electric motors. Remember that generators convert mechanical energy into electrical energy. The rotating armature method can only be used to produce small amounts of electrical power. The major disadvantage is that the ac voltage passes through the slip ring/brush assembly. High voltages would cause sparking or arc-over between the brushes and slip rings. The maintenance involved in replacing brushes and repairing slip-ring assemblies would be expensive. This method is used for alternators with low power outputs.

The rotating field method, shown in Fig. 5-16(b) is used for alternators with larger power outputs. The dc excitation voltage is used to develop the magnetic field. Dc voltage is applied to the rotor of the generator. The ac voltage output is induced into the stationary conductors of the machine. Since the dc excitation voltage is much smaller than the ac voltage output, maintenance problems are reduced. The conductors of the stationary part of the machine can be made larger. They will carry more current since they do not rotate.

THREE-PHASE AC GENERATORS

Most electrical power produced in the United States is three-phase power produced at power plants. Power distribution systems use many three-phase generators (alternators) connected in parallel. A simple diagram of a three-phase alternator and a three-phase voltage diagram are shown in Fig. 5-17. The alternator output windings are connected in

Figure 5-17. (a) Diagram of a three-phase alternator; (b) diagram of three-phase voltage.

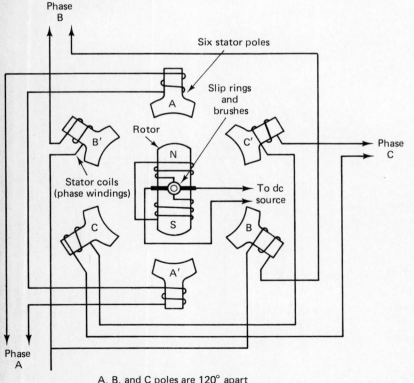

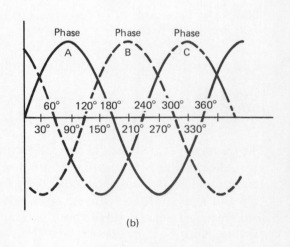

(b)

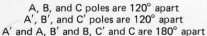

A, B, and C poles are 120° apart
A', B', and C' poles are 120° apart
A' and A, B' and B, C' and C are 180° apart

(a)

either of two ways. These methods are called the *wye* connection and the *delta* connection. These three-phase connections are shown with schematics in Fig. 5-18. These methods are also used for connecting the windings of three-phase transformers, three-phase motors, and other three-phase equipment.

In the wye connection of Fig. 5-18(a), the beginnings *or* ends of each winding are connected together. The other sides of the windings are the ac lines which extend from the alternator. The voltage across the power lines is called line voltage (E_L). Line voltage is higher than the voltage across each phase. Line voltage (E_L) is equal to the square root of 3 (1.73) multiplied by the voltage across the phase windings (E_p), or $E_L = E_p \times 1.73$. The line current (I_L) is equal to the phase current (I_p), or $I_L = I_p$.

Figure 5-18. Three-phase connections: (a) wye connection, sometimes called star connection; (b) delta connection.

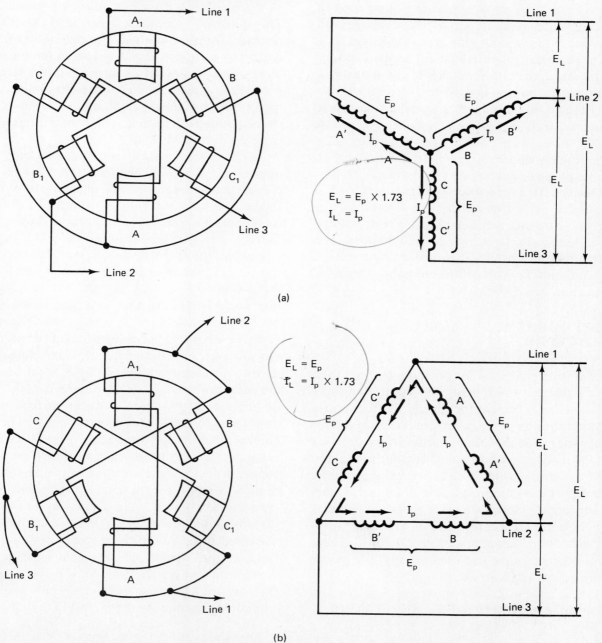

(a)

(b)

In the delta connection, shown in Fig. 5-18(b), the end of one phase winding is connected to the beginning of the next phase winding. Line voltage (E_L) is equal to the phase current (I_p) multiplied by 1.73, or $I_L = I_p \times 1.73$. The differences between voltages and currents in wye and delta systems should be remembered.

Three-phase power is used mainly for high-power industrial and commercial equipment. The power produced by three-phase generators is a more constant output than that of single-phase power. Three-phase power is more economical to supply energy to the large motors that are often used in industries. Three separate single-phase voltages can be delivered from a three-phase power system. It is more economical to distribute three-phase power from power plants to homes, cities, and industries. Three conductors are needed to distribute three-phase voltage. Six conductors would be necessary for three separate single-phase systems. Equipment that uses three-phase power is smaller in size than similar single-phase equipment. It saves energy to use three-phase power whenever possible.

One type of a three-phase alternator is used in automobiles. The three-phase ac it produces is converted to dc by a rectifier circuit. The dc voltage is then used to charge the automobile battery. The charging time and voltage is controlled by a voltage regulator circuit. The parts of an automobile alternator are shown in Fig. 5-19.

DIRECT CURRENT GENERATORS

Alternating current is used in greater quantities than direct current. However, many important operations depend on direct current (dc) power. Industries use direct current power for many operations. Electroplating and variable-speed motor drives are two examples of direct current use. Direct current energy is used to cause automobiles to start and for many types of portable equipment used in homes. Three-phase or single-phase ac power is easily converted to direct current. Direct current is available in the form of primary and secondary chemical cells. Direct current generators are also used to supply dc power for specialized applications.

Direct current generators are used to convert mechanical energy into direct current electrical energy. The parts of a simple direct current generator are shown in Fig. 5-20. The principle of operation of direct current generators is similar to alternating current generators. Electromagnetic induction causes a voltage to be generated. Armature coils rotate through a magnetic field. North and south polarities of permanent magnets or electromagnets are used as the field. As the coils rotate, electromagnetic induction causes a current to be induced into them. The current induced in the conductors is an *alternating current*. This ac is converted to a form of direct current. The conversion of ac to dc is done by a *split-ring* commutator. The split-ring commutator has segments that are insulated from one another. The ends of the armature conductors are connected to the commutator segments. The purpose of a split-ring commutator is to reverse the armature coil connection to the external load at the same time that the ac current induced into the armature coils reverse. This causes direct current to be applied to the load.

The current that flows from one coil would appear as shown in Fig. 5-21(a). This is called *pulsating direct current*. By using many turns of wire in the armature the voltage output is a smoother direct current. This type of output is shown in Fig. 5-21(b). The voltage developed depends on (1) the strength of the magnetic field, (2) the number of coils in the armature, and (3) the speed of rotation. By increasing any of these factors, voltage output is increased. This is the same as ac generators.

Direct current generators are made in several types. One type is a permanent-magnet dc generator. On this type of generator, permanent magnets are used to develop the magnetic field. Electromagnets are ordinarily used to develop the magnetic field of dc generators. A source of direct current must be applied to the electromagnetic coils. The most common method of developing a magnetic field is for part of the generator dc output to be used to supply the field. There are three major classifications of direct current generators: (1) permanent-magnet generators, (2) separately excited generators, and (3) self-excited generators. The self-excited types are classified according to the method used to connect the armature windings to the field windings. This is done in one of the following ways: (1) series, (2) parallel (shunt), or (3) compound. A shunt-wound dc generator is shown in Fig. 5-22.

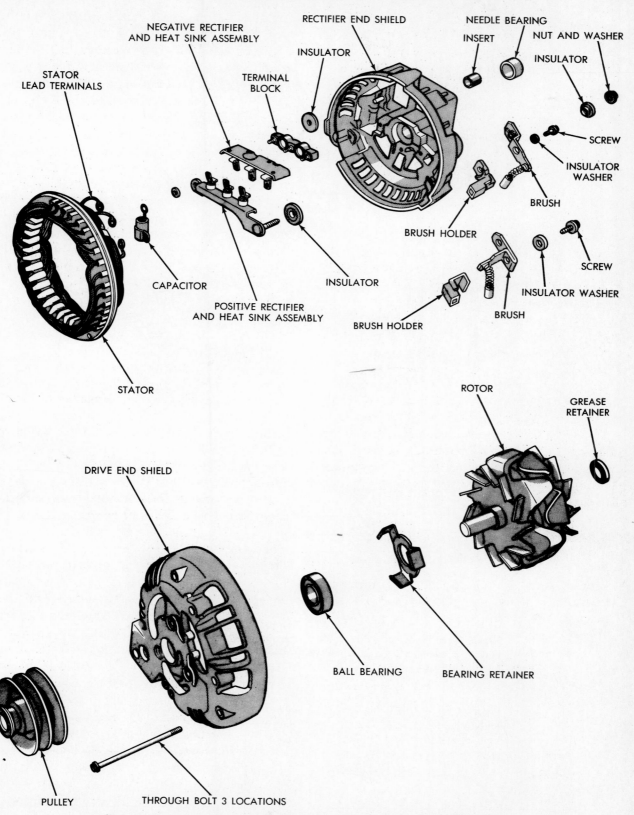

Figure 5-19. Parts of an automobile alternator. (Courtesy of Chrysler Corp.)

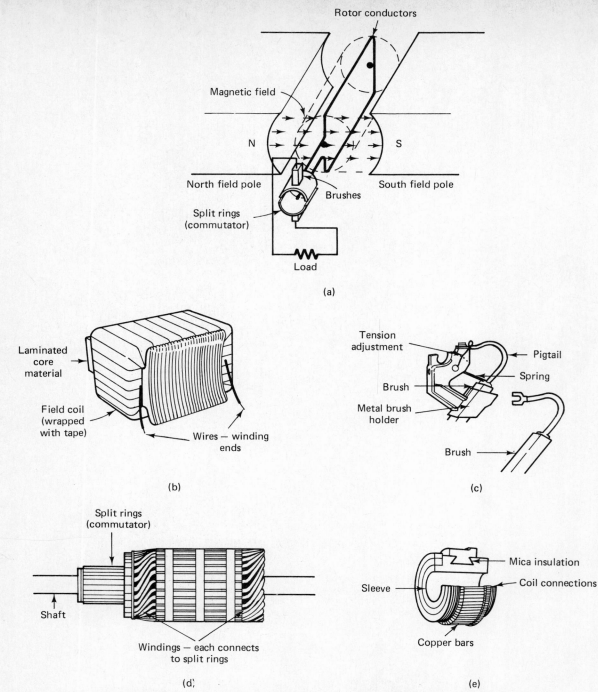

Figure 5-20. Parts of a direct current generator: (a) basic parts; (b) electro-magnetic pole piece; (c) brush assembly; (d) rotor; (e) split-ring detail.

Figure 5-21. Current flow from the coils of a dc generator: (a) single coil, pulsating direct current; (b) many coils, pure direct current.

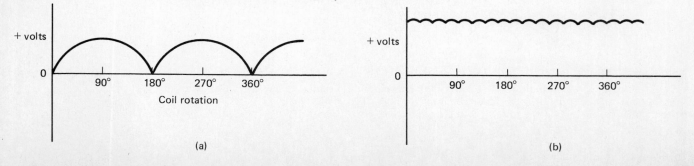

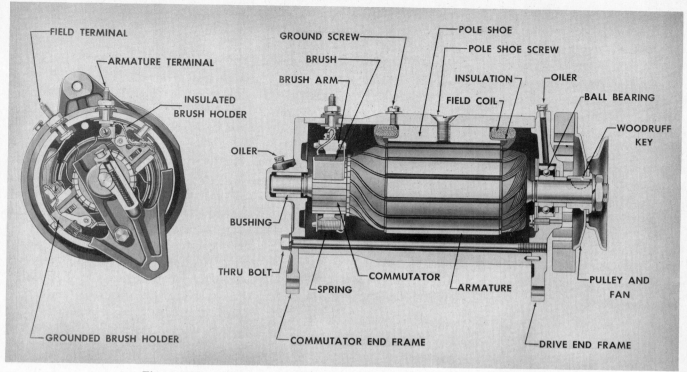

Figure 5-22. Cutaway of a shunt-wound dc generator. (Courtesy of Delco-Remy.)

Permanent-Magnet dc Generators

A diagram of a permanent-magnet dc generator is shown in Fig. 5-23. The rotor conductors are connected to a split-ring commutator and brushes. The magnetic field is developed by permanent magnets made of Alnico or some other alloy. Alnico is an alloy of *al*uminum, *n*ickel, and *i*ron, and *co*balt. Several permanent magnets can be used together to create a stronger magnetic field.

The armature of a permanent-magnet dc generator has many turns of insulated wires. The armature rotates inside the permanent-magnetic field. An induced voltage is then developed and applied to a load device. Applications for this type of dc generator usually require low amounts of power. A permanent-magnet dc generator is sometimes called a "magneto."

Separately Excited dc Generators

When large amounts of direct current electrical energy are needed, generators with electromagnetic fields are used. Strong magnetic fields are produced by electromagnets. It is also possible to control the

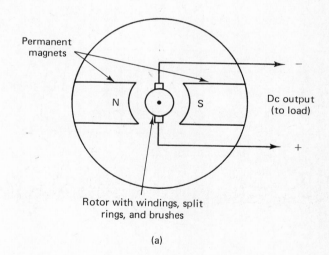

(a)

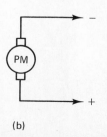

(b)

Figure 5-23. Permanent-magnet dc generator: (a) pictorial; (b) schematic.

strength of the field. This is done by varying the current through the field windings. The output of a generator is easily controlled in this way.

Direct current which is applied to the coils to develop an electromagnetic field is called *exciting current*. When dc exciting current comes from a source separate from the generator, it is called a *separately excited* dc generator. This type of generator is shown in Fig. 5-24. Storage batteries are sometimes used to supply dc exciting current to separately excited dc generators. The field circuit is not connected to the armature circuit. The separately excited dc generator has a very constant output voltage. Changes in load affect the armature current, but they do not change the strength of the magnetic field. The output voltage of a separately excited dc generator is varied by adjusting the current flow through the field coils. A large rheostat in series with

Figure 5-24. Separately excited dc generator: (a) pictorial; (b) schematic.

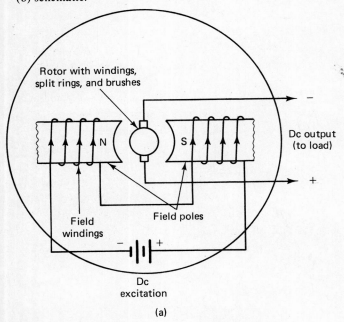

(a)

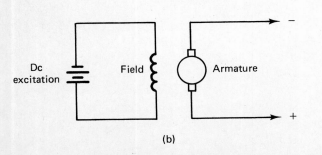

(b)

the field coils can be used to control current flow and adjust output voltage.

Separately excited dc generators are used when precise voltage control is needed. Certain industrial processes require this precision. The cost of separately excited dc generators is usually high. Another disadvantage is that a separate direct current electrical energy source is needed.

Self-Excited Generators

SELF-EXCITED, SERIES-WOUND DC GENERATORS. Dc generators produce direct current, so it is possible to take part of a generator's output to use as exciting current for the field coils. Generators that use part of their own output to supply dc exciting current are called *self-excited* dc generators. The method of connecting the armature windings and field windings together determines the type of generator. The armature and field windings may be connected in either series, parallel (shunt), or series-parallel (compound). These are the three types of self-excited dc generators.

Series-wound dc generators have armature windings connected in series with the field windings and the load as shown in Fig. 5-25. In the series-wound dc generator, the total current flows through the load and through the field coils and armature. The field coils are wound with low-resistance wire which has a few turns of large-diameter wire. An electromagnetic field is produced by the current flow through the coils. Remember that current flow is the *same* in all parts of a series circuit.

If the load is disconnected, no current would flow through the generator. The field coils retain a small amount of magnetism after they are de-energized. This is called *residual magnetism*. Due to residual magnetism, current begins to flow as soon as the generator operates again. As the current increases, there is an increase in the magnetic flux of the field. The output voltage rises as current flow increases. An output graph of a series-wound dc generator is shown in Fig. 5-26. The peak of the curve shows magnetic saturation of the field coils. At this point, the "domains" of the coils have maximum alignment. Beyond this point, an increase in load current causes a decrease in output voltage due to energy losses which occur. The output of a series-wound dc generator varies with changes in load current. Self-excited, series-wound generators have only a few applications.

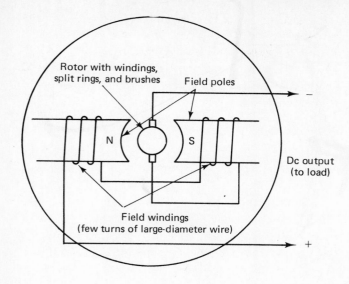

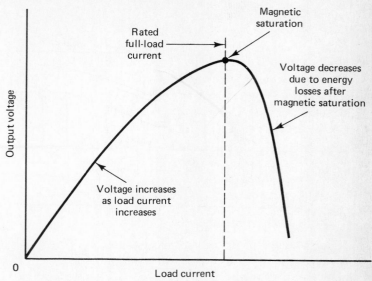

Figure 5-26. Output graph of a series-wound dc generator.

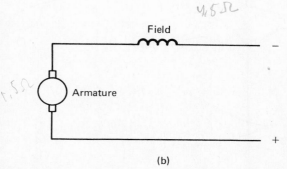

(b)

Figure 5-25. Series-wound dc generator: (a) pictorial; (b) schematic.

SELF-EXCITED, SHUNT-WOUND DC GENERATORS.
When the field coils, armature circuit, and load are connected in parallel, a shunt-wound dc generator is formed. Figure 5-27 shows a shunt-wound dc generator. The armature current developed by the generator (I_A) has two paths. One path is through the load (I_L) and the other is through the field coils (I_F). The shunt-wound dc generator is designed so that the field current is not more than 10% of the total armature current (I_A). This is so that most of the generated armature current (I_A) will flow to the load.

A strong electromagnetic field must be produced. Also, the field current must be low. The field coils are wound with many turns of wire. They rely very little on the amount of field current to produce a strong magnetic field. The small-diameter wires limit the field current to a low value due to their high resistance.

When no load is connected to a shunt-wound dc generator, a voltage is still generated. The voltage

supplies energy to the field coils. Residual magnetism in the field coils is important for shunt-wound dc generators also. When a shunt generator is turned on, current flows in the armature and field circuit due to residual magnetism. As current increases, the output voltage increases until magnetic saturation occurs.

When a load is connected to a dc shunt generator, the armature current (I_A) increases. The current increases the voltage ($I \times R$) drop of the armature. This causes a slightly smaller output voltage. Increases in load current cause slight decreases in output voltage as shown in the graph of Fig. 5-28. With load currents less than the rated value, the voltage is nearly constant. Large load currents cause the output voltage to drop sharply due to energy losses. Self-excited, shunt-wound dc generators are used when a fairly constant output voltage is needed.

SELF-EXCITED, COMPOUND-WOUND DC GENERATORS. Compound-wound dc generators have two sets of field windings. One set is made of low-resistance coils connected in series with the armature circuit. The other set is made of high-resistance coils connected in parallel with the armature circuit. A compound-wound dc generator is shown in Fig. 5-29.

The output voltage of a series-wound dc generator increases with increases in load current. The output voltage of a shunt-wound dc generator decreases slightly with increases in load current. A

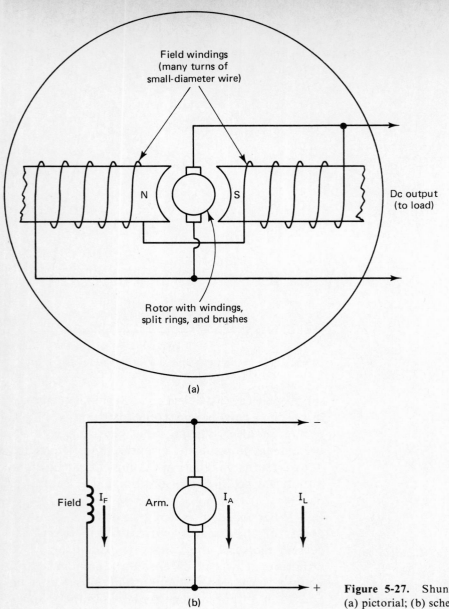

Field windings
(many turns of
small-diameter wire)

N   S

Dc output
(to load)

Rotor with windings,
split rings, and brushes

(a)

Field I_F Arm. I_A I_L

(b)

Figure 5-27. Shunt-wound dc generator:
(a) pictorial; (b) schematic.

Figure 5-28. Output graph of a shunt-wound dc generator.

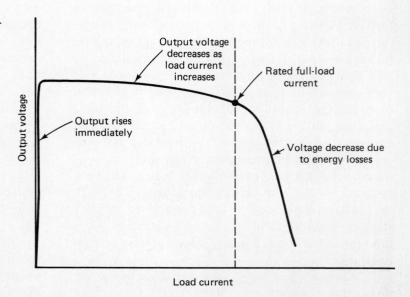

Output voltage
decreases as
load current
increases

Rated full-load
current

Output rises
immediately

Voltage decrease due
to energy losses

Output voltage

Load current

116

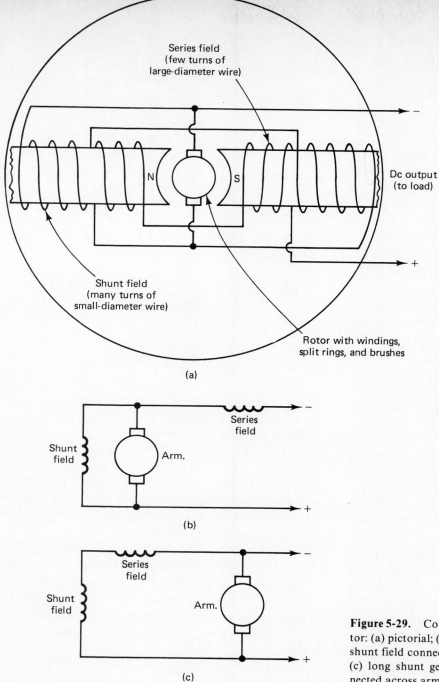

Series field
(few turns of
large-diameter wire)

N   S

Dc output
(to load)

Shunt field
(many turns of
small-diameter wire)

Rotor with windings,
split rings, and brushes

(a)

Shunt
field

Arm.

Series
field

−

+

(b)

Series
field

Shunt
field

Arm.

−

+

(c)

Figure 5-29. Compound-wound dc generator: (a) pictorial; (b) short shunt generator—shunt field connected across armature only; (c) long shunt generator—shunt field connected across armature and series field.

compound-wound dc generator has both series and shunt windings. Its output voltage is almost constant regardless of load current. The series field windings set up a magnetic field to counteract the voltage reduction caused by the voltage ($I \times R$) drop of the armature circuit. This produces a constant voltage.

A constant output voltage is produced by a *flat-compounded* dc generator. The no-load voltage is equal to the rated full-load voltage of a flat-compounded generator. No-load voltage is the output when there is no load connected to the generator.

Full-load voltage is the output when the rated value of load is connected to the circuit. A compound-wound dc generator with full-load voltage greater than no-load voltage is called an *overcompounded* generator. A generator with full-load voltage less than no-load voltage is called an *undercompounded* generator. Output graphs for the three types of compound generators are shown in Fig. 5-30.

Compound-wound dc generators can be made so that the series and shunt fields either aid or oppose each other. If the polarities of the coils on

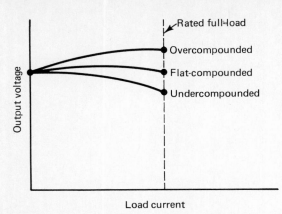

Figure 5-30. Output graphs for three types of compound-wound dc generators.

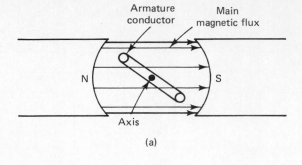

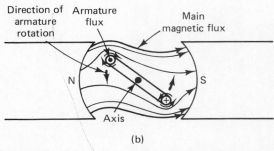

Figure 5-31. Effect of armature reaction in dc generators: (a) main magnetic field with no current flow through the armature windings; (b) distortion of main magnetic field with current flow through the armature windings.

one side are the same, the magnetic fields aid each other. This type is called a *cumulative* compound dc generator. A *differential* compound dc motor has opposite polarities on both sides. The cumulative generator is most used. Compound-wound dc generators are used for applications which require constant-voltage output.

Generator Operating Characteristics

Dc generators have a characteristic known as *armature reaction*. Current flow through the armature windings produces a circular magnetic field. These fields react with the main field as shown in Fig. 5-31. A magnetic field is produced which tends to distort the main magnetic field. As load current increases, armature current also increases. The increase in armature current causes more armature reaction to occur. Armature reaction causes sparking between the brushes and split rings. The theoretical switching point of the current to the load occurs at the generator's *neutral plane*. This is a plane or position where no voltage is induced into the armature conductor connected to the split ring. Also, no distortion of the main magnetic field takes place. The switching position is perpendicular to the main magnetic field. Armature reaction causes the main magnetic field to become distorted. The new switching position occurs when there is a small voltage induced into an armature coil. This new switching position is called the *running neutral plane*.

Armature reaction is reduced by using windings called *interpoles* between the main field coils. These coils are connected in series with the armature

circuit. An increase in armature current causes a stronger magnetic field around the interpoles. Its field counteracts the distortion of the main field caused by armature reaction.

Generator power output is usually rated in kilowatts. This rating is the electrical power generating capacity of a generator. Ratings are specified by the manufacturer on the nameplate of a generator. Other ratings include output voltage, speed, and temperature limits. Generators are made in many sizes.

As the load of a generator is increased, the generator tends to slow down. The output voltage then decreases. The amount of voltage change depends on the type of generator. The amount of change in output voltage from no-load value to rated full-load value is called *voltage regulation*. Voltage regulation is found by using the formula

$$\% \, VR = \frac{V_{NL} - V_{FL}}{V_{FL}} \times 100$$

where % VR is the voltage regulation, V_{NL} is the voltage with no load connected, and V_{FL} is the rated full-load voltage of the generator.

The efficiency of a generator is the ratio of its power output in watts and its power input in horsepower. The efficiency of a generator is found by using this formula

where P_{in} is the power input in horsepower, and P_{out} is the power output in watts. Horsepower must be converted to watts. Since 1 hp = 746 W, multiply the P_{in} by 746.

$$\% \text{ efficiency} = \frac{P_{out}}{P_{in}} \times 100$$

Review

1. List eight sources of electrical energy.

2. What are the two types of electrical current produced by energy sources called?

3. What are two classifications of chemical cells, and how are they different?

4. Discuss the operation of a carbon–zinc cell.

5. Discuss the operation of a lead–acid cell.

6. How do nickel–cadmium cells differ from carbon–zinc cells?

7. How is light energy used to produce electrical energy?

8. How is heat energy used to produce electrical energy?

9. How is pressure used to produce electrical energy?

10. Discuss electromagnetic induction.

11. What three factors determine the amount of voltage induced into a conductor inside a magnetic field?

12. What are the differences between slip rings and split rings used for electrical generators?

13. What is a sine wave?

14. If the maximum voltage of a generator is 50 V and the angle of rotation is 240°, what is the instantaneous voltage?

15. If the speed of rotation of a four-pole generator is 1800 rpm, what is its output frequency?

16. What are the two methods used to produce voltage output in generators, and how do they differ?

17. Discuss the two types of three phase connections used for stator windings in three-phase generators.

18. Discuss the construction of a dc generator.

19. What are the types of dc generators?

20. Discuss armature reaction.

21. If a generator has a no-load voltage output of 100 V and a full-load voltage output of 90 V, what is its voltage regulation percentage?

22. If a generator has a power output of 2000 W and a power input of 3 hp, what is its efficiency?

Student Activities

Chemical Cells (see Fig. 5-32)

1. Obtain a lemon, two pieces of metal (copper and zinc, if available), and a meter that measures low values of direct current.
2. Connect the meter as shown.
3. Squeeze the lemon until the opening where the metal strips are placed is juicy.
4. Exchange the positions of the two pieces of metal and watch the meter.
5. On a sheet of paper, discuss what occurs.
6. Turn in your paper to the teacher.

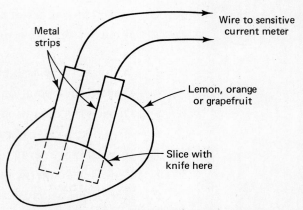

Figure 5-32.   Making a chemical cell.

Light Sources of Electrical Energy (see Fig. 5-8)

1. Obtain a solar cell.
2. Connect a VOM, set on a 1-V dc range, across the leads of the solar cell. Be sure to observe polarity.
3. Take a flashlight and shine onto the solar cell.
4. What is the maximum dc voltage output? Maximum voltage = _____ V dc.
5. On a sheet of paper, record the findings of this activity. Explain the relationship of light and voltage output and the method used to convert light energy into electrical energy.
6. Turn in your paper to the teacher.

Electromagnetic Induction (see Fig. 5-33)

1. Obtain the following materials:

    Electromagnet coil constructed in the third Student Activity in Chapter 4
    Galvanometer (zero-centered)
    Permanent magnet
    Piece of string (about 2 ft long)
    Wood to construct a platform

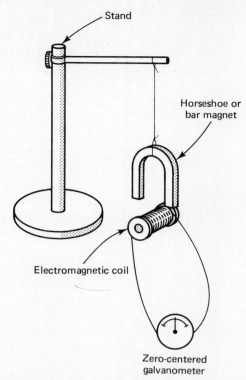

Figure 5-33. Setup for studying electromagnetic induction.

2. Connect the electromagnet coil to the terminals of the galvanometer. Place them on a flat surface (see Fig. 5-33).

3. Attach the permanent magnet to a string. Connect the other end of the string to the platform so that the magnet is about ¼ in. above the coil.

4. Push the magnet and allow it to move back and forth across the coil. Observe the galvanometer movement.

5. Let the teacher check your setup.

6. On a sheet of paper, discuss the electromagnetic induction. Be sure to discuss how speed and magnetic field strength affect current.

7. Turn in your paper to the teacher.

Generator Problems

1. **Work the following problems dealing with electrical generators, on a sheet of paper.**
   a. If the peak voltage of a single-phase ac generator is 320 V, what are the instantaneous induced voltages after an armature conductor has rotated the following number of degrees?
      (1) 25°   **(2)** 105°   **(3)** 195°   **(4)** 290°   **(5)** 335°
   b. A three-phase ac (wye-connected) generator has a phase voltage of 120 V and phase current of 10 A. What is its:
      (1) Line voltage   **(2)** Line current   **(3)** Total power
   c. A three-phase ac (delta-connected) generator has a line voltage of 480 V and a line current of 7.5 A. Find:
      (1) Phase voltage   **(2)** Phase current   **(3)** Total power

d. Find the operating frequencies of the following ac generators:
   (1) Single-phase, two-pole, 3600 rpm
   (2) Single-phase, four-pole, 1800 rpm

e. Find the voltage regulation of ac generators with the following no-load and full-load voltage values:
   (1) $E_{NL} = 120$ V, $E_{FL} = 115$ V
   (2) $E_{NL} = 220$ V, $E_{FL} = 210$ V
   (3) $E_{NL} = 480$ V, $E_{FL} = 478$ V

f. Find the efficiency of the following ac generators:
   (1) Power input = 20 hp, power output = 12 kW
   (2) Power input = 2.5 hp, power output = 1200 W
   (3) Power input = 4.25 hp, power output = 3 kW

g. A series-wound dc generator has a armature resistance of 1.5 Ω and a field resistance of 4.5 Ω. The voltage across a 12-Ω load is 100 V. What is the value of the following?
   (1) Armature current   (2) Power output

h. A shunt-wound dc generator has a armature resistance of 2.5 Ω and a field resistance of 40 Ω. Its terminal voltage is 120 V. Calculate:
   (1) Armature current   (2) Field current
   (3) Current delivered to the load   (4) Power output

i. A 120-V dc generator is rated at 20 kW. What is its maximum load current?

2. Turn in the answers to the teacher for grading.

Automobile Alternator (see Fig. 5-19)

1. Obtain an automobile alternator.

2. Disassemble the alternator to observe the parts.

3. Locate the stator, rotor, brushes, and slip rings.

4. On a sheet of paper, discuss the construction and operation of an automobile alternator.

5. Turn in the discussion to the teacher for grading.

Alternating-Current Electricity

Much of the electrical energy used today is called alternating current (ac). Most of the electrical equipment and appliances used in homes operate from the alternating-current energy which is delivered by power lines. Alternating-current electricity has many applications in homes, industries, and commercial buildings. Electrical power plants in our country produce alternating current or ac electricity. Most of our power plants have huge steam turbines that rotate ac generators. These generators produce three-phase ac which is distributed by long-distance power transmission lines to the places where the electrical power is used. Industries and large commercial buildings use *three-phase* ac. Homes use *single-phase* ac power. Alternating current is the most common form of electrical energy used in the United States.

IMPORTANT TERMS

In this chapter, alternating current (ac) electricity is studied. There are many devices and circuits which are used with ac voltage applied to them. Review these terms to gain an understanding of some of the topics of this chapter. There are many new terms introduced in Chapter 6.

Ac.   The abbreviation used for alternating current.

Admittance (Y).   The total ability of an ac circuit to conduct current, measured in siemens or mhos; the inverse of impedance (Z)—$Y = 1/Z$.

Air-core inductor.   A coil wound on an insulated core or a coil or wire which does not have a metal core.

Amplitude.   The vertical height of an ac waveform.

Angle of lead or lag.   The angle between applied voltage and current flow in an ac circuit, in degrees; in an inductive (L) circuit, voltage (E) leads current (I)—ELI; in a capacitive (C) circuit, current (I) leads voltage (E)—ICE.

Apparent power (*volt-amperes*).   The power *delivered* to an ac circuit; applied voltage times current.

Attenuation.   A reduction in value.

Average voltage (E_{AVG}).   The value of an ac sine-wave voltage which is found by this formula: $E_{Avg} = E_{Peak} \times 0.636$.

Bandpass filter.   A frequency-sensitive ac circuit which allows incoming frequencies within a certain band to pass through but attenuates frequencies which are below or above this band.

Bandwidth.   The band (range) of frequencies which will pass easily through a bandpass filter or resonant circuit.

Capacitance (*C*).   The property of a device to oppose changes in voltage due to energy stored in its electrostatic field.

Capacitive reactance (X_c).   The opposition to the flow of ac current caused by a capacitive device (measured in ohms).

Capacitor.   A device that has capacitance and is usually made of two metal plate materials separated by a dielectric material (insulator).

Center tap.   A terminal connection made to the center of a transformer winding.

Choke coil.   An inductor coil used to block the flow of ac current and pass dc current.

Condenser.   A term used occasionally to mean *capacitor*.

Conductance (*G*).   The ability of a resistance of a circuit to conduct current, measured in siemens or mhos; the inverse of resistance—$G = 1/R$.

Cycle.   A sequence of events that causes one complete pattern of alternating current from a zero reference, in a positive direction, back to zero, then in a negative direction and back to zero.

Decay.   A term used for a gradual reduction in value of a voltage or current.

Decay time.   The time for a capacitor to discharge to a certain percentage of its original charge or the time required for current through an inductor to reduce to a percentage of its maximum value.

Decibel.   A unit used to express an increase or decrease in power, voltage, or current in a circuit.

Dielectric.   An insulating material placed between the metal plates of a capacitor.

Dielectric constant.   A number that represents the ability of a dielectric to develop an electrostatic field, compared to air, which has a value of 1.0.

Effective voltage (E_{Eff}).   A value of an ac sine-wave voltage which has the same heating effect as an equal value of dc voltage: $E_{Eff} = E_{Peak} \times 0.707$.

Electrolytic capacitor.   A capacitor that has a positive plate made of aluminum and a dry paste or liquid used to form the negative plate.

Electrostatic field.   The field developed around a material due to the energy of an electrical charge.

ELI.   A term used to help remember that voltage (*E*) leads current (*I*) in an inductive (*L*) circuit.

Farad.   The unit of measurement of capacitance that is required to contain a charge of one coulomb when a potential of one volt is applied.

Filter.   A circuit used to pass certain frequencies and attenuate all other frequencies.

Frequency.   The number of ac cycles per second, measured in hertz (Hz).

Frequency response.   A circuit's ability to operate over a range of frequencies.

Henry (H).   The unit of measurement of inductance which is produced when a voltage of one volt is induced when the current through a coil is changing at a rate of one ampere per second.

Hertz (Hz).   The international unit of measurement of frequency equal to one cycle per second.

ICE.   A term used to help remember that current (*I*) leads voltage (*E*) in a capacitive (*C*) circuit.

Impedance (*Z*).   The total opposition to current flow in an ac circuit which is a combination of resistance (*R*) and reactance (*X*) in a circuit; measured in ohms; $Z = \sqrt{R^2 + X^2}$.

Inductance (*L*).   The property of a circuit to oppose changes in current due to energy stored in a magnetic field.

Inductive circuit.   A circuit that has one or more inductors or has the property of inductance, such as an electric motor circuit.

Inductive reactance (X_L).   The opposition to current flow in an ac circuit caused by an inductance (*L*), measured in ohms: $X_L = 2\pi \times f \times L$.

Inductor. A coil of wire which has the property of inductance and is used in a circuit for that purpose.

In phase. Two waveforms of the same frequency which pass through their minimum and maximum values at the same time and polarity.

Instantaneous voltage (E_i). A value at any instant (time) along a waveform.

Isolation transformer. A transformer with a 1:1 turns ratio used to isolate an ac power line from equipment with a chassis ground.

Lagging phase angle. The angle by which current *lags* voltage (or voltage *leads* current) in an inductive circuit.

Leading phase angle. The angle by which current *leads* voltage (or voltage *lags* current) in a capacitive circuit.

Maximum power transfer. A condition that exists when the resistance or impedance of a load (R_L) equals that of the source which supplies it (R_S).

Mho. Ohm spelled backward; the old unit of measurement for conductance, susceptance, and admittance; being replaced by the *seimen*.

Mica capacitor. A capacitor made of metal foil plates separated by a mica dielectric.

Mutual inductance (M). When two coils are located close together so that the magnetic flux of the coil affect one another in terms of their inductance properties.

Parallel resonant circuit. A circuit that has an inductor and capacitor connected in parallel to cause response to frequencies applied to the circuit.

Peak-to-peak voltage (E_{p-p}). The value of ac sine-wave voltage from positive peak to negative peak.

Peak voltage (E_{Peak}). The maximum positive or negative value of ac sine-wave voltage: $E_{Peak} = E_{Eff} \times 1.41$.

Period (time). The time required to complete one ac cycle: time = 1/frequency.

Phase angle (θ). The angular displacement between applied voltage and current flow in an ac circuit.

Power factor (PF). The ratio of power converted (true power) in an ac circuit and the power delivered (apparent power):

$$PF = \frac{\text{true power (watts)}}{\text{apparent power (volt-amps)}}$$

Primary winding. The coil of a transformer to which ac source voltage is applied.

Quality factor (Q). The "figure of merit" or ratio of inductive reactance and resistance in a frequency-sensitive circuit.

Reactance (X). The opposition to ac current flow due to inductance (X_L) or capacitance (X_c).

Reactive circuit. An ac circuit which has the property of inductance or capacitance.

Reactive power (VAR). The "unused" power of an ac circuit has inductance or capacitance, which is absorbed by the magnetic or electrostatic field of a reactive circuit.

Resistance. The opposition to current flow due to a resistive device.

Resistive circuit. A circuit whose only opposition to current flow is resistance; a nonreactive circuit.

Resonant circuit. *See* Parallel resonant circuit *and* Series resonant circuit.

Resonant frequency (f_r). The frequency which passes most easily through a frequency-sensitive circuit when $X_L = X_C$ in the circuit:

$$f_r = \frac{1}{2\pi \times \sqrt{L \times C}}$$

Root mean square voltage (rms). *See* Effective voltage.

Sawtooth waveform. An ac waveform shaped like the teeth of a saw.

Secondary winding. The coil of a transformer into which voltage is induced; energy is delivered to the load circuit by the secondary winding.

Selectivity. The ability of a resonant circuit to select a specific frequency and reject all other frequencies.

Series resonant circuit. A circuit that has an inductor and capacitor connected in series to cause response to frequencies applied to the circuit.

Siemen. *Same as* mho.

Signal. An electrical waveform of varying value which is applied to a circuit.

Sine wave. A waveform of one cycle of ac voltage.

Step-down transformer. A transformer that has a secondary voltage lower than its primary voltage.

Step-up transformer. A transformer that has a secondary voltage higher than its primary voltage.

Susceptance (B). The ability of an inductance (B_L) or a capacitance (B_c) to pass ac current; measured in siemens or mhos:

$$B_L = \frac{1}{X_L} \quad \text{and} \quad B_c = \frac{1}{X_c}.$$

Tank circuit. *See* Parallel resonant circuit.

Theta (θ). The Greek letter used to represent phase angle of an ac circuit.

Time constant (RC). The time required for the voltage across a capacitor in an *RC* circuit to increase to 63% of its maximum value or decrease to 37% of its maximum value; time = $R \times C$.

Time constant (RL). The time required for the current through an inductor in an *RL* circuit to increase to 63% of its maximum value or decrease to 37% of its maximum value; time = L/R.

Transformer. An ac power control device which transfers energy from its primary winding to its secondary winding by mutual inductance and is ordinarily used to increase or decrease voltage.

True power (W). The power actually *converted* by an ac circuit, as measured with a wattmeter.

Turns ratio. The ratio of the number of turns of the primary winding (N_p) of a transformer to the number of turns of the secondary winding (N_s).

Vector. A straight line whose length indicates magnitude and position indicates direction.

Volt-ampere (VA). The unit of measurement of apparent power.

Volt-amperes reactive (VAR). The unit of measurement of reactive power.

Waveform. The pattern of an ac frequency derived by looking at instantaneous voltage values which occur over a period of time; on a graph a waveform is plotted with instantaneous voltages on the vertical axis and time on the horizontal axis.

Wavelength. The distance from a point on a waveform to a corresponding point on an adjacent waveform.

Working voltage. A rating of capacitors which is the maximum voltage that can be placed across the plates of a capacitor without damage occurring.

ALTERNATING-CURRENT VOLTAGE

When an ac source is connected to some type of load, current direction changes several times in a given unit of time. Remember that direct current (dc) flows in one direction only. A diagram of one cycle alternating current is compared to a dc waveform in Fig. 6-1. This waveform is called an ac sine wave. When the ac generator shaft rotates one complete revolution, or 360°, one ac sine wave is produced. Ac voltage generation was studied in Chapter 5. Note that the sine wave has a positive peak at 90°, then decreases to zero at 180°. It then increases to a peak negative voltage at 270°, then decreases to zero at 360°. The cycle then repeats itself. Current flows in one direction during the positive part and in the opposite direction during the negative half-cycle.

Figure 6-2 shows five cycles of alternating current. If the time required for an ac generator to produce five cycles were 1 s, the frequency of the ac would be 5 cycles per second. Ac generators at power plants in the United States operate at a frequency of 60 cycles per second, or 60 Hz. Hertz is the international unit for frequency measurement. If 60 ac sine waves are produced every second, a speed of 60 revolutions per second is needed. This produces a frequency of 60 cycles per second.

Ac voltage is measured with a volt-ohm-milliammeter (VOM), and polarity of the meter leads *is not* important. This is because ac changes direction. Remember that polarity is important when measuring dc since direct current flows only in one direction. Many VOMs do not measure ac current. They have ranges for ac voltage only.

Shown in Fig. 6-3 are several voltage values associated with alternating current. Among these are peak positive, peak negative, and peak-to-peak ac values. Peak positive is the maximum positive

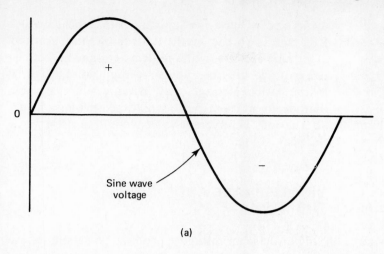

(a)

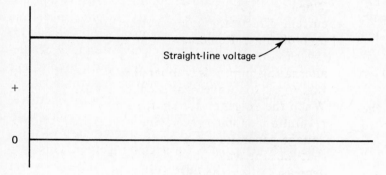

(b)

Figure 6-1. Comparison of (a) ac and (b) dc waveforms.

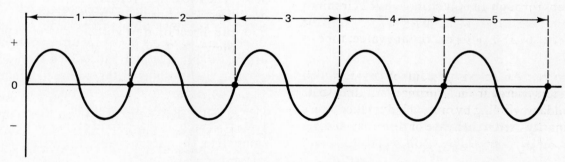

Figure 6-2. Five cycles of alternating current.

Figure 6-3. Voltage values of an ac waveform.

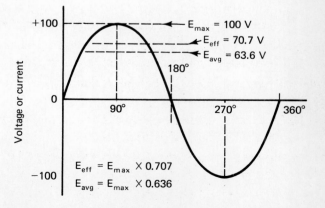

voltage reached during a cycle of ac. Peak negative is the maximum negative voltage reached. Peak-to-peak is the voltage value from peak positive to peak negative. These values are important to know when working with radio and TV amplifier circuits. For example, the most important ac value is called *effective* or measured value. This value is less than the peak positive value. A common ac voltage is 120 V, which is used in homes. This is an effective value voltage. Its peak value is about 170 V. Effective value of ac is defined as the ac voltage that will do the same amount of work as a dc voltage of the same value. For instance, in the circuit of Fig. 6-4, if the switch is placed in position 2, 10 V ac effective value is applied to the lamp. The lamp should produce the same amount of brightness with 10 V ac effective value as with 10 V dc applied. When ac voltage is measured with a meter, the reading indicated is effective value.

In some cases, it is important to convert one ac value to another. For instance, the voltage rating of electronic devices must be greater than the peak ac voltage applied to them. If 120 V ac is the measured voltage applied to a device, the peak voltage is about 170 V. So the device must be rated over 170 V rather than 120 V.

To determine peak ac, when the measured or effective value is known, the formula

$$peak = 1.41 \times effective\ value$$

is used. When 120 V is multiplied by the 1.41 conversion factor, the peak voltage is found to be about 170 V.

Two other terms that should be mentioned are *rms value* and *average value*. Rms stands for "root mean square" and is equal to 0.707 × peak value. Rms refers to the mathematical method used to determine effective voltage. Rms voltage and effective voltage are the same. Average voltage is the mathematical average of all instantaneous voltages that occur at each period of time throughout an alternation. The average value is equal to 0.636 times the peak value.

SINGLE-PHASE AND THREE-PHASE AC

Single-phase ac voltage is produced by single-phase ac generators or it can be obtained across two power lines of a three-phase system. A single-phase ac source has a hot wire and a neutral wire. The neutral is grounded to help prevent electrical shocks. Single-phase power is the type of power distributed to our homes. A three-phase ac source has three power lines. Three-phase voltage is produced by three-phase generators at power plants. Three-phase voltage is a combination of three single-phase voltages which are electrically connected together. This voltage is similar to three single-phase sine waves separated in phase by 120°. Three-phase ac is used to power large equipment in industry and commercial buildings. It is not distributed to homes. There are three power lines on a three-phase system. Some three-phase systems have a neutral connection and others do not. Chapter 5 discussed the voltage and current relationships of three-phase systems.

The term *phase* refers to time or the difference between one point and another. If two sine-wave voltages reach their zero and maximum values at the same time, they are *in phase*. Figure 6-5 shows two

Figure 6-4.   Comparison of effective ac voltage and dc voltage.

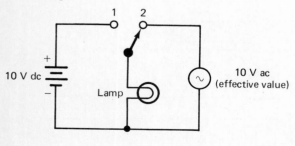

Figure 6-5.   Two ac voltages that are in phase.

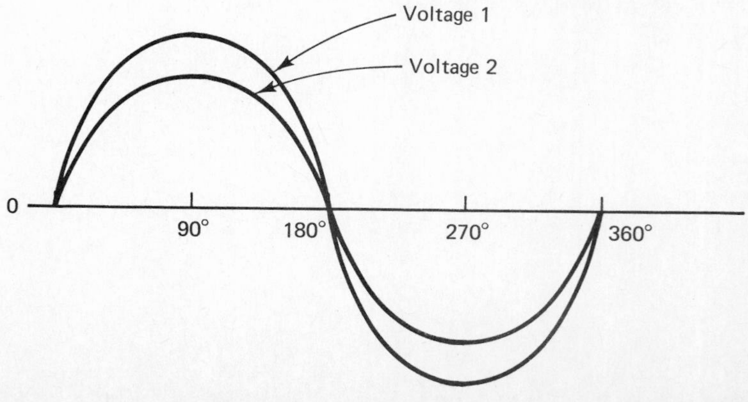

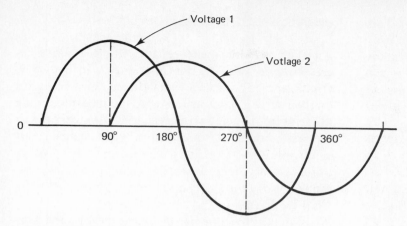

Figure 6-6. Two ac voltages that are out of phase by an angle of 90°.

alternating current voltages that are in phase. If two voltages reach their zero and maximum values at different times, they are *out of phase*. Figure 6-6 shows two alternating current voltages that are out of phase. Phase difference is given in degrees. The voltages shown are out of phase at an angle of 90°.

Remember that single-phase ac voltage is in the form of a sine wave. Single-phase ac voltage is used for low power applications, primarily in the home. Almost all electrical power is generated and transmitted over long distances as *three-phase* ac. Three coils are placed 120° apart in a generator to produce three-phase ac voltage. Most ac motors over 1 hp in size operate with three-phase ac power applied.

Figure 6-7. Comparison of (a) single-phase and (b) three-phase ac voltages.

Most industries and commercial buildings have three-phase equipment.

Three-phase ac systems have several advantages over single-phase systems. In a single-phase system, the power is said to be *pulsating*. The peak values along a single-phase ac sine wave are separated by 360°, as shown in Fig. 6-7(a). This is similar to a one-cylinder gas engine. A three-phase system is somewhat like a multicylinder gas engine. The power is more steady. One cylinder is compressing when the others are not. This is similar to the voltages in three-phase ac systems. The power of one separate phase is pulsating, but the total power is more constant. The peak values of three-phase ac are

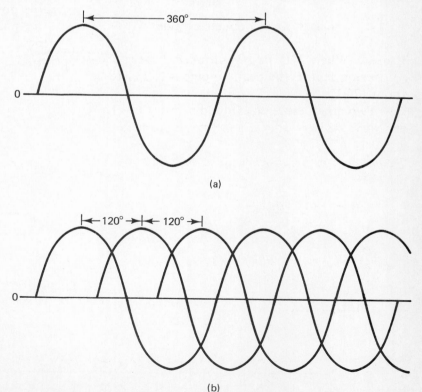

separated by 120°, as shown in Fig. 6-7(b). This makes three-phase ac power more desirable to use.

The power ratings of motors and generators is greater when three-phase ac power is used. For a certain frame size, the rating of a three-phase ac motor is almost 50% larger than a similar single-phase ac motor.

MEASURING AC VOLTAGE WITH A VOM

A VOM may be used to measure ac voltage. Ac voltage is measured in the same way as dc voltage with two exceptions:

1. When measuring ac voltage, proper polarity *does not* have to be observed.
2. When measuring ac voltage, the ac voltage ranges and scales of the meter must be used.

Figure 6-8 shows the scale of a VOM with those used to measure ac voltage marked. The scales used to measure ac voltage are identical to those used to measure dc voltage.

USING AN OSCILLOSCOPE

Another way to measure ac voltage is with an *oscilloscope*, such as the one shown in Fig. 6-9. Oscilloscopes provide a visual display of waveforms on their screen. They are also used to measure a wide range of frequencies with precision. Oscilloscopes or "scopes" are used to examine wave shapes. For electronic servicing, it is necessary to be able to observe the voltage waveform while troubleshooting. Oscilloscopes are discussed in more detail in Chapter 8.

An oscilloscope permits various voltage waveforms to be visually analyzed. It produces an image on its screen. The controls must be properly adjusted. The image, called a *trace*, is usually a line on the screen or cathode-ray tube (CRT). A stream of electrons strikes the phosphorescent coating on the inside of the screen, causing the screen to produce light.

The oscilloscope displays voltage waveforms on two axes, like a graph. The horizontal axis on the screen is the *time* axis. The vertical axis is the voltage axis. An ac waveform is displayed on the CRT as shown in Fig. 6-10. For the CRT to display a trace properly, the internal circuits of the scope must be

Figure 6-8. Scale of a VOM with ac voltage section shown.

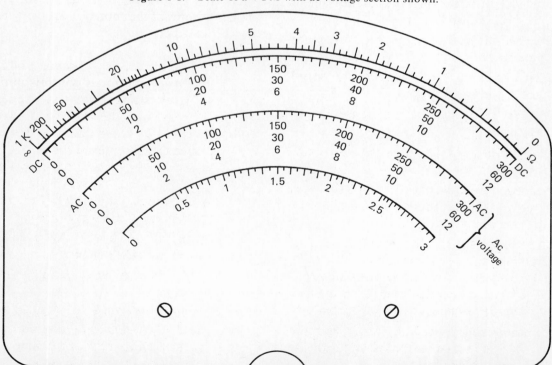

Figure 6-9. Oscilloscope used to measure ac voltage. (Courtesy of Hickok Electrical Instrument Co.)

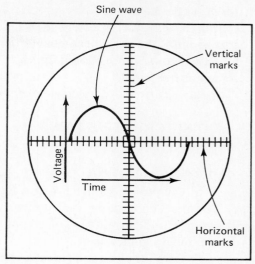

Figure 6-10. Ac waveform displayed on the screen of an oscilloscope.

properly adjusted. These adjustments are made by controls on the front of the oscilloscope. Oscilloscopes are slightly different, but most scopes have some of the following controls:

1. *Intensity*: Controls the brightness of the trace, sometimes the on–off control also.
2. *Focus*: Adjusts the thickness of the trace so that it is clear and sharp.
3. *Vertical position*: Adjusts the entire trace up or down.
4. *Horizontal position*: Adjusts the entire trace to the left or right.
5. *Vertical gain*: Controls the height of the trace.
6. *Horizontal gain*: Controls the horizontal size of the trace.
7. *Vertical attenuation or variable volts/cm*: Acts as a "coarse" adjustment to reduce the trace vertically.
8. *Horizontal sweep or variable time/cm*: Controls the speed at which the trace moves across (sweeps) the CRT horizontally. This control determines the number of waveforms displayed on the screen.
9. *Synchronization select*: Controls how the input to the scope is "locked in" with the circuitry of the scope.

10. *Vertical input*: External connections used to apply the input to the vertical circuits of the scope.
11. *Horizontal input*: External connections used to apply an input to the horizontal circuits of the scope.

The following procedure is used to adjust the oscilloscope controls to measure ac voltage. The names of some of the controls vary on different types of oscilloscopes.

1. Turn on the oscilloscope and adjust the *intensity* and *focus* controls until a bright, narrow, straight-line trace appears on the screen. Use the *horizontal position* and *vertical position* controls to position the trace in the center of the screen. Adjust the *horizontal gain* and the *variable time/cm* until the trace extends from the left of the screen to the right side of the screen. This allows the entire waveform to be displayed.
2. Connect the proper *test probes* into the oscilloscope's *vertical input* connections.
3. The scope is now ready to measure ac voltage.
4. After a waveform is displayed, adjust the *vertical attenuation* (volts/cm) and *vertical gain* controls until the height of the trace equals about 2 in. or 4 centimeters (cm).

Most scopes have scales that are marked in centimeters. Adjust the *vernier* or stability control until the trace becomes stable. One ac waveform or more should appear on the screen of the scope.

OHM'S AND KIRCHHOFF'S LAWS FOR AC CIRCUITS

In Chapter 3, Ohm's law was used with dc circuits. As discussed in Chapter 3, Ohm's law can be used with ac circuits containing resistance *only*. Kirchhoff's voltage and current laws also apply to ac circuits containing resistance *only*. The use of Ohm's and Kirchhoff's laws for ac circuits is influenced by the effects of two circuit elements which are not present in dc circuits. Before learning to apply basic electrical laws to ac circuits, it is important to learn about *inductance* and *capacitance*.

INDUCTANCE

When energized with dc voltage, a magnetic field is produced around a coil. Dc current flow produces a *constant* magnetic field around a coil. When the same coil is supplied with ac voltage, the constantly changing ac current produces a constant *changing* magnetic flux. This changing flux sets up a magnetic field with a constantly reversing polarity and changing strength. It also induces a counter EMF or counter voltage. This counter electromotive force (CEMF) opposes the source voltage. CEMF limits current flow from the source.

The opposition to the flow of ac current by a magnetic field is due to a property called *inductance* (L). The opposition to current flow of an inductive device depends on the resistance of the wire and the magnetic properties of the circuit. This opposition due to the magnetic effect is called *inductive reactance* (X_L). X_L varies with the applied frequency and is found by using the formula $X_L = 2\pi \times f \times L$, where $2\pi = 6.28$, f is the applied frequency in hertz, and L is the inductance in *henries*. The basic unit of inductance is the henry (H).

At zero frequency (or dc), there is no opposition due to inductance. Only a coil's resistance limits current flow. As frequency increases, the inductive effect becomes greater. Many ac machines use magnetic circuits in one form or another. The inductive reactance of an ac circuit usually has more effect on current flow than resistance. An ohmmeter measures *dc resistance* only. Inductive reactance must be calculated or determined experimentally.

Ohm's law *does not* apply to ac circuits with inductance as it did with dc circuits. Ohm's law *does* apply if the resistance (R) *and* inductive reactance (X_L) are considered. Both values limit current flow. The total opposition to current flow is called *impedance* (Z). Impedance is a combination of resistance and reactance. In ac circuits with both types of opposition, the Ohm's law relationship becomes $I = E/Z$.

Figure 6-11 shows some symbols used to indicate various types of inductors. Figure 6-12 shows some types of inductors. They are all coils of wire designed for specific functions. Often they are called "choke" coils. Choke coils are used to pass dc current and block ac current flow.

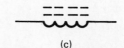

(a)

Figure 6-11.   Symbols used for inductors: (a) air core; (b) iron core; (c) powdered metal core.

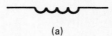

(b)

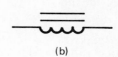

(c)

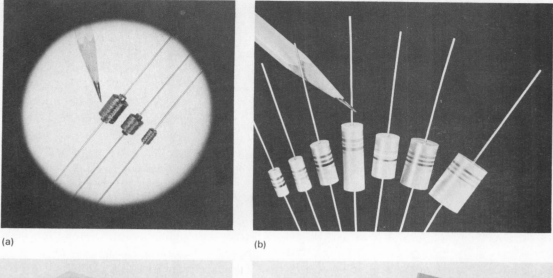

Figure 6-12. Some types of inductors: (a) subminiature inductors used with high frequencies; (b) inductors that use a color-code value; (c) inductors used in equipment power supplies; (d) inductors for various electronic applications. [(a), (b), and (d) Courtesy of J. W. Miller Div./Bell Industries; (c) courtesy of TRW/UTC Transformers.]

CAPACITANCE

When two conductors are separated by an insulator (dielectric), an *electrostatic* charge may be set up on the conductors. This charge becomes a source of stored energy. The strength of the charge depends on the applied voltage, the size of the conductors, which are called *plates*, and the quality or *dielectric strength* of the insulation. The closer the two plates are placed together, the more charge may be set up on them. This type of device is called a *capacitor*. The size of capacitors is measured in units called

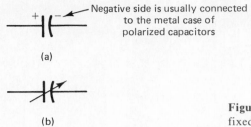

(a)

(b)

Figure 6-13. Capacitor symbols: (a) fixed capacitor; (b) variable capacitor.

Figure 6-14. (a) Small capacitors used in electronic equipment; (b) variable trimmer capacitors used in communication circuits; (c) variable capacitor used in tuning circuits; (d) capacitors used for power factor correction on electrical power systems. [(a) Courtesy of Centralab, Inc., a North American Philips Co.; (b) courtesy of Johanson Mfg. Corp.; (c) courtesy of J. W. Miller Div./Bell Industries; (d) courtesy of McGraw-Edison Co.]

(a)   (b)

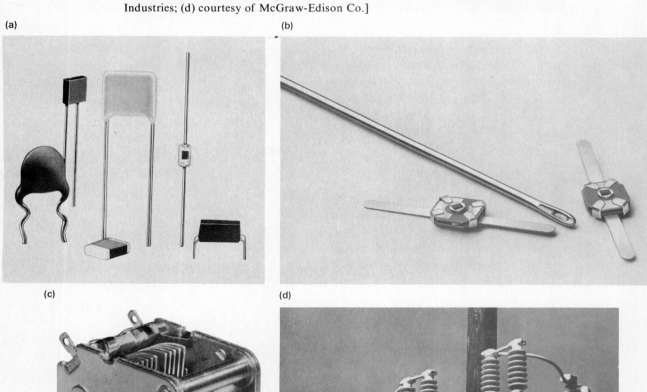

(c)   (d)

farads, *microfarads*, and *picofarads*. Some capacitor symbols are shown in Fig. 6-13. Various types of capacitors are shown in Fig. 6-14.

When a dc voltage is applied to the plates of a capacitor, it will charge to the value of the source voltage. The dielectric between the plates then stops current flow. The capacitor remains charged to the value of the dc voltage source. When the voltage source is removed, the capacitor charge will leak away. Dc current flows to or from a capacitor only when the source voltage is turned on or off.

If the same capacitor is connected to an ac voltage source, the voltage constantly *changes*. The capacitor receives energy from the source during one-quarter cycle and very little current flows through the dielectric of a capacitor. Ac voltage applied to a capacitor causes it to constantly change its amount of charge. Capacitors have the ability to pass ac current because of their charging and discharging action. They can be used to pass ac current and block dc current flow. The opposition of a capacitor to a source voltage depends on frequency. The faster the applied voltage changes across a capacitor, the easier the capacitor passes current.

INDUCTIVE EFFECTS IN CIRCUITS

Inductance is the property of a circuit that opposes changes of current flow. Remember that inductance is the characteristic of an electrical circuit that opposes *changes* in current flow. Any coil of wire has inductance. Coils oppose current changes by producing a counter voltage (CEMF). The term *reactive* is used due to a coil's reaction to changes in applied voltage. The opposition of a coil is called *reactance* (X) and is measured in ohms. The subscript L is added to represent inductive reactance (X_L). Inductive reactance depends on the "speed" or frequency of the ac source.

Inductance *does not* change with frequency changes. Coils have inductance due to their ability to oppose current change. This property stays the same for any coil. The factors that determine inductance are the number of turns of the coil, area and length of the core material, and the type of core used. The main factor is the number of coil turns. The *inductive reactance* (X_L) *does* change with changes in frequency. This can be verified by looking at the

formula: $X_L = 2\pi \times f \times L$. Notice that as frequency is increased, inductive reactance increases. At high frequencies, inductors approach the characteristics of an open circuit. At low frequencies, inductors approach the characteristics of short circuits.

Mutual Inductance

When inductors are connected together, a property called *mutual inductance* (M) must be considered. Mutual inductance is the magnetic field interaction or *flux linkage* between coils. The amount of flux linkage is called the coefficient of coupling (k). If all the lines of force of one coil cut across a nearby coil, it is called *unity coupling*. There are many possibilities, determined by coil placement, of coupling between coils. The amount of mutual inductance between coils is found by using the formula

$$\text{mutual inductance } (M) = k \times \sqrt{L_1 \times L_2}$$

The term k is the coefficient of coupling, which gives the amount of coupling. L_1 and L_2 are the inductance values of the coils. Mutual inductance should also be considered when two or more coils are connected together.

Inductors in Series and Parallel

Inductors may be connected in series or parallel. When inductors are connected to prevent the magnetic field of one from affecting the others, these formulas are used to find total inductance (L_T).

1. Series inductance:

$$L_T = L_1 + L_2 + L_3 + \cdots$$

2. Parallel inductance:

$$\frac{1}{L_T} = \frac{1}{L_1} + \frac{1}{L_2} + \frac{1}{L_3} + \cdots$$

L_1, L_2, L_3, and so on, are inductance values in henries.

When inductors are connected so that the magnetic field of one affects the other, mutual inductance increases or decreases the total inductance. The effect of mutual inductance depends on the physical positioning of the inductors. Their distance apart and the direction in which they are wound affects mutual inductance. Inductors are connected in series or parallel with an aiding or opposing mutual inductance (M). The formulas used to find total inductance (L_T) are:

1. Series aiding:

$$L_T = L_1 + L_2 + 2M$$

2. Series opposing:

$$L_T = L_1 + L_2 - 2M$$

3. Parallel aiding:

$$\frac{1}{L_T} = \frac{1}{L_1 + M} + \frac{1}{L_2 + M}$$

4. Parallel opposing:

$$\frac{1}{L_T} = \frac{1}{L_1 - M} + \frac{1}{L_2 - M}$$

L_1 and L_2 are the inductance values and M is the value of mutual inductance.

Inductive Phase Relationships

A change in current through a coil causes a change in the magnetic flux around the coil. Voltage changes at its maximum rate when passing through its zero value at $0°$, $180°$, and $360°$, as shown in Fig. 6-15. The magnetic flux changes are also greatest at these times. The CEMF of the coil is at *maximum*

value at these times also. Since CEMF opposes source voltage, it is directly opposite applied voltage in the diagram.

Lenz's law states that the counter voltage (CEMF) always opposes a change in current. At $0°$, the counter voltage is opposite in polarity to the applied voltage. This opposes the rise in current and causes current to equal zero. When the values of the counter voltage and the applied voltage are at $90°$, the current is maximum. This causes the current flow to *lag* behind the applied voltage by $90°$.

The dc resistance of a coil is small. The largest opposition to ac current flow through a coil is the counter voltage (CEMF). The applied voltage is slightly higher than the counter voltage. If they were equal, no current could flow. Ohm's law can be used with inductive circuits to find current as follows:

$$I_L = \frac{E_L}{X_L}$$

where I_L is the current through the coil, E_L is the voltage across the coil, and X_L is the inductive reactance of the coil (in ohms). The resistance of the coil is usually not taken into account.

CAPACITIVE EFFECTS IN CIRCUITS

Inductance is defined as the property of a circuit to oppose changes in *current*. *Capacitance* is the property of a circuit to oppose changes in *voltage*. Inductance stores energy in an *electromagnetic field* while capacitance stores energy in an *electrostatic field*. A capacitor is measured by a unit called the *farad*. One farad is the amount of capacitance that permits a current of one ampere to flow when the

Figure 6-15.   Inductive phase relationships.

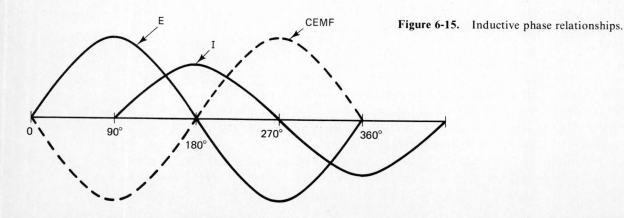

voltage change across the plates of a capacitor is one volt per second. The farad is too large for practical use. The microfarad (one-millionth of a farad, abbreviated μF or MFD) is the most common subunit of capacitance. For high-frequency circuits, the microfarad is also too large. The subunit micro-microfarad (one-millionth of a microfarad, abbreviated $\mu\mu F$ or MMFD) is then used. This subunit is called the picofarad (pF) to avoid confusion.

The three factors that determine the capacitance of a capacitor are:

1. *Plate area*: Increasing plate area increases capacitance.

2. *Distance between plates*: Capacitance is decreased when the distance between the plates increases.

3. *Dielectric material*: Dielectrics, including air, are used as insulators between capacitor plates. They are rated with a dielectric constant (K). Capacitors with higher dielectric constants have higher capacitance.

Capacitors in Series

Adding capacitors in series has the same effect as increasing the distance between plates. This reduces total capacitance. The total capacitance of capacitors connected in series is less than any individual capacitance. Total capacitance (C_T) is found in the same way as parallel resistance. The reciprocal formula is used:

$$\frac{1}{C_T} = \frac{1}{C_1} + \frac{1}{C_2} + \frac{1}{C_3} + \cdots$$

Capacitors in Parallel

Capacitors in parallel are similar to one capacitor with its plate area increased. Doubling the plate area would double capacitance. Capacitance in parallel is found by adding individual values, just as with series resistors, or

$$C_T = C_1 + C + C_3$$

LEADING AND LAGGING CURRENTS IN AC CIRCUITS

Inductors and capacitors cause ac circuits to react differently than circuits with resistance only. Inductors oppose changes in *current*, and capacitors oppose changes in *voltage*. In dc circuits, voltage and current change in step or "in phase" with each other. Ac circuits have different effects because the voltage is constantly changing. The current in an inductive circuit *lags* behind the ac source voltage. Both voltage and current are in the form of sine waves, as shown in Fig. 6-15. The current *lags* the applied voltage by 90°.

The current in a capacitive circuit is maximum when the applied voltage is zero. This is true with both dc and ac circuits. The capacitive current *leads* the applied voltage. The phase relationships of voltage and current in a capacitive circuit is shown in Fig. 6-16. The 90° phase difference between current and voltage is the maximum theoretically possible. In actual ac circuits, it is not possible to have purely inductive or purely capacitive circuits. Resistance is always present in actual circuits. Voltage and current in a resistive circuit are always in phase.

Figure 6-16. Capacitive phase relationships.

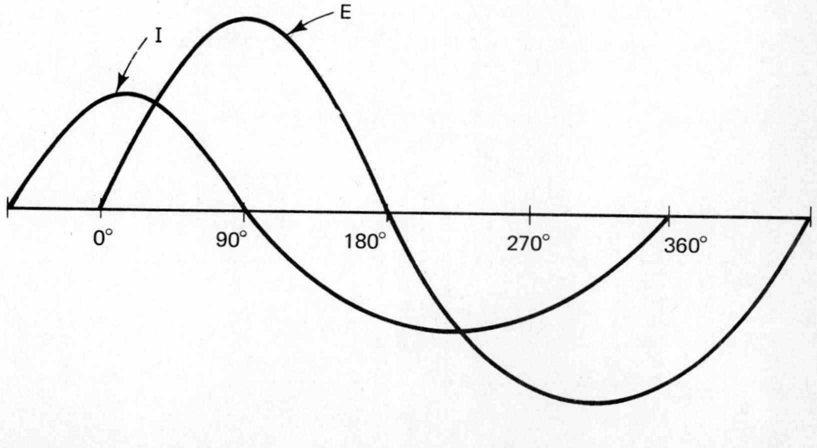

Combinations of resistance, inductance, and capacitance cause current flow to vary in phase between 90° lagging and 90° leading. An easy way to remember phase relationships in inductive and capacitive circuits is to use the terms "ELI" and "ICE." "ELI" means: voltage (E) leads current (I) in an inductive (L) circuit. "ICE" means: current (I) leads voltage (E) in a capacitive (C) circuit.

CAPACITOR CHARGING AND DISCHARGING

Refer to the circuit of Fig. 6-17. A capacitor is shown connected to a two-position switch. With the switch in position 1, the capacitor is uncharged. No voltage is applied to the capacitor. Each plate of the capacitor is neutral since no voltage is applied. When a voltage is placed across the capacitor, an electrostatic field or charge is developed between its plates. When the switch is placed in position 2, this places the capacitor across a 6-V battery. This causes a displacement of electrons in the circuit. Electrons move away from the negative polarity and toward the positive polarity of the battery. There is a surge of current flow as the capacitor charges to the value of the battery voltage. Electrons accumulate on the negative plate of the capacitor. At the same time, electrons leave the positive plate. This causes a difference of potential to develop across the capacitor. When electrons move onto the negative plate, it becomes more negative. Also, as electrons leave the positive plate, it becomes more positive.

The polarity of the potential that exists across the capacitor opposes the source voltage. As the capacitor continues to charge, the voltage across the capacitor increases. Current flow stops when the voltage across the capacitor is equal to the source voltage. At this time, the two voltages cancel each other out. No current actually flows through the

capacitor. This is true since the material between the plates of a capacitor is an insulator.

An important safety factor to remember is that a capacitor may hold a charge for a long period of time. Capacitors can be an electrical shock hazard if not handled properly.

When a capacitor is discharged, the charges on the plates become neutralized. A current path between the two plates must be developed. When the switch in Fig. 6-17 is moved to position 1, the electrons on the negative plate move to the positive plate and neutralize the charge. The capacitor releases the energy it has absorbed during its charging as it discharges through the resistor.

TYPES OF CAPACITORS

Capacitors are classified as either fixed or variable. Fixed capacitors have one value of capacitance. Variable capacitors are constructed to allow capacitance to be varied over a range of values. Variable capacitors often use air as the dielectric. The capacitance is varied by changing the position of the movable plates. This changes the plate area of the capacitor. When the movable plates are fully meshed together with the stationary plates, the capacitance is maximum. Several types of capacitors are shown in Fig. 6-14. Capacitor color codes are contained in Appendix 7.

Fixed capacitors come in many types. Some types of fixed capacitors are as follows:

1. *Paper capacitors*: Paper capacitors use paper as their dielectric. As shown in Fig. 6-18, they are made of flat strips of metal foil plates separated by a dielectric, which is

Figure 6-18.   Paper capacitor construction.

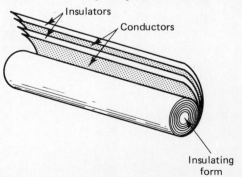

Figure 6-17.   Capacitor charge and discharge circuit.

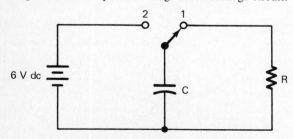

usually waxed paper. Paper capacitors have values in the picofarad and low-microfarad ranges. The voltage ratings are usually less than 600 V. Paper capacitors are usually sealed with wax to prevent moisture problems.

The voltage rating of capacitors is very important. A typical set of values marked on a capacitor might be "10 μF, 50 DCWV." This capacitor would have a capacitance of 10 μF and a "dc working voltage" of 50 V. This means that a voltage in excess of 50 V could damage the plates of the capacitors.

2. *Mica capacitors*: Mica capacitors have a layer of mica and then a layer of plate material. Their capacitance is usually small (in the picofarad range). They are small in physical size but have high voltage ratings.

3. *Oil-filled capacitors*: Oil-filled capacitors are used when high capacitance and high voltage ratings are needed. They are like paper capacitors immersed in oil. The paper, when soaked in oil, has a high dielectric constant.

4. *Ceramic capacitors*: Ceramic capacitors use a ceramic dielectric. The plates are thin films of metal deposited on ceramic material or made in the shape of a disk. They are covered with a moisture-proof coating and have high voltage ratings.

5. *Electrolytic capacitors*: Electrolytic capacitors are used when very high capacitance is needed. Electrolytic capacitors contain a paste electrolyte. They have two metal plates with the electrolyte between them and are usually housed in a cylindrical aluminum can. The aluminum can is the negative terminal of the capacitor. The positive terminal (or terminals) is brought out of the can at the bottom. The size and voltage rating are usually printed on the capacitor. Electrolytic capacitors often have two or more capacitors housed in one unit. They are called multisection capacitors.

The positive plate of an electrolytic capacitor is aluminum foil covered with a thin oxide film. The film is formed by an electrochemical reaction and acts as the dielectric. A strip of paper that contains a paste electrolyte is placed next to the positive plate. The electrolyte is the negative plate of the capacitor. Another strip of aluminum foil is placed next to the electrolyte. These three layers are then coiled up and placed into a cylinder.

Electrolytic capacitors are said to be "polarized." If the positive plate is connected to the negative terminal of a voltage source, the capacitor will become short-circuited. Capacitor polarity is marked on the capacitor to prevent this from happening. Some special high-value nonpolarized electrolytic capacitors are used for ac applications. Motor-starting capacitors are an example.

ALTERNATING CURRENT CIRCUITS

Ac circuits are similar in many ways to dc circuits. They have a source, a load, a path, and usually controls and indicators. Ac circuits are classified by the electrical characteristics they have (resistive, inductive, or capacitive). All ac circuits are either resistive, inductive, capacitive, or a combination of these. The operation of each type of electrical circuit is different. The nature of alternating current causes certain electrical circuit properties to exist.

Resistive Circuits

The simplest type of ac circuit is a resistive circuit, such as that illustrated in Fig. 6-19. A resistive circuit is the same with ac applied as it is with dc applied. In dc circuits, the following formulas are used:

$$E = I \times R$$

$$I = \frac{E}{R}$$

$$R = \frac{E}{I}$$

These show that when the voltage applied to a circuit is increased, the current will increase. Also when the resistance of a circuit is increased, the current decreases. The waveforms of Fig. 6-20 show the relation of the voltage and current in a resistive ac circuit. Voltage and current are *in phase*. In phase means that the minimum and maximum values of voltage and current occur at the same time. The

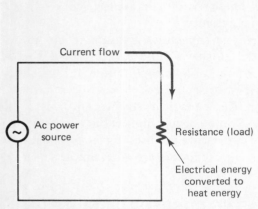

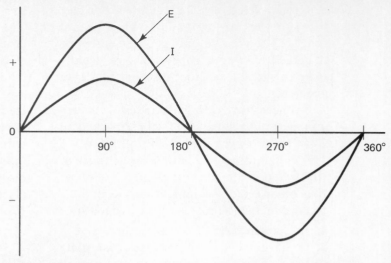

Figure 6-19. Resistive ac circuit.

Figure 6-20. Voltage and current waveforms of a resistive ac circuit.

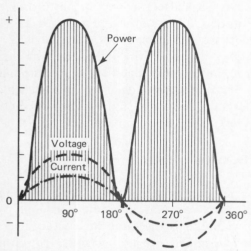

Figure 6-21. Power curve for a resistive ac circuit.

power converted by the resistance is found by multiplying voltage times current ($P = E \times I$). A power curve for a resistive ac circuit is shown in Fig. 6-21. When an ac circuit has only resistance, it is very similar to a dc circuit.

Inductive Circuits

The property of inductance (L) is very common. It adds more complexity to the relationship between voltage and current in an ac circuit. All motors, generators, and transformers have inductance. Inductance is due to the counter electromotive force (CEMF) produced when a magnetic

field is developed around a coil of wire. The magnetic field produced around coils affects a circuit. The CEMF produced by a magnetic field offers opposition to change in the current of a circuit. In an inductive circuit, voltage *leads* the current. If the circuit were purely inductive (containing no resistance), the voltage would lead the current by 90° (see Fig. 6-22).

A purely inductive circuit does not convert any power in the load. All power is delivered back to the source. Refer to points A and B on the waveforms of Fig. 6-23. At the peak of each waveform the value of the other waveform is zero. The power curves are equal and opposite. They will cancel each other out. Where voltage and current are positive, the power is also positive, since the product of two positive values is positive. When voltage is positive and current is negative, the product of the two is negative. The power converted is also negative. Negative power means that electrical energy is returned from one load to the source without being converted to another form. The power converted in a purely inductive circuit is equal to zero.

Resistive–Inductive (*RL*) Circuits

Since all actual circuits have some resistance, the inductance in a circuit might typically cause the condition shown in Fig. 6-24. The voltage leads the current by 30°. The angular separation between voltage and current is called *phase angle*. The phase angle increases as the inductance of the circuit

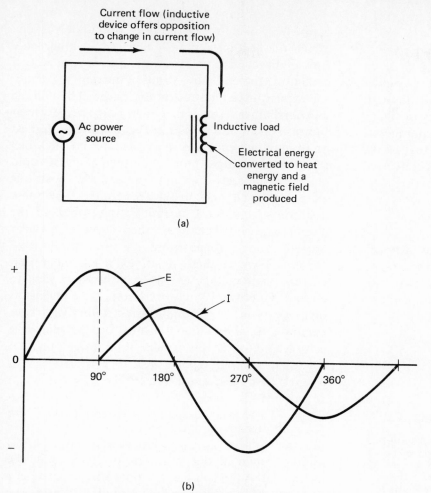

(a)

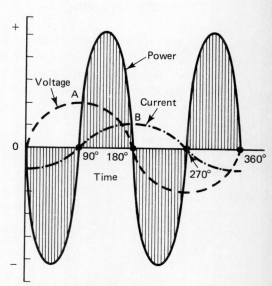

(b)

Figure 6-22. Voltage and current waveforms of a purely inductive ac circuit. (a) circuit; (b) waveforms.

Figure 6-23. Power curve for a purely inductive circuit.

Figure 6-24. Resistive–inductive (*RL*) circuit and its waveforms: (a) *RL* ac circuit; (b) voltage and current waveforms.

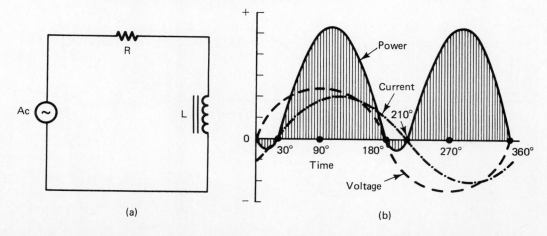

(a)   (b)

increases. This type of circuit is called a resistive–inductive (*RL*) circuit.

Compare the purely inductive circuit's waveforms of Figs. 6-22 and 6-23 to those of Fig. 6-24. In a resistive–inductive (*RL*) circuit, part of the power supplied from the source is converted in the load. During the intervals from 0 to 30° and from 180 to 210°, negative power is produced. The remainder of the ac cycle produces positive power. Most of the electrical energy supplied by the source is converted to another form of energy in the circuit.

Capacitive Circuits

Figure 6-25 shows a capacitive circuit. Capacitors have the ability to store an electrical charge. They have many applications in electrical circuits. The operation of a capacitor in a circuit depends on its ability to charge and discharge.

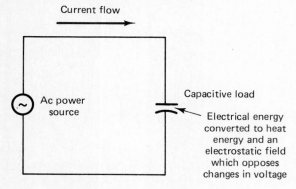

Figure 6-25. Capacitive ac circuit.

If dc voltage is applied to a capacitor, the capacitor will charge to the value of that dc voltage. When the capacitor is fully charged, it blocks the flow of direct current. However, if ac is applied to a capacitor, the changing value of current will cause the capacitor to charge and discharge. The voltage and current waveforms of a purely capacitive circuit (no resistance) are shown in Fig. 6-26. The highest amount of current flows in a capacitive circuit when the voltage changes more rapidly. The most rapid change in voltage occurs at the 0° and 180° positions as polarity changes from (−) to (+). At these positions, maximum current is developed in the circuit. The rate of change of the voltage is slow near the 90° and 270° positions. A small amount of current flows at these positions. Remember that current leads voltage by 90° in a purely capacitive circuit. No power is converted in this circuit, just as no power is developed in the purely inductive circuit. Figure 6-27 shows that the positive and negative power waveforms cancel each other out.

Resistive–Capacitive (*RC*) Circuits

All circuits contain some resistance. A more practical circuit is the resistive–capacitive (*RC*) circuit shown in Fig. 6-28. In an *RC* circuit, the current leads the voltage by a phase angle between 0 and 90°. If capacitance in a circuit increases, the phase angle increases. The waveforms of Fig. 6-29 shows an *RC* circuit in which the current leads the

Figure 6-26. Voltage and current waveforms of a purely capacitive ac circuit.

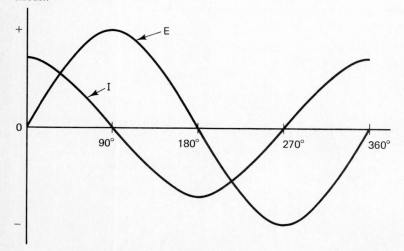

Figure 6-27. Power curves for a purely capacitive ac circuit.

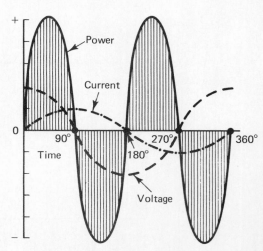

voltage by 30°. This circuit is similar to the *RL* circuit of Fig. 6-24(a). No power is converted in the circuit during the 0 to 30° and the 180 and 210° intervals. In this *RC* circuit, most of the electrical energy supplied by the source is converted to another form of energy in the circuit.

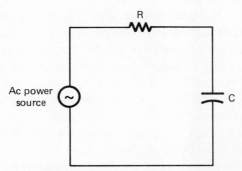

Figure 6-28. Resistive–capacitive (*RC*) circuit.

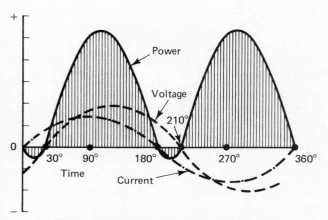

Figure 6-29. Waveforms of an *RC* circuit.

VECTOR DIAGRAMS

A vector diagram for each ac circuit discussed in this chapter is shown in Fig. 6-30. The understanding of the vector diagram is helpful when working with ac circuits. Rather than using waveforms to show phase relations, it is easier to use vector diagrams. Vectors are straight lines that have specific direction and length (*magnitude*). They are used to represent voltage and current values in ac circuits.

A horizontal line is drawn when beginning a vector diagram. Its left end is the reference point. In the diagrams of Fig. 6-30, the voltage vector line is the reference. For the inductive circuits, the current vectors are drawn in a *clockwise* direction from the voltage vector. This indicates that voltage leads current in an inductive circuit. For the capacitive circuits, the current vectors are drawn in a *counter-clockwise* direction from the voltage vectors. This indicates that current *leads* voltage in a capacitive circuit.

SERIES AC CIRCUITS

In any series ac circuit, the current is the same in all parts of the circuit. The voltages must be added by using a *voltage triangle*. Impedance (*Z*) of a series ac circuit is found by using an *impedance triangle*. Power values are found by using a *power triangle*.

Series *RL* Circuits

Series *RL* circuits are often found in electronic equipment. When an ac voltage is applied to a series *RL* circuit, the current is the same through each part.

Figure 6-30. Vector diagrams showing voltage and current relationships of ac circuits: (a) resistive (*R*) circuit; (b) purely inductive (*L*) circuit; (c) *RL* circuit; (d) purely capacitive (*C*) circuit; (e) *RC* circuit.

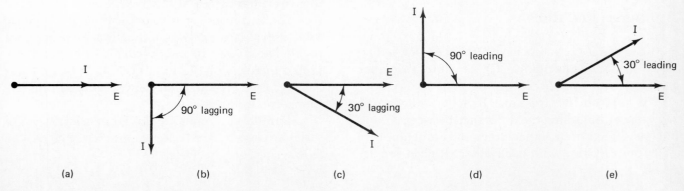

(a)   (b)   (c)   (d)   (e)

The voltage drops across each component distribute according to the values of resistance (R) and inductive reactance (X_L) in the circuit.

The total opposition to current flow in any ac circuit is called impedance (Z). Both resistance and reactance in an ac circuit oppose current flow. The impedance of a series RL circuit is found by using either of these formulas:

$$Z = \frac{E}{I} \quad \text{or} \quad Z = \sqrt{R^2 + X_L^2}$$

The impedance of a series RL circuit is found by using an *impedance triangle*. This right triangle is formed by the three quantities that oppose ac. A triangle is also used to compare voltage drops in series RL circuits. Inductive voltage (E_L) leads resistive voltage (E_R) by 90°. E_A is the voltage applied to the circuit. Since these values form a right triangle, the value of E_A may be found by using the formula

$$E_A = \sqrt{E_R^2 + E_L^2}$$

An example of a series RL circuit is shown in Fig. 6-31. Appendix 6 should be reviewed, if necessary, to gain a better understanding of trigonometry for ac circuit problems.

Series *RC* Circuits

Series RC circuits also have many uses. This type of circuit is similar to the series RL circuit. In a capacitive circuit, current leads voltage. The reactive values of RC circuits act in the opposite directions to RL circuits. An example of a series RC circuit is shown in Fig. 6-32.

Series *RLC* Circuits

Series RLC circuits have resistance (R), inductance (L), and capacitance (C). The total reactance (X_T) is found by subtracting the smaller reactance (X_L or X_C) from the larger one. Reactive voltage (E_X) is found by obtaining the difference between E_L and E_C. The effects of the capacitance and inductance are 180° out of phase with each other. Right triangles are used to show the simplified relationships of the circuit values. An example of a series RLC circuit is shown in Fig. 6-33.

PARALLEL AC CIRCUITS

Parallel ac circuits have many applications. The basic formulas used with parallel ac circuits are different from those of series circuits. Impedance (Z) of a parallel circuit is less than individual branch values of resistance, inductive reactance, or capacitive reactance. There is no impedance triangle for parallel circuits since Z is smaller that R, X_L, or X_C. A right triangle is drawn to show the currents in the branches of a parallel circuit.

The voltage of a parallel ac circuit is the same across each branch. The currents through the branches of a parallel ac circuit are shown by a right triangle called a *current triangle*. The current through the capacitor (I_C) is shown leading the current through the resistor (I_R) by 90°. The current through the inductor (I_L) is shown lagging I_R by 90°. I_L and I_C are 180° out of phase. They are subtracted to find the total reactive current (I_X). Since these values form a right triangle, the total current may be found by using the formula

$$I_T = \sqrt{I_R^2 + I_X^2}$$

This method is used to find currents in parallel RL, RC, or RLC circuits.

When components are connected in parallel, finding impedance is more difficult. An impedance triangle cannot be used. A method that can be used to find impedance is to use an *admittance triangle*. The following quantities are plotted on the triangle: (1) admittance: $Y = 1/Z$; (2) conductance: $G = 1/R$; (3) inductive susceptance: $B_L = 1/X_L$; and (4) capacitive susceptance: $B_C = 1/X_C$. These quantities are the inverse of each type of opposition to ac. Since total impedance (Z) is the smallest quantity in a parallel ac circuit, its reciprocal ($1/Z$) becomes the largest quantity on the admittance triangle (just as $\frac{1}{2}$ is larger than $\frac{1}{4}$). The values are in *siemens* or *mhos* (ohms spelled backward).

Parallel ac circuits are similar in several ways to series ac circuits. Study the examples of Fig. 6-34 to 6-36.

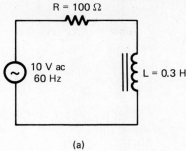

(a)

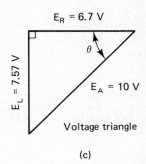

R = 100 Ω

Impedance triangle

E_R = 6.7 V

E_A = 10 V

Voltage triangle

(c)

Finding circuit values:

1. Find inductive reactance (X_L):

$$X_L = 2\pi \cdot f \cdot L$$
$$= 6.28 \times 60 \times 0.3$$
$$= 113 \; \Omega$$

2. Find impedance (Z):

$$Z = \sqrt{R^2 + X_L^2}$$
$$= \sqrt{100^2 + 113^2}$$
$$= \sqrt{10,000 + 12,769}$$
$$= \sqrt{22,769}$$
$$= 150 \; \Omega$$

3. Check to be sure that Z is larger than R or X_L.

4. Find total current (I_T):

$$I_T = \frac{E}{Z} = \frac{10 \; V}{150 \; \Omega} = 0.067 \; A$$

5. Find the voltage across R(E_R):

$$E_R = I \times R$$
$$= 0.067 \; A \times 100 \; \Omega$$
$$= 6.7 \; V$$

6. Find the voltage across L (E_L):

$$E_L = I \cdot X_L$$
$$= 0.067 \; A \times 133 \; \Omega$$
$$= 7.57 \; V$$

7. Check to see that

$$E_A = \sqrt{E_R^2 + E_L^2}$$
$$10 \; V = \sqrt{6.7 \; V^2 + 7.57 \; V^2}$$
$$= \sqrt{44.89 + 57.3}$$
$$= \sqrt{102.2}$$
$$\approx 10.1 \; V^*$$

(approximately equal to)

8. Find circuit phase angle (θ):

$$\text{cosine } \theta = \frac{\text{adjacent}}{\text{hypotenuse}} = \frac{E_R}{E_A} = \frac{6.7}{10}$$
$$= 0.67$$
$$= \text{inverse cosine } 0.67$$
$$= 48°$$

(b)

*Slight difference is due to rounding off of numbers as they are calculated.

Figure 6-31. Series *RL* circuit example: (a) circuit; (b) procedure for finding circuit values; (c) circuit triangles.

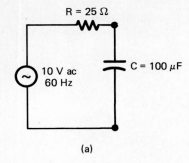

(a)

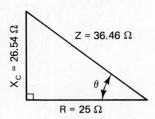

Impedance triangle

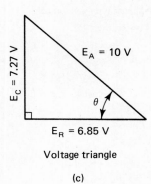

Voltage triangle

(c)

*Any trig function can
be used with either
triangle.

Finding circuit values:

1. Find capacitive reactance (X_C):

$$X_C = \frac{1}{2\pi \cdot f \cdot C} \text{ or } \frac{1,000,000}{2\pi \cdot f \cdot C}$$
$$\text{(in farads)} \qquad \text{(in } \mu\text{F)}$$

$$= \frac{1,000,000}{6.28 \times 60 \times 100}$$

$$= 26.54 \ \Omega$$

2. Find impedance (Z):

$$Z = \sqrt{R^2 + X_C^2}$$

$$= \sqrt{25^2 + 26.54^2}$$

$$= \sqrt{625 + 704.37}$$

$$= \sqrt{1329.37}$$

$$= 36.46 \ \Omega$$

3. Check to be sure that Z
is larger than R or X_C.

4. Find total current (I_T):

$$I_T = \frac{E}{Z} = \frac{10 \text{ V}}{36.46 \ \Omega} = 0.274 \text{ A}$$

5. Find the voltage across R (E_R):

$$E_R = I \times R$$

$$= 0.274 \text{ A} \times 25 \ \Omega$$

$$= 6.85 \text{ V}$$

6. Find the voltage across C (E_C):

$$E_C = I \cdot X_C$$

$$= 0.274 \text{ A} \times 26.54 \ \Omega$$

$$= 7.27 \text{ V}$$

7. Check to see that

$$E_A = \sqrt{E_R^2 + E_C^2}:$$
$$10 \text{ V} = \sqrt{6.85 \text{ V}^2 + 7.27 \text{ V}^2}$$

$$= \sqrt{46.92 + 52.88}$$

$$= \sqrt{99.8}$$

$$\approx 9.99 \text{ V}$$

8. Find the circuit phase angle (θ):

$$*\text{sine } \theta = \frac{\text{opposite}}{\text{hypotenuse}} = \frac{X_C}{Z} = \frac{26.54}{36.46}$$

$$= 0.727$$

$$\theta = \text{inverse sine } 0.727$$

$$= 47°$$

(b)

Figure 6-32.   Series *RC* circuit example: (a) circuit; (b) procedure for finding
circuit values; (c) circuit triangles.

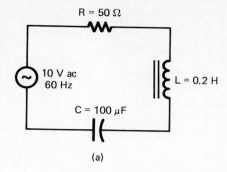

(a)

Finding circuit values:

1. Find inductive reactance (X_L):

$$X_L = 2\pi \cdot f \cdot L$$
$$= 6.28 \times 60 \times 0.2$$
$$= 75.36\ \Omega$$

2. Find capacitive reactance (X_C):

$$X_C = \frac{1,000,000}{2\pi \cdot f \cdot C\ (\mu F)}$$
$$= \frac{1,000,000}{6.28 \times 60 \times 100}$$
$$= 26.54\ \Omega$$

3. Find total reactance (X_T):

$$X_T = X_L - X_C$$
$$= 75.36\ \Omega - 26.54\ \Omega$$
$$= 48.82\ \Omega$$

4. Find impedance (Z):

$$Z = \sqrt{R^2 + X_T^2}$$
$$= \sqrt{50^2 + 48.82^2}$$
$$= \sqrt{2500 + 2383.4}$$
$$= \sqrt{4,883.4}$$
$$= 69.88\ \Omega$$

5. Check to be sure that Z is larger than X_T or R.

6. Find total current (I_T):

$$I_T = \frac{E}{Z} = \frac{10\ V}{69.88\ \Omega} = 0.143\ A$$

7. Find voltage across R (E_R):

$$E_R = I \times R$$
$$= 0.143\ A \times 50\ \Omega$$
$$= 7.15\ V$$

8. Find voltage across L (E_L):

$$E_L = I \cdot X_L$$
$$= 0.143\ A \times 75.36\ \Omega$$
$$= 10.78\ V$$

9. Find voltage across C (E_C):

$$E_C = I \cdot X_C$$
$$= 0.143\ A \times 26.54\ \Omega$$
$$= 3.8\ V$$

10. Find total reactive voltage (E_X):

$$E_X = E_L - E_C$$
$$= 10.78\ V - 3.8\ V$$
$$= 6.98\ V$$

11. Check to see that

$$E_A = \sqrt{E_R^2 + E_X^2}$$
$$= \sqrt{7.15\ V^2 + 6.98\ V^2}$$
$$10\ V = \sqrt{51.12 + 48.72}$$
$$= \sqrt{99.84}$$
$$\approx 9.99\ V$$

12. Find circuit phase angle (θ):

$$\text{tangent } \theta = \frac{\text{opposite}}{\text{adjacent}} = \frac{X_T}{R} = \frac{48.82}{50}$$
$$= 0.976$$
$$= \text{inverse tangent } 0.976$$
$$= 44°$$

(b)

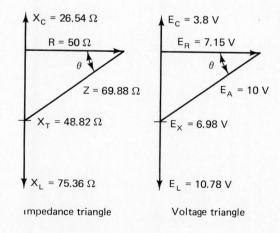

impedance triangle   Voltage triangle

(c)

Figure 6-33. Series *RLC* circuit example: (a) circuit; (b) procedure for finding circuit values; (c) circuit triangles.

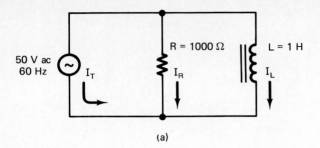

(a)

Finding circuit values:

1. Find inductive reactance (X_L):

$$X_L = 2\pi \cdot f \cdot L$$
$$= 6.28 \times 60 \times 1$$
$$= 376.8\ \Omega$$

2. Find current through $R(I_R)$:

$$I_R = \frac{E}{R} = \frac{50\ V}{1000\ \Omega} = 0.05\ A$$

3. Find current through L (I_L):

$$I_L = \frac{E}{X_L} = \frac{50\ V}{376.8\ \Omega} = 0.133\ A$$

4. Find total current (I_T):

$$I_T = \sqrt{I_R^2 + I_L^2}$$
$$= \sqrt{0.05^2 + 0.133^2}$$
$$= \sqrt{0.0025 + 0.0177}$$
$$= \sqrt{0.0202}$$
$$= 0.142\ A$$

5. Check to see that I_T is larger than I_R or I_L.

6. Find impedance (Z):

$$Z = \frac{E}{I_T} = \frac{50\ V}{0.142\ A} = 352.1\ \Omega$$

7. Check to see that Z is less than R or X_L.

8. Find circuit phase angle (θ):

$$\text{sine}\ \theta = \frac{\text{opposite}}{\text{hypotenuse}} = \frac{I_L}{I_T} = \frac{0.133}{0.142}$$
$$= 0.937$$
$$\theta = \text{inverse sine } 0.937$$
$$= 70°$$

(b)

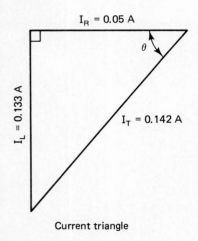

Current triangle

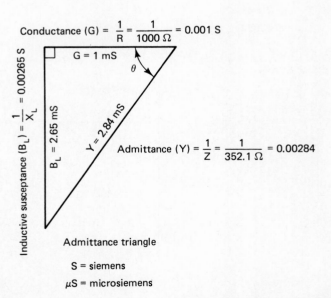

Admittance triangle

S = siemens

μS = microsiemens

(c)

Figure 6-34. Parallel *RL* circuit example: (a) circuit; (b) procedure for finding circuit values; (c) circuit triangles.

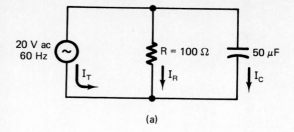

(a)

Finding circuit values:

1. Find capacitive reactance (X_C):

$$X_C = \frac{1,000,000}{2\pi \cdot f \cdot C \, (\mu F)}$$

$$= \frac{1,000,000}{6.28 \times 60 \times 50}$$

$$= 53 \, \Omega$$

2. Find current through R (I_R):

$$I_R = \frac{E}{R} = \frac{20 \, V}{100 \, \Omega} = 0.2 \, A$$

3. Find current through C (I_C):

$$I_C = \frac{1}{X_C} = \frac{20 \, V}{53 \, \Omega} = 0.377 \, A$$

4. Find total current (I_T):

$$I_T = \sqrt{I_R^2 + I_C^2}$$

$$= \sqrt{0.2^2 + 0.377^2}$$

$$= \sqrt{0.04 + 0.142}$$

$$= \sqrt{0.182}$$

$$= 0.427 \, A$$

5. Check to see that I_T is larger than I_R or I_C.

6. Find impedance (Z):

$$Z = \frac{E}{I_T} = \frac{20 \, V}{0.427 \, A} = 46.84 \, \Omega$$

7. Check to see that Z is less than R or X_C.

8. Find circuit phase angle (θ):

$$\text{cosine } \theta = \frac{\text{adjacent}}{\text{hypotenuse}} = \frac{G}{Y} = \frac{10}{21.3}$$

$$= 0.469$$

$$\theta = \text{inverse cosine } 0.469$$

$$= 62°$$

(b)

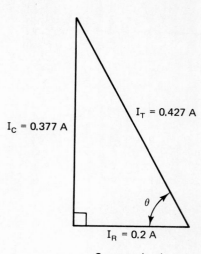

$I_C = 0.377$ A

$I_T = 0.427$ A

θ

$I_R = 0.2$ A

Current triangle

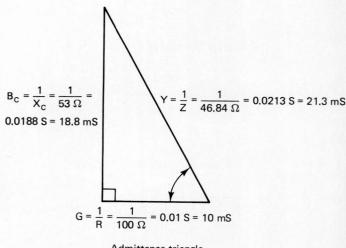

$B_C = \dfrac{1}{X_C} = \dfrac{1}{53 \, \Omega} =$ 0.0188 S $= 18.8$ mS

$Y = \dfrac{1}{Z} = \dfrac{1}{46.84 \, \Omega} = 0.0213$ S $= 21.3$ mS

$G = \dfrac{1}{R} = \dfrac{1}{100 \, \Omega} = 0.01$ S $= 10$ mS

Admittance triangle

(c)

Figure 6-35. Parallel *RC* circuit example: (a) circuit; (b) procedure for finding circuit values; (c) circuit triangles.

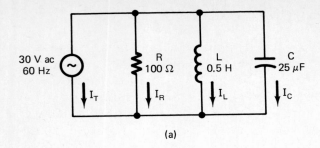

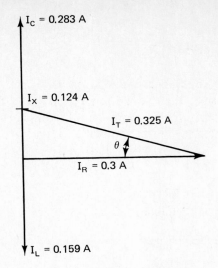

(a)

Finding circuit values:

1. Find inductive reactance (X_L):

$$X_L = 2\pi \cdot f \cdot L$$
$$= 6.28 \times 60 \times 0.5$$
$$= 188.4 \ \Omega$$

2. Find capacitive reactance (X_C):

$$X_C = \frac{1,000,000}{6.28 \times 60 \times 25}$$
$$= 106 \ \Omega$$

3. Find current through R (I_R):

$$I_R = \frac{E}{R} = \frac{30 \text{ V}}{100 \ \Omega} = 0.3 \text{ A}$$

4. Find current through L (I_L):

$$I_L = \frac{E}{X_L} = \frac{30 \text{ V}}{188.4 \ \Omega} = 0.159 \text{ A}$$

5. Find current through C (I_C):

$$I_C = \frac{E}{X_C} = \frac{30 \text{ V}}{106 \ \Omega} = 0.283 \text{ A}$$

6. Find total reactive current (I_X):

$$I_X = I_C - I_L$$
$$= 0.283 - 0.159$$
$$= 0.124$$

7. Find total current (I_T):

$$I_T = \sqrt{I_R^2 + I_X^2}$$
$$= \sqrt{0.3^2 + 0.124^2}$$
$$= \sqrt{0.9 + 0.0154}$$
$$= \sqrt{0.1054}$$
$$= 0.325 \text{ A}$$

8. Check to see that I_T is larger than I_R or I_X.

9. Find impedance (Z):

$$Z = \frac{}{I_T} = \frac{30 \text{ V}}{0.325 \text{ A}} = 92.3 \ \Omega$$

10. Check to see that Z is less than R or X_L or X_C.

11. Find circuit phase angle (θ):

$$\text{tangent } \theta = \frac{\text{opposite}}{\text{adjacent}} = \frac{I_X}{I_R} \frac{0.124}{0.3}$$
$$= 0.413$$
$$\theta = \text{inverse tangent } 0.413$$
$$= 22°$$

(b)

Current triangle

$B_C = \dfrac{1}{X_C} = \dfrac{1}{106 \ \Omega} = 0.0094 \text{ S} = 9.4 \text{ mS}$

$B_X = B_C - B_L = 9.4 - 5.3 = 4.1 \text{ mS}$

$Y = \dfrac{1}{Z} = \dfrac{1}{92.3 \ \Omega} = 0.0108 \text{ S} = 10.8 \text{ mS}$

4.1 μs

$G = \dfrac{1}{R} = \dfrac{1}{100 \ \Omega} = 0.1 \text{ S} = 10 \text{ mS}$

$B_L = \dfrac{1}{X_L} = \dfrac{1}{188.4 \ \Omega} = 0.053 \text{ S} = 5.3 \text{ mS}$

Admittance triangle

(c)

Figure 6-36. Parallel *RLC* circuit example: (a) circuit; (b) procedure for finding circuit values; (c) circuit triangles.

POWER IN AC CIRCUITS

Power values in ac circuits are found by using *power triangles*, as shown in Fig. 6-37. The power *delivered* to an ac circuit is called *apparent power*. It is equal to voltage times current. The unit of measurement is the volt-ampere (VA) or kilovolt-ampere (kVA).

Meters are used to measure apparent power in ac circuits. Apparent power is the voltage applied to a circuit multiplied by the total current flow. The power *converted* to another form of energy by the load is measured with a wattmeter. This actual power converted is called *true power*. The ratio of true power converted in a circuit to apparent power

Figure 6-37. Power triangles for ac circuits: (a) *RL* circuits; (b) *RC* circuits; (c) *RLC* circuit with X_C larger than X_L; (d) *RLC* circuit with X_C larger than X_L. (e) Formulas for calculating power values.

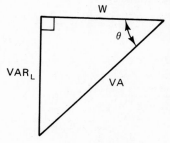

VA   = volt-amperes (apparent power)
W   = watts (true power)
VAR_L = volt-amperes reactive (inductive)
VAR_C = volt-amperes reactive (capacitive)
VAR_T = volt-amperes reactive (total)

(a)

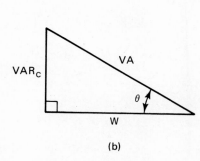

(b)

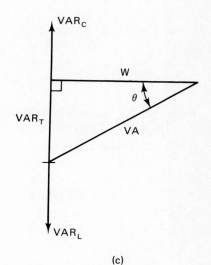

(c)

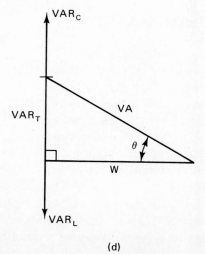

(d)

Calculating power values:

1. In dc circuits:

$$P = E \times I = I^2 \times R = \frac{E^2}{R}$$

2. In ac circuits:

a. $VA = E_A \times I_T = I_T^2 \times Z = \dfrac{E_A^2}{Z}$

b. $W = E_R \times I_R = I_R^2 \times R = \dfrac{E_R^2}{R}$

c. $VAR_L = E_L \cdot X_L = I_L^2 \cdot X_L = \dfrac{E_L^2}{X_L}$

d. $VAR_C = E_C \cdot X_C = I_C^2 \cdot X_C = \dfrac{E_C^2}{X_C}$

e. $VAR_T = VAR_L - VAR_C$ or $VAR_C - VAR_L$

(e)

delivered to an ac circuit is called the *power factor* (PF). Power factor is found by using the formula

$$PF = \frac{\text{true power } (W)}{\text{apparent power (VA)}}$$

or

$$\% \text{ PF} = \frac{W}{VA} \times 100$$

The term W is the true power in watts and VA is the apparent power in volt-amperes. The highest power factor of an ac circuit is 1.0 or 100%. This value is called *unity* power factor.

The phase angle (θ) of an ac circuit determines the power factor. Remember that the phase angle is the angular separation between voltage applied to an ac circuit and current flow through the circuit. Also, more inductive or capacitive reactance causes a larger phase angle. In a purely inductive or capacitive circuit, a 90° phase angle causes a power factor of *zero* degrees. The power factor varies according to the values of resistance and reactance in a circuit.

There are two types of power that affect power conversion in ac circuits. Power conversion in the resistive part of the circuit is called *active power* or true power. True power is measured in watts. The other type of power is caused by inductive or capacitive loads. It is 90° out of phase with the true power and is called *reactive power*. It is a type of "unused" power. Reactive power is measured in volt-amperes reactive (VAR).

The power triangle of Fig. 6-37 has true power (watts) marked as a horizontal line. Reactive power (VAR) is drawn at a 90° angle from the true power. Volt-amperes or apparent power (VA) is the longest side (hypotenuse) of the right triangle. This right triangle is similar to the impedance triangle and the voltage triangle for series ac circuits and the current triangle and admittance triangle for parallel ac circuits. Each type of right triangle has a horizontal line used to mark the *resistive* part of the circuit. The vertical lines are used to mark the *reactive* part of the circuit. The longest side of the triangle (hypotenuse) is used to mark *total* values of the circuit. The size of the hypotenuse depends on the amount of resistance and reactance in the circuit. Vector diagrams and right triangles are extremely important for understanding ac circuits.

Power in Three-Phase Circuits

In Chapter 5, three-phase wye and delta connections were discussed. A review of wye and delta connections might be helpful. The formulas for three-phase power relationships are:

Apparent power (volt-amperes) per phase:

$$VA_P = E_P \times I_P$$

Total apparent power (volt-amperes):

$$VA_T = 3 \times E_P \times I_P \quad \text{or} \quad 1.73 \times E_L \times I_L$$

Power factor:

$$PF = \frac{\text{true power (watts)}}{1.73 \times E_L \times I_L} \quad \text{or} \quad \frac{\text{watts}}{3 \times E_P \times I_P}$$

True power per phase:

$$P_P = E_P \times I_P \times PF$$

Total true power:

$$P_T = 3 \times E_P \times I_P \times PF$$

or

$$1.73 \times E_L \times I_L \times PF$$

Remember that the symbols used above are as follows: E_L is the line voltage in volts, I_L is the line current in amperes, E_P is the phase voltage in volts, and I_P is the phase current in amperes.

FREQUENCY-SENSITIVE AC CIRCUITS

Some types of ac circuits are designed to respond to ac frequencies. Circuits that are used to pass some frequencies and block others are called frequency-sensitive circuits. There are two types of frequency-sensitive circuits: (1) filter circuits and (2) resonant circuits. Each type of circuit uses reactive devices to respond to different ac frequencies. These circuits have *frequency response curves*. Frequency is graphed on the horizontal axis and voltage output on the vertical axis. Sample frequency response curves for each type of filter and resonant circuit are shown in the examples that follow.

Filter Circuits

The three types of filter circuits are shown in Fig. 6-38. Filter circuits are used to separate one range of frequencies from another. Low-pass filters pass low ac frequencies and block higher frequencies. High-pass filters pass high frequencies and block lower frequencies. Bandpass filters pass a midrange of frequencies and block lower and higher frequencies. All filter circuits have resistance and capacitance *or* inductance.

Figure 6-39 shows the circuits used for low-pass, high-pass, and bandpass filters and their frequency response curves. Many low-pass filters are series *RC* circuits, as shown in Fig. 6-39(a). Output voltage (E_{out}) is taken across a capacitor. As frequency increases, capacitive reactance (X_C) decreases, since

$$X_C = \frac{1}{2\pi \times f \times C}$$

The voltage drop across the output is equal to I times X_C. So as frequency increases, X_C decreases, and voltage output decreases. Series *RL* circuits may

Figure 6-38. Three types of filter circuits: (a) low-pass filter—passes low frequencies and blocks high frequencies; (b) high-pass filter—passes high frequencies and blocks low frequencies; (c) bandpass filter—passes a midrange of frequencies and blocks high and low frequencies.

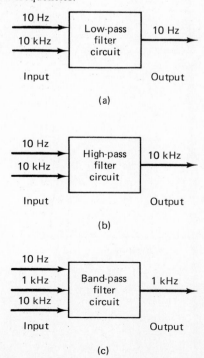

(a)

(b)

(c)

also be used as low-pass filters. As frequency increases, inductive reactance (X_L) increases, since $X_L = 2\pi \times f \times L$. Any increase in X_L reduces the circuit's current. The voltage output taken across the resistor is equal to $I \times R$. So when I decreases, E_{out} also decreases. As frequency increases, X_L increases, I decreases, and E_{out} decreases.

Figure 6-39(b) shows two types of high-pass filters. The series *RC* circuit is a common type. The voltage output (E_{out}) is taken across the resistor (R). As frequency increases, X_C decreases. A decrease in X_C causes current flow to increase. The voltage output across the resistor (E_{out}) is equal to $I \times R$. So as I increases, E_{out} increases, As frequency increases, X_C decreases, I increases, and E_{out} increases. A series *RL* circuit may also be used as a high-pass filter. The E_{out} is taken across the inductor. As frequency increases, X_L increases. E_{out} is equal to I times X_L. So as X_L increases, E_{out} also increases. In this circuit, as frequency increases, X_L increases and E_{out} increases.

The bandpass filter of Fig. 6-39(c) is a combination of low-pass and high-pass filter sections. It is designed to pass a midrange of frequencies and block low and high frequencies. R_1 and C_1 form a low-pass filter and R_2 and C_2 form a high-pass filter. The range of frequencies to be passed is determined by calculating the values of resistance and capacitance.

Resonant Circuits

Resonant circuits are designed to pass a range of frequencies and block all others. They have resistance, inductance, *and* capacitance. Figure 6-40 shows the two types of resonant circuits: (1) series resonant and (2) parallel resonant and their frequency response curves.

SERIES RESONANT CIRCUITS. Series resonant circuits are a series arrangement of inductance, capacitance, and resistance. A series resonant circuit offers a small amount of opposition to some ac frequencies and much more opposition to other frequencies. They are important for selecting or rejecting frequencies.

The voltage across inductors and capacitors in ac series circuits are in direct opposition to each other (180° out of phase). They tend to cancel each other out. The frequency applied to a series resonant circuit affects inductive reactance and capacitive reactance. At a specific input frequency, X_L will

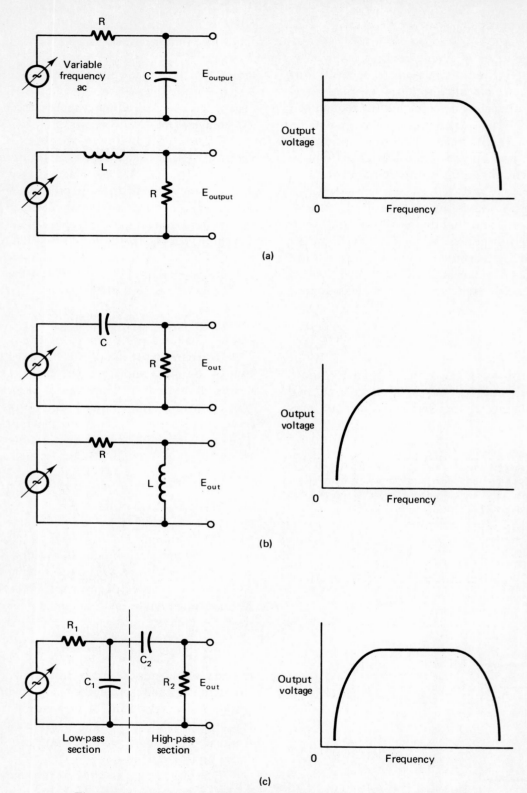

Figure 6-39. Circuits used to filter ac frequencies and their response curves: (a) low-pass filters; (b) high-pass filters; (c) bandpass filter.

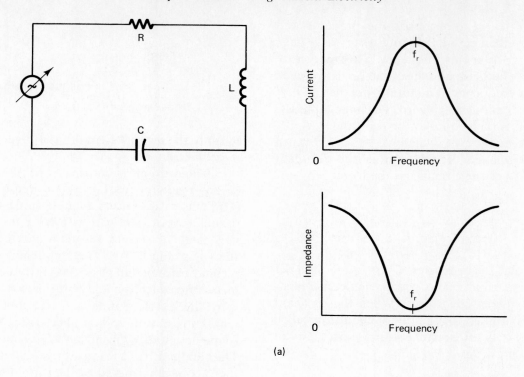

(a)

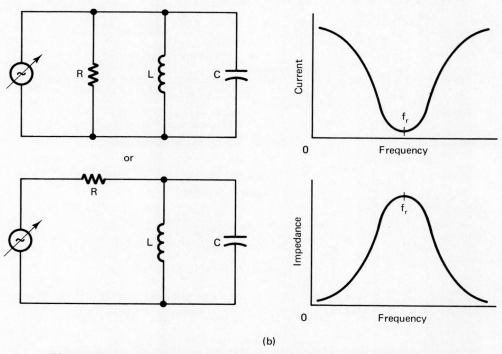

(b)

Figure 6-40. (a) Series and (b) parallel resonant circuits and their frequency response curves.

equal X_C. The voltages across the inductor and capacitor are then equal. The total reactive voltage (E_X) is zero volts at this frequency. The opposition offered by the inductor and the capacitor cancel each other at this frequency. The total reactance (X_T) of the circuit (X_L minus X_C) is zero. The impedance (Z) of the circuit is then equal to the resistance (R).

The frequency at which $X_L = X_C$ is called *resonant frequency.* To determine the resonant frequency (f_r) of the circuit, use the formula

$$f_r = \frac{1}{2\pi \times \sqrt{L \times C}}$$

In the formula, L is in henries, C is in farads, and f_r is in hertz. As either inductance or capacitance increases, resonant frequency decreases. When the resonant frequency is applied to a circuit, a condition called *resonance* exists. Resonance for a series circuit causes the following:

1. $X_L = E_C$.
2. X_T is equal to zero.

3. $E_L = E_C$
4. Total reactive voltage (E_X) is equal to zero.
5. $Z = R$.
6. Total current (I_T) is maximum.
7. Phase angle (θ) is $0°$.

Refer to the series *RLC* circuit of Fig. 6-33 to clarify these values.

The ratio of reactance (X_L or X_C) to resistance (R) at resonant frequency is called *quality factor* (Q). This ratio is used to determine the range of frequencies or *bandwidth* (BW) a resonant circuit will pass. A sample resonant circuit problem is shown in Fig. 6-41. The frequency range that a resonant circuit will pass (BW) is found by using the procedure of steps 5 and 6 in Fig. 6-41.

The cutoff points are at about 70% of the maximum output voltage. These are called the low-frequency cutoff (f_{lc}) and high-frequency cutoff (f_{hc}). The bandwidth of a resonant circuit is determined by the Q. Q is determined by the ratio of X_L and X_C to R. Resistance mainly determines bandwidth. This effect is summarized as follows:

Figure 6-41. Sample resonant circuit problem: (a) circuit; (b) procedure to find circuit values.

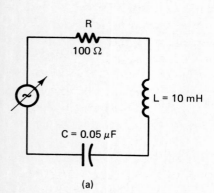

(a)

Finding circuit values:

1. Find resonant frequency (f_r):

$$f_r = \frac{1}{2\pi \times \sqrt{L \times C}} = \frac{1}{6.28 \times \sqrt{(10 \times 10^{-3}) \times (0.05 \times 10^{-6})}} =$$

$$= \frac{1}{6.28 \times \sqrt{0.5 \times 10^{-9}}} = \frac{1}{6.28 \times (2.23 \times 10^{-5})} = \frac{1}{1.4 \times 10^{-4}} = 7121 \text{ Hz}$$

2. Find X_L^* and X_C at resonant frequency:

$$X_L = 2\pi \cdot f \cdot L = 6.28 \times 7121 \times (10 \times 10^{-3}) = 447 \ \Omega$$

*Easier to calculate.

3. Find quality factor (Q):

$$Q = \frac{X_L}{R} = \frac{447 \ \Omega}{100 \ \Omega} = 4.47$$

4. Find bandwidth (BW):

$$BW = \frac{f_r}{Q} = \frac{7121 \text{ Hz}}{4.47} = 1593 \text{ Hz}$$

5. Find low-frequency cutoff (f_{lc}):

$$f_{lc} = f_r - \tfrac{1}{2} BW = 7121 \text{ Hz} - 797 \text{ Hz} = 6324 \text{ Hz}$$

6. Find high-frequency cutoff (f_{hc})

$$f_{hc} = f_r + \tfrac{1}{2} BW = 7121 + 797 = 7918 \text{ Hz}$$

(b)

1. When R is increased, Q decreases, since $Q = X_L/R$.

2. When Q decreases, BW increases, since BW $= f_r/Q$.

3. When R is increased, BW increases.

Two curves are shown in Fig. 6-42 which show the effect of resistance on bandwidth. The curve of Fig. 6-42(b) has high *selectivity*. This means that a resonant circuit with this response curve would select a small range of frequencies. This is very important for radio and television tuning circuits.

Figure 6-42. Effect of resistance on bandwidth of a series resonant circuit: (a) high resistance, low selectivity; (b) low resistance, high selectivity.

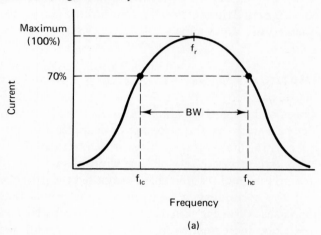

(a)

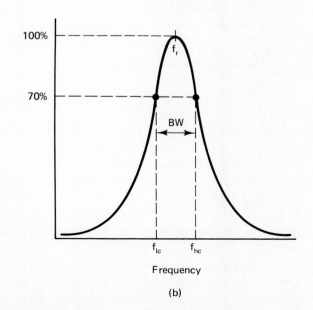

(b)

PARALLEL RESONANT CIRCUITS. Parallel resonant circuits are similar to series resonant circuits. Their electrical characteristics are somewhat different, but they accomplish the same purpose. Another name for parallel resonant circuits is *tank circuit*. A tank circuit is a parallel combination of L and C used to select or reject ac frequencies.

With the resonant frequency applied to a parallel resonant circuit, the following occurs:

1. $X_L = X_C$.

2. X_T equals 0.

3. $I_L = I_C$.

4. I_X equals zero, so the circuit current is minimum.

5. $Z = R$ and is maximum.

6. Phase angle $(\theta) = 0°$.

Refer to the parallel RLC circuit of Fig. 6-36 to clarify these values.

The calculations used for parallel resonant circuits are similar to those for series circuits. There is one exception. The quality factor (Q) is found by using this formula for parallel circuits: $Q = R/X_L$. Otherwise, the example of Fig. 6-41 is the same for parallel resonant circuits.

TIME CONSTANT CIRCUITS

Another important type of circuit is the time constant circuit. These circuits use the properties of inductance and capacitance to operate as timing circuits to control loads. Remember that inductance (L) opposes changes in current and capacitance (C) opposes changes in voltage. The reaction time of inductors and capacitors to oppose changes of current and voltage depends on resistance. The time (in seconds) for a capacitor or an inductor to react is called a *time constant*.

In Fig. 6-43(a), a resistor and an inductor are connected in series to a voltage source. Current rises from zero to maximum after a certain time period due to the CEMF of the magnetic field. The time required for the current to reach maximum is controlled by the values of R and L. Resistance opposes current flow and inductance opposes changes in current. The change in current occurs from zero to

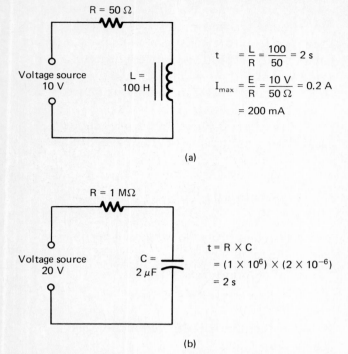

$$t = \frac{L}{R} = \frac{100}{50} = 2\ s$$

$$I_{max} = \frac{E}{R} = \frac{10\ V}{50\ \Omega} = 0.2\ A$$

$$= 200\ mA$$

(a)

$$t = R \times C$$
$$= (1 \times 10^6) \times (2 \times 10^{-6})$$
$$= 2\ s$$

(b)

Figure 6-43. Time constant circuits: (a) *RL* circuit; (b) *RC* circuit.

maximum. The time required for the current to reach about 63% of maximum is found by using the formula $t = L/R$. Time constant (t) is in seconds, L is in henries, and R is in ohms.

After one time constant, current through the inductor is about 63% of maximum. After five time constants, current through the inductor is maximum.

The maximum current (I_{max}) in the circuit is 200 mA. The time required for the current to reach 63% of 200 mA is 2.0 s. The time required for the current to reach the maximum (200 mA) is 10 s ($t \times 5 = 2.0 \times 5 = 10$ s).

A resistor and capacitor are connected in series to a voltage source in Fig. 6-43(b). The *time* for the voltage across the capacitor to reach maximum is controlled by the values of R and C. Resistance opposes the current flow. Capacitance opposes changes in voltage. The change in voltage is from zero to maximum. The amount of time for a capacitor to charge to about 63% of the applied voltage is found by using this formula: $t = R \times C$. Time (t) is in seconds, R is in ohms, and C is in farads.

After *one time constant*, the voltage across the capacitor is about 63% of the source voltage. In *five*

time constants, the voltage across the capacitor equals the source voltage of 20 V. The time for the capacitor to charge to 63% of the source voltage is 2 s. The time for the voltage across the capacitor to equal the source voltage is 10 s ($t \times 5 = 2 \times 5 = 10$ s).

The discharge of the capacitor is similar to its charging action. When the source voltage is removed, the capacitor discharges about 63% of the source voltage in *one time constant*. In five time constants, a capacitor discharges 100% of its voltage. Charge and discharge curves for capacitors are shown in Fig. 6-44. The time for current changes in inductive circuits is similar. The curves are called *universal time constant curves*. The vertical axis is marked as percent of capacitor voltage (E_C) or percent of inductor current (I_L) The horizontal axis is marked in time constant values. Sometimes the charging curve is called the *rise time* and the discharging curve is called the *decay time*.

TRANSFORMERS

Transformers are very important electrical devices. They are used to either increase or decrease alternating current (ac) voltage. Transformers are made by using two separate sets of wire windings which are wound on a metal core. These are called the primary and the secondary windings. A transformer that increases voltage is called a step-up transformer and one that decreases voltage is called a step-down transformer. Several types of transformers are shown in Fig. 6-45.

Transformer Operation

Transformers are electrical control devices used to either increase or decrease ac voltage. They do not operate with dc voltage applied. Notice in Fig. 6-46 that ac voltage is applied to the primary winding of the transformer. There is no connection of the primary and secondary windings. The transfer of energy from the primary to the secondary winding is due to magnetic coupling or mutual inductance. The transformer relies on electromagnetism to operate.

The primary and secondary windings of transformers are wound around a laminated iron core. The iron core is used to transfer the magnetic energy from the primary winding to the secondary winding.

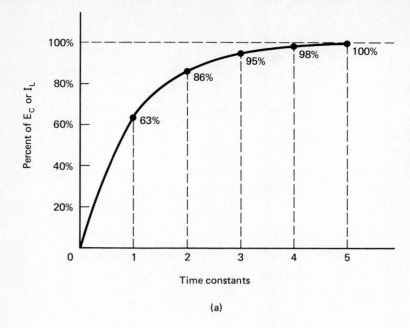

(a)

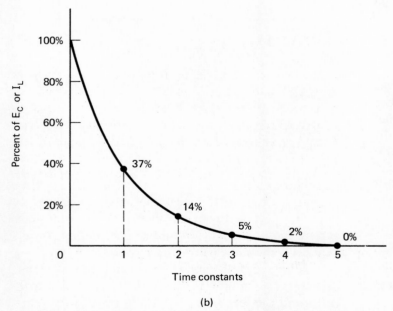

(b)

Figure 6-44. Universal time constant curves: (a) charging or rise time curve; (b) discharging or decay time curve.

Figure 6-45. Transfomers. (Courtesy of TRW/UTC Transformers.)

Figure 6-46. Transformer: (a) pictorial; (b) schematic symbol.

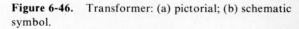

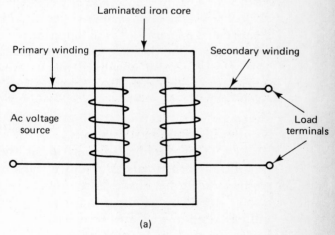

(a)

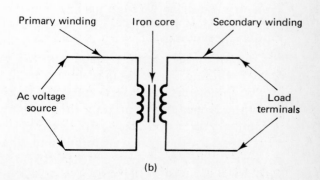

(b)

Remember that ac voltage is applied to the primary winding and the secondary winding is connected to an electrical load.

There are many different types and sizes of transformers. However, the same basic principles of operation apply to all transformers. The operation of a transformer relies on the expanding and collapsing of the magnetic field around the primary winding. When current flows through a conductor, a magnetic field is developed around the conductor. When ac voltage is applied to the primary winding, it causes a constantly changing magnetic field around the winding. During times of increasing ac voltage, the magnetic field around the primary winding expands. After the peak value of the ac cycle is reached, the voltage decreases toward zero. When the ac voltage decreases, the magnetic field around the primary winding collapses. The collapsing magnetic field is transferred to the secondary winding. Transformers will not operate with dc voltage applied since dc voltage does not change in value.

Types of Transformers

There are three basic transformer types. These types are air core, iron core, and powered metal core transformers. The core material used to construct transformers depends on the application of the transformers. Radios and televisions use several types of air-core and powdered metal-core transformers. Iron-core transformers are sometimes called power transformers. They are often used in the power supplies of electrical equipment and at electrical power distribution substations.

Two other classifications of transformers are step-up transformers and step-down transformers. Step-up transformers have more turns of wire used for their secondary windings than for their primary windings. Step-down transformers have more turns of wire on their primary windings than on their secondary windings. An example of a step-up transformer is shown in Fig. 6-47(a). A step-up transformer has a higher voltage across the secondary winding than the input voltage applied to the primary winding. The step-up is caused by the *turns ratio* of the primary and secondary windings. The turns ratio is the ratio of primary turns of wire (N_P) to secondary turns of wire (N_S). If there are 100 turns of wire used for the primary winding and 200 turns for the secondary winding, the turns ratio is 1:2. The

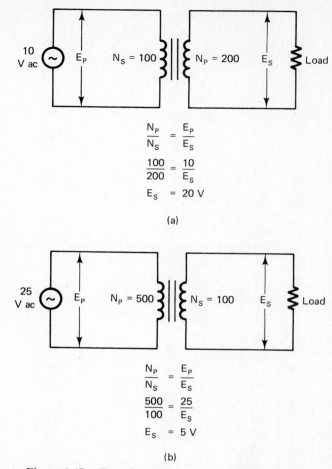

$$\frac{N_P}{N_S} = \frac{E_P}{E_S}$$

$$\frac{100}{200} = \frac{10}{E_S}$$

$$E_S = 20 \text{ V}$$

(a)

$$\frac{N_P}{N_S} = \frac{E_P}{E_S}$$

$$\frac{500}{100} = \frac{25}{E_S}$$

$$E_S = 5 \text{ V}$$

(b)

Figure 6-47. Transformers: (a) step-up; (b) step-down.

secondary voltage will be two times as large as the primary voltage. If 10 V ac is applied to the primary, 20 V ac is developed across the load connected across the secondary winding. The *voltage ratio* (E_P/E_S) is the same as the turns ratio.

Figure 6-47(b) shows an example of a step-down transformer. The primary winding has 500 turns and the secondary has 100 turns of wire. The turns ratio (N_P/N_S) is 5:1. This means that the primary voltage (E_P) is five times greater than the secondary voltage (E_S). If 25 V is applied to the primary, 5 V is developed across the secondary. Remember that a step-up transformer increases voltage and a step-down decreases voltage. The voltage formula shown in Fig. 6-47 shows the relationship between the primary and secondary voltages and the turns ratio of the primary and secondary windings.

Transformers are *power control* devices. Power is equal to voltage (E) multiplied by current (I). In a transformer, primary power (P_P) is equal to secondary power (P_S). Therefore, primary voltage (E_P) times primary current (I_P) is equal to secondary voltage (E_S) times secondary current (I_S)—$P_P = P_S$ or ($E_P \times I_P$) = ($E_S \times I_S$). The *current ratio* of transformers is an inverse ratio: $N_P/N_S = I_S/I_P$. An application of the current ratio is shown in Fig. 6-48. If voltage is increased, current is decreased in the same proportion. The secondary voltage is decreased by 10 times and secondary current is increased by 10 times in the example shown. This is a step-down transformer, since voltage decreased.

Some transformers have more than one secondary winding. These are called *multiple secondary transformers*. Some are step-up and some are step-down windings, depending on the application. There are many other special types of transformers. *Autotransformers* have only one winding. They may be either step-up or step-down transformers, depending on how they are connected. *Variable autotransformers*, such as the one shown in Fig. 6-49, are used in power supplies to provide variable ac voltages.

Current transformers, such as the one shown in Fig. 6-50, are used to measure large ac currents. A conductor is placed inside the hole in the center. A current-carrying conductor produces a magnetic field. This magnetic field is transferred to the winding of the secondary of the current transformer. The high current is reduced by the transformer so that it can be measured.

An important application of transformers is in automobile ignition systems. Figure 6-51 shows the type of step-up transformer used to supply high voltage to the spark plugs of automobiles.

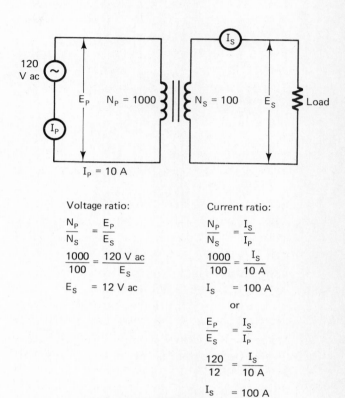

Voltage ratio:

$$\frac{N_P}{N_S} = \frac{E_P}{E_S}$$

$$\frac{1000}{100} = \frac{120 \text{ V ac}}{E_S}$$

$$E_S = 12 \text{ V ac}$$

Current ratio:

$$\frac{N_P}{N_S} = \frac{I_S}{I_P}$$

$$\frac{1000}{100} = \frac{I_S}{10 \text{ A}}$$

$$I_S = 100 \text{ A}$$

or

$$\frac{E_P}{E_S} = \frac{I_S}{I_P}$$

$$\frac{120}{12} = \frac{I_S}{10 \text{ A}}$$

$$I_S = 100 \text{ A}$$

Figure 6-48. Transformer current ratio.

Figure 6-50. Current transformer. (Courtesy of F. W. Bell Co.)

Figure 6-49. Variable auto transformer. (Courtesy of The Superior Electric Co.)

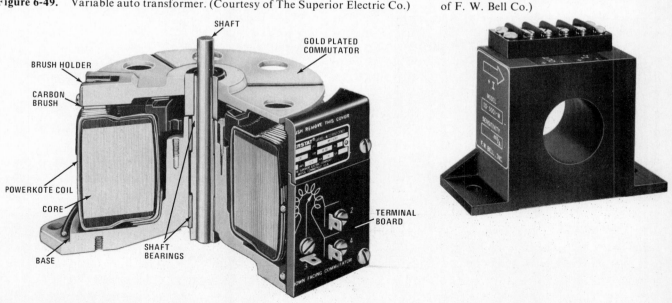

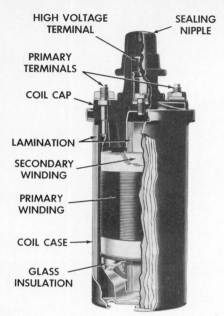

HIGH VOLTAGE TERMINAL

SEALING NIPPLE

PRIMARY TERMINALS

COIL CAP

LAMINATION

SECONDARY WINDING

PRIMARY WINDING

COIL CASE

GLASS INSULATION

Figure 6-51. Step-up transformer used in automobile ignition system. (Courtesy of Delco-Remy.)

Many transformers are used to control three-phase voltages. Some small three-phase transformers are shown in Fig. 6-52(a). Large three-phase transformers are used at electrical power distribution substations to step-down voltages. A large three-phase power transformer is shown in Fig. 6-52(b).

Another type of transformer is called an *isolation transformer*. When making tests with any equipment that plugs into an ac power outlet, an isolation transformer should be used. Its purpose is to isolate the power source of the test equipment from the equipment being tested. An isolation transformer is plugged into a power outlet. The equipment being tested is plugged into the secondary winding of the isolation transformer. There is no connection between the primary and secondary windings of the transformer. This "isolates" the hot and neutral wires of the two pieces of equipment. The possibility of electrical hot to ground shorts is eliminated by using an isolation transformer.

Transformer Efficiency

Transformers are very efficient electrical devices. A typical efficiency rating for a transformer is around 98%. Efficiency of electrical equipment is found by using the formula

$$\text{efficiency } (\%) = \frac{P_{\text{out}}}{P_{\text{in}}} \times 100$$

where P_{out} is the secondary power ($E_S \times I_S$) in watts and P_{in} is the primary power ($E_P \times I_P$) in watts.

Figure 6-52. (a) Small three-phase transformers; (b) large three-phase power transformer used at electrical substations. [(a) Courtesy of TRW/UTC Transformers; (b) courtesy of Kuhlman Corp.]

(a)

(b)

Review

1. What type of alternating current is used in homes?
2. What type of alternating current is used in industries?
3. What is a sine wave?
4. What are the formulas used to find the following ac values when effective value ac is known?
    a. Peak
    b. Peak-to-peak
5. What is the difference between "in phase" and "out of phase"?
6. What is the difference between measuring ac voltage and dc voltage?
7. What are some advantages of using oscilloscopes?
8. List and describe three controls used on oscilloscopes.
9. What is Ohm's law for ac circuits?
10. What is inductance?
11. What is inductive reactance?
12. What is impedance?
13. What is capacitance?
14. What is capacitive reactance?
15. What is the phase relationship between voltage and current in an inductive circuit?
16. What is the phase relationship between voltage and current in a capacitive circuit?
17. What are three factors that affect capacitance?
18. How does a capacitor charge and discharge?
19. What are five types of capacitors?
20. Why are vector diagrams used?
21. Draw a voltage triangle and label each side with $E_R = 10$ V, $E_L = 20$ V, and $E_C = 15$ V.
22. Draw a power triangle for an ac circuit with $VAR_L = 15$, $VAR_C = 32$, and, $W = 12$.
23. Draw a current triangle for a parallel ac circuit with $I_C = 8$ A, $I_L = 4$ A, and $I_R = 9$ A.
24. What is the power factor of a circuit?
25. What is the phase angle of a circuit?
26. What are the three types of filter circuits and how do they differ?
27. What are two types of resonant circuits?
28. What is the difference between a filter circuit and a resonant circuit?
29. What is a time constant?
30. Discuss the operation of a transformer.
31. Why are isolation transformers used?
32. What are some types of transformers?

34. How is transformer efficiency determined?

Student Activities

Oscilloscope Use (see Fig. 6-9)

1. Obtain an oscilloscope.
2. On a sheet of paper, record the following information:
   a. The manufacturer of the oscilloscope.
   b. The model number of the oscilloscope.
   c. In the following, match the oscilloscope control with its function. If the control on your scope has a different name, put it on the paper beside the control name listed. Put the letter of the correct function for each control on the paper. (All of those listed will not be used.)

(1)	Intensity	(a)	Selects horizontal frequency setting
(2)	Focus	(b)	Provides up and down adjustments of beam
(3)	Vertical position	(c)	Determines amount of horizontal deflection
(4)	Horizontal position	(d)	Controls size of electron beam
(5)	Vertical gain	(e)	Provides left and right adjustments of beam
(6)	Horizontal gain	(f)	Controls brightness of beam
(7)	Vertical attenuation (V/cm)	(g)	Acts as a stability control
(8)	Sweep select	(h)	External voltages are applied to this control
(9)	Vernier	(i)	Determines the precise frequency of the horizontal movement
(10)	Synchronization select	(j)	Determines the amount of vertical deflection
(11)	Horizontal input	(k)	Setting for calibration voltage
(12)	Vertical input	(l)	Controls external signal to the vertical amplifier
		(m)	Selects type of synchronization desired

3. Turn in the answers to the teacher for grading.

Alternating-Current Values (see Fig. 6-3)

There are several ac values that are commonly used. These units are ac sine-wave relationships. The following ac sine-wave values are often converted from one value to the other.

Effective value (rms) = 0.707 × peak value

Average value = 0.636 × peak value

Peak value = 1.41 × rms value

Peak-to-peak value = 2.82 × rms value

1. Solve the following problems. Place all answers on a sheet of paper.
   a. Find the following effective values.

Peak voltage	Rms voltage
4 V ac =	_____ V rms
12 V ac =	_____ V rms
6 V ac =	_____ V rms
10 V ac =	_____ V rms
1.5 V ac =	_____ V rms
2 V ac =	_____ V rms
5 V ac =	_____ V rms
9 V ac =	_____ V rms

   b. Find the following peak and peak-to-peak (p–p) values.

Rms voltage	Peak voltage	P–P voltage
3 V ac =	_____ V peak;	_____ V p–p
8 V ac =	_____ V peak;	_____ V p–p
7 V ac =	_____ V peak;	_____ V p–p
9 V ac =	_____ V peak;	_____ V p–p
10 V ac =	_____ V peak;	_____ V p–p
15 V ac =	_____ V peak;	_____ V p–p
11 V ac =	_____ V peak;	_____ V p–p
18 V ac =	_____ V peak;	_____ V p–p

2. Turn in the answers to the teacher for grading.

Inductance and Inductive Reactance

1. Solve the following problems. Place all answers on a sheet of paper.
   a. Total inductance in series: $L_1 = 2$ H, $L_2 = 3$ H, $L_3 = 2$ H.
   b. Total inductance in parallel: $L_1 = 2$ H, $L_2 = 3$ H, $L_3 = 8$ H.
   c. Inductive reactance ($X_L = 2\pi fL$) for the following:
      (1) 30 Hz, 8 H, $X_L = $ _____ Ω.
      (2) 50 Hz, 2 H, $X_L = $ _____ Ω.
      (3) 60 Hz, 1.0 H, $X_L = $ _____ Ω.
      (4) 40 Hz, 3 H, $X_L = $ _____ Ω.
   d. Mutual inductance in series to increase inductance (fields aiding)
      (1) $L_1 = 2$ H, $L_2 = 5$ H, $M = 0.55$
         $L_T = L_1 + L_2 + 2M = $ _____ H.
      (2) $L_1 = 3$ H, $L_2 = 2$ H, $M = 0.35$
         $L_T = $ _____ H.

    e. Mutual inductance in series to decrease inductance (fields opposing)

        (1) $L_1 = 4$ H, $L_2 = 3$ H, $M = 0.85$

            $L_T = L_1 + L_2 - 2M = $ _____ H.

        (2) $L_1 = 3$ H, $L_2 = 3$ H, $M = 0.4$

            $L_T = $ _____ H.

    f. Compute problems d and e when the inductors are connected in parallel:

        (d1) $\dfrac{1}{L_T} = \dfrac{1}{L_1 + M} + \dfrac{1}{L_2 + M}$

        (d2) $L_T = $ _____ H.

                $L_T = $ _____ H.

        (e1) $\dfrac{1}{L_T} = \dfrac{1}{L_1 - M} + \dfrac{1}{L_2 - M}$

        (e2) $L_T = $ _____ H.

                $L_T = $ _____ H.

2. Turn in the answers to the teacher for grading.

Capacitance and Capacitive Reactance

1. Solve the following problems. Place all answers on a sheet of paper.

    a. Compute the capacitance for the following capacitors, using the formula

$$C = \frac{0.0885\,KA}{D}$$

    where C is capacitance in picofarads, K is the dielectric constant, A is the plate area in square centimeters, and D is the dielectric thickness in centimeters.

    (1) Dielectric constant $= 3$

    (2) Size of plate $= 2 \times 200$ cm

    (3) Thickness of dielectric $= 0.03$ cm

$$C = \text{_____} \text{ pF}$$

    (1) Dielectric constant $= 3$

    (2) Size of plate $= 3 \times 200$ cm

    (3) Thickness of dielectric $= 0.04$ cm

$$C = \text{_____} \text{ pF}$$

    (1) Dielectric constant $= 2$

    (2) Size of plate $= 0.25 \times 275$ cm

    (3) Thickness of dielectric $= 0.02$ cm

$$C = \text{_____} \text{ pF}$$

    b. Compute the following for the total capacitance in a series circuit, using the formula

$$\frac{1}{C_T} = \frac{1}{C_1} + \frac{1}{C_2} + \frac{1}{C_3}$$

(1) $C_1 = 10 \ \mu F$
$C_2 = 10 \ \mu F$
$C_3 = 30 \ \mu F$

$$C_T = \underline{\hspace{1.5cm}} \mu F$$

(2) $C_1 = 40 \ \mu F$
$C_2 = 20 \ \mu F$
$C_3 = 20 \ \mu F$

$$C_T = \underline{\hspace{1.5cm}} \mu F$$

(3) $C_1 = 80 \ \mu F$
$C_2 = 60 \ \mu F$
$C_3 = 80 \ \mu F$

$$C_T = \underline{\hspace{1.5cm}} \mu F$$

c. Compute the capacitive reactance for an ac circuit of 60 Hz with a 30-μF capacitor, using the formula

$$X_C = \frac{1}{2\pi f C} = \underline{\hspace{1.5cm}} \Omega$$

d. Compute the following for total capacitance in parallel, using the formula

$$C_T = C_1 + C_2 + C_3$$

(1) $C_1 = 80 \ \mu F$
$C_2 = 60 \ \mu F$
$C_3 = 80 \ \mu F$

$$C_T = \underline{\hspace{1.5cm}} \mu F$$

(2) $C_1 = 50 \ \mu F$
$C_2 = 20 \ \mu F$
$C_3 = 30 \ \mu F$

$$C_T = \underline{\hspace{1.5cm}} \mu F$$

(3) $C_1 = 70 \ \mu F$
$C_2 = 50 \ \mu F$
$C_3 = 50 \ \mu F$

$$C_T = \underline{\hspace{1.5cm}} \mu F$$

2. Turn in your answers to the teacher.

Alternating-Current Circuit Problems

1. For more practice in solving ac circuit problems, complete this activity. It is important to have an understanding of basic ac circuit problems. Examples of several types of ac circuit problems are done.

    a. Series inductance: $L_T = L_1 + L_2 + L_3$
    $L_1 = 2$ H, $L_2 = 5$ H, $L_3 = 6$ H
    What is the inductance of this circuit? $L_T = $ _____ H.

    b. Parallel inductance: $\dfrac{1}{L_T} = \dfrac{1}{L_1} + \dfrac{1}{L_2} + \dfrac{1}{L_3}$
    $L_1 = 2$ H, $L_2 = 4$ H, $L_3 = 8$ H
    What is the inductance of this circuit? $L_T = $ _____ H.

    c. Inductive reactance: $X_L = 2\pi f L$.
    $L = 2$ H, $F = 60$ Hz
    What is the inductive reactance? $X_L = $ _____ Ω.

    d. Series capacitance: $\dfrac{1}{C_T} = \dfrac{1}{C_1} + \dfrac{1}{C_2} + \dfrac{1}{C_3}$
    $C_1 = 2$ μF, $C_2 = 4$ μF, $C_3 = 20$ μF
    What is the capacitance? $C_T = $ _____ μF.

    e. Parallel capacitance: $C_T = C_1 + C_2 + C_3$
    $C_1 = 20$ μF, $C_2 = 40$ μF, $C_3 = 65$ μF
    What is the capacitance? $C_T = $ _____ μF.

    f. Capacitive reactance: $X_C = \dfrac{1}{2\pi f C}$
    $C = 20$ μF, $F = 20$ Hz
    What is the capacitive reactance? $X_C = $ _____ Ω.

    g. RL time constant: t (seconds) $= L/R$
    $R = 4$ Ω, $L = 20$ H
    What is the time constant? $t = $ _____ s.

    h. RC time constant: $t = RC$
    $R = 50$ Ω, $C = 1$ μF
    What is the time constant? $t = $ _____ s.

    i. Impedance in RC or RL circuit: $Z = \sqrt{R^2 + (X_c)^2}$ or $\sqrt{R^2 + (X_L)^2}$
    $R = 10$ Ω, $X_C = 10$ Ω
    What is the impedance? $Z = $ _____ Ω.

    j. Impedance in RLC circuits:

    $$Z = \sqrt{R^2 + (X_L - X_C)^2}$$

    $R = 30$ Ω, $X_L = 250$ Ω, $X_C = 210$ Ω
    What is the impedance? $Z = $ _____ Ω.

2. Turn in the answers to the teacher.

Transformer Operation

1. Obtain transformer and a VOM. Record all data on a sheet of paper.
2. Identify the primary windings of the transformer. The primary will be connected to an ac voltage source. Plug it in when you are ready to make

the necessary measurements. Remove the power cord when the transformer is not in use. Be very aware of safety. It is best to make measurements with a low ac voltage applied to the primary (from 6 to 15 V). If a 120-V ac source is used, be very careful.

3. Some transformers have one secondary winding. Many transformers have three or more secondary windings. Count the number of secondary windings on the transformers. Number of secondary windings = _____ .

4. Make sure that none of the wires from the secondary windings touch.

5. Prepare the VOM to measure ac voltage. Begin with the highest ac voltage range for each secondary voltage measurement.

6. Carefully connect the primary winding to an ac source. Measure the voltage across each secondary winding. Record these voltages.

7. Disconnect the ac voltage from the transformer primary. With no voltage applied, make the following resistance measurements with a VOM.
   a. Resistance of primary winding = _____ Ω.
   b. Resistance of each secondary winding = _____ Ω.

8. Turn in the data to the teacher for grading.

Transformer Problems

1. Solve the following problems which deal with transformer operation. Record your answers on a sheet of paper.
   a. A 1500-turn primary winding has a 120 V applied. The secondary voltage is 20 V. How many turns of wire does the secondary winding have?
   b. When a 200-Ω resistance is placed across a 240-V secondary winding, the primary current is 30 A. What is the value of the primary voltage of the transformer?
   c. A transformer has a 20:1 turns ratio. The primary voltage is 4800 V. What is the secondary voltage?
   d. A transformer has a 2400-V primary winding and a 120-V secondary. The primary winding has 1000 turns. How many turns of wire does the secondary winding have?
   e. A 240- to 4800-V transformer has a primary current of 95 A and a secondary current of 4 A. What is its efficiency?
   f. A transformer has a 20-kVA power rating. Its primary voltage is 120 V and its secondary voltage is 480-V. What are the values of its maximum primary current and secondary current?

2. Turn in the answers to the teacher for grading.

Series and Parallel AC Circuit Problems

1. Solve each of the following series and parallel ac circuit problems. Put all answers and calculations on a sheet of paper.
   a. A 20-μF capacitor and a 1000-Ω resistor are connected in series with a 120-V 60-Hz ac source. Find:

(1)	X_C	(6)	True power
(2)	Z	(7)	Apparent power
(3)	I	(8)	Reactive power
(4)	E_C	(9)	Power factor
(5)	E_R		

    b. Draw an impedance triangle, a voltage triangle, and a power triangle for the circuit for problem (a) and label each value.

    c. A parallel ac circuit converts 12,000 W of power. The applied voltage is 240 V and the total current is 72 A. Find:

       (1) Apparent power

       (2) Power factor

       (3) Phase angle of circuit

    d. Draw a power triangle for the circuit of problem (c) and label each value.

    e. A series circuit with 20 V applied has a resistance of 100 Ω, a capacitance of 40 μF, and an inductance of 0.15 H. Find:

       (1) X_C **(6)** E_C

       (2) X_L **(7)** E_L

       (3) X_T **(8)** E_R

       (4) Z **(9)** Phase angle

       (5) I

    f. Draw an impedance triangle and a voltage triangle for the circuit for problem (e) and label each value.

    g. Use the same values that were given for R, C, and L in problem (e). Connect them in a parallel circuit that has 10 V applied to it. Find:

       (1) I_R **(6)** Conductance

       (2) I_C **(7)** Susceptance

       (3) I_L **(8)** Admittance

       (4) I_T **(9)** Phase angle

       (5) Z

    h. Draw a current triangle and an admittance triangle and label each value for the circuit for problem (g).

  2. Turn in the answers to the teacher for grading.

7

Electrical Energy Conversion

Electrical energy conversion takes place in all electrical systems. The part of the system where energy is converted to another form is called the *load*. There are many different types of loads which are part of electrical circuits. A load can be a resistance, inductance, or capacitance as discussed in Chapter 6. In this chapter the function of the load is considered. Electrical energy is commonly converted to light energy, heat energy, and mechanical energy. Many types of electrical lights are used today. Electrical heating has also become very popular in homes and other buildings. Electrical motors are loads that convert electrical energy into mechanical energy. Several different types of electrical energy conversion will be studied in this chapter.

IMPORTANT TERMS

Several important terms are defined below. These should be studied as needed.

Armature. The rotating coils of a motor which has a wound rotor.

Back EMF. *See* Counter electromotive force.

Ballast. An inductive coil placed in a fluorescent light circuit to develop a high voltage, thereby causing the light to operate.

Brush. A sliding contact made of carbon and graphite, connected between a split-ring commutator and the power source of a dc motor.

Capacitive heating. A method of heating nonconductive materials by placing them between two metal plates and applying a high-frequency ac voltage.

Commutation. The process of applying direct current of the proper polarity at the proper time to the rotor windings of a dc motor through split rings and brushes.

Copper losses. The power losses that occur when the heat produced as electrical current flows through the copper windings of an electric machine; equal to $I^2 \times R$.

171

Core. Magnetic iron or steel materials used to form the laminated parts of electrical motors, generators, and other electromagnetic equipment.

Counter electromotive force (CEMF). "Generator action" which takes place in motors when voltage (EMF) is induced into the rotor conductors which opposes the source voltage (EMF).

Counter EMF. A shortened term which means *counter electromotive force* (CEMF).

Delta connection. A method of connecting the stator windings of three-phase machines in which the beginning of one winding is connected to the end of the next winding; power input lines come into the junctions of the windings.

Efficiency. The ratio of output power to input power; for motors,

$$\text{efficiency} = \frac{P_{out}}{P_{in}} = \frac{\text{horsepower output} \times 746}{\text{watts input}}$$

Field coils. Electromagnetic coils that develop the magnetic fields of motors and generators.

Field poles. Laminated metal pieces that are used to hold the field coils of motors and generators.

Fluorescent lighting. A lighting method that uses lamps which are tubes filled with mercury vapor to produce a bright white light.

Horsepower. One horsepower (hp) = 33,000 foot-pounds of work per minute or 550 foot-pounds per second; it is the unit of measuring mechanical power; 1 hp = 746 W.

Incandescent lighting. A method of lighting which uses bulbs that have tungsten filaments which, when heated by electrical current, produce light.

Induction motor. An ac motor which has a solid squirrel cage rotor and operates on the induction (transformer action) principle.

Inductive heating. A method of heating conductors in which the material to be heated is placed inside a coil of wire and high-frequency ac voltage is applied.

Laminations. Thin pieces of sheet metal used to construct the metal parts of motors and generators.

Motor. A rotating machine which converts electrical energy into mechanical energy.

Repulsion motor. A single-phase ac motor which has a wound rotor and brushes which are shorted together.

Resistive heating. A method of heating that relies on the heat produced when electrical current moves through a conductor.

Right-hand motor rule. The rule applied to find the direction of motion of the rotor conductors of a motor.

Rotor. The part of a motor that rotates.

Self-starting. The ability of a motor to begin rotation when electrical power is applied to it.

Shaded-pole motor. A single-phase ac motor that uses copper shading coils around its poles to produce rotation.

Single-phase motor. Any motor that operates with single-phase ac voltage applied to it.

Slip rings. Solid metal rings mounted on the end of a rotor shaft and connected to the brushes and the rotor windings; used with some types of ac motors.

Speed regulation. The ability of a motor to maintain a steady speed with changes in load.

Split phase motor. A single-phase ac motor that has start windings and run windings and operates on the induction principle.

Split rings. A group of copper bars mounted on the end of a rotor shaft of a dc motor and connected to the brushes and the rotor windings.

Squirrel-cage rotor. A solid rotor used in ac induction motors which has round metal conductors joined together at the ends and placed inside the laminated metal motor core.

Starter. A resistive network used to limit starting current of motors.

Stator. The stationary part of a motor.

Synchronous motor. An ac motor that operates at a constant speed regardless of the load applied.

Torque. Mechanical energy in the form of rotary motion.

Universal motor. A motor that operates with either ac or dc voltage applied.

Vapor lighting. A method of lighting that uses

lamps which are filled with certain gases that produce light when electrical current is applied.

Wye connection. A method of connecting the stator windings of three-phase machines in which the three beginnings *or* ends of the windings are joined together to form a common (neutral) point and the other wires connect to the power lines.

LIGHTING SYSTEMS

Lighting systems convert electrical energy to light energy. There are several types of lighting systems used today. Among these are incandescent, fluorescent, mercury vapor, and several other special-purpose lighting systems.

Incandescent Lighting

For many years, incandescent lighting was by far the most widely used lighting method. Today, fluorescent lamps are used extensively. However, incandescent lamps are still used for most home applications. The construction of an incandescent lamp is shown in Fig. 1-5(a). This lamp is simple to install and maintain. It costs less to buy than most other types of light bulbs. They have lower efficiency and a shorter life span than many other lighting systems.

Incandescent lamps have thin tungsten filaments. They are in the shape of coiled wires. The filament is connected directly to a voltage source (usually 120 V ac) through the lamp base. When electric current is passed through the filament, the temperature of the filament raises to 3000 to 5000° F. At this temperature range, the tungsten filament produces a high-intensity white light. When incandescent lamps are made, the air is removed from inside the glass. This prevents the filament from burning out instantly.

Fluorescent Lighting

Fluorescent lighting is also used in many places today. Fluorescent lamps are tube-shaped bulbs with a filament on each end. There is no electrical connection between the two filaments. The operating principle of the fluorescent lamp is shown in Fig. 7-1. The tube is filled with mercury vapor. High-speed electrons passing between the filaments collide with the mercury atoms. This collision produces ultraviolet radiation. The inside of the lamp tube has a phosphor coating. The coating reacts with the ultraviolet radiation to produce visible light. Several types of phosphors may be used to produce different colors of light.

The circuit for fluorescent lamp is shown in Fig. 7-2. A thermal starter is used. It is a bimetallic strip and a heater connected in series with the filaments. The bimetallic strip remains closed long enough for the filaments to heat. Heat causes the mercury in the tube to become a vapor and conduct electrical current. The bimetallic strip then bends due to the heat produced by current flow through the

Figure 7-1. Operating principle of a fluorescent lamp: (1) filament glows when ac current flows through it; (2) when starter opens, the ballast discharges to ionize the mercury (Hg) atoms; (3) ultraviolet radiation is given off by Hg atoms; (4) visible light is produced.

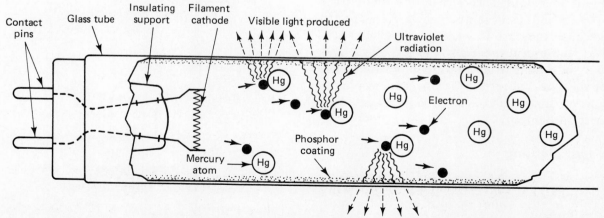

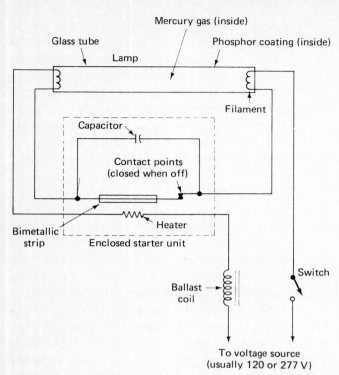

Figure 7-2. Circuit for a fluorescent lamp. (From Patrick/ Fardo, *Energy Management and Conservation*, Prentice-Hall, Englewood Cliffs, N.J., 1982.)

heater. The contact points then separate. This opens the filament circuit. A capacitor is connected across the switch to reduce contact sparking. When the contacts of the starter open, a high voltage is placed across the lamp filaments. The high voltage is caused by action of the ballast coil. The ballast coil has many turns of fine wire. It produces a strong magnetic field. When the contacts of the starter separate, the magnetic field collapses. The high voltage across the filaments causes the mercury inside the lamp to ionize. A current flow then starts through the lamp.

Fluorescent lamps produce more light than the same-size incandescent lamp. They are cheaper to operate since they use less energy. The light produced by fluorescent tubes also has less glare. This light is very similar to natural light. Various sizes and shapes of fluorescent lamps are available. Fluorescent lamps have several advantages over incandescent lighting.

Vapor Lighting

Another type of electrical lighting is the vapor type. The mercury vapor lamp is one of the most common. Another type is the sodium vapor lamp.

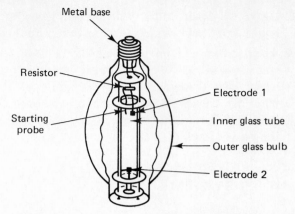

Figure 7-3. Cutaway view of a mercury vapor lamp.

These lamps are filled with a gas that produces a color. For instance, mercury produces a greenish-blue light and argon produces a bluish-white light. Gases can be mixed to produce other color combinations.

A mercury vapor lamp is shown in Fig. 7-3. It is made of a glass arc tube placed inside the bulb. The inner tube contains mercury. A voltage is applied between the starting probe and an electrode. This causes an arc to start between them. The arc strength is limited by a resistor. The arc causes the mercury in the inner tube to ionize. When the mercury has ionized, greenish-bluish light is produced.

Mercury vapor lamps are long-lasting and easy to maintain. They are used in industry and commercial buildings to provide intense light outputs. Mercury vapor lamps may be used for outdoor lighting.

The sodium vapor lamp shown in Fig. 7-4 is popular for outdoor industrial and commercial lighting and for highway lighting. This lamp contains low-pressure neon gas and sodium. When

Figure 7-4. Sodium vapor lamp.

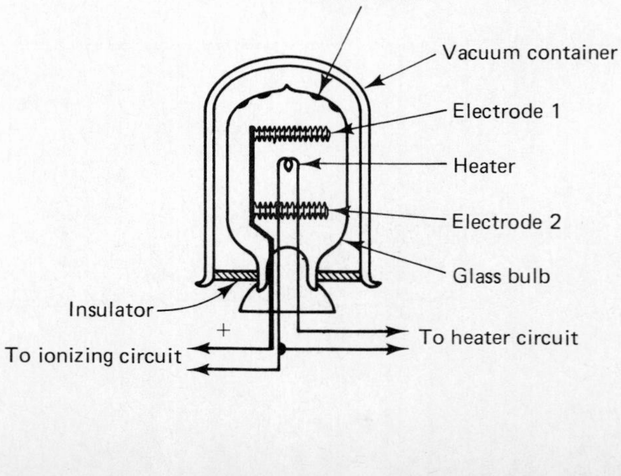

electric current is passed through a heater, electrons are given off. An ionizing circuit causes positive charges to be placed on the positive electrodes. Electrons pass from the heater to the positive electrodes. This causes the neon gas to ionize. The neon produces enough heat to cause the sodium to ionize. A yellowish light is then produced by the sodium vapor. Sodium vapor lamps can produce about three times as much light as the same-size incandescent lamp.

The types of lighting loads discussed in this section are three of the most used. Incandescent, fluorescent, and vapor lamps are used for many lighting applications. A few other specialized types of lighting are used.

HEATING SYSTEMS

Most electrical loads produce a certain amount of heat. Heating is caused when current flows through a resistive device. In some equipment, heat causes a power loss. Lamps, for example, produce heat energy as well as light energy. There are several types of loads that are mainly heating loads. Their main purpose is to convert electrical energy into heat energy. These load devices include resistance heating, inductive heating, and dielectric (capacitive) heating.

Resistance Heating

Heat energy is produced when electrical current flows through a resistive material. In some cases, the heat energy produced is undesirable. Resistance heating produces useful heat that can be transferred from a resistive material to where it is used. The heating element has coiled resistance wire designed as a heat-conducting material. The element is contained in a metal sheath. This principle can be used to heat water, oil, or a room. Resistive heaters may be used in open air or immersed in a material that is to be heated. The heat energy produced depends on the amount of current flow and the resistance of the element. A typical application is shown in Fig. 7-5, where resistance heating coils are used as supplemental elements for a heat pump unit.

Induction Heating

The principle of induction heating is shown in Fig. 7-6. Heat is produced in magnetic materials when alternating current is applied. In the example, current is induced into the material to be heated. Electromagnetic induction is caused by the application of alternating current to the heating coil. The material to be heated must be a conductor. Current is induced into the conductor. A high-frequency ac source is used to produce higher heat output. The high-speed magnetic field created by the high-frequency ac source moves across the material to be heated. The induced voltage causes circulating currents called *eddy currents* to flow in the material. Heat is produced due to the materials resistance to the flow of eddy currents. The heat produced by this method is used for many industrial processes, such as heating metals.

Figure 7-5. Resistance heating element. (Courtesy of Williamson Co.)

Figure 7-6. Principle of induction heating.

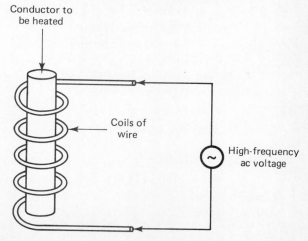

Dielectrical (Capacitive) Heating

Induction heating is used only for conductive materials. Another method must be used to heat nonconductive materials. Dielectric or capacitive heating is shown in Fig. 7-7. Nonconductors may be heated by placing them in electrostatic field between two metal electrodes. The electrodes are supplied by a high-frequency ac source. The material to be heated is like the dielectric of a capacitor. The metal electrodes are like two plates of a capacitor.

High-frequency ac is applied to the dielectric heater. The high-speed AC voltage causes the atomic structure of the dielectric material to become distorted. As frequency increases, the amount of distortion also increases. Internal friction is caused which produces a large amount of heat in the nonconductive material. Dielectric heating produces rapid heat that spreads throughout the heated material. Common applications of dielectric heating are for making plywood and bonding plastic sheets.

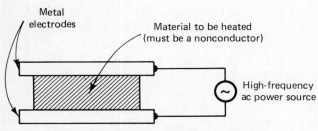

Figure 7-7.   Dielectric (capacitive) heating.

MECHANICAL LOADS (MOTORS)

Equipment that converts electrical energy into mechanical energy is another type of electrical load. Electric motors are the main type of mechanical load. There are many types of motors used today. Motors are our major energy-consuming load. Motors of various sizes are used for many applications.

Basic Principles of Motors

All motors have several basic parts in common. Their basic parts include (1) a stator, which is the frame and other stationary parts; (2) a rotor, which is a rotating shaft and other parts that rotate; and (3)

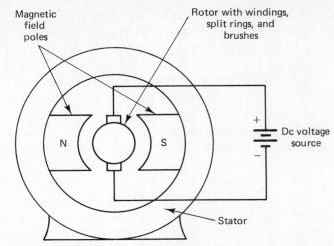

Figure 7-8.   Basic parts of a dc motor.

other equipment, such as a brush/commutator assembly for dc motors, and a starting circuit for certain ac motors. The basic parts of a dc motor are shown in Fig. 7-8. Dc motors are constructed in the same way as dc generators. Their basic parts are the same. The function of a motor is to convert electrical energy into mechanical energy. Motors produce rotary motion. A motor has an electrical input and a mechanical output. Mechanical energy is produced by a motor due to the interaction of a magnetic field and a set of conductors.

The motor principle is shown in Fig. 7-9. In Fig. 7-9(a) no current is flowing through the conductors. This is due to the position of the brushes against the commutator. No motion is produced at this time. When current flows through the conductors, a circular magnetic field is developed around the conductors. The direction of current flow determines the direction of the circular magnetic fields, as shown in Fig. 7-9(b). The magnetic field around the conductors interacts with the main magnetic field. The interaction of these two magnetic fields results in rotary motion. The circular magnetic field around the conductors causes a compression of the main magnetic field at points *A* and *B* in Fig. 7-9(c). This compression causes the magnetic field to react in the opposite direction. Motion is produced away from points *A* and *B*. Rotary motion is produced in clockwise direction. To change the direction of rotary motion, the direction of current flow through the conductors is reversed. This can be observed by using the right-hand rule shown in Fig. 7-10.

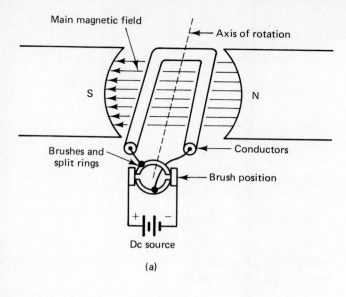

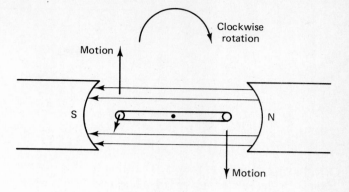

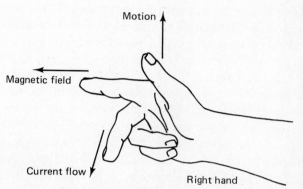

Figure 7-10. Right-hand motor rule.

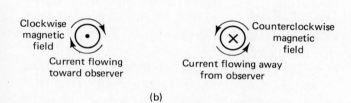

(b)

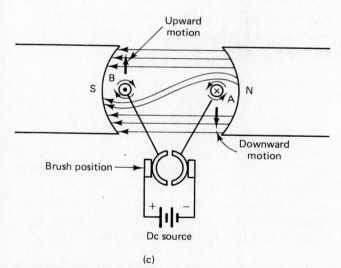

(c)

Figure 7-9. Motor principle: (a) no current flow through rotor conductor; (b) direction of circular magnetic fields around the rotor conductors when current flows; (c) compression points produced at *A* and *B* when current flows through rotor conductors—causing rotation.

The rotary motion produced by the interaction of two magnetic fields is called *torque* or *motor action*. The amount of torque produced by a motor depends on the strength of the main magnetic field and the amount of current flowing through the rotor conductors. If the magnetic field or the current through the conductors increases, the amount of torque will increase.

Direct Current Motors

Motors that operate from direct-current power sources are often used. Direct-current motors are classified as series, shunt, or compound machines. The method of connecting the armature and field determines the type. Also, permanent-magnet dc motors are used for some applications.

The operation of dc motors can be explained by referring to Fig. 7-11. Some important terms are used here. These terms are load, speed, counter electromotive force (CEMF), line current, and torque. The amount of mechanical load applied to the shaft of a motor depends on how it is used. Load applied to the motor affects its operation. As the

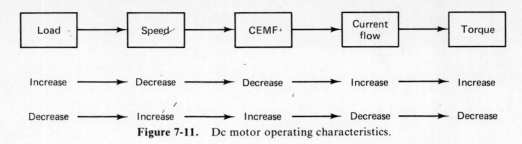

Figure 7-11. Dc motor operating characteristics.

mechanical load increases, the speed of a motor tends to decrease. As speed decreases, voltage induced into the rotor conductors decreases. This is reduced due to generator action or CEMF. This generated voltage (counter electromotive force) depends on the number of rotating conductors and the speed of rotation, just like a generator. As motor speed decreases, CEMF decreases. The counter electromotive force produced by a motor opposes the applied voltage. The actual voltage applied to a motor increases as the CEMF decreases. When the working voltage increases, more current flows from the power source. The torque of a motor depends on current flow. Torque will increase as current increases. The opposite situation occurs when the mechanical load connected to the shaft of a motor decreases. The speed of the motor tends to increase. Increased speed causes increased CEMF. The CEMF is in opposition to the applied voltage, so the CEMF increases. Current decreases as CEMF increases. Decreased current causes a decrease in torque. Torque varies with changes in load. As the load on a motor is increased, its torque also increases. The current drawn from the power source also increases when the load is increased.

CEMF is very important in motor operation. When a motor is started a very large starting current flows. The starting current is much larger than the current when full speed is reached. Current flow at full speed is called running. Maximum current flows when there is no CEMF (when the rotor is not turning). As the CEMF increases, the current flow decreases. A resistance in series with the armature circuit is often used to reduce the starting current of a motor. After the motor has reached full speed, this resistance is bypassed by automatic or manual switching systems. This allows the motor to produce maximum torque.

The torque of a motor is measured in foot-pounds. Motors are rated in terms of their horsepower. The horsepower rating of a motor is based on the amount of torque produced. Horsepower can be found by using the formula

$$hp = \frac{\text{speed} \times \text{torque}}{5252}$$

The speed is in revolutions per minute (rpm) and the torque is in foot-pounds.

Dc motors, like generators, have *armature reaction*. Armature reaction was discussed in Chapter 5. This effect *distorts* the main magnetic field in dc motors. The brushes of a dc motor can be moved to counteract the effect of armature reaction. Coils called *interpoles* are used to control armature reaction in dc motors.

The speed of dc motors can be easily controlled. By adjusting the applied dc voltage with a rheostat, motor speed can be varied from zero to maximum rpm. Some dc motors have better speed control than others. This can be understood by studying the different types of motors. The types of dc motors are (1) permanent-magnet dc motors, (2) series-wound dc motors, (3) shunt-wound dc motors, and (4) compound-wound dc motors. Each has different characteristics.

PERMANENT-MAGNET DC MOTORS. The permanent-magnet dc motor shown in Fig. 7-12 is constructed like a permanent-magnet dc generator. The permanent-magnet motor is used only where a low amount of torque is needed. The dc power supply is connected directly to the armature conductors through the brush/commutator assembly. The main magnetic field is produced by permanent magnets mounted on the stator. Its speed can be changed by adjusting the supply voltage.

SERIES-WOUND DC MOTORS. The way in which the armature and field circuits of a dc motor are connected affects its operation. Each type of dc motor is similar in construction to that of a corresponding type of dc generator. The only difference is

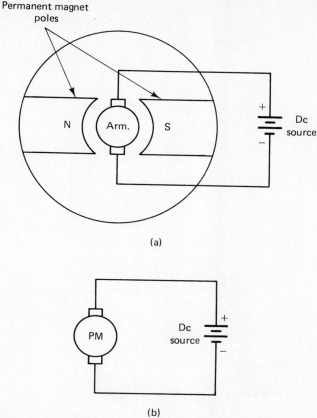

(a)

(b)

Figure 7-12. Permanent-magnet dc motor: (a) pictorial; (b) schematic.

that a generator is a voltage source and the motor is a *load*.

The series-wound motor, shown in Fig. 7-13, has its armature and field circuits connected in series. There is only one path for current flow. The field coils are wound with a few turns of a large-diameter wire. This gives the field circuit a low resistance. Changes in mechanical load applied to the motor shaft cause charges in the current flow through the field. If the load increases, the current flow also increases. Increased current causes a stronger magnetic field. The speed of a series-wound motor varies from very fast with no load to very slow with a heavy load applied. A high value of current flows through the low-resistance field. The series-wound motor produces a high starting torque due to the large current flow. Series-wound motors are used where heavy loads must be moved.

SHUNT-WOUND DC MOTORS. Shunt-wound dc motors are used more than any other type of dc motor. As shown in Fig. 7-14, the shunt-wound dc motor has its field coils connected in parallel with the armature. The field coils are wound with many turns of small-diameter wire. This gives the field a high resistance. The field coils are a high-resistance parallel path. This causes a small amount of current

Figure 7-13. Series-wound dc motor: (a) pictorial; (b) schematic.

(a)

(b)

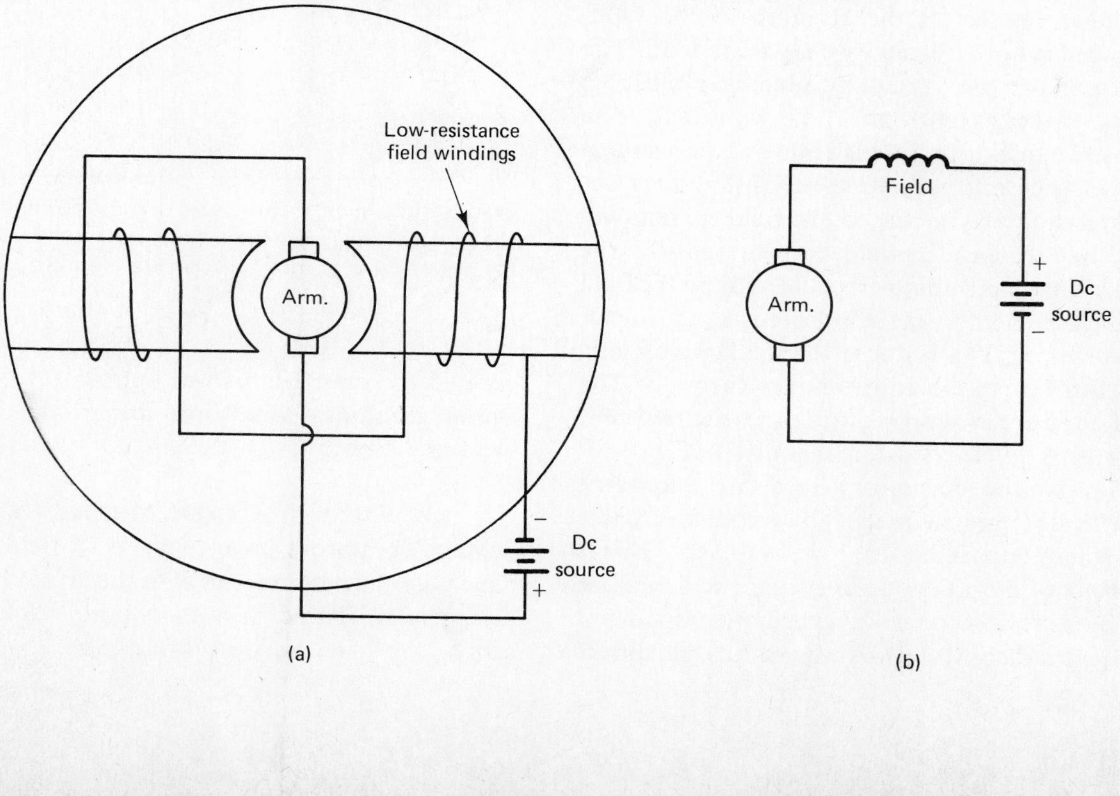

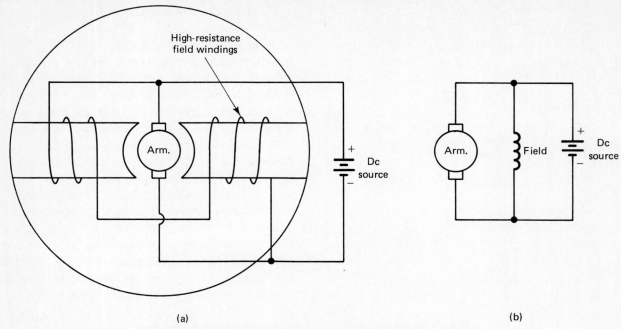

(a)   (b)

Figure 7-14.   Shunt-wound dc motor: (a) pictorial; (b) schematic.

flow through the field. A strong electromagnetic field is developed due to the many turns of wire that the field windings have. Remember that the amount of current flow and the number of turns affect the strength of an electromagnet.

Most of the current drawn by shunt motor flows in the armature circuit path. The armature current does not affect the strength of the field. Motor speed is not affected by changes in load. The field current can be varied by placing a variable resistance in series with the field windings. The current in the field is low. This allows a low-wattage rheostat to be used to vary the speed of the motor. As the field resistance is increased, the field current will decrease. A decrease in field current reduces the strength of the electromagnetic field around the field coils. When the field strength is decreased, the motor will rotate faster. This is due to the weaker magnetic lines of force that the rotor moves through. The speed of dc shunt-wound motors can be varied very easily by using a field rheostat, as shown in Fig. 7-15.

Shunt-wound dc motors have fairly constant speed with changes in load. The speed decreases slightly when the mechanical load increases. This is due to increased voltage drop across the armature circuit. Because of its constant speed under load and its ease of speed control, the shunt-wound dc motor

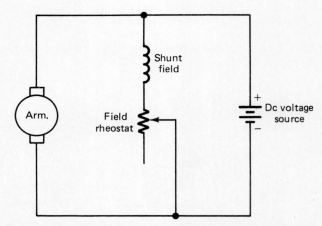

Figure 7-15.   Field rheostat used to control dc motor speed. As rheostat resistance increases, field current decreases, causing magnetic field strength to decrease and motor speed to increase.

is used for many industrial applications. Variable-speed machines and tools are driven by shunt-wound dc motors.

COMPOUND-WOUND DC MOTORS. Compound-wound dc motors have two sets of field windings. One is connected in series with the armature and one in parallel. These motors combine the desirable aspects of the series-wound and shunt-wound motors. They have high starting torque, similar to a series-wound motor. They also have constant speed, similar to a shunt-wound motor. When high torque and very constant speed are needed, the compound-wound dc motor can be used. Compound-wound motors are more expensive than series-wound or shunt-wound dc motors. Dc motors are designed so that their shaft will rotate in either direction. It is simple to reverse the direction of rotation of a dc motor. By reversing the connections of the armature windings or field windings, reversal of direction is achieved. This is done by changing the terminal connections. Four terminals are ordinarily used. They are labeled A_1 to A_2 for the armature connections, and F_1 and F_2 for the field connections. If either the armature connection or the field connections are reversed, the rotation of the motor will reverse (see Fig. 7-16). If both the armature and field connections are reversed, the motor shaft will rotate in the same direction. The relative direction of current flow through the armature and the field circuits would still be the same.

Single-Phase AC Motors

Another type of mechanical load is the single-phase ac motor. This type of motor is common for industrial, commercial, and home use. They operate from a single-phase ac power source. Many are plugged into 120-V power outlets. There are three types of single-phase ac motors: (1) universal motors, (2) induction motors, and (3) synchronous motors.

UNIVERSAL MOTORS. Universal motors can operate with either ac or dc voltage applied. A universal motor is shown in Fig. 7-17. It is constructed in the same way as a series-wound dc motor. The series-wound arrangement is the only type of motor that will operate with ac voltage applied. Series motors have windings with low inductance (few turns of large-diameter wire). They offer low opposition to the flow of alternating current. The universal motor has concentrated field windings which are similar to those of dc motors.

Figure 7-16. Reversing the direction of rotation of dc motors.

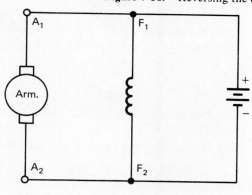

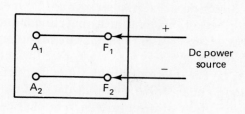

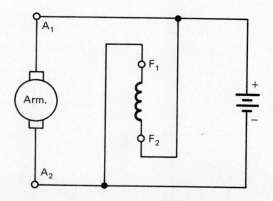

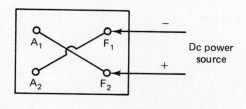

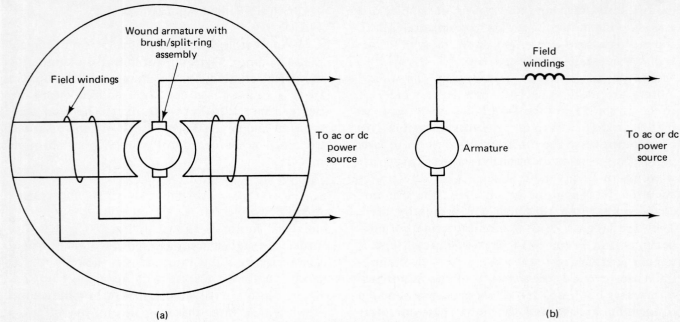

(a)   (b)

Figure 7-17.   Universal motor: (a) pictorial; (b) schematic.

The operating principle of the universal motor with ac applied involves the change of field and armature polarities at the same time. A reversal of field and armature polarity is brought about by the changing nature of alternating current. The current changes direction 120 times per second! Reversal of current direction through the armature conductors at the proper time causes the armature to rotate. The principle is the same as for dc motors. In fact, the universal motor operates in the same manner as a series-wound dc motor except that the field polarity and the direction of armature current change at a

rate of 120 times per second with 60-Hz ac voltage applied. The speed and torque of universal motors are similar to dc series-wound motors. Applications of universal motors are for many portable tools and small motor-driven equipment. Appliances such as mixers and blenders and tools such as drills and saws use universal motors.

INDUCTION MOTORS. The operation of a single-phase ac induction motor is shown in Fig. 7-18. The coil symbols around the stator represent

Figure 7-18.   Operation of a single-phase ac induction motor.

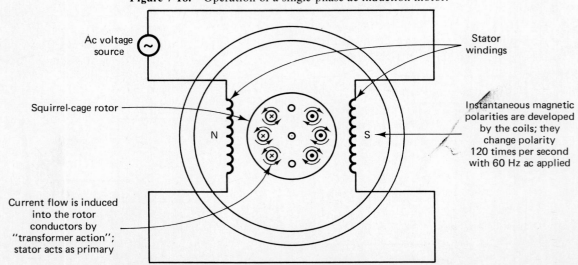

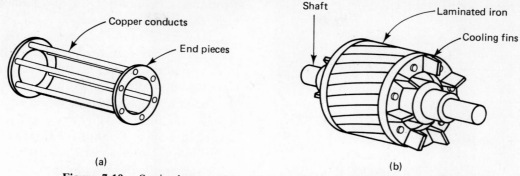

Figure 7-19. Squirrel-cage rotor: (a) internal construction; (b) assembled rotor.

the field coils. The coils are connected to an ac voltage source. The polarity of the coils will change 120 times per second when 60 Hz ac is applied. Induction motors have a solid rotor, called a squirrel-cage rotor. Remember that dc motors have wound rotors. The squirrel-cage rotor shown in Fig. 7-19 has large copper conductors soldered at each end to a round connecting plate. The plates actually short-circuit the copper conductors. When current flows through the stator windings, a current flow is induced into the rotor conductors. This current is caused by "transformer action" between the stator and rotor. The stator has ac voltage and acts as a transformer primary winding. The rotor is similar to a transformer secondary winding since current is induced into it.

The stator polarity changes with the applied ac voltage. It develops a changing or "rotating" magnetic field. The rotor conductors change polarity due to the induced current flow through the copper conductors. The rotor rotates in step with the changing magnetic field of the stator. Some method of starting the rotor to move must be used. Due to inertia, a rotor will not move until it is put into motion. Most ac motors have some method for starting rotation.

The operating speed of an ac induction motor is based on the speed of the rotating magnetic field. The number of stator poles also affects speed. The speed of the rotor will never be as high as the speed of the changing stator field. If the two speeds were equal, there would be no relative motion between the rotor and stator conductors. No induced rotor current could flow and no torque would be developed. The operating speed (rotor speed) of an ac induction motor is always less than the speed of the stator field.

The speed of the stator may be found by using the formula

$$\text{speed (rpm)} = \frac{\text{frequency (Hz)} \times 120}{\text{number of stator poles}}$$

A motor with two stator poles that operate on a 60-Hz source would have a stator speed of 3600 rpm:

$$\frac{60 \times 120}{2} = \frac{7200}{2} = 3600 \text{ rpm}$$

The stator speed is also called the *synchronous speed* of a motor.

The difference between the stator speed of an ac induction motor and the rotor speed is called *slip*. The rotor speed must lag behind the stator speed. Slip is found by using the formula

$$\% \text{ slip} = \frac{\text{stator speed} - \text{rotor speed}}{\text{stator speed}} \times 100$$

If rotor speed is close to stator speed, the percentage of slip is small. For example, a two-pole motor has a stator speed of 3600 rpm and a rotor speed of 3450 rpm. The percent slip is

$$\frac{3600 - 3450}{3600} \times 100 = \frac{150}{3600} \times 100 = 4.17\%$$

Single-phase ac induction motors are classified according to the method used for starting. Common types include (1) split-phase motors, (2) capacitor motors, and (3) shaded-pole motors.

Split-Phase Motors. A split-phase ac induction motor is shown in Fig. 7-20. It has two sets of

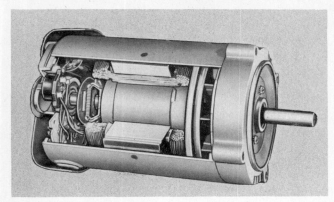

Figure 7-20. Split-phase ac induction motor. (Courtesy of Dayton Electric Mfg. Co.)

stator windings. One set is called a run winding. The run winding is connected across the ac power source. The other winding is called the start winding. The start winding is also connected across the ac line. A centrifugal switch mounted on the shaft of the motor is connected in series with the start winding. The centrifugal switch is in the closed position when the motor is turned off.

To understand the operation of a split-phase ac motor, refer to Fig. 7-21(a). This figure shows a two-pole motor with single-phase ac applied. A permanent magnet is placed inside the stator for simplicity. Remember that a squirrel-cage rotor is used in ac induction motors. At time t_0 of the ac sine wave, no stator magnetic field is produced. At time t_1, a north magnetic polarity is produced on the right pole of the stator and a south polarity on the left pole. These polarities will cause the rotor to align horizontally. Remember that unlike magnetic poles attract. At time t_2 the stator poles are demagnetized. They then magnetize in the opposite direction, at time t_3. The rotor will now align horizontally in the opposite direction. This effect will continue at a rate of 120 polarity changes per second, with 60 Hz ac applied to the stator. The rotor would not move unless it is positioned so that it will be drawn toward one of the poles. Some starting method must be used for single-phase ac motors. They are not "self-starting."

A two-phase ac voltage applied to a motor is shown in Fig. 7-21(b). Two sets of stator windings are used. One phase is connected to each set. Two-phase voltage is not produced by power companies, but this example shows the operating principle of a split-phase motor. Two-phase voltage has one phase

at minimum value while the other phase is at maximum value.

At time t_1, phase 1 is maximum and phase 2 is minimum. The right stator pole is now a north polarity and the left stator pole is a south polarity. The rotor aligns itself horizontally. No polarities are developed by the vertical poles at this time. At time t_2, phase 1 is minimum and phase 2 is maximum. The top stator pole is a south polarity and the bottom stator pole is now a north polarity. The rotor will now align itself vertically, by moving 90°. At time t_3, phase 1 is maximum in the opposite direction and phase 2 is minimum. A north pole is now developed on the left and a south pole on the right. The rotor moves 90° in a clockwise direction. This effect will continue as two-phase voltage is applied to the stator poles. The direction of rotation is established by the relation of phase 1 to phase 2. An ac motor with two-phase voltage applied would be self-starting.

The same is true with three-phase voltage applied to a motor as shown in Fig. 7-21(c). Movement of the rotor results due to 120° phase separation of three-phase ac voltage. The voltage is applied to six stator poles. Three-phase induction motors are "self-starting." No starting method needs to be built into three-phase motors.

Look at the split-phase ac motor diagram in Fig. 7-22(a). The purpose of the two sets of windings is to produce a type of two-phase ac voltage to start the motor. The single-phase ac voltage applied to the motor is said to be "split" into a two-phase current. The magnetic field of the stator is created by "phase splitting." The start winding of the split-phase motor is made of a few turns of small-diameter wire. This gives the winding a high resistance (small-diameter wire) and a low inductance (few turns of wire). The run winding is wound with many turns of large-diameter wire. It has a lower resistance (large-diameter wire) and a higher inductance (many turns of wire). Inductance in an ac circuit causes the current flow to lag behind the applied voltage. With more inductance there is a greater lag in current flow.

Single-phase ac voltage waveform of a split-phase ac induction motor is shown in Fig. 7-22(b). The current flow in the start winding lags the applied ac voltage due to the windings inductance. The current flow in the run winding lags behind by a greater amount due to its higher inductance. The phase separation of the currents in the start and run

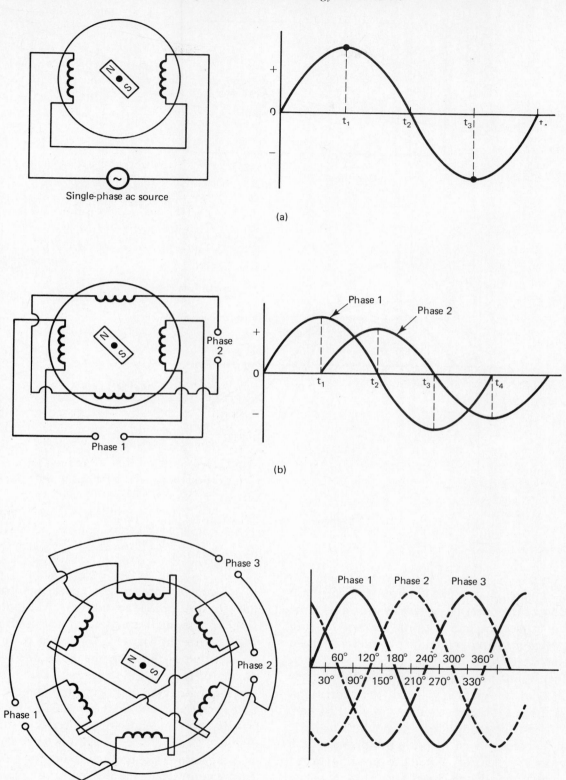

Figure 7-21. Operation of a split-phase motor: (a) two-pole motor with single-phase ac applied; (b) motor with two-phase ac applied; (c) motor with three-phase ac applied.

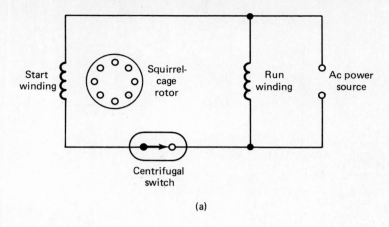

(a)

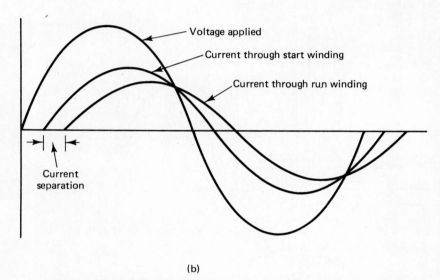

(b)

Figure 7-22. Split-phase ac motor: (a) schematic; (b) waveforms.

windings creates a two-phase current. The phase separation of the currents is usually around 30° or less. This causes the motor to have a low starting torque. Higher torque would be produced if phase separation were closer to 90°.

When a split-phase ac induction motor reaches about 75% of its operating speed, the centrifugal switch will open the start-winding circuit. The start winding is no longer needed. Removing the start winding reduces power losses and prevents the motor from overheating. When the motor is turned off the centrifugal switch closes to connect the start winding across the ac voltage source again. Split-phase motors are used where low torque is required to drive mechanical loads such as small machinery and appliances in the home.

Capacitor Motors. Capacitor motors are similar to split-phase ac motors. A capacitor-start, single-phase ac induction motor and its wiring diagram are shown in Fig. 7-23. The wiring diagram is the same as the split-phase motor except that a capacitor is placed in series with the start winding. The capacitor causes the current in the start winding to lead (rather than lag) the applied voltage. Remember that current leads voltage in a capacitive circuit. Figure 7-24 shows the voltage and current waveforms in a capacitor-start motor circuit. The current in the start winding leads the applied ac voltage. This is due to the high capacitance in the start-winding circuit. The run winding has only inductance, so the current in the run winding lags the applied voltage. The amount of phase separation of the current is close to

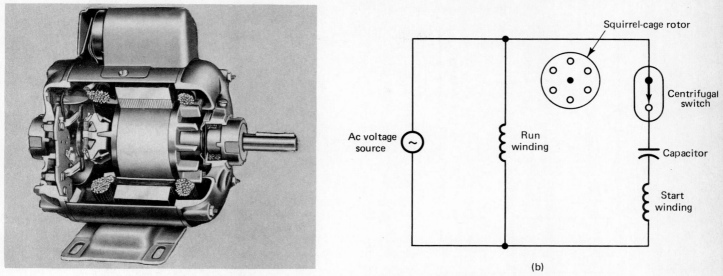

Figure 7-23. Capacitor-start motor: (a) cutaway view; (b) wiring diagram. [(a) Courtesy of Dayton Electric Mfg. Co.]

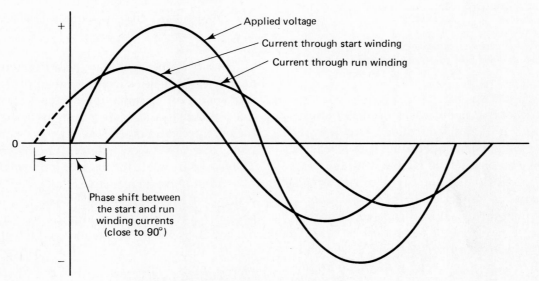

Figure 7-24. Waveforms of a capacitor-start motor.

90°. It is almost like a two-phase ac voltage. The starting torque of a capacitor-start induction motor is greater than that of a split-phase motor. This type of motor can be used where greater torque is needed. Most capacitor motors, as well as split-phase motors, are used in fractional-horsepower sizes (less than 1 hp). Common fractional-horsepower sizes are $\frac{1}{4}$, $\frac{1}{2}$, and $\frac{3}{4}$ horsepower.

Another type of capacitor motor is called a capacitor-start, capacitor-run, or two-value capacitor motor. Its circuit is shown Fig. 7-25(a). This type of motor has two capacitors. One capacitor of higher value is in series with the start winding and centrifugal switch. The other capacitor is lower in value. It is connected in series with the start winding. The smaller capacitor remains in the circuit during

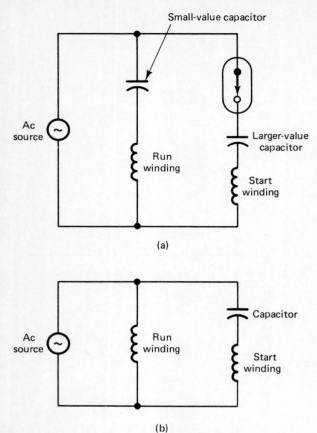

(a)

(b)

Figure 7-25.   Other types of capacitor motors: (a) capacitor-start, capacitor-run motor; (b) permanent capacitor motor.

Figure 7-26.   Shaded-pole ac induction motor.

operation. The larger capacitor is used to increase starting torque. It is removed from the circuit by the centrifugal switch when operating speed reaches about 75%. The smaller capacitor and the start winding are part of the motor circuit during operation. The smaller capacitor helps to produce a constant running torque. It also reduces vibration.

Another type of capacitor motor is called a *permanent capacitor motor.* Its circuit is shown in Fig. 7-25(b). This motor has no centrifugal switch. The capacitor is connected into the circuit at all times. These motors are used when very low torque is needed. They are made in small, fractional-horsepower sizes.

Split-phase motors and capacitor motors may have their direction of rotation reversed easily. The connection of the start winding or run winding must be changed. By reversing either the start-winding connections or the run-winding connections (but not both) the direction of rotation is reversed.

Shaded-Pole Motors.   Another method of producing a two-phase current is called "pole shading." A shaded-pole ac induction motor is shown in Fig. 7-26. Shaded-pole motors are used for very low torque applications such as fans and small blower units. They are low-cost, rugged, and reliable motors. They are only made in very low horsepower ratings.

The operating principle of a shaded-pole motor is shown in Fig. 7-27. A single-phase ac alternation is shown to represent the applied ac voltage. The dashed lines show voltage induced into the shaded section of the field poles. The shading coils are shown in the upper right and lower left of the field poles. The shaded pole is surrounded by a heavy copper conductor. It is actually part of the main field pole. This closed-looped copper conductor has current induced into it when ac voltage is applied to the main field poles.

Single-phase ac voltage is applied to the stator windings. The magnetic field of the main field poles induces a current flow into the shaded sections of the main poles. The shaded sections act like transformer secondary windings. The voltage induced into a

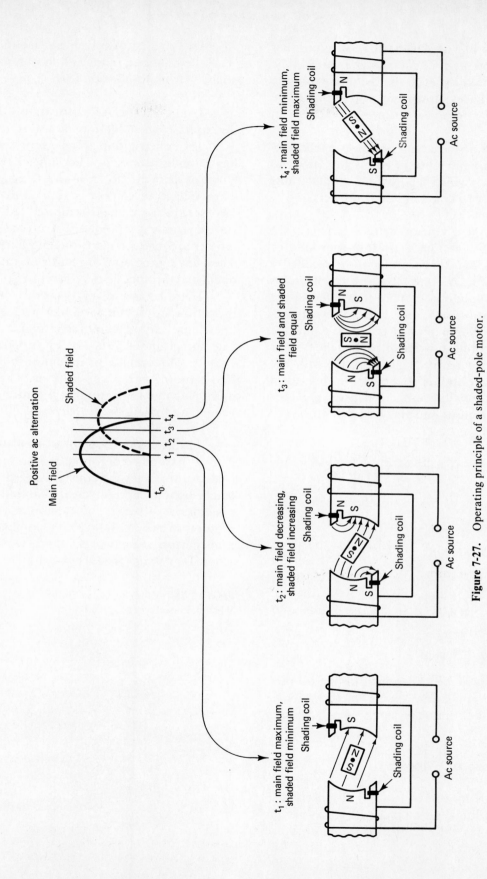

Figure 7-27. Operating principle of a shaded-pole motor.

189

shaded section is out of phase with the main field voltage. Four time periods are shown in Fig. 7-27. The voltage induced into the shaded pole from the main field pole causes movement of the rotor. The movement of the rotor at each time period is shown in the figure.

Shaded-pole ac induction motors are inexpensive. They use a squirrel-cage rotor and have no starting winding or centrifugal switch. Applications are limited mainly to small fans and blowers.

SINGLE-PHASE SYNCHRONOUS MOTORS. In timing or clock applications a constant-speed motor must be used. A constant-speed motor is called a synchronous motor. A single-phase ac synchronous motor has stator windings that are connected across the ac voltage source and a rotor made of permanent-magnet material. Some method of starting must be used to start the rotor's movement. When the rotor is started, it will rotate in "synchronism" with the changing ac voltage applied to the stator. This motor does not operate on the induction principle. The speed of a synchronous motor is based on the speed formula:

$$\text{speed (rpm)} = \frac{\text{frequency} \times 120}{\text{number of poles}}$$

At 60 Hz frequency, these are the synchronous speeds:

1. Two-pole = 3600 rpm
2. Four-pole = 1800 rpm
3. Six-pole = 1200 rpm
4. Eight-pole = 900 rpm
5. Ten-pole = 720 rpm
6. Twelve-pole = 600 rpm

Single-phase ac synchronous motors are used for low-torque applications, which include clocks, phonographs, and timing systems which require constant speed.

Three-Phase AC Motors

Three-phase motors are used in industry and commercial buildings. Three-phase ac voltage is not supplied to homes. Most three-phase motors are *integral* horsepower sizes. These are ratings of 1 or more horsepower. They may be as large as 10,000 hp!

Two types of three-phase ac motors are (1) three-phase ac induction motors and (2) three-phase ac synchronous motors.

THREE-PHASE AC INDUCTION MOTORS. A three-phase ac induction motor is shown in Fig. 7-28. The construction of the motor is very simple. It has wound stator coils. They are connected in either a wye or delta circuit. A squirrel-cage rotor is used. Three-phase ac voltage is applied to the stator. A 120° separation makes the motor "self-starting." No starting winding is needed. Three-phase ac induction motors are made in many integral horsepower sizes. They have good starting and running torque. The operational principle is shown in Fig. 7-29.

The direction of rotation of a three-phase ac induction motor can be changed easily. Two power lines coming into the stator windings can be reversed. The direction of rotation of the shaft will then change. Three-phase ac induction motors are used for many industrial applications. They are used to drive machine tools, pumps, elevators, hoists, conveyors, and many other machines.

THREE-PHASE AC SYNCHRONOUS MOTORS. Three-phase ac synchronous motors are very specialized motors. They are constant-speed motors which can be used to "correct" power factor of three-phase systems. A diagram of three-phase synchronous motor is shown in Fig. 7-30. Three-phase ac synchronous motors are constructed like three-phase alternates. Dc voltage produces an electromagnetic field.

Figure 7-28. Three-phase ac induction motor. (Courtesy of Westinghouse Electric Corp.)

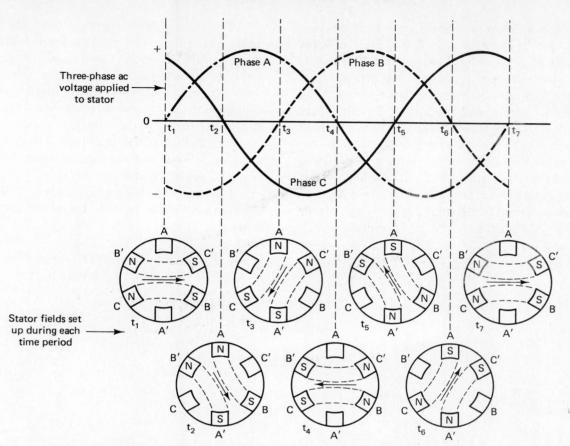

Figure 7-29. Operational principle of a three-phase induction motor.

Figure 7-30. Diagram of a three-phase ac synchronous motor.

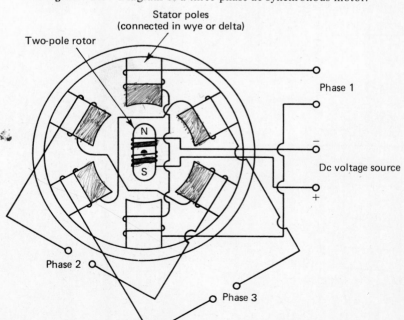

Notice the north and south polarities. The stator windings are connected in either a wye or delta circuit. Three-phase voltage is applied to the stator.

The three-phase ac synchronous motor is different from the three-phase ac induction motor. The rotor has wire windings. It is connected by slip-rings and brushes to a dc power source. Three-phase ac synchronous motors are not self-starting. Some method must be used to start the motor. Synchronous motors are designed so that they will rotate at one speed regardless of the load. This speed is called *synchronous speed*. The speed of an ac synchronous motor may be found by using the formula

$$\text{speed (rpm)} = \frac{\text{ac frequency} \times 120}{\text{number of poles} \div 3}$$

This is the same formula that was used for single-phase motors except that the number of poles must be divided by 3 (the number of phases). A three-phase ac motor with 12 stator poles has four poles per phase (12 ÷ 3 = 4). Its speed is 1800 rpm. Synchronous ac motors have speeds that are based on the number of stator poles and the ac frequency.

One method of starting a three-phase ac synchronous motor is to use a small dc machine connected to its shaft, as shown in Fig. 7-31(a).

It is one of the most complex types of motors in use. Listed below are the steps for starting this motor.

1. Dc power is applied to the auxiliary motor. This causes it to increase in speed.

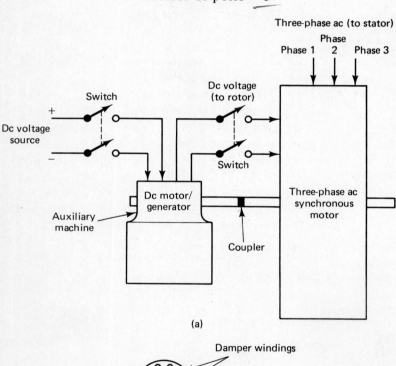

(a)

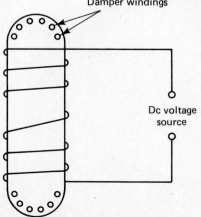

(b)

Figure 7-31. Methods of starting three-phase synchronous motors: (a) auxiliary dc motor/generator used to accelerate motor; (b) squirrel cage or damper windings placed in rotor.

2. Three-phase ac power is applied to the stator of the synchronous motor.

3. When motor speed nears the synchronous speed of the motor, the dc power circuit is opened. At the same time, the line terminals of the auxiliary machine are connected across the slip ring/brush assembly of the synchronous motor.

4. The auxiliary machine now converts to generator operation. It supplies dc current to the rotor of the synchronous motor. The synchronous motor now acts as a prime mover to turn the auxiliary machine.

5. When the rotor is magnetized, it will "lock in step" or "synchronize" with the revolving stator field.

6. The speed of the motor will remain the same even with changes in load as long as the rotor is magnetized.

Another starting method is shown in Fig. 7-31(b). This method uses "damper" windings. They are similar to the conductors of a squirrel-cage rotor. A damper winding is placed inside the laminated iron of the rotor. No auxiliary machine is required to start a synchronous motor when damper windings are used. The steps of operation are now simplified as follows:

1. Three-phase ac power is applied to the stator.

2. The motor operates like a three-phase induction motor due to the damper windings.

3. The motor speed builds up to slightly less than the speed. Remember that the speed of an induction motor is slower than synchronous.

4. Dc power is applied to the slip ring/brush assembly.

5. The rotor becomes magnetized. The motor then builds up speed until synchronous speed is reached.

6. The speed of the motor remains the same regardless of the load placed on the shaft of the motor.

An advantage of the three-phase ac synchronous motor is that it can be connected to a three-phase power system to increase power factor of ac circuits. Three-phase ac synchronous motors are sometimes used only to increase power factor. If no load is connected to the shaft of a three-phase ac synchronous motor, it is called a synchronous *capacitor*. In many cases, this motor is used for its constant-speed capability.

As load increases on a three-phase synchronous motor, the stator current will increase. The motor will remain "synchronized" (same speed) unless the load causes "pull-out" to take place. The motor would then stop rotating due to excessive load. Many three-phase synchronous motors are larger than 100 hp. They are used mainly for industrial applications that require constant speed.

Synchro Systems and Servo Systems

Synchro systems cause rotary motion to control angular position. Rotation of one shaft causes positioning of the second device as shown in Fig. 7-32. Synchro systems are used to control movement. It is possible to transfer rotary motion from one point to another without using mechanical linkage.

A synchro system is two or more motor/generator units connected together. When an operator turns the "generator" shaft of the unit to a certain position it automatically rotates the "motor" shaft to the same direction. The motor may be at a remote location. It is possible to have accurate control over long distances. An antenna rotor for a TV set is an example of a synchro system.

In synchro systems that require high torque, a servo system is used. A servo system is a special type of motor that controls movement of equipment. Synchro systems have two or more electromagnetic devices that look like small electric motors. They are connected together so that the position of the generator shaft determines the position of the motor shaft. Figure 7-32(a) shows a schematic diagram of a synchro unit. The generator and motor units look identical. The motor unit has a metal flywheel attached to its shaft to prevent shaft vibration. The letters G and M inside the electrical symbol are used for the generator and motor units.

Figure 7-32(b) shows the circuit of a synchro system. Single-phase ac voltage is used to power the system. The ac voltage is applied to the rotors of both the generator and motor. The stator coils are connected together in a wye connection.

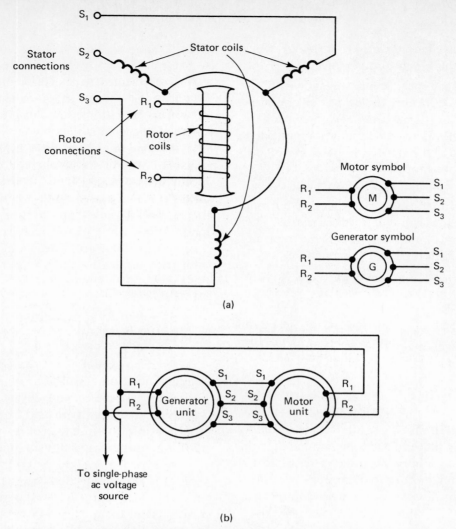

Figure 7-32. Synchro systems: (a) schematic diagram; (b) circuit. R_1, R_2, rotor connections; S_1, S_2, S_3, stator connections.

When power is applied, the motor will lock into position according to the location of the generator shaft. Both units remain in the same position until the generator shaft is turned. Turning the generator shaft 30° in a clockwise direction will cause the motor unit to move 30° clockwise. If calibrated dials are attached to the shaft of each unit, they would show the same movement.

The stator coils of a synchro unit are wound inside a metal housing. The coils act like the secondary windings of a transformer to transfer energy. The rotor of a synchro unit is wound on a laminated core. Slip rings and brushes on the shaft are used to supply ac power to the rotor. The rotor acts like the primary winding of a transformer.

When ac voltage is applied to the rotor of a synchro unit, an alternating magnetic field is produced. By transformer action this field cuts across the stator coils. This induces a voltage in the stator windings. The position of the rotor determines the amount of voltage induced in the stator coils. The stator coils of the generator and motor are connected together as shown in Fig. 7-32(a). Voltage induced in the stator coils of the generator causes the same current flow in the stator coils of the motor. This causes the same magnetic field to develop the motor stator. The rotor of the motor unit will align itself with the magnetic field of the stator coils.

A change in rotor position of the generator unit affects the induced voltage of the stator coils. In this way, movement can be transferred to the motor through three small wires which connect the stators.

Synchro and servo systems are used to change the position or speed of an object. Position changes are needed for numerical controlled machinery and process control equipment in industry. Speed control

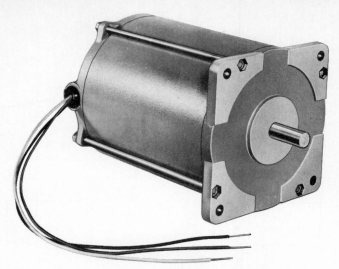

Figure 7-33. Servomotor. (Courtesy of The Superior Electric Co.)

is used for conveyor belt units, operating speed of machine tools, and disk or magnetic-tape drives for computers. A servomotor is shown in Fig. 7-33.

DC STEPPING MOTORS. Dc stepping motors are servomotors used to drive many types of computer-controlled machines. They convert electrical pulses into rotary motion. The shaft of a dc

Figure 7-34. Dc stepping motor construction. (Courtesy of The Superior Electric Co.)

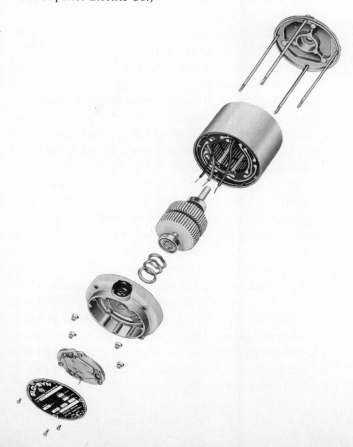

stepping motor rotates a specific number of degrees when a pulse is applied to it. The amount of movement caused by each pulse can be repeated. The shaft of a dc stepping motor is used to position industrial machinery accurately. The speed, distance, and direction of a machine can be controlled by a dc stepping motor. The stator construction and coil placement are shown in Fig. 7-34. The rotor is made of permanent-magnet material. The permanent-magnet rotor is placed between two series-connected stator coils as shown in Fig. 7-35. With power

Figure 7-35. Stator coil placement of a dc stepping motor: (a) one set of stator coils—rotor can move in either direction; (b) two sets of stator coils—rotor will move in a clockwise direction.

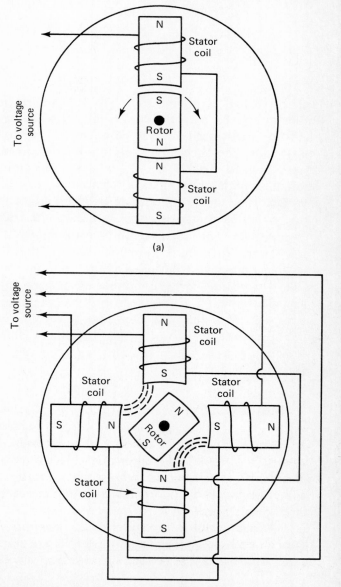

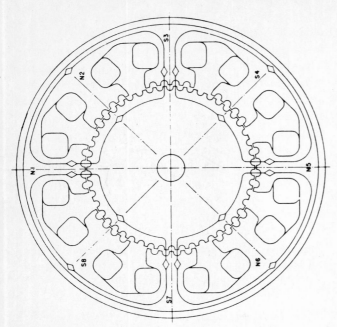

Figure 7-36. Dc stepping-motor rotor. (Courtesy of The Superior Electric Co.)

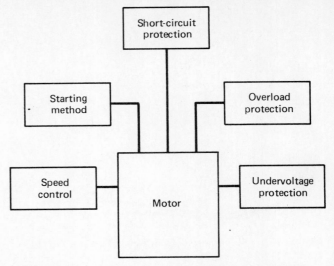

Figure 7-37. Block diagram of a motor control circuit.

applied to the stator, the rotor can be repelled in either direction. With the stator polarities shown, the rotor aligns itself midway between two pairs of stator coils. The direction of rotation is determined by the polarities of the stator coils. Adding more stator coils to this motor makes positioning the rotor very accurate. Operation of a dc stepping motor is achieved by electrical pulses from a computerized power supply. The stepping-motor rotor shown in Fig. 7-36 has 50 teeth. It takes 200 pulses or "steps" for each complete revolution. The amount of movement is determined by the number of teeth on the rotor and the electrical pulse sequence. A dc stepping motor that takes place 200 steps for one revolution will move 1.8 degrees per step (360° divided by 200). Dc stepping motors are very important for use in industry.

Motor Control Basics

A block diagram of a motor control circuit is shown in Fig. 7-37. The components used to control motors include circuit breakers, switches, magnetic contactors, protective overloads, limit switches, and many other devices.

On many motor control circuits, *pushbutton switches*, such as those shown in Fig. 7-38, are used. A pushbutton may be pressed to cause a motor to start. Another pushbutton may be used to cause it to stop. Pushbuttons either apply power from a source or open the power circuit to stop a motor.

Limit switches may also be used to control a motor's on or off operation. Limit switches are used to limit the movement or travel of a piece of equipment. For example, they could be used in an elevator to cause it to stop at the right floor.

Fuses or circuit breakers are also part of a motor control circuit. Fuses are made so that they will open a circuit by melting. If the current flowing through a circuit exceeds a safe limit, the fuse element will melt. A fuse is used to prevent excessive current form flowing through a motor. It is a form of protective control. *Circuit breakers* have the same function. A circuit breaker must be reset after it is opened. If a circuit breaker is rated at 20 A, currents greater than that amount will disconnect the motor power source. By automatically disconnecting the power source, circuit breakers act as a protective control. They are usually preferred over fuses since they do not have to be replaced.

Magnetic contactors are another type of motor control device. Contactors are magnetically closed by a solenoid mechanism. They are opened by a spring or by gravity. They require only a small current to operate and usually control a much larger current flow through a motor.

Another type of motor-starting device is called a manual starter. They are similar to magnetic contactors except that no magnetic circuit is used. A manual motor starter, such as the one shown in Fig. 7-39, is used to turn a motor on and off.

(a)

(b)

Figure 7-38. Pushbutton switches. [(a) Courtesy of Cutler-Hammer Co.; (b) courtesy of Furnas Electric Co.]

Figure 7-39. Manual motor starter. (Courtesy of Furnas Electric Co.)

Review

1. How do incandescent, fluorescent, and vapor lighting systems differ?
2. How do resistive, inductive, and capacitive heating methods differ?
3. Discuss basic motor action, which is how torque is produced by a motor.
4. What is meant by (a) horsepower and (b) torque?
5. Discuss the following types of motors: (a) permanent magnet, (b) series, (c) parallel, and (d) compound.
6. Discuss the induction principle of single-phase motor operation.
7. How is the speed of an ac motor calculated?
8. Discuss the operation of the following types of ac motors:
    a. Universal motor
    b. Split-phase motor
    c. Capacitor-start, induction motor
    d. Shaded-pole motor
    e. Single-phase synchronous motor
    f. Three-phase induction motor
    g. Three-phase synchronous motor

9.   How are three-phase synchronous motors commonly started?

10.   What are three-phase ac synchronous motor applications?

Student Activities

AC Induction Motor Analysis (see Fig. 7-20)

In this activity a split-phase ac motor will be analyzed. These motors are used for many applications.

1.   Obtain an ac induction motor, an ac ammeter, and a VOM. Record all data on a sheet of paper.

2.   The run windings and start winding are connected in parallel. A centrifugal switch is in series with the start windings. Dual-voltage motors (120/240 V, for example) have two set of run windings. They are connected in parallel for low-voltage operation and in series for high-voltage operation. What *type* of motor is being tested?

3.   Observe the arrangement of the start and run windings around the stator assembly. How many run-winding coil sets (poles) does your motor have? How many starting-coil sets?

4.   Trace the external connecting wires of the motor to their origin within the stator. Momentarily disconnect the necessary wires to make the following resistance measurements. Use a low-resistance range of the meter. Measure and record run-winding resistance and start-winding resistance.

5.   Make sure that all wires are properly connected so that the motor can be operated. Connect an ac ammeter in series with ac line.

6.   Apply the rated voltage to the motor from the power source. Observe and record the *starting current* drawn by the motor when it is initially turned on.

7.   Record the current flow when the motor is running.

8.   Turn off the power and perform the necessary procedure to reverse the direction of rotation of the motor.

9.   Turn on the motor and test the motor for operation in the reverse direction.

10.   Discuss the operation of an ac induction motor on a sheet of paper.

11.   Turn in the data and discussion to the teacher.

Universal Motor Analysis (see Fig. 7-17)

In this activity the operation of a universal motor will be studied. It will operate with either alternating current or direct current applied. Universal motors are one of the most used types of motors. Applications of the universal motor include electric motors, electric saws, and household items such as mixers, blenders, and sewing machines. The universal motor is one of the few types of variable-speed motors which operate on alternating curent. By varying the voltage, the speed of the motor will change.

1. Obtain a universal motor and a VOM.

2. Look at the motor construction. Note that the armature and field coils are connected in series with the power source.

3. Connect the power terminals of the universal motor to an ac voltage source of its rated value.

4. Turn on the voltage and measure the starting current. The current should be measured with the meter in series with either one of the power terminals.

Starting current = _____ A

5. Now record the running current.

Running current = _____ A

6. Look toward the shaft of the motor and observe the direction of rotation.

7. Disconnect the motor and the direction of the motor.

8. Plug the motor into the power source and check for proper operation.

9. Discuss the operation of a universal motor on a sheet of paper.

10. Turn in the data and discussion to the teacher.

Motor Repair and Troubleshooting

Motor troubleshooting should be a systematic process. For good troubleshooting one should have a good understanding of the type of motor that is being repaired. With this knowledge of operation and some test equipment, many repair jobs can be completed.

Some of the following instruments are used to make tests on motors: (1) tachometers—speed tests; (2) meggers—insulation tests; (3) VOMs—resistance, current, and voltage tests; and (4) growlers—wound armature tests.

In this activity, a motor will be tested and repaired. Many parts of this procedure listed below may not apply. However, try to follow the steps.

1. Obtain a motor that does not function properly, and a VOM. Record all data on a sheet of paper.

2. Examine the construction of the motor.

3. Find the motor nameplate and record the following information:
   a. Manufacturing company
   b. Motor type
   c. Identification number
   d. Model number
   e. Frame type
   f. Number of phases (ac)
   g. Horsepower
   h. Cycles (ac)
   i. Speed (rpm)
   j. Voltage rating
   k. Current ratings (amperes)
   l. Thermal protection
   m. Temperature rating (°C)
   n. Time rating
   o. Other information

4. Determine the type of motor, such as ac induction, shaded-pole, universal, or capacitor-start. Consult with the teacher to determine the specific type. What is the specific type?

5. Draw a schematic of the motor. This will provide a general reference for later tests.

6. Determine the proper type of power source and voltage. Apply rated voltage to the motor for a very brief period of time. Attempt to find the specific problem.

7. What problem seems to exist in the motor?

8. What are some possible remedies to stop the problem?

9. If the motor must be disassembled, remove the end plates carefully. Be careful with the connecting wires and bearings. Remove the rotor assembly at this time. Check the bearings.

10. If connecting wires are removed, use tape labels or some other means to label the proper terminal connections.

11. It may be necessary to make some resistance tests with a VOM to find the condition of the stator windings. (Example: Ac single-phase induction motor; start winding, 20 Ω resistance; run windings, 5 Ω resistance.) Of course, those resistances will vary. Record your resistance ratings.

Winding resistances = _____ .

12. Now inspect all wiring connections, splices, and other mechanical connections for possible opens or shorts.

13. Inspect the stator of the motor for excessive dirt, oil, or other damage.

14. Check the rotor for possible troubles.

15. If the rotor is a squirrel-cage type, it may be in need of cleaning with fine sandpaper.

16. If a wound rotor is used, check the commutator for high mica insulation, roughness, irregular wear, or other damage. It may be necessry to turn the commutator on a metal lathe, clean it with sandpaper, or remove high mica with an undercutting tool.

17. Inspect the rotor windings for damaged insulation. The rotor windings may be tested for damaged insulation with a megger. Wound rotors may be tested for shorts, grounds, or open circuits with a growler.

18. If the motor has brushes, check for extensive wear, oil, grease, chipping, or improper contact.

19. Perform any other tests that may be needed to locate the trouble. Consult with your teacher.

20. Perform the necessary procedure to repair the motor.

21. Check with your teacher when you feel you have completed the troubleshooting procedure and have repaired the motor malfunction.

22. On a sheet of paper, write a detailed discussion of the troubleshooting process used in this activity.

23. Turn in the discussion to the teacher for grading.

Motor Problems

1. Solve the following problems which deal with several types of electric motors. Record all your answers on a sheet of paper.

    a. A shunt-wound dc motor converts 300 W of electrical power when connected to a 120-V dc source. Its field resistance is 85 Ω. Find:

        (1) Line current   **(3)** Armature current

        (2) Field current

    b. A shunt-wound dc motor draws 40 A of current from a 120-V source. It has a field resistance of 30 Ω and a armature resistance of 1 Ω. Find:

        (1) Field current   **(3)** Total power

        (2) Armature current

    c. What is the horsepower of a running motor at 3450 rpm that has a torque of 35 ft-lb?

    d. The no-load speed of an induction motor is 1790 rpm and full-load speed is 1725 rpm. What is its synchronous speed?

    e. What is the % slip of the motor given in problem d?

    f. What is the synchronous speed of a 12-pole 60-Hz three-phase ac motor?

    g. What is the synchronous speed of a 12-pole 60-Hz single-phase ac motor?

2. Turn in your answers to the teacher for grading.

Electrical Instruments

Instruments of many types are in use today. They measure many different quantities. Instruments are used to measure electrical quantities, as well as other physical quantities.

All instruments have some common characteristics. Some quantity is monitored either periodically or continously. Some type of visual display of the quantity must be presented. Several types of instruments for measuring electrical quantities are used. The basic type of instruments may be classified as: (1) hand-deflection instruments, (2) comparison instruments, (3) cathode-ray tube instruments, (4) numerical readout instruments, and (5) chart recording instruments.

IMPORTANT TERMS

The following terms should be reviewed before studying electrical instruments in this chapter.

Ammeter. A meter used to measure current flow.

Bridge circuit. A circuit with groups of components that are connected by a "bridge" in the center and used for precision measurements.

Cathode-ray tube (CRT). A large vacuum tube in which electrons emitted from a cathode are formed into a narrow beam and accelerated so that upon striking a phosphorescent screen, a visible pattern of light energy is produced.

D'Arsonval meter movement. The internal portion of a meter which has a stationary permanent magnet, an electromagnetic coil, and a pointer that moves in direct proportion to the current flow through the coil.

Galvanometer. A meter which measures very small current values.

Kilowatt-hour (kWh). 1000 watts per hour, a unit of measurement for electrical energy.

Loading. The effect caused when a meter is connected into a circuit that causes the meter to draw current from the circuit.

Megger. A meter used to measure very high resistances.

Moving coil. *See* D'Arsonval meter movement.

Multiplier. A resistance connected in series with a meter movement to extend the range of voltage it will measure.

Ohmmeter. A meter used to measure resistance.

Ohms per volt rating (Ω/V). The rating that indicates the loading effect a meter has on a circuit when making a measurement.

Oscilloscope. An instrument that has a cathode-ray tube to allow a visual display of voltages.

Sensitivity. *See* Ohms per volt rating.

Shunt. A resistance connected in parallel with a meter movement to extend the range of current it will measure.

Voltmeter. A meter used to measure voltage.

Watt-hour. A unit of energy measurement equal to one watt per hour.

Watt-hour meter. A meter that measures the rate of power conversion.

Wattmeter. A meter used to measure power conversion.

Wheatstone bridge. A bridge circuit that makes precision resistance measurements by comparing an "unknown" resistance with a known or standard resistance value.

X axis. A horizontal line.

Y axis. A vertical line.

HAND-DEFLECTION INSTRUMENTS

Instruments that rely on the motion of a hand or pointer are called hand-deflection instruments. The volt-ohm meter (VOM) is one type of hand-deflection instrument. The VOM is an instrument used for measuring several electrical quantities. Single-function meters, such as those shown in Fig. 8-1, are also used to measure electrical quantities. They measure only one quantity.

The basic part of an electrical hand-deflection meter is called a *meter movement*. Physical quan-

(a)

(b)

Figure 8-1. (a) Dc milliammeter; (b) dc microammeter. (Courtesy of Triplett Corp.)

tities such as airflow or fluid pressure are also measured by hand-deflection meters. The movement of the hand or pointer over a calibrated scale indicates the quantity being measured.

Many meters use the *D'Arsonval* or *moving-coil* type of meter movement. The construction details of this meter movement are shown in Fig. 8-2. The hand or pointer of the movement stays on the left side of the calibrated scale. A moving coil is located inside a horseshoe magnet. Current flows through the coil from the circuit being tested. A reaction occurs between the electromagnetic field of the coil and the permanent magnetic field of the horseshoe magnet. This reaction causes the hand to move toward the right side of the scale. This moving-coil meter movement operates on the same principle as an electric motor. It can be used for single-function meters that measure only one quantity. It can also be used for multifunction meters, such as VOMs, VTVMs (vacuum-tube voltmeters), or TVMs (transistorized voltmeters), which measure more than one quantity. The D'Arsonval meter movement can be used to measure voltage, current, or resistance. Resistors of proper value are connected to the meter movement for making these measurements.

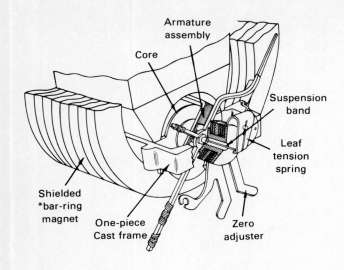

Armature
assembly

Core

Suspension
band

Leaf
tension
spring

Shielded
*bar-ring
magnet

One-piece
Cast frame

Zero
adjuster

SUSPENSION DRAWING

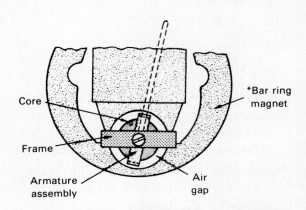

Core

Frame

Armature
assembly

*Bar ring
magnet

Air
gap

PARTIAL VIEW SHOWING AIR GAP

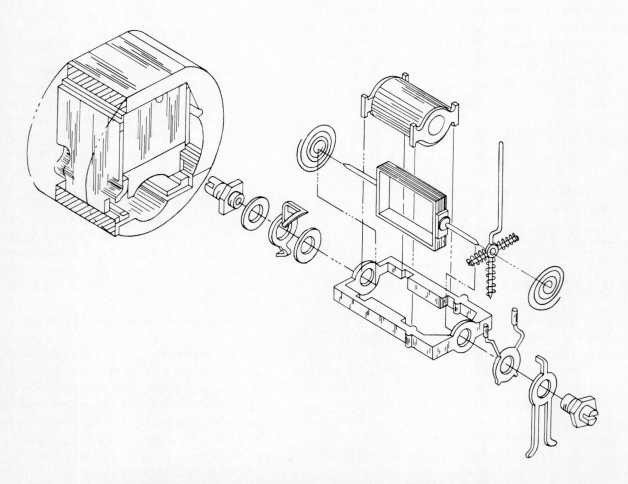

EXPLODED VIEW OF DC PIVOT AND JEWEL TYPE METER

Figure 8-2. D'Arsonval meter movement.

Measuring Direct Current

The moving-coil meter movement can be used to measure any value of direct current by using *shunt resistors* in parallel with the movement. A diagram of a basic movement with a shunt resistor is shown in Fig. 8-3. The coil of the meter movement will not handle high currents. The typical current rating of a meter movement ranges from 10 μA to 10 mA. This current rating (I_M) is the amount of current needed to move the pointer to the extreme right side of the scale. This is called full scale. When this current value is smaller, more turns of wire are used for the coil. More turns allow a strong electromagnetic field to be developed. As the current rating (I_M) decreases, the resistance (R_M) of the movement increases due to the greater length of wire. The values of I_M and R_M must be known to design a meter circuit that will measure currents higher than the value of I_M.

A shunt resistor is placed in parallel with the meter movement, as shown in Fig. 8-3. Some of the current flowing through the external circuit is "shunted" through the resistor. For this reason the resistor is called a *shunt resistor* (R_{SH}) The value of shunt resistance must be precisely calculated. Its value creates conditions so that ranges of current above the value of the meter movement can be measured. The value of shunt resistance is found by using the formula

$$R_{SH} = \frac{I_M \times R_M}{I_{SH}}$$

where R_{SH} is the shunt resistance in ohms, I_M is the full-scale current of rating of the meter movement in amperes, R_M is the resistance of the meter movement in ohms, and I_{SH} is the current flow through the shunt resistor in amperes.

For example, a 1.0-mA, 100-Ω meter movement could be used to measure 10 mA. The shunt resistance value is found as follows:

$$R_{SH} = \frac{I_M \times R_M}{I_{SH}} \qquad 1 \text{ mA} = 0.001 \text{ A}$$
$$= \frac{0.001 \times 100}{0.009} \qquad 9 \text{ mA} = 0.009 \text{ A}$$
$$= 11.11 \ \Omega$$

The value of I_{SH} is 9 mA (0.009 A) or 10 mA − 1 mA = 9 mA. This amount of current must be shunted through the shunt resistor. At the maximum current of the 10 mA range, I_M equals 1.0 mA. A *multirange* ammeter is designed by using several values of shunt resistance and a switching arrangement as shown in Fig. 8-4. Multirange meters usually have only one scale. They are either directly read or have a multiplying factor so that the scale may be easily read for each range. Scales were discussed in Chapter 2. This type of scale is called a *linear* scale. It has the same distance between all division marks.

Measuring DC Voltage

Hand-deflection meter movements are also used to measure dc voltage. Meter movements respond to current flow through the electromagnetic coil. As voltage increases, current increases in the same proportion. Voltage values also produce accurate readings on a calibrated meter scale. To measure voltage, a resistor is placed in series with the meter movement. This resistor, shown in Fig. 8-5, is called a *multiplier* resistor. The purpose of the multiplier is to adjust the value of current across the

Figure 8-3. Meter movement with a shunt resistor to measure current.

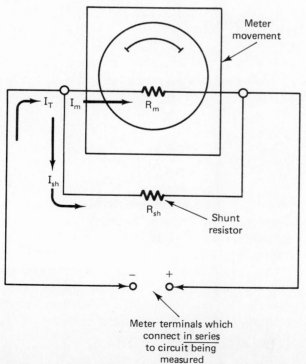

Meter terminals which connect in series to circuit being measured

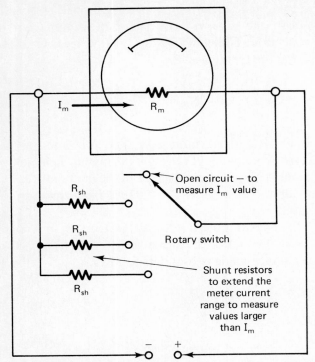

Figure 8-4.   Switching arrangement for a multirange current meter.

Figure 8-5.   Meter movement with a multiplier used to measure voltage.

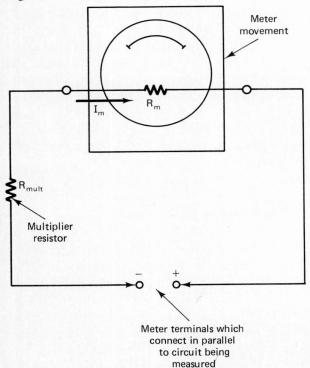

meter movement. At a certain voltage value ($I_M \times R_M$), full-scale deflection (E_{FS}) on the meter scale will result. The formula to find full-scale deflection is

$$E_{FS} = I_M \times R_M$$

A range of dc voltage from 0 to 10 V could be measured using 1.0 mA, 100-Ω meter movement. With 1.0 mA flowing through the 100-Ω meter movement (full scale), 0.1 V is applied across the meter movement ($E_{FS} = 1\,\text{mA} \times 100\,\Omega = 0.1\,\text{V}$). To measure 10 V at full scale, a multiplier resistance is added in series with the meter movement to drop 9.9 V. This value is found by using the formula

$$R_{\text{Mult}} = \frac{E_{\text{Mult}}}{I_T}$$
$$= \frac{9.99}{0.001}$$
$$= 9900\ \Omega$$

Another way to find the value of a multiplier resistance is to calculate the total resistance in the circuit. The values for the circuit with 10 V applied are found as follows:

$$R_T = \frac{E_T}{I_T}$$
$$= \frac{10\ \text{V}}{0.001\ \text{A}}$$
$$= 10{,}000\ \Omega$$

The total resistance of the circuit is $R_T = R_M + R_{\text{Mult}}$, so

$$R_{\text{Mult}} = R_T - R_M$$
$$= 10{,}000\ \Omega - 100\ \Omega$$
$$= 9900\ \Omega$$

A switching arrangement is used for several voltage ranges that use the same meter movement. A switching arrangement for a multirange voltmeter is shown in Fig. 8-6.

An important characteristic of meters is called *sensitivity*. The sensitivity of a meter increases as the current rating (I_M) decreases. Voltmeters are connected in parallel with a circuit to measure voltage.

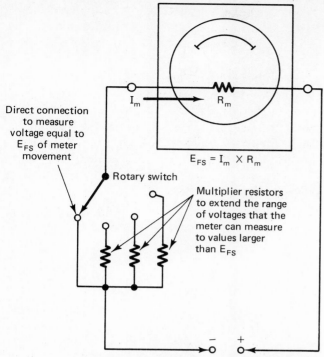

Direct connection to measure voltage equal to E_{FS} of meter movement

I_m R_m

$E_{FS} = I_m \times R_m$

Rotary switch

Multiplier resistors to extend the range of voltages that the meter can measure to values larger than E_{FS}

− +

Figure 8-6. Switching arrangement for a multirange voltmeter.

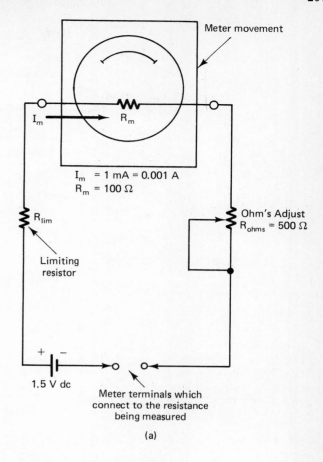

Meter movement

I_m R_m

$I_m = 1 \text{ mA} = 0.001 \text{ A}$
$R_m = 100 \ \Omega$

R_{lim}

Limiting resistor

Ohm's Adjust
$R_{ohms} = 500 \ \Omega$

+ −

1.5 V dc

Meter terminals which connect to the resistance being measured

(a)

1. Find total resistance (R_T) which is needed in the circuit to cause full-scale deflection of the meter when the meter terminals are touched together:

$$R_T = \frac{E_T}{I_m} = \frac{1.5 \text{ V}}{1 \text{ mA}} = \frac{1.5 \text{ V}}{0.001 \text{ A}} = 1500 \ \Omega$$

2. Find the value of limiting resistor (R_{lim}) needed in the circuit:

$$R_{lim} = R_T - R_{Ohm's \ Adjust} - R_m$$
$$= 1500 \ \Omega - 500 \ \Omega - 100 \ \Omega$$
$$= 900 \ \Omega$$

(b)

Figure 8-7. (a) Ohmmeter circuit with a meter movement used to measure resistance. (b) Ohmmeter circuit calculations.

Part of the circuit flows through the meter to make the needle deflect. The current that flows through the meter should be very small. More sensitive meters draw less current from the circuit. Sensitivity is measured in ohms per volt and is equal to $1/I_M$. Remember that I_M is the current required for full-scale deflection of the meter. A 1-mA movement would have an ohms per volt rating of 1/0.001 or 1000 Ω/V. This means that the meter would have 1000 Ω of resistance for its 1-V range. Sensitivities of meter movements range from as low as 100 Ω/V to as high as 200,000 Ω/V. Low-sensitivity meters should not be used for making accurate measurements.

Measuring Resistance

Resistance is also measured by using a meter movement. A ohmmeter circuit is shown in Fig. 8-7(a). The resistance to be measured is connected in series with the meter circuit. A 1.5-V cell is used to supply current to the meter circuit. The scale of the meter is calibrated so that the movement of the pointer indicates a value of resistance. When the meter probes are touched together, the meter pointer

is adjusted to full scale. An "ohms" scale is shown in Fig. 8-8(a). The full-scale mark (right side) indicates zero resistance. As higher values of resistance are measured, less current flows through the meter circuit. The ohm-meter pointer deflects less for higher resistance values. The "Ohms Adjust" control is a potentiometer used to adjust the meter pointer. Remember that an ohmmeter must be "zeroed" before making a measurement. The pointer is ad-

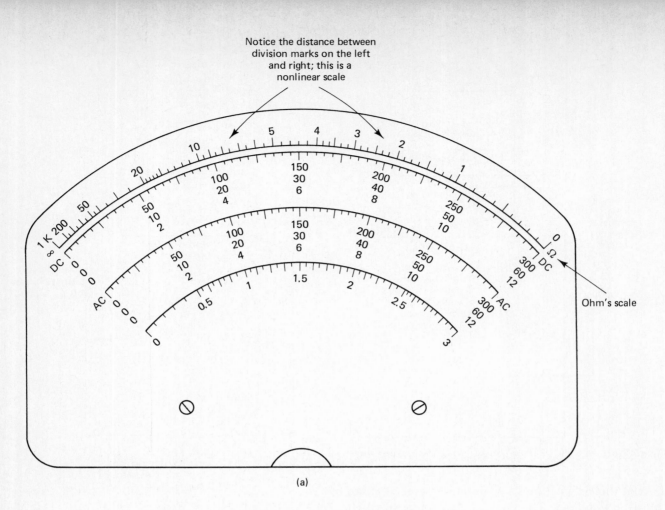

(a)

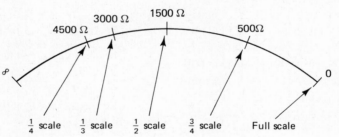

(b)

Meter deflection	Current through meter	Resistance indicated
Full-scale	1 mA	0 Ω
One-half scale	0.5 mA	1500 Ω
One-third scale	0.33 mA	3000 Ω
One-fourth scale	0.25 mA	4500 Ω
Three-fourths scale	0.75 mA	500 Ω

Figure 8-8. (a) Ohm's scale of a meter; (b) marking or calibrating an Ohm's scale.

justed to the zero on the scale when the probes are touched together. The purpose of the ohms adjust control is to allow accurate measurements as the voltage of the battery in the circuit changes. The left side of the scale indicates *infinite* resistance. This is an open circuit which has no current flow.

In the ohmmeter circuit of Fig. 8-7(a), a 1.0-mA 100-Ω meter movement is used. A value of 900 Ω for the current-limiting resistor (R_{Lim}) is calculated in Fig. 8-7(b). The total resistance of the meter circuit will be $R_M + R_{Lim} + R_{Ohms}$, or 100 Ω + 900 Ω + 500 Ω = 1500 Ω. When the meter probes are touched together, the total current in the circuit will be 1 mA:

$$I_T = \frac{E_T}{R_T}$$

$$= \frac{1.5 \text{ V}}{1500 \text{ }\Omega}$$

$$= 0.001 \text{ A (1.0 mA)}$$

One milliampere is the current (I_M) for full-scale deflection of the meter. The "ohms" scale is calibrated to read zero ohms at the full-scale deflection point of the meter.

The measured resistance that causes half-scale deflection of the meter pointer is equal to the total resistance of the ohmmeter circuit. To measure a 1500-Ω resistance, it is connected between the meter probes. The total circuit resistance is then 3000 Ω. The meter pointer deflects to only one-half of full scale. With twice as much resistance in the circuit, there is only half as much current. The center of the ohms scale is marked "1500" Ω, as shown in Fig. 8-8(b).

If the 3000-Ω resistance is measured, the total circuit resistance is 4500 Ω. This is three times the resistance of the meter circuit. The current in the circuit is then only one-third of the full-scale current (33 mA). The meter pointer deflects to only one-third of full scale when 3000 Ω is measured. The point on the "ohms" scale for 3000 Ω is then marked. The amount of current can be calculated for any value of measured resistance. The "ohms" scale is marked according to the current value. The table in Fig. 8-8(b) shows some values used to calibrate the "ohms" scale of a meter. This type of scale is called a *nonlinear* scale. It has division marks which are farther apart on the right than on the left side.

There are limits to the amount of resistance an ohmmeter can measure. Several values can be measured by using a multirange ohmmeter that has a small battery (usually 1.5 V) for the ×1 and ×100 ranges, and a larger battery (usually 9 or 30 V) for the ranges of ×1000 and higher.

Multifunction meters are usually designed to measure voltage, current, and resistance. Most hand-deflection meters used today have one meter movement with internal circuits to make many types of measurements. A rotary switch is often used for multifunction meters to select the quantity to be measured and the proper range.

Measuring Electrical Power

Electrical power is measured with a *wattmeter*. A meter movement called a *dynamometer* is used for most wattmeters. Figure 8-9 shows a meter movement that has two electromagnetic coils. One coil, called the current coil, is connected in series with the circuit to be measured. The other coil, called the potential coil, is connected in parallel with the circuit. The strength of each electromagnetic field affects the movement of the meter pointer. The operating principle of this movement is like the moving-coil type. The main difference is that an electromagnetic field rather than a permanent-magnet field surrounds the moving coil. Dc power is found by multiplying voltage and current ($P = E \times I$). Ac power is found by multiplying voltage and current and power factor of the load ($P = E \times I \times$ PF). The *true power* of an ac circuit is read with a wattmeter. When a load is either inductive or capacitive, remember that the true power is less than apparent power ($E \times I$). The true power is the actual power converted by the load.

Measuring Electrical Energy

Wattmeters with dynamometer movements are used to measure electricity power. They monitor the voltage and current in a circuit. They are used to measure the electrical power converted by some type of load. The amount of electrical energy converted into mechanical energy by a motor load is measured by connecting a wattmeter to the motor circuit.

The amount of electrical energy used *over a period of time* is measured by a *watt-hour meter*.

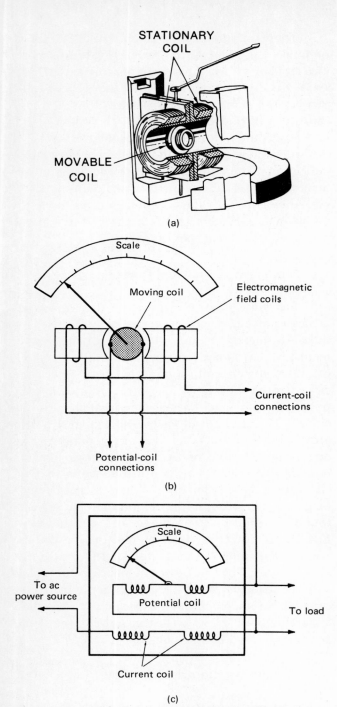

Figure 8-9 labels:
STATIONARY COIL
MOVABLE COIL
(a)

Scale
Moving coil
Electromagnetic field coils
Current-coil connections
Potential-coil connections
(b)

Scale
To ac power source
Potential coil
To load
Current coil
(c)

Figure 8-9. Meter used to measure electrical power: (a) dynamometer movement; (b) meter movement schematic; (c) circuit. (From Patrick/Fardo, *Energy Management and Conservation*. Prentice-Hall, Englewood Cliffs, N.J., 1982.)

The watt-hour meter shown in Fig. 8-10 has a small motor inside. The speed of the motor is increased as more current passes through it. The rotor is an aluminum disk that is connected to a numerical display. The display indicates the number of *kilowatt-hours* (kWh) of electrical energy used. Figure 8-10

Figure 8-10. Watt-hour meter. (Courtesy of Sangamo-Schlumberger.)

shows a dial display that is often used on watt-hour meters. Other types have a numerical display of the actual kilowatt-hours used.

Watt-hour meters are connected between the power lines coming into a building and the branch circuits inside the building. The electrical energy used by a building must pass through the kWh meter. The operation of a watt-hour meter is similar to a wattmeter. A voltage coil is connected in parallel with power lines to monitor voltage. A current is placed in series with one line to measure current. The voltage and current of the system affect the speed of the aluminum disk. A watt-hour meter is also similar to an ac induction motor. This stator is an electromagnet that has two sets of windings: the voltage windings and the current windings. The field developed in the voltage windings causes current to be induced into the aluminum disk. The disk's speed increases when the voltage or current of the system increases. Since voltage usually stays the same, the disk's speed increases as current increases. This causes the number display to increase also. Watt-hour meters monitor the true power converted in a system over a period of time.

Measuring Three-Phase Electrical Power

It is often necessary to monitor the three-phase power used by industrial and commercial buildings. A combination of single-phase wattmeters can be used to measure three-phase power, as shown in Fig.

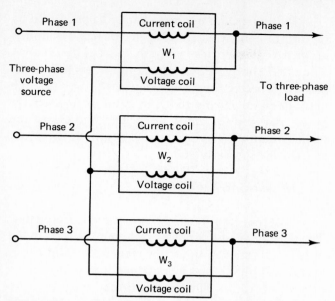

Figure 8-11. Using single-phase wattmeters to measure three-phase power.

8-11. The method shown is not too practical. The sum of the meter readings is found to calculate the total power of a three-phase power system. A meter called a *three-phase wattmeter* is used to measure the true power of a three-phase system. The power indicated on the meter is dependent on the voltage and current of all three phases of the system. Three-phase watt-hour meters are also available. They use three aluminum disks to monitor the three-phase power used by a system over a period of time. The rotation of the shaft is caused by a combination of the power of each phase. The kWh display increases as the three-phase power converted by the system increases.

Measuring Power Factor

Power factor is the ratio of the measured true power of a system to the apparent power (volts × amperes). To find the power factor of a system, a wattmeter, a voltmeter, and an ammeter may be used. The power factor could be found as PF = W/VA. It is more convenient to use a *power factor meter* when power factor must be measured.

The circuit of a power factor meter is shown in Fig. 8-12. A power factor meter is similar to a wattmeter. It has two armature coils that rotate due to their electromagnetic field strengths. The armature coils are mounted on the same shaft. They are placed about 90° apart. One coil is connected across

Figure 8-12. Circuit of a power factor meter.

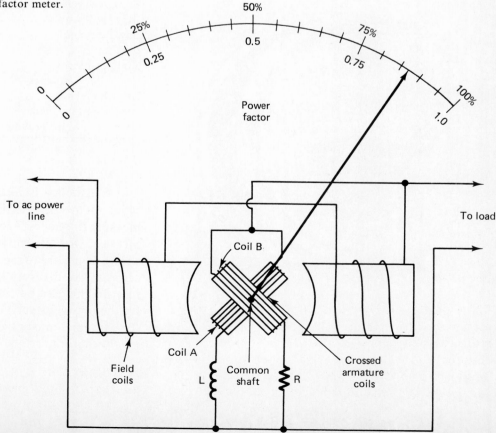

the power line in series with a resistance (R). The other coil is connected across the line in series with an inductance (L). The resistor in series with the coil produces a magnetic field due to the *in-phase* part of the power. The inductor in series with the other coil produces a magnetic field due to the out-of-phase part of the power. The scale of the meter is calibrated to measure power factors of 0 to 1.0, or 0 to 100%.

Measuring Power Demand

Figure 8-13 shows a power demand meter. This is an important instrument for industries and large commercial buildings. Power demand is found by using the formula

$$\text{power demand} = \frac{\text{peak power used (kW)}}{\text{average power used (kW)}}$$

Power demand is important since it shows the ratio of average power used to the peak value that a utility company must supply. Power demand is usually calculated over 15-, 30-, or 60-min intervals. It may be converted into longer periods of time.

Figure 8-13. Power demand meter. (Courtesy of Sangamo-Schlumberger.)

Industries and businesses may be penalized by the utility company if their peak power demand is far greater than their average demand. This is called a demand charge. Power demand meters help companies to better use their electrical power. A high peak demand means that the equipment used for power distribution must be larger. The closer the peak demand is to the average demand, the more efficient the power system.

Measuring Frequency

Another important measurement is frequency. The frequency of the power system must remain the same at all times. Frequency refers to the number of cycles of voltage or current that occur in a given period of time. The international unit used to measure frequency is the hertz (Hz). One hertz equals one cycle per second. A table of frequencies is shown in Table 8-1. The standard power frequency in the United States is 60 Hz. Some countries use 50 Hz.

Table 8-1
FREQUENCY VALUES*

Band	Frequency range
Extremely low frequency	30 Hz to 300 Hz
Voice frequency	300 Hz to 3 kHz
Very low frequency	3 kHz to 30 kHz
Low frequency	30 kHz to 300 kHz
Medium frequency	300 kHz to 3 MHz
High frequency	3 MHz to 30 MHz
Very high frequency	30 MHz to 300 MHz
Ultra-high frequency	300 MHz to 3 GHz
Super-high frequency	3 GHz to 30 GHz
Extremely high frequency	30 GHz to 300 GHz

*Power frequency = 60 Hz; AM radio band = 550–1600 kHz; FM radio band = 88–108 MHz.

Frequency is measured with several different types of meters. An electronic counter is used to measure frequency. Vibrating-reed frequency meters are used for measuring frequencies. Figure 8-14 shows some frequency meters. An oscilloscope is also used to measure frequency. Graphic recording instruments can be used to give a visual display of frequency over a period of time. The power industry, for example, must at all times monitor frequency of its alternators.

Figure 8-14. Frequency meters. (Courtesy of J-B-T Instruments.)

Ground-Fault Indicators

A ground-fault indicator is used to locate faulty grounding of equipment or power systems. Equipment must be properly grounded. A ground-fault indicator, shown in Fig. 8-15, is used to check for faulty grounding. Several grounding conditions exist that might be dangerous. Faulty conditions include (1) hot and neutral wires reversed, (2) open equipment ground wires, (3) open neutral wires, (4) open hot wires, (5) hot and equipment grounds reversed, and (6) hot wires on neutral terminals. Each of these conditions presents a serious problem. Proper wiring eliminates most of these problems. A check with a ground-fault indicator assures safe and efficient electrical wiring in a building.

Measuring High Resistance

A megohmmeter such as the one shown in Fig. 8-16 is used to measure very high resistances. These resistances are beyond the range of most ohmmeters. Megohmmeters are used to check the quality of insulation on electrical equipment in industry. The quality of insulation of equipment varies with age, moisture content, and applied voltage. Megohmmeters are similar to most ohmmeters. They use a hand-crank, permanent-magnet dc generator as a voltage source rather than a battery. The dc generator is cranked by the operator while making a test. Insulation tests should be made on all power equipment. Insulation breakdown causes equipment to fail. Insulation resistance value can be used to

Figure 8-15. GFI circuit tester. (Courtesy of Leviton Co.)

Figure 8-16. Megohmmeter used to measure high resistance. (Courtesy of James G. Biddle Instruments.)

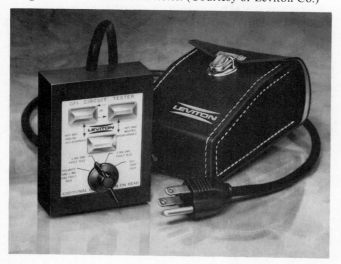

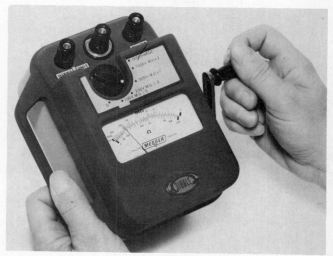

determine when equipment needs to be replaced or repaired. A decrease in insulation resistance over a period of time means that a problem could soon exist.

Clamp-On Meters

Clamp-on meters, such as the one shown in Fig. 8-17, are used to measure current in power lines. They are used to check current by clamping around a power line. They are easy to use for maintenance and testing of equipment. The meter is simply clamped around a conductor. Current flow through a conductor creates a magnetic field around the conductor. The magnetic field induces a current into the iron core of the clamp-on part of the meter. The meter scale is calibrated to indicate current flow. An increase in the current flow through a power line causes the current induced into the clamp-on part of the meter to increase. Clamp-on meters usually have voltage and resistance ranges to make them more versatile.

Figure 8-17. Clamp-on current meter. (Courtesy of VIZ Mfg. Co.)

COMPARISON INSTRUMENTS

Another type of indicator is called a comparison instrument. A comparison instrument compares a *known value* to an *unknown value*. The accuracy of a comparison instrument is much greater than hand-deflection instruments. They can be used to make very accurate measurements.

Wheatstone Bridge

The Wheatstone bridge is a comparison instrument. The circuit of a Wheatstone bridge is shown in Fig. 8-18. A voltage source is used with a sensitive zero-centered movement and a circuit called a resistance bridge. The resistance circuit has an unknown external resistance (R_x) which is the resistance to be measured. Resistor R_s is the known "standard" resistance. R_s is adjusted so that the current path of R_x and R_s is the same value as the path formed by R_1 and R_2. No current flows through the meter at this time. The meter indicates zero current. This is called a *null* condition. The bridge is said to be balanced. The value of R_s is marked on the instrument to compare to the value of R_x. Resistors

Figure 8-18. Circuit of a Wheatstone bridge.

Ratio arm

R_1 R_2

Sensitive current meter

R_x R_s

(resistance to be measured)

(standard or known value)

Dc voltage source

Circuit switch

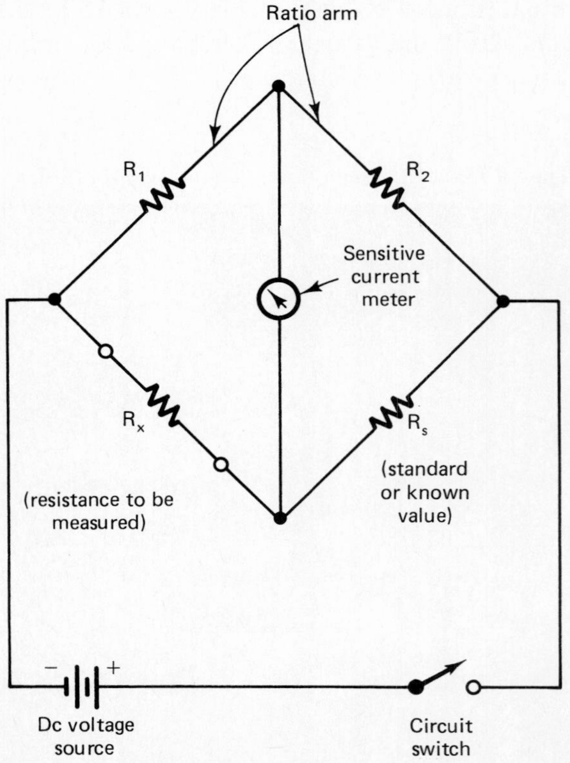

R_1 and R_2 are called the *ratio arm* of the bridge circuit. The value of the unknown resistance (R_x) is found by using the formula

$$R_x = \frac{R_1}{R_2} \times R_s$$

Wheatstone bridges measure resistance with precise accuracy. Other comparison instruments use the Wheatstone bridge principle. Comparison instruments compare unknown quantities with known quantities in the circuit of the instruments.

CATHODE-RAY-TUBE INSTRUMENTS

Cathode-ray-tube (CRT) instruments are usually called oscilloscopes. The use of oscilloscopes was discussed in Chapter 6. Oscilloscopes monitor voltages of a circuit visually. The basic part of the oscilloscope is called a cathode-ray tube.

Figure 8-19 shows the construction of the CRT and its internal electron gun assembly. A beam of electrons is produced by the cathode of the tube. Electrons are given off when the filament is heated by a filament voltage. The electrons have a negative

Figure 8-19. Construction of a CRT: (a) cutaway view; (b) electron beam detail.

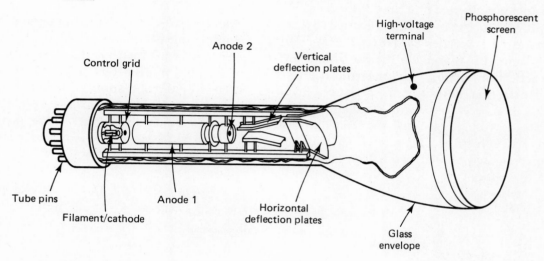

(a)

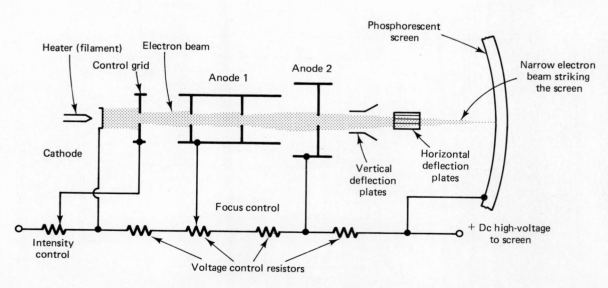

(b)

(−) charge. They are attracted to the positive (+) potential of anode 1. The number of electrons which pass to anode 1 is changed by the amount of negative (−) voltage applied to the control grid. Anode 2 has a higher positive voltage applied. The higher voltage accelerates the electron beam toward the screen of the CRT. The difference in voltage between anodes 1 and 2 sets the point where the beam strikes the CRT screen. The screen has a phosphorescent coating. When electrons strike the CRT, light is produced. The phosphorescent coating of the screen allows an electrical charge to produce light. The horizontal and vertical movement of the electron beam is controlled by *deflection plates*. With no voltage applied to either set of plates, the electron beam appears as a dot in the center of the CRT screen. The movement of the electron beam caused by the change of the plates is called *electrostatic deflection*. When a potential is placed on the horizontal and vertical deflection plates, the electron beam is moved. Horizontal deflection is produced by a circuit called a *sweep oscillator circuit* inside the oscilloscope. A voltage "sweeps" the electron beam back and forth across the CRT screen. The horizontal setting of an oscilloscope is adjusted to match the

frequency of the voltage being measured. Vertical deflection is caused by the voltage being measured. The voltage to be measured is applied to the vertical deflection of the oscilloscope. A block diagram of an oscilloscope circuit is shown in Fig. 8-20. Oscilloscopes are used to measure ac and dc voltages, frequency, phase relationships, distortion in amplifiers, and various timing and special-purpose applications. They also have some important medical uses, such as to monitor heartbeat.

NUMERICAL READOUT INSTRUMENTS

Many instruments have numerical readouts to simplify measurement. They make very accurate measurements. Instruments such as digital counters and digital multimeters (Fig. 8-21) are commonly used. Numerical readout instruments have internal circuits used to produce a ditigal display of the quantity being measured. These instruments are easy to read since a scale does not have to be interpreted.

Figure 8-20. Block diagram of an oscilloscope.

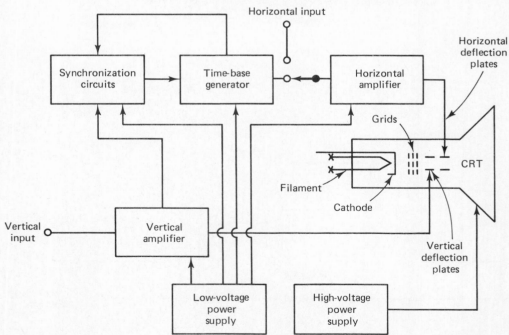

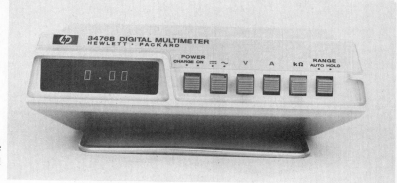

Figure 8-21. Autoranging digital multimeter—range adjusts automatically. (Courtesy of Hewlett-Packard Co.)

CHART RECORDING INSTRUMENTS

Most instruments are used in applications where no permanent record of the measured quantity is needed. Sometimes instruments must provide a permanent record of some quantity over a period of time. Typical chart recording instruments are shown in Fig. 8-22. Some chart recorders use pen and ink recorders and inkless recorders.

Pen and ink recorders have a pen that touches a paper chart. The ink leaves a permanent record of the measured quantity on the chart. The charts are either roll charts that revolve on rollers under the pen mechanism, or circular charts that rotate under the pen. Chart recorders sometimes use more than one pen to record several different quantities at the same time.

The pen of a chart recorder is connected to the meter movement. The pen is supplied with a source of ink. It is moved by the meter movement in the same way as the pointer of a hand-deflection meter. The charts used ordinarily have lines that indicate the amount of pen movement. Spaces on the chart are marked according to time periods. Charts are moved under the pen at a constant speed. Spring-drive mechanisms, synchronous ac motors, or dc servomotors are used to drive charts.

Inkless recorders may have voltage applied to the pen point. Heat is produced which causes a mark to be made on a sensitive paper chart. The advantage

Figure 8-22. Chart recording instruments. (Courtesy of Gould Inc., Instruments Division, Cleveland, Ohio.)

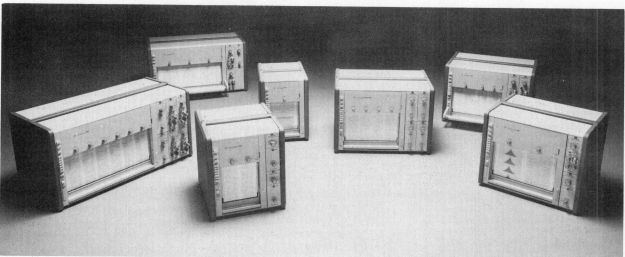

Figure 8-23. X-Y recorder. (Courtesy of Esterline Angus Instruments Div.)

of inkless recorders is that ink is not required. Ink must be replaced and it is often messy to use.

A special type of chart recording instrument is the X-Y recorder shown in Fig. 8-23. X-Y recorders are used to plot the relationship of two quantities. Often some variable quantity is plotted versus time. One variable quantity could also be plotted versus another variable quantity. One input is applied to cause vertical (*Y* axis) movement of the indicating pen. The other input causes horizontal (*X* axis) movement of the pen. The plot produced on the chart shows the relationship of the two quantities connected to the *X* and *Y* inputs.

Review

1. What are the five basic types of instruments?
2. Discuss the construction of a meter movement.
3. What is a single-function meter?
4. What is a multifunction meter?
5. How is the range of a meter movement extended to measure higher current values?
6. What is meant by the following abbreviations? (a) I_M, (b) R_M, (c) R_{SH}, and (d) I_{SH}
7. How is a meter movement used to mesure voltages?
8. What is meant by the following abbreviations: (a) E_{FS} and (b) R_{Mult}
9. What is meter sensitivity?
10. How is a meter movement used to measure resistances?
11. What is the difference between a linear and a nonlinear scale?
12. What is the purpose of the following in an ohmmeter circut? (a) R_{Lim}, (b) voltage source, and (c) Ohm's adjust resistance?
13. Discuss the method used to calibrate an Ohm's scale.
14. Describe the construction of an electrical power meter or wattmeter.

15. What is the difference between dc power and ac power measurement?

16. Describe a kilowatt-hour (kWh) meter.

17. How is three-phase ac power mesured?

18. How could power factor of an ac circuit be determined if a power factor meter is not available?

19. Why is power demand important to industries?

20. What is the purpose of a ground-fault indicator?

21. What is the purpose of a megger?

22. What is the purpose of a Wheatstone bridge?

23. Discuss the operation of an oscilloscope.

24. What is an advantage of using a numerical readout instrument?

25. Why are chart recording instruments used?

Student Activities

Measuring Meter Movement Resistance

1. Obtain a meter movement of any value, two potentiometers (1000 to 25,000 Ω values will usually work), a VOM, and a 6-V lantern battery.

2. Meter movement resistance *cannot* be measured with a VOM. It would possibly damage the moving coil since it has a low current rating.

3. Connect the circuit shown in Fig. 8-24. Record all data on a sheet of paper.

4. Disconnect the wire at point *A*, and adjust R_1 until meter reads *full scale*. R_1 should first be set to maximum resistance.

5. Connect the wire at point *A*. Adjust R_2 until the meter reads *half-scale*. The resistance of R_2 (from point *A* to point *B*) will now equal the resistance of the meter movement.

Figure 8-24. Circuit for measuring meter movement resistance.

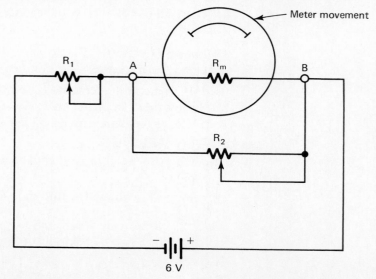

6. Do not disturb the setting of R_2. Remove it from the circuit. Measure the resistance from point A to point B with a VOM. $R =$ _____ Ω.

7. On a sheet of paper, discuss the construction and operation of a moving-coil meter movement.

8. Turn in the data and discussion to the teacher for grading.

Current Meter Design (see Fig. 8-3)

In this activity, any meter movement with known value of I_M and R_M can be used. If no meter movement are available, values of I_M and R_M can be selected by the teacher. In either case, the following procedure is used to design a current meter to measure a given value of current flow.

1. Obtain a meter movement, a 6-V battery, and a resistor (50 to 250 Ω values will work well). Record all data on a sheet of paper.

2. Using any meter movement with known resistance (R_M) and current (I_M), calculate the values of shunt resistance (R_{SH}) needed to extend its range to measure 200 mA of current. Use the following formula:

$$R_{SH} = \frac{I_M \times R_M}{I_{SH}}$$

3. Use the necessary resistor combinations connected together to construct the shunt.

4. Connect the shunt across the meter movement. Have the teacher check the R_{SH} value.

5. Connect the resistor (50 to 250 Ω) to the 6-V battery. Then put the meter which was designed *in series* with this circuit to test the meter.

6. Describe the construction of the meter movement on a sheet of paper.

7. If a 200-Ω 1-mA meter movement is used, what are the values of shunt resistance required to extend its range to measure the following currents? Show your calculations on a sheet of paper.
    a. 5 mA   c. 50 mA   e. 9.5 mA
    b. 10 mA   d. 100 mA   f. 1 A

8. Turn in the data to your teacher for grading.

Voltmeter Design (see Fig. 8-5)

In this activity, any meter movement with known values of I_M and R_M can be used. If no meter movements are available, values of I_M and R_M can be selected by the teacher. In either case, the following procedure is used to design a voltmeter to measure a given value of voltage. In this activity, 10 V will be used as the voltage range.

1. Obtain a meter movement and a 6-V battery. Record all data on a sheet of paper.

2. Find the voltage for full-scale deflection of the meter movement.

$$E_{FS} = I_M \times R_M = \text{_____} \text{ V}$$

3. Find the voltage that must be dropped across the multiplier resistor to measure 10 V.

$$E_{\text{Mult}} = 10 \text{ V} - E_{\text{FS}} = \underline{\hspace{1.5cm}} \text{ V}$$

4. Find the value of multiplier resistor needed in series with the meter movement to measure 10 V.

$$R_{\text{Mult}} = \frac{E_{\text{Mult}}}{I_M} = \underline{\hspace{1.5cm}} \Omega$$

5. Use the necessary resistor combination connected together to construct the multiplier.

6. Connect the multiplier in *series* with the meter movement.

7. Have the teacher check the R_{Mult} value.

8. Connect a 6-V battery across the meter which was designed to test the meter.

9. Calculate the values of multiplier resistance needed to extend the range of a 1-mA 100-Ω meter movement to measure the following voltages.
    a. 1 V, $R_{\text{Mult}} = \underline{\hspace{1.5cm}} \Omega$.
    b. 5 V, $R_{\text{Mult}} = \underline{\hspace{1.5cm}} \Omega$.
    c. 15 V, $R_{\text{Mult}} = \underline{\hspace{1.5cm}} \Omega$.
    d. 150 V, $R_{\text{Mult}} = \underline{\hspace{1.5cm}} \Omega$.

10. Turn in the data to the teacher for grading.

Ohmmeter Design [see Fig. 8-7(a)]

In this activity, any meter movement with known values of I_M and R_M can be used. If no meter movements are available, values of I_M and R_M can be selected by the teacher. In either case, the following procedure is used to design an ohmmeter.

1. Obtain a meter movement and a 1.5-V battery. Record all data on a sheet of paper.

2. Find the total resistance needed in the ohmmeter circuit using a 1.5-V battery.

$$R_T = \frac{1.5 \text{ V}}{I_M}$$

3. Find the value of limiting resistor (R_{Lim}) needed.

$$R_{\text{Lim}} = R_T - R_M$$

4. Use the necessary resistor combination connected together to construct the limiting resistor (R_{Lim}).

5. Connect R_{Lim} in series with the meter movement and a 1.5-V battery.

6. Have the teacher check the R_{Lim} value.

7. Draw a half-circle on a sheet of paper to represent the ohmmeter scale.

8. Mark the following points on the scale.
   a. Full-scale on the right side (0 Ω)
   b. Infinite (∞) on the left side ($\infty \Omega$)
   c. Half-scale ($R_{Lim} + R_M$ value) _____ Ω.

9. Take some known values of resistors (1 to 100 kΩ) and test the meter which was designed.

10. Find the value of R_{Lim} needed for a 1-mA 100-Ω meter movement using a 9-V battery.

11. Turn in the data to the teacher for grading.

Electronic Basics

Electronics is a unique part of the electrical field. It has made possible such things as radio, television, computers, pocket calculators, and stereo amplifiers. The term "electronics" takes its name from the electron. Electrons are the tiny, negatively charged particles of an atom. Electronics deals with the flow of these particles through a number of rather specialized devices. One division of this field deals with devices that respond to electrons released from a metal surface. This is called electron emission. Vacuum tubes and photoelectric devices operate on this principle. Solid-state electronics is another division. It deals with the flow and control of electrons through semiconductor materials. Transistors, diodes, and integrated circuits are solid-state devices. Electronics deals with devices of both types.

Vacuum tubes or electron tubes have been responsible for the main thrust of electronics for a number of years. In 1948, scientists at the Bell Telephone Laboratories were able to make pieces of solid material amplify an electric current. This has brought a significant change to electronics. A number of unique solid-state devices have been developed. Solid-state electronic applications have surpassed the vacuum tube. Complete electronic systems are now built on a single integrated-circuit chip. These systems have caused a significant change in the electronics field.

Electronic basics deals with the things that a person should know about electronic devices in order to use them. Solid-state basics will be discussed first. This deals with the structure of a semiconductor, crystal formation, covalent bonding, and current flow. Material theory of this type applies to all solid-state devices. Electron emission is then discussed. It is important today because of applications in communications and electronic control circuitry. Solid-state electronics and emission electronics both play an essential role in the field of electronics.

IMPORTANT TERMS

A number of new, and to some extent, unusual terms are used in this chapter. In general, these terms will assist the reader in understanding the material. Review each item very carefully before proceeding with the chapter. Refer back to a specific item if its meaning is not clear in the text.

Acceptor impurity. Atoms joined together by an electron-sharing process that causes a molecule to be stable.

Cathode. The electron-emitting element of a device.

Compound. Two or more elements combined together chemically.

Conduction band. The outermost energy level of an atom.

Covalent bonding. An element with more than four valence electronics that is used as a dopant for semiconductors. Acceptor atoms mixed with silicon makes an N material.

Donor impurity. An element with less than four valence electrons that is used as a dopant for semiconductors. Donor atoms mixed with silicon makes a P material.

Dopant. An impurity added to silicon or germanium.

Doping. The process of adding impurities to a pure semiconductor.

Electrode. A specific component or part of a unit such as the cathode of a vacuum tube.

Electron volt (eV). A measure of energy acquired by an electron passing through a potential of one volt.

Electrostatic force. An interaction between negative and positive charged particles.

Emission. The release of electrons from the surface of a conductive material.

Extrinsic. A purified material that has a controlled amount of impurity added to it when manufactured.

Forbidden gap. An area between the valence and conduction band on an energy-level diagram.

Intrinsic. A semiconductor crystal in a high level of purification.

Ion. An atom that has gained or lost an electron.

Ionic bond. A binding force that joins one atom to another in a way that fills the valence band of an atom.

Kinetic energy. Energy due to motion.

Light-dependent resistor (LDR). A photoconductive device that changes resistance with light energy.

Majority current carrier. Electrons of an N material and holes of a P material.

Metallic bond. A force such as a floating cloud of ions that holds atoms loosely together in conductor.

Minority current carrier. A conduction vehicle opposite to the majority current carrier, such as holes in an N material and electrons in the P material.

Photon. A small packet of energy or quantum of energy associated with light.

Quanta. A discrete amount of energy required to move an electron into a higher energy level.

Quantum theory. A theory based on the absorption and emission of discrete amounts of energy.

Secondary emission. The electrons released by bombardment of a conductive material with force.

Solid state. An area of electronics dealing with the conduction of current through semiconductors.

Stabilized atom. An atom that has a full complement of electrons in its valence band.

Thermionic emission. The release of electrons from a conductive material by the application of heat.

Valence band. The part of an energy-level diagram that deals with the outermost electrons of an atom.

Valence electrons. Electrons in the outer shell or layer of an atom.

SEMICONDUCTOR THEORY

Nearly any study of semiconductor devices begins with an investigation of atomic theory. This is purposely done to familiarize the reader with a number of ideas related to semiconductor operation.

In Chapter 1 we looked at electrons, protons, and neutrons. Each of these parts had a specific role in the construction of an atom. Elements are classified according to the number of particles they possess. No two elements have the same physical structure.

We also pointed out in Chapter 1 that the electrons of a particular atom are not an equal distance from the nucleus. They rotate in well-defined orbits. Each shell or orbit can only hold a certain number of electrons. Outer-shell electrons have a great deal to do with the electrical conductivity of an atom. Uncontrolled movement of these electrons causes certain atoms to be good electrical conductors. A material composed of this type of atom has many free electrons. An insulating material, by comparison, has a relatively small number of free electrons. Insulators tend to hold outer-shell electrons firmly in place. Very little electrical conductivity takes place in an insulating material.

Midway between conductors and insulators is a third classification of atoms known as semiconductors. Silicon and germanium are the most common semiconductor elements. Semiconductor compounds such as copper oxide, cadmium–sulfide, and gallium arsenide are also frequently used. Semiconductor materials are generally classified as type IVB elements. See the periodic table of elements in Appendix 1. This type of atom has four valence electrons. Atoms of this type are indecisive about their status. If they were to give up four valence electrons, stability could be achieved. If it could gain four electrons, stability could also be achieved. The conductive status of a semiconductor is subject to some degree of change.

Stability of an atom is an important concept in status of semiconductor materials. Atoms cannot possess more than eight electrons in the valence band. When exactly eight electrons appear in the valence band, an atom is considered to be stable. The valence electrons in a stable atom become tightly bound together. Atoms of this type are excellent insulators. They do not have any free electrons for electric conduction. Gases such as neon, argon, krypton, and xenon are examples of stabilized elements. They will not mix with other materials because of this condition. A stabilized gas is generally called inert.

Atoms having fewer than eight valence electrons are unstable. They have a tendency to seek electrons from neighboring atoms to become stable. Atoms with five, six, or seven valence electrons tend to borrow electrons from other atoms to seek stability. Atoms with one, two, or three valence electrons try to release these electrons to other atoms.

ATOM COMBINATIONS

The valence electrons of different atoms are frequently connected together to form stabilized molecules or compounds. The process of linking or interconnection is generally called *bonding*. Several kinds of bonding occur in atom combinations. *Ionic bonding*, *covalent bonding*, and *metallic bonding* are of particular importance in semiconductor theory.

When atoms bond together to form molecules, each atom is seeking stability. A stabilized condition occurs when the valence band contains eight electrons. In ionic bonding, the valence electrons of one atom join together with those of a different atom to become stable. When an atom has more than four valence electrons it is seeking additional electrons. This type of atom is often called an *acceptor*. Atoms containing less than four valence electrons try to rid themselves of these electrons. These atoms are called *donors*. Donor and acceptor atoms frequently combine together in ionic bonding. When this occurs the combination becomes stabilized.

Table salt is a common example of ionic bonding. Figure 9-1 shows an example of independent atoms and ionic bonding. The sodium (Na) atom donates its one valence electron to a chlorine (Cl) atom with seven valence electrons. Upon acquiring the extra electron, Cl immediately becomes overbalanced negatively. This causes the atom to become a *negative ion*. At the same time, the sodium atom loses its valence electron. The sodium atom then becomes a *positive ion*. Since unlike charges attract, the Na and Cl are bonded together by an ionizing force.

Copper oxide (Cu_2O) is also an example of ionic bonding. Two atoms of pure copper (Cu) combine with one atom of oxygen (O). Copper is a good electrical conductor. It has one valence electron in its atomic structure. Oxygen has six valence electrons. The combination of two copper atoms and one oxygen atom forms a stable copper oxide compound. The stability of this compound makes it a good insulating material.

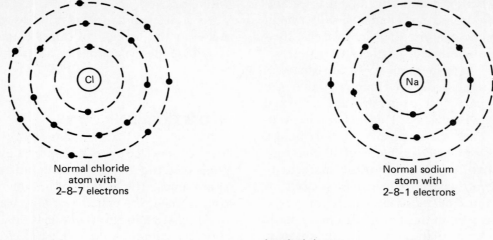

Normal chloride
atom with
2-8-7 electrons

Normal sodium
atom with
2-8-1 electrons

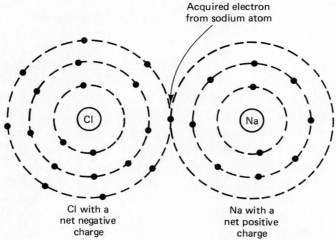

Acquired electron
from sodium atom

Cl with a
net negative
charge

Na with a
net positive
charge

Figure 9-1.   Ionic bonding of sodium chloride atoms.

Covalent bonding takes place when the valence electrons of neighboring atoms are shared with other atoms. When an atom contains four valence electrons, it may share one electron with four neighboring atoms. There is a unique difference in covalent bonding and ionic bonding. No ions are formed in covalent bonding. A *covalent force* is established between the two connecting electrons. These electrons alternately shift orbits between the atoms. A polarity shifting field alternates back and forth between each atom. This covalent force bonds the individual atoms together.

An illustration of covalent bonding is shown in Fig. 9-2(a). A single silicon atom is shown independently in Fig. 9-2(b). A simplified version of the silicon atom is used in the covalent bonding structure. Only the nucleus and valence electrons of each atom are shown in this arrangement. Individual atoms are bonded together by electron pairs. Five atoms are

needed to complete the bonding function. The bonding process extends out in all directions. This links each atom together in a *lattice* network. A crystal structure is formed by the lattice network.

Metallic bonding occurs in good electrical conductors. Copper, for example, has one electron in its valence band. This electron has a tendency to wander around the material between different atoms. Upon leaving one atom, it immediately enters the orbit of another atom. The process is repeated on a continuous basis. When an electron leaves an atom, the atom becomes a positive ion. The process takes place in a random manner. This means that one electron is always associated with an atom but is not in one particular orbit. As a result, all atoms tend to share all the valence electrons.

In metallic bonding there is a type of electrostatic force between positive ions and electrons. In a sense, electrons float around in a cloud that covers

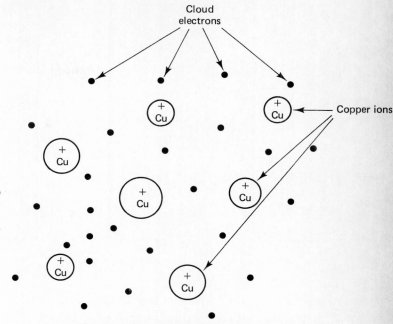

Figure 9-3. Metallic bonding of copper.

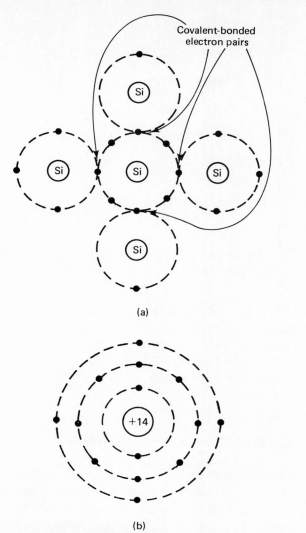

(a)

(b)

Figure 9-2. (a) Covalent-bonded silicon atoms; (b) a single silicon atom (Si).

the positive ions. This floating cloud bonds the electrons randomly to the ions. Figure 9-3 shows an example of the metallic bonding of copper.

CONDUCTION IN SOLID MATERIALS

Solid materials are primarily used in electrical devices to achieve electron conduction. These materials can be divided into conductors, semiconductors, and insulators. For discussion purposes, assume that we have a cubic centimeter block of these three materials. If the resistance is measured between opposite faces of each block, it shows some

very unique differences. The insulator cube measures several million ohms. The conductor cube, by comparison, measures only 0.000001 Ω. The semiconductor cube measures approximately 600 Ω. In effect, this shows a rather wide range of resistance differences in the three materials. A comparison of this type shows only the general boundaries of the respective materials.

Scientifically, conductors, semiconductors, and insulators are evaluated by *energy-level diagrams*. This takes into account the amount of energy needed to cause an electron to leave its valence band and go into conduction. The diagram is a composite of all atoms within the material.

Energy-level diagrams of insulators, semiconductors, and conductors are shown in Fig. 9-4. The valence band is located at the bottom and the conduction band is at the top. There are other energy levels in the structure of the atom. Each electron shell has a discrete energy level. In electronics we are interested primarily in the valence and conduction bands.

The valence band shows the energy levels of the valence electrons of each atom. A certain amount of energy must be added to the valence electrons to cause them to go into conduction. The *forbidden gap*, when it exists, separates the valence and conduction bands. A certain amount of energy is

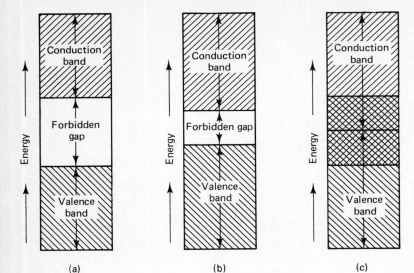

Figure 9-4. Energy-level diagrams for (a) insulators, (b) semiconductors, and (c) conductors.

needed to cross the forbidden gap. If it is insufficient, electrons are not released for conduction. They will remain in the valence band. The width of the forbidden gap indicates conduction status of a particular material. In atomic theory the width of the gap is expressed in *electron volts* (eV). An electron volt is defined as the amount of energy gained or lost when an electron is subjected to a potential difference of 1 V. The atoms of each element have a distinct energy-level value that permits conduction.

Insulating materials have a very wide forbidden gap. It takes a very large amount of energy to cause an insulator to go into conduction. Thyrite is a material commonly used as a lightning arrester. At normal voltages it is an ideal insulator. When it is subjected to high voltage, electrons cross the forbidden gap and go into conduction. This shunts the high voltage to ground, thus protecting a device from lightning. When good insulators are operated at high temperatures, the increased heat energy causes valence electrons to go into conduction.

The forbidden gap of a semiconductor is much smaller than that of an insulator. Silicon, for example, needs to gain 1.12 eV of energy to go into conduction. In many cases, the addition of heat energy at room temperature may be sufficient to cause conduction in a semiconductor. This particular characteristic is extremely important in solid-state electronic devices.

In a conductor, the conduction band and valence band overlap one another. In a sense, the valence band and conduction band are one and the same. This means that valence band electrons are readily released to become free electrons. Some electrical conduction takes place within this material through the application of heat at normal room temperature.

SEMICONDUCTOR MATERIALS

Materials such as silicon, germanium, and carbon are found naturally in crystalline form. Instead of being in a random mass, these atoms are arranged in an orderly manner. A crystal of silicon or germanium forms a definite geometrical pattern. Figure 9-5 shows a simplification of the silicon (Si) crystal. Germanium (Ge) would have a similar type of crystal formation. Note that the individual atoms are covalently bonded together. This display shows a two-dimensional layout of these atoms. The drawing is also simplified by displaying only the nucleus and valence electrons of each atom. The geometrical pattern of silicon causes its crystal to be in the shape of a cube.

Crystals of silicon and germanium can be manufactured by melting the natural elements. The process is somewhat complex and rather expensive to achieve. Manufactured crystals must be made extremely pure to be usable in semiconductor devices. A very pure semiconductor crystal is called an *intrinsic* material. Germanium is considered to be intrinsic when only one part of impurity exists in 10^8

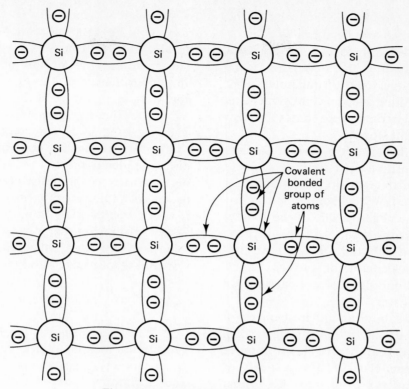

Figure 9-5. Intrinsic crystal of silicon.

parts of germanium. Silicon is intrinsic when the impurity ratio is $1:10^{13}$. In more sophisticated solid-state devices the ratio may even be higher.

An intrinsic crystal of silicon would appear as the structure in Fig. 9-5. Covalent bonding changes the electrical conductivity of the material, causing each group of atoms to become stable. Thus only a limited number of free electrons are available for conduction. Silicon and germanium crystals respond to some extent as an insulator.

Intrinsic silicon at absolute zero (-273° C) is considered to be a perfect insulator. The valence electrons of each atom are firmly bound together in perfect covalent bonds. No free electrons are available for conduction. In actual circuit operation, the absolute zero condition is not very meaningful. It cannot be readily attained. Any temperature above -273° C causes silicon to become somewhat conductive. The insulating quality of a semiconductor material is therefore dependent on its operating temperature.

At room temperature, such as 25° C, silicon atoms receive enough energy from heat to break their bonding. A number of free electrons become available for conduction. At room temperature, intrinsic silicon becomes somewhat conductive.

An important solid-state concept takes place when intrinsic silicon goes into conduction. For every electron that is freed from its covalent bonding a void spot is created. This spot, which is normally called a *hole*, represents an electron deficiency. A hole in a covalent bonding group occurs when an electron is released. An increase in the temperature of a piece of silicon causes the number of free electrons and holes to increase.

It should be remembered that when a neutral atom loses an electron, it acquires a positive charge. The atom then becomes a positive ion. Since an atom acquires a hole at the same time as the positive charge, the hole bears a positive charge. It should be noted, however, that the crystal remains electrically neutral. For every free electron there is an equivalent hole. These two balance out the overall charge of the crystal.

When a valence electron leaves its covalent bond to become a free electron, a hole appears in its place. This hole can then attract a different electron from a nearby bonded group. Upon leaving its bonded group, this electron creates a new hole. The original hole is then filled and becomes void. Each electron that leaves its bonding to fill a hole creates a new hole in its original group. In a sense, this means that electrons move in one direction and holes in the opposite direction. Since electron movement is considered to be an electric current, holes are also representative of current flow.

Electrons are called negative current carriers and holes are called positive current carriers. These current carriers move in opposite directions in a semiconductor material. When voltage is applied holes move toward the negative side of the source. Electron current flows toward the positive side of the source. This condition only takes place in a semiconductor material. In a conductor, such as copper, there is no covalent bonding. Current carriers are restricted to only electrons.

Figure 9-6 shows how an intrinsic piece of silicon will respond at room temperature when voltage is applied. Note that the free electrons move toward the positive terminal of the battery. Electrons leaving the semiconductor flow into the copper connecting wire. Each electron that leaves the material creates a hole in its place. The holes appear to be moving by jumping between covalent bonded groups. Holes are attracted by the negative terminal connection. For each electron that flows out of the material a new electron enters at the negative connection point. In effect, the new electron fills a hole at this point. The process of hole flow and electron flow through the material is continuous as long as energy

is supplied. Current flow is the resulting carrier movement. An intrinsic semiconductor has an equal number of current carriers moving in each direction. The resulting current flow of a semiconductor is limited primarily to the applied voltage and the operating temperature of the material.

Pure silicon or germanium in its intrinsic state is rarely used as a semiconductor. Usable semiconductors must have controlled amounts of impurities added to them. The added impurities change the conduction capabilities of a semiconductor. The process of adding an impurity to an intrinsic material is called *doping*. The impurity is called a *dopant*. Doping a semiconductor causes it to be an extrinsic material. Extrinsic semiconductors are the operational basis of nearly all solid-state devices.

N-Type Material

When an impurity is added to silicon or germanium without altering the crystal structure, an N-type material is produced. Atoms of arsenic (As) and antimony (Sb) have five electrons in their valence band (see the periodic element table in Appendix 1). Adding either impurity to silicon must not alter the crystal structure or bonding process. Each impurity atom has an extra electron that does not take part in a covalent bonding group. These electrons are loosely held together by their parent atoms. Figure 9-7 shows how the crystal is altered with the addition of an impurity atom.

When arsenic is added to pure silicon the crystal becomes an N-type material. It has extra electrons or negative (N) charges that do not take part in the covalent bonding process. Impurities that add electrons to a crystal are generally called *donor* atoms. An N-type material has more extra or free electrons than an intrinsic piece of material. A piece of N material is not negative charged. Its atoms are all electrically neutral. There is, however, a number of extra electrons that do not take part in the covalent bonding process. These electrons are free to move about through the crystal structure.

An extrinsic silicon crystal of the N-type will go into conduction with only 0.05 eV of energy applied. An intrinsic crystal requires 0.12 eV to move electrons from the valence band into the conduction band. Essentially, this means that an N material is a fairly good electrical conductor. In this type of crystal, electrons are considered to be the *majority current carriers*. Holes are the minority current carriers. The

Figure 9-6.   Current carrier movement in silicon.

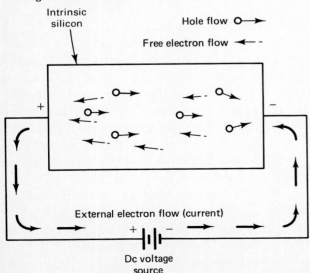

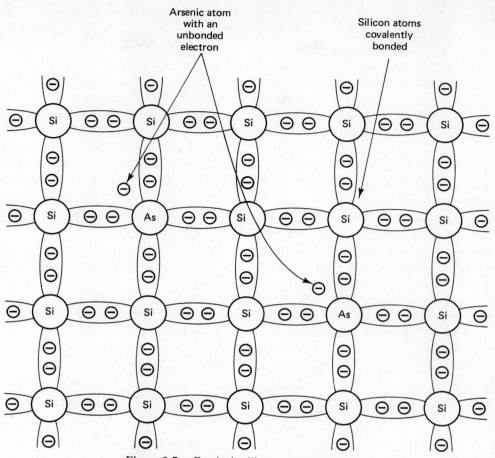

Figure 9-7. Extrinsic silicon crystal structure.

amount of donor material added to silicon determines the number of majority current carriers in its structure.

The number of electrons in a piece of N-type silicon is a million or more times greater than the electron–hole pairs of a piece of intrinsic silicon. At room temperature, there is a decided difference in the electrical conductivity of this material. Extrinsic silicon becomes a rather good electrical conductor. There is a larger number of current carriers to take part in conduction. Current flow is achieved primarily by electrons in this material. Figure 9-8 shows how the current carriers respond in a piece of N material. There is a larger number of electrons indicated than holes. This shows that electrons are the majority current carriers and holes are the minority carriers.

If the voltage source of Fig. 9-8 is reversed, the current flow would reverse its direction. Essentially, this means that N-type silicon will conduct equally as well in either direction. The flow of current

carriers is simply reversed. This is an important consideration in the operation of a device that employs N material in its construction.

Figure 9-8. Current carriers in an N-type material.

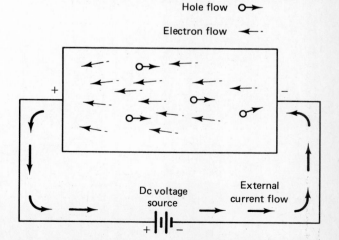

P-Type Material

P-type material is formed when indium (In) or gallium (Ga) is added to intrinsic silicon. Indium or gallium are type III elements according to the periodic table of Appendix 1. These elements are often called *acceptors*. They are readily seeking a fourth electron. In effect, this type of dopant material has three valence electrons. Each covalent bond that is formed with an indium atom will have an electron deficiency or hole. This represents a positive charge area in the covalent bonding structure. A positive charge area crystal or P material describes this type of structure. Each hole in the P material can be filled with an electron. Electrons from neighboring covalent bond groups require very little energy to move in and fill a hole.

The ratio of doping material to silicon is typically in the range 1 to 10^6. This means that P material will have a million times more holes than the heat-generated electron–hole pairs of pure silicon. At room temperature, there is a very decided difference in the electrical conductivity of this material. Figure 9-9 shows how the crystal structure of silicon is altered when doped with an acceptor element—in this case, indium.

A piece of P material is not positively charged. Its atoms are primarily all electrically neutral. There are, however, holes in the covalent structure of many atom groups. When an electron moves in and fills a hole, the hole becomes void. A new hole is created in the bonded group where the electron left. Hole movement in effect is the result of electron movement.

A P material will go into conduction with only 0.05 eV of energy applied. Intrinsic silicon, by comparison, requires 0.7 eV for conduction. Extrinsic silicon is therefore considered to be a rather good electrical conductor. In this material, holes are the majority carriers and electrons are the minority

Figure 9-9.   P-type crystal material.

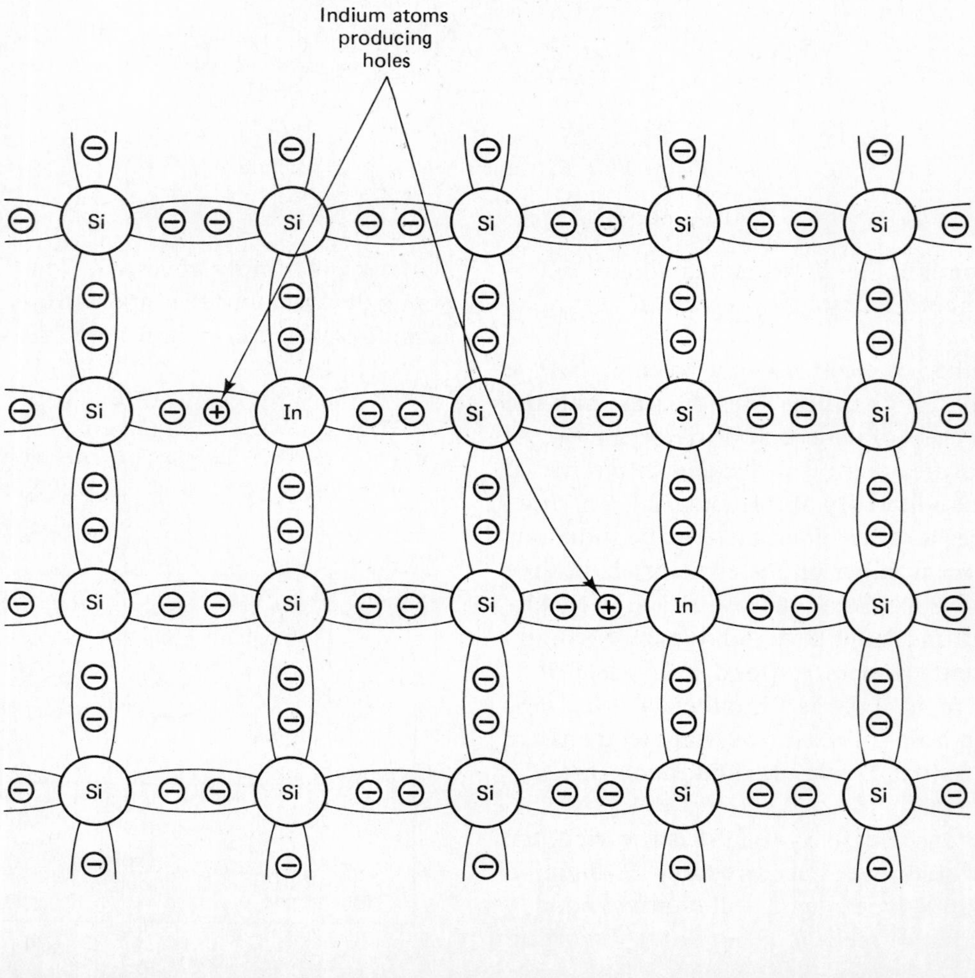

Indium atoms
producing
holes

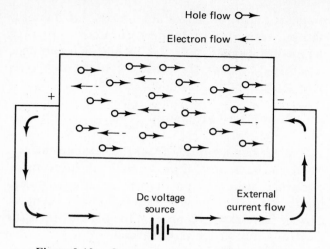

Figure 9-10. Current carriers in a P-type material.

carriers. The amount of acceptor material added to silicon determines the majority carrier content of the crystal.

Figure 9-10 shows how a P-type crystal will respond when connected to a voltage source. Note that there is a larger number of holes than electrons. With voltage applied, electrons are attracted to the positive battery terminal. Holes move, in a sense, toward the negative battery terminal. An electron is picked up at this point. The electron immediately fills a hole. The hole then becomes void. At the same time, an electron is pulled from the material by the positive battery terminal. Holes therefore move toward the negative terminal due to electrons shifting between different bonded groups. With energy applied, hole flow is continuous.

ELECTRON EMISSION

Electronics is a unique part of the electrical field that deals with the flow of electrons. These tiny particles can be made to perform a variety of work functions when they are placed into motion. A number of electronic devices operate by causing electrons to be released from the surface of a conductive material. This function is called *emission*. Emitted electrons travel through space, where they are eventually attracted to a positively charged electrode. A number of other electrodes may be included in the physical makeup of this device. Electrons must pass through these electrodes in order to be controlled.

Electron *emission* can be achieved in several different ways. Emission devices are classified according to the way electrons are released from a material. The application of energy in some form causes electron emission. Essentially, this energy is applied to the valence electrons of the emitting electrode. The added energy causes these electrons to increase their permissive energy level. When enough energy is supplied, the valence band electrons cross over the forbidden gap and move into the conduction band. See the energy-level diagram of Fig. 9-4. Continued energy increases may cause electrons to escape the retaining forces of the parent atom. This effect is called electron emission.

Thermionic Emission

The process of freeing electrons from a metal or metal substance by heat is called *thermionic emission*. Ordinarily, electrons would not escape from the surface of a solid substance. In the structure of an atom, electrons are in a continuous state of motion. If the material is heated, individual electrons begin to move faster. *Kinetic energy,* which is a result of motion, increases with electron speed. Valence electrons eventually gain enough energy to escape from the surface of the material. The amount of heat needed to produce emission is usually quite high. For materials such as oxided barium, 1000°C is needed. Tungsten emits electrons at 2300°C. The emitting material is normally heated by electricity. Tungsten, which has a rather high resistance, produces heat when current flows through it.

When electricity is applied directly to a piece of metal, electrons are emitted from its surface. The emitting electrode, in this case, is called a filament. Tubes with this type of emitter are called directly heated filaments. Battery-powered tube circuits and high-power transmitting tubes employ directly heated filaments. Figure 9-11 shows a directly heated filament and its schematic symbol.

If alternating current is used to energize the tube filament, electron emission will vary according to the frequency of the source. These fluctuations will produce an ac hum in the stream of emitted electrons. In most applications this causes a form of interference. Indirect heating of the emitting surface isolates it from the ac filament source. The emitting electrode of this tube is called a cathode. Essentially,

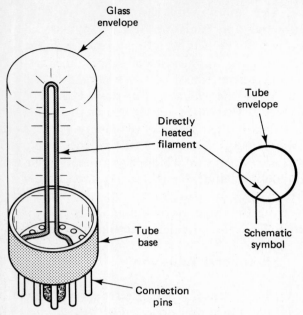

Figure 9-11. Directly heated tube.

the emitter consists of a cathode and a heater. The heater or filament is placed inside the cathode. Heat from the filament is transferred to the cathode. Electrons are emitted from the surface of the cathode after it has been heated.

Indirectly heated cathodes do not reach emission temperatures as quickly as directly heated surfaces. Heat takes time to be transferred to the cathode. The cathode usually has a rather large surface area compared with a directly heated filament. Once the surface is brought up to its emitting temperature, it retains this heat for a short period of

time. Its temperature will, therefore, not fluctuate with ac heater voltage. Indirectly heated tubes have a very constant emission level while in operation. Figure 9-12 shows the structure of an indirectly heated tube and its schematic symbol.

Secondary Emission

After being emitted from the cathode surface, electrons have very little velocity. A large metal electrode, with a high positive charge, is often placed inside the tube. This electrode is called a plate or anode. The positive charge on the plate attracts the electrons emitted from the cathode. A high positive charge will cause electrons to move to the plate with a great deal of velocity. When striking the plate, these electrons will transfer some of their energy to the atoms of the plate material. If the energy level is great enough, it will cause electrons to be released from the surface of the plate. This effect is called *secondary emission*. Figure 9-13 shows an illustration of secondary emission.

In normal circuit operation, each primary electron may cause the release of several secondary

Figure 9-13. Secondary emission.

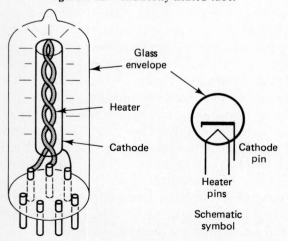

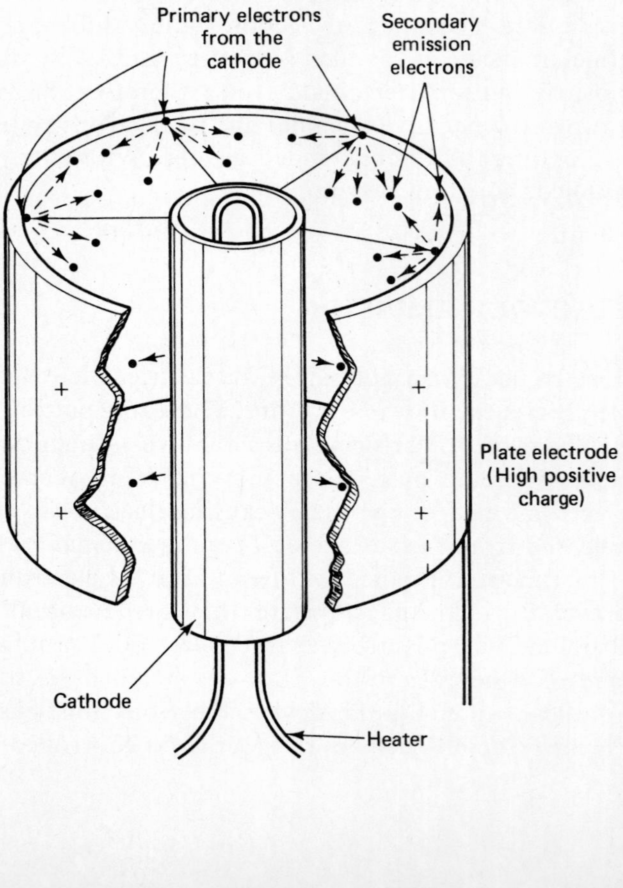

Figure 9-12. Indirectly heated tube.

electrons. This means that a cloud of secondary electrons may form around the surface of the plate. In effect, this can restrict the flow of primary electrons. As a general rule, secondary emission is not a desirable condition. In some tubes special provisions are made to reduce or eliminate secondary emission. In some phototubes secondary emission is used to achieve electron multiplication. This action permits the tube to have greater output capabilities.

Photoemission

Materials that emit or give off electrons when light strikes them are considered to be photoemissive. These materials are sensitive to light. Devices to make use of this effect are called photoelectric tubes or photocells. These devices are used to detect light in various control applications.

The operating principle of a photoemissive device is best explained by the *quantum theory* of light. According to this theory, light is made up of tiny packages of energy called *photons*. A small packet of energy is called a *quanta*, hence the name quantum theory. These packages contain uncharged particles that possess certain levels of energy. This energy depends on the frequency of light.

A device whose electrical conductivity changes with light is considered to be *photoconductive*. Essentially, this kind of device changes its internal resistance with light. These devices are also called *photoresistive cells* or light-dependent resistors (LDRs). Semiconductor materials such as silicon, selenium, and cadmium compounds are used in the construction of these devices.

Figure 9-14 shows the construction of a cadmium sulfide (CdS) photoresistive cell. A thick film of cadmium sulfide is deposited on a glass *substrate*. Metal leads are attached to each end of film pattern. A transparent window is placed over the top of the cell. The entire assembly is then covered with protective coating. This seals the cell and protects it from air and foreign material.

Operation of a photoemissive cell is based on the amount of light energy striking its working surface. Photons of light striking this surface transfer their energy to the valence electrons of the film material. If the transfer of energy is great enough, valence electrons will cross the forbidden gap and go

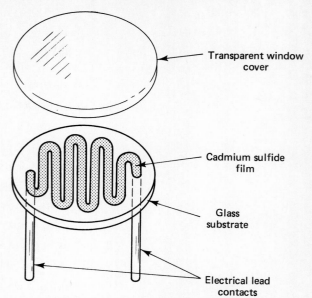

Figure 9-14.   Structure of a photoresistive cell.

into the conduction band. These electrons then become free and respond as current carriers. Each electron that goes into the conduction band leaves a corresponding hole in the crystal structure. The amount of applied light energy therefore determines the electron–hole production of the material. An increase in the number of current carriers causes a decrease in material resistance.

A light-detection circuit using cadmium sulfide cell is shown in Fig. 9-15. The CdS cell responds in this circuit as a variable resistor. In a dark area the cell is high resistant. Very little current will be indicated by the meter. An increase in light intensity will cause a decrease in cell resistance. This, in turn, will cause an increase in circuit current. The meter shows how the cell responds to different light levels.

Figure 9-15.   Light detection circuit, General Electric GE-X6 or equivalent.

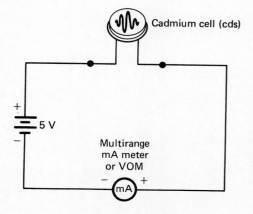

Review

1. Explain the difference between conductors, semiconductors, and insulators.
2. What is ionic bonding?
3. Explain the meaning of covalent bonding.
4. Explain the energy-level concept of solid material conduction.
5. What is an intrinsic material and an extrinsic material?
6. How does electric conduction occur in a piece of intrinsic silicon?
7. Explain how the current carriers of a piece of N material respond when a voltage source is connected to it.
8. How is current conduction achieved in P material?
9. Explain how thermionic emission is achieved.
10. What is secondary emission?
11. What is a photon?
12. How is photoemission achieved?
13. How does the current conduction of a light-dependent resistor change when exposed to light?

Student Activities

N-Material Conduction

1. Select either an N channel or P channel junction field-effect transistor or JFET.
2. With an ohmmeter measure the resistance between any two of the three leads. If a resistance value of a few hundred ohms is measured, reverse the two leads. A reading of the same resistance in the reverse direction indicates the N channel of the device. Find the two leads of the JFET where the resistance value is the same in each direction. Record the measured resistance.
3. Connect the two JFET leads into the circuit as indicated in Fig. 9-16.
4. Turn on the voltage source. Measure and record the value of current.
5. Turn off the voltage source. Reverse the two selected leads.
6. Turn on the voltage source. Measure and record the value of current. This shows how current passes through an N or P material.

Vacuum Tube Heater Testing

1. Select any low-heater-voltage vacuum tube for this test.
2. With an ohmmeter, find two leads that indicate resistance between them. Record the measured resistance and note the location of the leads.

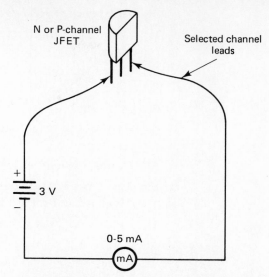

Figure 9-16. Semiconductor conduction circuit.

3. If a vacuum tube data manual is available, look up the specific tube being used. Note the pin location of heater leads. Also note the voltage rating of the heater. Generally, the number of a tube denotes the heater voltage. A 6C4, for example, uses a 6-V heater. A 12AX7 uses 12 V.

4. Connect the heater of a selected tube in the circuit as indicated in Fig. 9-17. The specific pin connections will depend on the tube selected for the circuit.

5. Turn on the filament transformer. Measure and record the filament current value.

6. Dc can also be used as an energy source for the circuit.

7. If the heater voltage is smaller than the voltage source, a dropping resistor may be added in series with the heater. Values can be calculated if needed.

8. With the heater source applied, locate the position of the glowing filament within the tube.

Figure 9-17. Vacuum-tube heater testing.

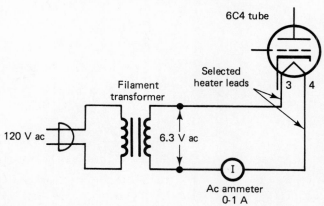

Light-Dependent Resistor Testing

1. Connect an ohmmeter to the two leads of a CdS cell. Use an $R \times 1$ kΩ range.

2. Cover the cell so that no light is on the lens. The dark resistance is _____ Ω.

3. In normal room light the resistance is _____ Ω.

4. Exposed to bright sunlight or near an incandescent lamp of 100 W, the resistance is _____ Ω.

5. Connect the light-detection circuit of Fig. 9-15.

6. Test the operation of the cell for dark, normal room light, and bright light. Measure and record the resulting current values for these light levels.

10

Electronic Diode

Diodes are used rather extensively in nearly all phases of the electronic field today. Radio, TV, industrial control circuits, home entertainment equipment, computers, and electrical appliances are only a few of the applications. Diodes are probably the simplest of all electronic components. Only two leads or electrodes are used in their construction. Functionally, this device has unidirectional conductivity. This means that it conducts well in only one direction. AC can be easily converted into dc with a diode. In this application, a diode is used as a rectifier. In addition to this, other functions can be performed. Some of these are the operational bases of more complex electronic devices. Diode operation is therefore extremely important. A person working in electronics must be very familiar with the characteristic operation of this device.

IMPORTANT TERMS

In the study of diode devices one frequently encounters a number of rather unusual terms in common usage. As a rule, these terms play a key role in the presentation of new material. Review these terms very carefully before proceeding with the text of this chapter.

Anode. A terminal of an electronic device that collects electrons.

Barrier potential or voltage. The voltage that is developed across a P-N junction due to the diffusion of holes and electrons.

Bias forward. Voltage applied across a P-N junction that causes the conducting of majority current carriers.

Bias reverse. Voltage applied across a P-N junction that causes little or no current flow.

Cathode. The terminal of an electronic device that is responsible for electron emission.

Covalent bond. Atoms joined together that share electrons to form a stable molecule.

Depletion zone. An area near the P-N junction where it is void of current carriers.

Dielectric constant. A number that compares the insulating ability of a material to that of air. Air has a value of 1.

Doping. The process of adding an impurity to a semiconductor material.

Efficiency. A ratio between output and input.

Electron. A negatively charged atomic particle.

Energy. Something that is capable of producing work, such as heat, light, chemical, and mechanical action.

Filament. The heating element of a vacuum tube.

Hole. A charge area or void where an electron is missing.

Ion. An atom that has either lost an electron (positive ion) or gained an electron (negative ion).

Junction. The point where P and N semiconductor materials are joined.

Photon. A small bundle of light energy.

Thermionic emission. The release of electrons from the surface of a metal conductor due to the application of heat.

Valence electrons. The outermost shell or layer of electrons in the structure of an atom.

JUNCTION DIODE

The term "junction diode" is often used to describe a crystal structure made of P and N materials. A diode is generally described as a two-terminal device. One terminal is attached to P material and the other to N material. The common connecting point where these materials are joined is called a junction. A junction diode permits current carriers to flow readily in one direction and blocks the flow of current in the opposite direction.

Figure 10-1 shows the crystal structure of a junction diode. Notice the location of the P and N materials with respect to the junction. The crystal structure is continuous from one end to the other. The junction serves only as a dividing line that marks the ending of one material and the beginning of the other. This type of structure permits electrons to move readily through the entire structure.

Figure 10-2 shows two pieces of semiconductor material before they are formed into a P-N junction. As indicated, each piece of material has majority and minority current carriers. The number of carrier symbols shown in each material indicates the minority or majority function. Electrons are the majority carriers in the N material and are the minority carriers in the P material. Holes are the majority carriers of the P material and are in the minority in the N material. Both holes and electrons have freedom to move about in their respective materials.

Depletion Zone

When a junction diode is first formed, there is a unique interaction between current carriers. Electrons from the N material move readily across the junction to fill holes in the P material. This action is commonly called *diffusion*. Diffusion is the result of high concentration of carriers in one material and a lower concentration in the other. Only those current carriers near the junction take part in the diffusion process.

The diffusion of current carriers across the junction of a diode causes a change in its structure. Electrons leaving the N material cause positive ions to be generated in their place. Upon entering the P material to fill holes, these same electrons create negative ions. The area on each side of the junction then contains a large number of positive and negative ions. The number of holes and electrons in this area becomes depleted. The term *depletion zone* is

Figure 10-1. Crystal structure of a junction diode.

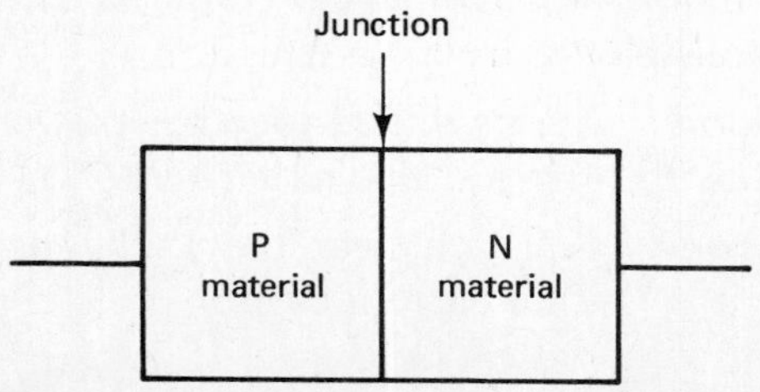

Figure 10-2. Semiconductor materials.

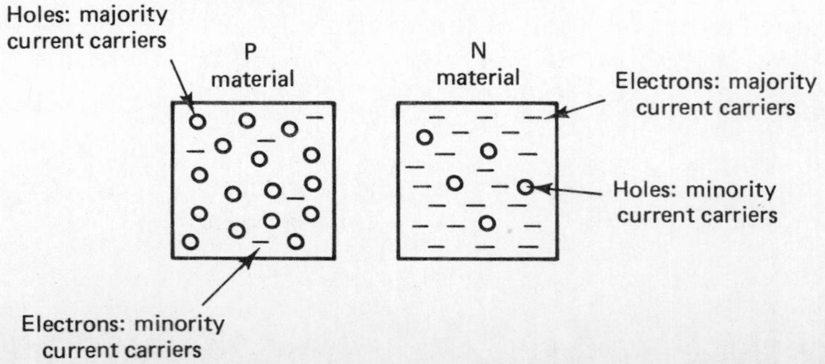

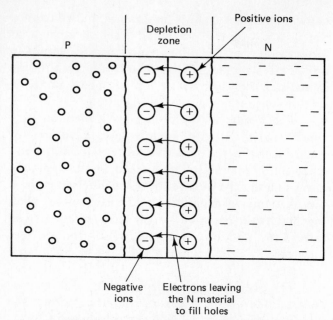

Figure 10-3. Depletion-zone formation.

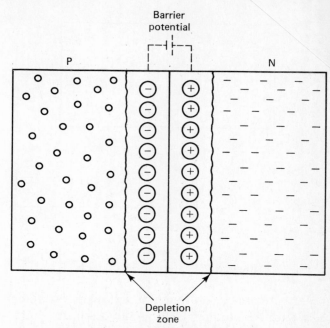

Figure 10-4. Barrier potential of a diode.

used to describe this area. It represents an area that is void of majority current carriers. All P-N junctions develop a depletion zone when they are formed. Figure 10-3 shows the depletion zone of a junction diode.

Barrier Potential

Before N and P materials are joined together at a common junction they are considered to be electrically neutral. After being joined, however, diffusion takes place immediately. Electrons crossing the junction to fill holes cause negative ions to appear in the P material. This action causes the area near the junction to take on a negative charge. Similarly, electrons leaving the N material cause it to produce positive ions. This, in turn, causes the N side of the junction to take on a net positive charge. These respective charge creations tend to drive remaining electrons and holes away from the junction. This action makes it somewhat difficult for additional charge carriers to diffuse across the junction. The end result is a charge buildup or *barrier potential* appearing across the junction.

Figure 10-4 shows the resulting barrier potential as a small battery connected across the P-N junction. Note the polarity of this potential with respect to P and N material. This voltage will exist even when the crystal is not connected to an outside source of energy. For germanium the barrier poten-

tial is approximately 0.3 V, and 0.7 V for silicon. These voltage values cannot, however, be measured directly. They appear only across the space charge region of the junction. The barrier potential of a P-N junction must be overcome by an outside voltage source in order to produce current conduction.

JUNCTION BIASING

When an external source of energy is applied to a P-N junction it is called a bias voltage or simply biasing. This voltage either adds to or reduces the barrier potential of the junction. Reducing the barrier potential causes current carriers to return to the depletion zone. Forward biasing is used to describe this condition. Adding external voltage of the same polarity to the barrier potential causes an increase in the width of the depletion zone. This action hinders current carriers from entering the depletion zone. The term "reverse biasing" is used to describe this condition.

Reverse Biasing

When a battery is connected across a P-N junction as in Fig. 10-5 it causes reverse biasing. Note that the negative terminal of the battery is connected to the P material and the positive to the N material. Connection in this way causes the battery

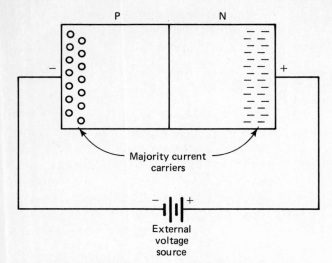

Figure 10-5.   Reverse-biased diode.

polarity to oppose the material polarity of the diode. Since unlike charges attract, the majority charge carriers of each material are pulled away from the junction. Reverse biasing of a diode normally causes it to be nonconductive.

Figure 10-6 shows how the majority current carriers are rearranged in a reverse-biased diode. As

shown, electrons of the N material are pulled toward the positive battery terminal. Each electron that moves or leaves the diode causes a positive ion to appear in its place. This causes a corresponding increase in the width of the depletion zone on the N side of the junction.

The P side of the diode has a reaction very similar to the N side. In this case, a number of electrons leave the negative battery terminal and enter the P material. These electrons immediately move in and fill a number of holes. Each filled hole then becomes a negative ion. These ions are then repelled by the negative battery terminal and driven toward the junction. As a result of this, the width of the depletion zone increases on the P side of the junction.

The overall depletion zone width of a reverse-biased diode is directly dependent on the value of the supply voltage. With a wide depletion zone a diode cannot effectively support current flow. The charge buildup across the junction will increase until the barrier voltage equals the external bias voltage. When this occurs a diode is effectively a noncon-ductor.

Figure 10-6.   Reverse-biased P-N junction.

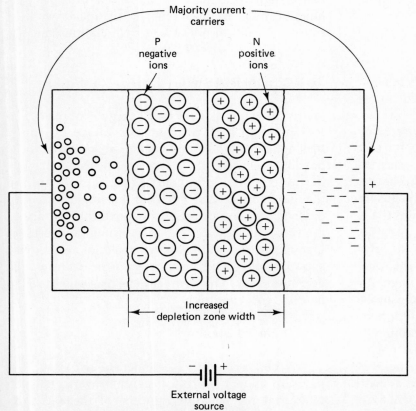

LEAKAGE CURRENT. When a diode is reverse biased its depletion zone increases in width. Normally, we would expect this condition to restrict current carrier formation near the junction. The depletion zone in effect represents an area that is primarily void of majority current carriers. Because of this, the depletion zone will respond as an insulator. Ideally, current carriers do not pass through an insulator. In a reverse-biased diode some current actually flows through the depletion zone. This is called leakage current. Leakage current is dependent on minority current carriers.

Remember that minority carriers are electrons in the P material and holes in the N material. Figure 10-7 shows how these carriers respond when a diode is reverse biased. It should be noted that the minority carriers of each material are pushed through the depletion zone to the junction. This action causes a very small leakage current to occur. Normally, leakage current is so small that it is often considered negligible.

The minority current carrier content of a semiconductor is primarily dependent on temperature. At normal room temperatures of 25°C or 77°F there are a rather limited number of minority carriers present in a semiconductor. However, when the surrounding temperature rises it causes considerable increase in minority carrier production. This causes a corresponding increase in leakage current.

Leakage current occurs to some extent in all reverse-biased diodes. In germanium diodes, leakage current is only a few microamperes. Silicon diodes normally have a lower minority carrier content. This in effect means less leakage current. Typical leakage current values for silicon are a few nanoamperes. The construction material of a diode is an important consideration to remember. Germanium is much more sensitive to temperature than is silicon. Germanium therefore has a higher level of leakage current. This factor is largely responsible for the widespread use of silicon in modern semiconductor devices.

Forward Biasing

When voltage is applied to a diode as in Fig. 10-8 it causes forward biasing. In this example, the positive battery terminal is connected to the P

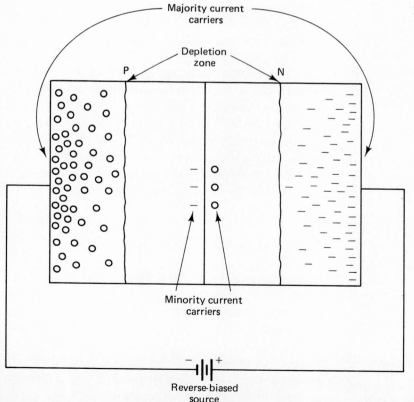

Figure 10-7. Minority current carriers.

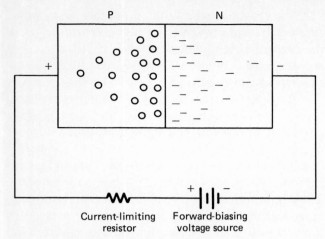

Figure 10-8.   Forward-biased diode.

constant supply of electrons. Large arrows are used in the diagram to show the direction of current flow outside the diode. Inside the diode, smaller arrows show the movement of majority current carriers. Remember that electron flow and current refer to the same thing.

Starting at the negative battery terminal, assume now that electrons flow through a wire to the N material. Upon entering this material, they flow immediately to the junction. At the same time, an equal number of electrons are removed from the P material and are returned through a resistor to the positive battery terminal. This action generates new holes and causes them to move toward the junction. When these holes and electrons reach the junction they combine and effectively disappear. At the same time new holes and electrons appear at the outer ends of the diode. These majority carriers are generated on a continuous basis. This action continues as long as the external voltage source is applied.

It is important to realize that electrons flow through the entire diode when it is forward biased. In the N material this is quite obvious. In the P material, however, holes are the moving current carriers. Remember that hole movement in one direction must be initiated by electron movement in the opposite direction. Therefore, the combined flow of holes and electrons through a diode equals the total current flow.

The current-limiting resistor of Fig. 10-9 is essential in a forward-biasing diode circuit. This resistor is needed to keep the current flow at a safe operating level. The maximum current (I_{max}) rating of a diode is representative of this value. As long as diode current does not exceed this value it can operate satisfactorily. Resistors are used to limit current flow to a reasonable operating level.

material and the negative terminal to the N material. This voltage repels the majority current carriers of each material. A large number of holes and electrons therefore appear at the junction. On the N side of the junction, electrons move in to neutralize the positive ions in the depletion zone. In the P material, electrons are pulled from negative ions which causes them to become neutral again. This action means that forward biasing causes the depletion zone to collapse and the barrier potential to be removed. The P-N junction will therefore support a continuous current flow when it is forward biased.

Figure 10-9 shows how the current carriers of a forward-biased diode respond. Since the diode is connected to an external voltage source, it has a

DIODE CHARACTERISTICS

Now that we have seen how a junction diode operates, it is time to examine some of its electrical characteristics. This takes into account such things as voltage, current, and temperature. Since these characteristics vary a great deal in an operating circuit, it is best to look at them graphically. This makes it possible to see how the device will respond under different operating conditions.

Figure 10-9.   Current carrier flow in a forward-biased diode.

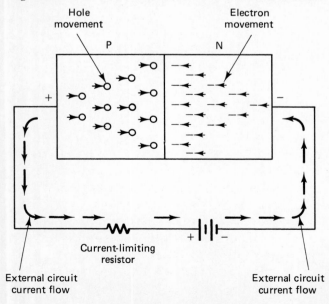

Forward Characteristic

When a diode is connected in the forward bias direction it conducts forward current I_F. The value of I_F is directly dependent on the amount of forward voltage V_F. The relationship of forward voltage and forward current is called the ampere-volt or $I\text{–}V$ characteristic of a diode. A typical diode forward $I\text{–}V$ characteristic is shown in Fig. 10-10(a) with its test circuit in Fig. 10-10(b). Note in particular that V_F is measured across the diode and that I_F is a measure of what flows through the diode. The value of the source voltage V_S does not necessarily compare in value with V_F.

Figure 10-10. (a) $I\text{–}V$ characteristics of a diode; (b) diode characteristic test circuit.

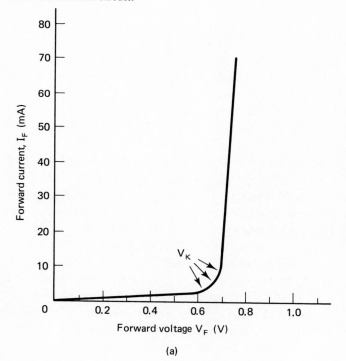

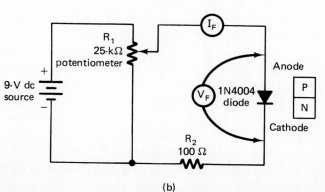

When the forward voltage of the diode equals 0 V, the I_F equals 0 mA. This value starts at the origin point (0) of the graph. If V_F is gradually increased in 0.1-V steps, I_F begins to increase. When the value of V_F is great enough to overcome the barrier potential of the P-N junction, a substantial increase in I_F occurs. The point at which this occurs is often called the knee voltage V_K. For germanium diodes V_K is approximately 0.3 V and 0.7 V for silicon.

If the value of V_F increases much beyond V_K, the forward current becomes quite large. This, in effect, causes heat to develop across the junction. Excessive junction heat can destroy a diode. To prevent this from happening a protective resistor is connected in series with the diode. This resistor limits the I_F to its maximum rated value. Diodes are rarely operated in the forward direction without a current-limiting resistor.

Reverse Characteristic

When a diode is connected in the reverse bias direction it has an $I_R\text{–}V_R$ characteristic. Figure 10-11(a) shows the reverse $I\text{–}V$ characteristic of a diode and its test circuit. This characteristic has different values of I_R and V_R. Reverse current is usually quite small. The vertical I_R line in this graph has current values graduated in microamperes. The number of minority current carriers that take part in I_R is quite small. In general, this means that I_R remains rather constant over a large part of V_R. Note also that V_R is graduated in 100-V increments. Starting at zero when the reverse voltage of a diode is increased there is a very slight change in I_R. At the voltage breakdown V_{BR} point current increases very rapidly. The voltage across the diode remains fairly constant at this time. This constant-voltage characteristic leads to a number of reverse bias diode applications. Normally, diodes are used in applications where the V_{BR} is not reached.

The physical processes responsible for current conduction in a reverse-biased diode are called zener breakdown and avalanche breakdown. Zener breakdown takes place when electrons are pulled from their covalent bonds in a strong electric field. This occurs at a rather high value of V_R. When large numbers of covalent bonds are broken at the same time there is a sudden increase in I_R.

Avalanche breakdown is an energy-related condition of reverse biasing. At high values of V_R

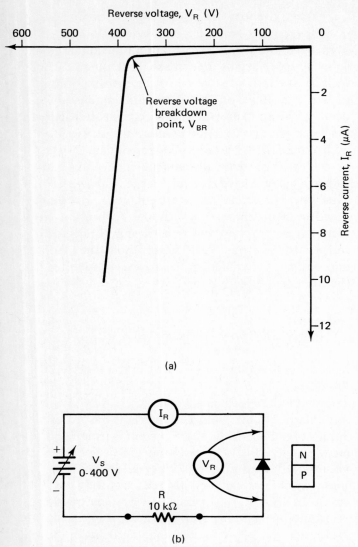

Figure 10-11. (a) Reverse *I–V* characteristics of a diode; (b) reverse characteristic test circuit.

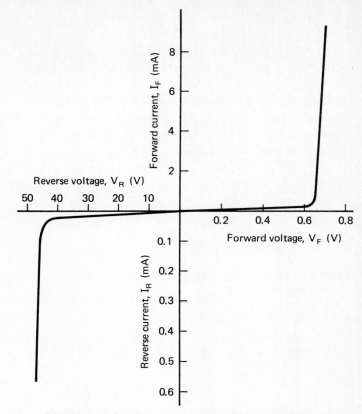

Figure 10-12. Diode current–voltage characteristics.

minority carriers gain a great deal of energy. This gain may be great enough to drive electrons out of their covalent bonding. This action creates new electron-hole pairs. These carriers then move across the junction and produce other ionizing collisions. Additional electrons are produced. The process continues to build up until an avalanche of current carriers takes place. When this occurs the process is irreversible.

Combined *I–V* Characteristics

The forward and reverse *I–V* characteristics of a diode are generally combined on a single characteristic curve. Figure 10-12 shows a rather standard

method of displaying this curve. Forward and reverse bias voltage V_F and V_R are usually plotted on the horizontal axis of the graph. V_F extends to the right and V_R to the left. The point of origin or zero value is at the center of the horizontal line. Forward and reverse current values are shown vertically on the graph. I_F extends above the horizontal axis with I_R extending downward. The origin point serves as a zero indication for all four values. This means that combined V_F and I_F values are located in the upper right part of the graph and V_R and I_R in the lower left corner. Different scales are normally used to display forward and reverse values.

A rather interesting comparison of silicon and germanium characteristic curves is shown in Fig. 10-13. Careful examination shows that germanium requires less forward voltage to go into conduction than silicon. This characteristic is a distinct advantage in low-voltage circuits. Note also that a germanium diode requires less voltage drop across it for different values of current. This means that germanium has a lower resistance to forward current flow. Germanium therefore appears to be a better conductor than silicon. Silicon is more widely used, however, because of its low leakage current and cost.

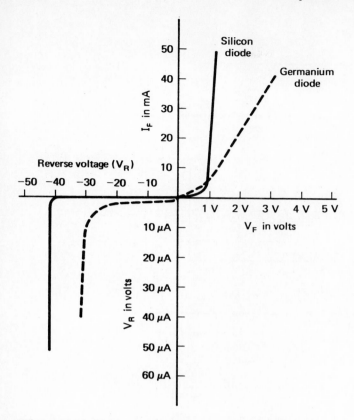

Figure 10-13.   Silicon and germanium diode characteristics.

The reverse bias characteristics of silicon and germanium diodes can also be compared in Fig. 10-13. The I_R of a silicon diode is very small compared to that of a germanium diode. Reverse current is determined primarily by the minority current content of the material. This condition is influenced primarily by temperature. For germanium diodes, I_R doubles for each $10°$ C rise in temperature. In a silicon diode the change in I_R is practically negligible for the same rise in temperature. As a result of this, silicon diodes are preferred over germanium diodes in applications where large changes in temperature occur. Comparisons of this type are quite obvious through the study of characteristic curves.

DIODE SPECIFICATIONS

When selecting a diode for a specific application some consideration must be given to specifications. This type of information is usually made available through the manufacturer. Diode specifications usually include such things as absolute maximum ratings, typical operating conditions, mechanical

data, lead identification, mounting procedures, and characteristic curves. Figure 10-14 shows a representative diode data sheet. Some of the important specifications are explained in the following:

1. *Maximum reverse voltage V_{RM}*: The absolute maximum or peak reverse bias voltage that can be applied to a diode. This may also be called the peak inverse voltage (PIV) or peak reverse voltage (PRV).

2. *Reverse breakdown voltage V_{BR}*: The minimum steady-state reverse voltage at which breakdown will occur.

3. *Maximum forward current I_{FM}*: The absolute maximum repetitive forward current that can pass through a diode at 25°C (77° F). This is reduced for operation at higher temperatures.

4. *Maximum forward surge current I_{FM} (Surge)*: The maximum current that can be tolerated for a short interval of time. This current value is much greater than I_{FM}. This represents the increase in current that occurs when a circuit is first turned on.

5. *Maximum reverse current I_R*: The absolute maximum reverse current that can be tolerated at device operating temperature.

6. *Forward voltage V_F*: Maximum forward voltage drop for a given forward current at device operating temperature.

7. *Power dissipation P_D*: The maximum power that the device can safely absorb on a continuous basis in free air at 25° C (77° F).

8. *Reverse recovery time T_{rr}*: The maximum time that it takes the device to switch from its on to its off state.

Diode Temperature

The operation of a diode is influenced by a great deal by temperature. All semiconductor materials are similar in this respect. The primary effect of temperature is the generation of additional electron–hole pairs. When valence electrons are excited by heat they cause a change in covalent bonding. This means that more electrons become available for conduction with an increase in temperature. These additional current carriers cause a decided change in the *I–V* characteristics of a diode.

TYPES 1N4001 THROUGH 1N4007
DIFFUSED-JUNCTION SILICON RECTIFIERS

***electrical characteristics at specified ambient† temperature**

PARAMETER		TEST CONDITIONS	MAX	UNIT
I_R	Static Reverse Current	V_R = Rated V_R, T_A = 25°C	10	μa
		V_R = Rated V_R, T_A = 100°C	50	μa
$I_{R(avg)}$	Average Reverse Current	V_{RM} = Rated V_{RM}, I_O = 1 a, f = 60 cps, T_A = 75°C	30	μa
V_F	Static Forward Voltage	I_F = 1 a, T_A = 25°C to 75°C	1.1	v
$V_{F(avg)}$	Average Forward Voltage	V_{RM} = Rated V_{RM}, I_O = 1 a, f = 60 cps, T_A = 25°C to 75°C	0.8	v
V_{FM}	Peak Forward Voltage	V_{RM} = Rated V_{RM}, I_O = 1 a, f = 60 cps, T_A = 25°C to 75°C	1.6	v

*Indicates JEDEC registered data.

THERMAL INFORMATION

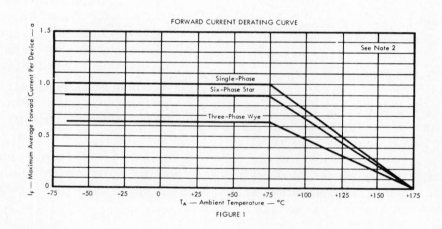

FORWARD CURRENT DERATING CURVE

FIGURE 1

†The ambient temperature is measured at a point 2 inches below the device. Natural air cooling shall be used.

NOTE 2: This rectifier is a lead-conduction-cooled device. At (or above) ambient temperatures of 75°C, the lead temperature 3/8 inch from case must be no higher than 5°C above the ambient temperature for these ratings to apply.

TEXAS INSTRUMENTS RESERVES THE RIGHT TO MAKE CHANGES AT ANY TIME IN ORDER TO IMPROVE DESIGN AND TO SUPPLY THE BEST PRODUCT POSSIBLE.

Figure 10-14. Data sheet for a silicon diode. (Courtesy of Texas Instruments.)

TYPES 1N4001 THROUGH 1N4007
DIFFUSED-JUNCTION SILICON RECTIFIERS

TYPES 1N4001 THROUGH 1N4007
BULLETIN NO. DL-S 657366, FEBRUARY 1965

50-1000 VOLTS • 1 AMP AVG

- **MINIATURE MOLDED PACKAGE**
- **INSULATED CASE**
- **IDEAL FOR HIGH-DENSITY CIRCUITRY**

*mechanical data

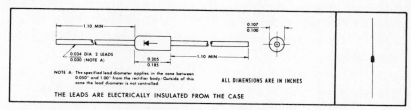

1.10 MIN

0.034 DIA 2 LEADS
0.030 (NOTE A)

0.205
0.185

1.10 MIN

0.107
0.100

NOTE A: The specified lead diameter applies in the zone between 0.050" and 1.00" from the rectifier body. Outside of this zone the lead diameter is not controlled.

ALL DIMENSIONS ARE IN INCHES

THE LEADS ARE ELECTRICALLY INSULATED FROM THE CASE

*absolute maximum ratings at specified ambient† temperature

		1N4001	1N4002	1N4003	1N4004	1N4005	1N4006	1N4007	UNIT
V_{RM}	Peak Reverse Voltage from −65°C to +175°C (See Note 1)	50	100	200	400	600	800	1000	v
V_R	Steady State Reverse Voltage from 25°C to 75°C	50	100	200	400	600	800	1000	v
I_O	Average Rectified Forward Current from 25°C to 75°C (See Notes 1 and 2)	1							a
$I_{FM(rep)}$	Repetitive Peak Forward Current, 10 cycles, at (or below) 75°C (See Note 3)	10							a
$I_{FM(surge)}$	Peak Surge Current, One Cycle, at (or below) 75°C (See Note 3)	30							a
$T_{A(opr)}$	Operating Ambient Temperature Range	−65 to +175							°C
T_{stg}	Storage Temperature Range	−65 to +200							°C
	Lead Temperature ⅜ Inch from Case for 10 Seconds	350							°C

NOTES: 1. These values may be applied continuously under single-phase, 60-cps, half-sine-wave operation with resistive load. Above 75°C derate I_O according to Figure 1.

2. This rectifier is a lead-conduction-cooled device. At (or above) ambient temperatures of 75°C, the lead temperature ⅜ inch from case must be no higher than 5°C above the ambient temperature for these ratings to apply.

3. These values apply for 60-cps half sine waves when the device is operating at (or below) rated values of peak reverse voltage and average rectified forward current. Surge may be repeated after the device has returned to original thermal equilibrium.

* Indicates JEDEC registered data.

† The ambient temperature is measured at a point 2 inches below the device. Natural air cooling shall be used.

TEXAS INSTRUMENTS
INCORPORATED
POST OFFICE BOX 5012 • DALLAS, TEXAS 75222

Figure 10-14. *(Continued.)*

Figure 10-15 shows how the *I–V* characteristics of a silicon diode change with temperature. These operating temperatures are indicated in Celsius degrees (°C). Three main differences should be noted. The first of these is in the forward bias region. This shows that less voltage is needed to produce conduction at elevated temperatures. The other differences are in the reverse bias area. One of these shows that there is a great deal more leakage current at higher temperatures. The third difference is in breakdown voltage. This indicates that lower temperatures produce higher breakdown voltages. These considerations must all be taken into account when selecting a diode for a circuit application.

Electronic circuits that employ diodes are called upon to operate at a rather wide range of temperatures. Consumer-grade diodes are usually rated for a range of −50 to +100° C. The extremes of this range are more meaningful if equated to the boiling point of water 212° F (+100° C) and the freezing of mercury −58° F (−50° C). Military-grade diodes are rated at −65 to 125° C. This type of diode is much more expensive than consumer-grade devices.

Figure 10-15. *I–V* characteristics of a silicon diode at 100, 50, and 25° C.

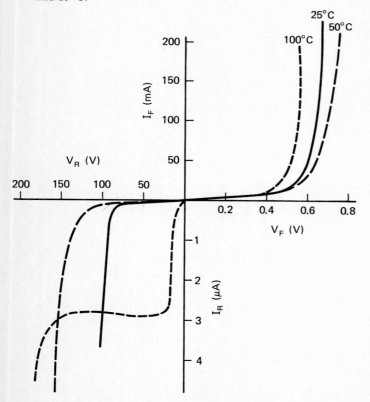

Junction Capacitance

A capacitor is defined as two or more conductive plates separated by an insulating material. A reverse-biased diode has a structure that responds as a capacitor. The two independent crystal materials serve as conductor plates with the depletion zone responding as a dielectric material. Junction capacitance is used to describe this effect.

Junction capacitance is a variable value that is dependent on bias voltage. Small amounts of reverse bias, and in some cases forward bias, produce the largest values of junction capacitance. This, in effect, reduces the width of the depletion zone. A decrease in dielectric thickness causes a corresponding increase in capacitance.

When the reverse bias voltage of a diode is increased, there is a decrease in junction capacitance. This condition causes a corresponding increase in the width of the depletion zone. With a thicker dielectric material, there is a smaller junction capacitance.

The dielectric constant of a silicon depletion zone is approximately 12. This means that the depletion zone is 12 times better than air as a dielectric. As a result, some rather significant capacitance values appear across a junction diode. This value is extremely important in high-frequency circuit applications. A reverse-biased diode can therefore respond as a voltage-controlled capacitor. Special devices known as varicap diodes are designed to perform this function.

Switching Time

One rather important characteristic for some diodes is switching time. This refers to the time it takes for a diode to switch from one state to the other. This condition is important in computer applications that involve rapid turn-on and turn-off times.

When a junction diode is forward biased, its depletion zone begins to fill with current carriers. This action produces a large number of electron–hole recombinations near the junction. If the bias voltage were suddenly removed, the recombination process would not stop instantly. Current carriers have a type of inertia that causes them to continue moving once they are placed in motion. The nonconduction state cannot be reached until all the current carriers have cleared the depletion zone. This means that the depletion zone has a current carrier storage function.

State changing from reverse bias to forward bias is also time dependent in a diode. The delay involved in this characteristic is due to junction capacitance. This capacitance tends to absorb the initial forward bias current when it is being charged. After the charging operation has been completed, current carriers become available for normal conduction. Special switching diodes are now available for most high-speed control operations.

DIODE PACKAGING

Diodes are manufactured today in a wide range of case styles and packages. A person working with these devices must be familiar with some of the common methods of packaging. Element identification and lead marking techniques are essential for proper installation and testing. A diode improperly connected in a circuit may be damaged or cause damage to other parts. Figure 10-16 shows three general classes of diodes according to their current-handling capability. These represent some of the more popular package styles in use today.

Low-current diodes [Fig 10-16(a)] are the smallest of all packages. Body length rarely exceeds 0.3 cm in length. The cathode is usually denoted by a color band. Glass-packaged diodes often have two or three color bands to indicate specific number types. This group of diodes is generally capable of passing forward current values of approximately 100 mA. The peak reverse voltage rating rarely ever exceeds 100 V. Low reverse current values for these devices are typically 5 μA at 25°C.

Medium-current diodes [Fig. 10-16(b)] are slightly larger in size than the low-current devices. Body size is approximately 0.5 cm with larger connecting leads. Diodes in this group can pass forward current values up to 5 A. Peak reverse voltage ratings generally do not exceed 1000 V. The anode and cathode terminals may be identified by a diode symbol on the body of the device. A color band near one end of the case is also used to identify the cathode. Low- and medium-current diodes are usually mounted by soldering. Any heat generated

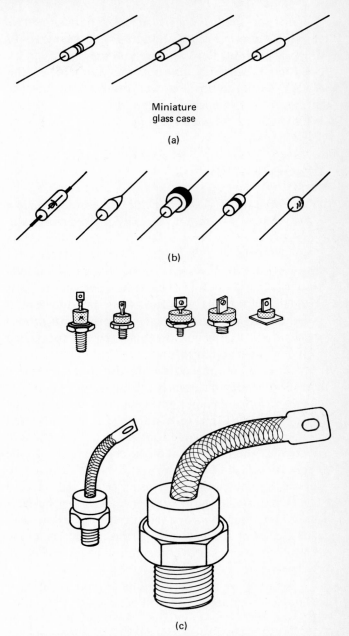

Figure 10-16. Diode packages: (a) low-power diodes; (b) medium-power diodes; (c) high-power diodes.

during operation is carried away by air or lead conduction.

High-current or power diodes are the largest of all diode types [see Fig. 10-16(c)]. These devices normally generate a great deal of heat. Air convection of heat is generally not adequate for most installations. These devices are designed to be mounted on metal heat sinks. Heat is conducted away from the diode by metal. Diodes of this classification can pass hundreds of amperes of forward current. Peak reverse voltage ratings are in the 1000 V range. Numerous packaging types and styles are used to house power diodes. As a rule, the rating of the device and method of installation usually dictate its package type.

DIODE TESTING

Diode operation is determined by its connection in a circuit. Current conduction is directly dependent on the polarity of the device. Heavy conduction occurs when it is forward biased and very little current flows during reverse biasing. Connecting a diode backward in a circuit reverses its conduction. This can destroy the diode and possibly damage several other circuit parts. A person working with diodes must be absolutely certain that they are connected properly into a circuit.

The schematic symbol of a diode is commonly used to identify its polarity. Figure 10-17 shows a diode symbol, element names, and the polarity of the crystal material. The term anode is commonly used to describe the electrode that attracts or gathers in electrons. The P material of the crystal serves as the anode in a diode. The term "cathode" is used to denote the electrode that gives off or emits electrons. The N material of the crystal serves as the cathode. It should be noted that forward current passes through a diode from the cathode to the anode. A diode symbol shows this moving from the bar to the point of the arrow.

Diodes are available today in many different package types and styles. Because of the many differences in these devices there is often some degree of confusion about lead identification. A

person working with these devices is frequently called on to test diodes and identify leads. Volt-ohm-milliameters (VOMs) or vacuum-tube voltmeters (VTVMs) are used to perform this operation. The ohmmeter section of the instrument is used to perform this test.

Figure 10-18 shows how an ohmmeter is used to test a diode. In preparation for this test the meter should be placed in the $R \times 100$ or 1 kΩ range. The

Figure 10-18. Diode testing with an ohmmeter: (a) forward bias test connection; (b) reverse bias test connection.

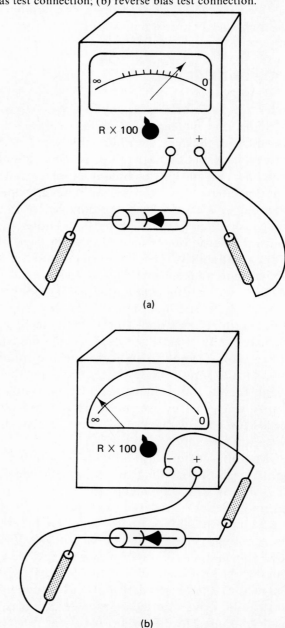

(a)

(b)

Figure 10-17. Diode: (a) crystal structure; (b) symbol and element names.

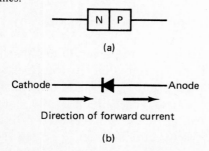

(a)

Cathode ——————|◀|—————— Anode

Direction of forward current

(b)

two meter leads are then connected to the diode. Exact resistance readings are not particularly important. Only high- or low-resistance indications are used. Figure 10-18(a) shows how an ohmmeter is connected to forward bias a diode. This test will normally cause the meter to show a low resistance. Reverse biasing of a diode is shown in Fig. 10-18(b). An infinite resistance is indicated by the ohmmeter during this test.

A good diode will normally show low resistance when forward biased and high resistance when reverse biased. If a diode does not respond in this manner, it is probably defective. A diode that shows low resistance in both directions is considered to be shorted. Shorting usually occurs when the maximum current or voltage ratings of the device are exceeded. Shorted diodes often cause damage to other circuit components. Blown fuses, tripped circuit breakers, or overheated resistors are generally good indicators of a shorted diode. It is a good practice to test other circuit components before replacing a shorted diode.

When a diode shows infinite resistance in both directions it is considered to be open. Open diodes generally cause a circuit to be nonconductive. This condition rarely causes damage to other circuit parts. Operation simply stops and there is no current flow through the diode. Open diodes do not occur very often in electronic circuits. A quick ohmmeter check will readily detect this type of fault should it occur.

An ohmmeter can also be used to identify the leads of a diode. The polarity of the ohmmeter voltage source must be known for this test to be effective. Straight polarity ohmmeters have the black or common lead negative and the red lead positive. A diode connected to the ohmmeter will show low resistance when forward biased and infinite resistance when reverse biased. Forward biasing occurs when the polarity of the ohmmeter matches the crystal polarity of the diode. The red lead is therefore connected to the anode and the black lead to the cathode. Lead identification is simply a matter of finding the forward bias direction of a diode and noting the polarity of the ohmmeter leads.

Ohmmeters with reverse polarity have the black lead positive and the red lead negative. Lead identification with this instrument is achieved by connecting the ohmmeter to the diode and deter-

mining the direction of forward biasing. Forward biasing will again cause the ohmmeter to read low resistance. The lead polarity of the diode, however, is a reverse of the ohmmeter lead polarity. The positive lead must therefore be connected to the cathode with the negative lead to the anode. Lead identification is simply a matter of finding the forward bias direction of a diode and then noting the reverse polarity of the ohmmeter.

It is essential that the polarity of an ohmmeter be known before it can effectively be used to identify diode leads. An unfamiliar meter can be easily checked with a separate dc voltmeter. Simply connect the voltmeter to the ohmmeter probes and observe the voltage indication. An up-scale voltage indication shows that the ohmmeter voltage matches the voltmeter polarity. A straight polarity ohmmeter will respond in this manner. Reverse polarity ohmmeters will cause an off-scale voltage indication to occur when the probe polarity is matched. As a rule, this polarity test should be made on all ranges of the ohmmeter. In some instruments, there may be a reverse polarity on only the higher resistance ranges.

The voltage value of an ohmmeter supply source is also an important consideration when testing diodes. If the supply voltage is less than 0.2 V, it will not be large enough to forward bias all diodes. This could cause a good diode to appear to be open. The ohmmeter would therefore show an infinite resistance in both directions. You may recall that it takes at least 0.3 V to cause conduction in germanium diodes and 0.7 V for silicon. Modern ohmmeters are often designed with a low supply voltage so that they will not turn on a P-N junction.

The supply voltage of some ohmmeters may also be extremely large. This type of meter could actually develop enough current or voltage to damage certain diodes. As a rule, high-frequency detection diodes are very susceptible to being damaged in this way. To avoid this kind of problem, do not attempt to test high-frequency diodes with high-voltage ohmmeters.

An ohmmeter is probably the most common instrument used to evaluate the status of a diode. The testing procedure is easy to follow and it can be performed very quickly. It is important to know the meter's voltage source polarity and value for these tests to be meaningful. Essentially, you must know your ohmmeter in order to use it effectively.

SEMICONDUCTOR DIODE DEVICES

A number of rather unusual semiconductor devices have been developed as a result of P-N junction theory. These devices also employ two electrodes and are classified as diodes. As a rule, there is some unique difference in each device that distinguishes it from the junction diode. A person working in electronics will find numerous applications of these devices. Some of the more importent of these devices will be discussed here.

Zener Diodes

The characteristics of a regular junction diode show that it is designed primarily for operation in the forward direction. Forward biasing will cause a large I_F with a rather small value of V_F. Reverse biasing will generally not cause current conduction until higher values of reverse voltage are reached. If V_R is great enough, however, breakdown will occur and cause a reverse current flow. Junction diodes are usually damaged when this occurs. Special diodes are now designed to operate in the reverse direction without being damaged. Zener diodes are manufactured specifically for this purpose.

Figure 10-19 shows a test circuit and the volt-ampere characteristic of a typical zener diode. In the forward bias direction this device behaves like an ordinary silicon diode. In the reverse bias direction, there is practically no reverse current flow until the breakdown voltage is reached. When this occurs there is a sharp increase in reverse current. Varying amounts of reverse current can pass through the

Figure 10-19. Zener diode: (a) $I-V$ characteristics; (b) test circuit.

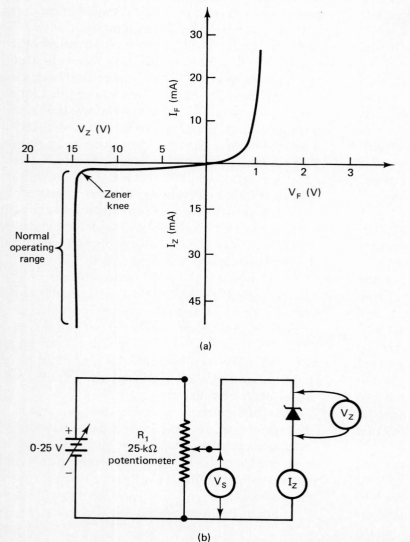

(a)

(b)

diode without damaging it. The breakdown voltage or zener voltage (V_Z) across the diode remains relatively constant. The maximum reverse current is limited, however, by the wattage rating of the diode.

Manufacturers rate zener diodes according to their V_Z value and the maximum power dissipation (P_D) at 25° C. This gives an indication of the maximum reverse current or I_R that a diode can safely conduct. A 1.0-W 15-V zener diode can, for example, conduct an I_R of 0.066 A or 66 mA. This is determined by the basic power equation $P = IE$, by solving for I. Thus

$$\text{maximum } I_R = \frac{P_D}{V_Z}$$

$$= \frac{1\text{ W}}{15\text{ V}}$$

$$= 0.066\text{ A or } 66\text{ mA}$$

Zener diodes are available commercially in a wide range of breakdown voltages. V_Z values range from 1.4 to 200 V with accuracies between 5 and 20%. Each V_Z value is generally specified at minimum zener current I_Z. Figure 10-19 shows that 1 mA of I_Z is needed to enter the knee of the I_R curve. After this value has been reached, the V_Z remains fairly constant over a wide range of I_Z. Power dissipation values are therefore used to indicate the safe operating range. Typical P_D ratings are from 150 mW to 50 W. When purchasing this kind of a device, one usually asks for a 1-W zener diode at 8.2 V_Z with an accuracy of 10%.

The symbol and crystal structure of a zener diode is shown in Fig. 10-20. Notice that it is very similar to that of a regular diode. The cathode is, however, drawn with a bent line to distinguish it from a regular diode symbol. This is supposedly done to represent the letter Z. Zener diodes are normally used only in the reverse bias direction. This means that the anode must be connected to the

negative side of the voltage source and the cathode to the positive.

An important difference in zener diodes and regular silicon diodes is in the way they are used in circuits. Because of the constant V_Z characteristic of a zener diode, it is used primarily to regulate voltages. A large change in I_R will cause only a small change in V_Z. This means that a zener diode can be used as an alternate current path. The constant V_Z developed across the diode can then be applied to a load device. The load voltage thus remains at a constant value by altering the current flow through the zener diode.

Figure 10-21 shows a representative zener diode regulator circuit. Notice that the circuit contains a series resistor, zener diode, load device, and a dc power supply. The primary function of this circuit is to maintain the load voltage at a constant 15 V. With 25 V applied, current through R_S will cause a voltage drop of 10 V. The output voltage or V_Z across the diode is therefore 15 V. An increase in the supply voltage to 30 V would simply cause the zener to conduct more current. The voltage drop across R_S would therefore increase to 15 V. This in turn would cause V_Z and the load voltage to remain at 15 V. A decrease in supply voltage to 20 V would cause a reduction in total current. The voltage drop across R_S would now drop to only 5 V. The load voltage across the diode would continue to be 15 V. This means that the regulated output will remain at 15 V for a change in supply voltage from 20 to 30 V. Voltage regulation is an important function of many electronic circuits today.

Light-Emitting Diodes

The light-emitting diode (LED) is another important diode device. It contains a P-N junction that emits light when passing forward current. A schematic symbol and typical package of the LED

Figure 10-20. Zener diode symbol and crystal structure.

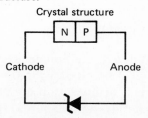

Crystal structure

Cathode　　　Anode

Figure 10-21. Zener diode regulator circuit.

Ac input | Unregulated dc power supply | 25 V dc | Series resistor R_S | R_L | 15-V regulated dc

15-V 1-W zener diode

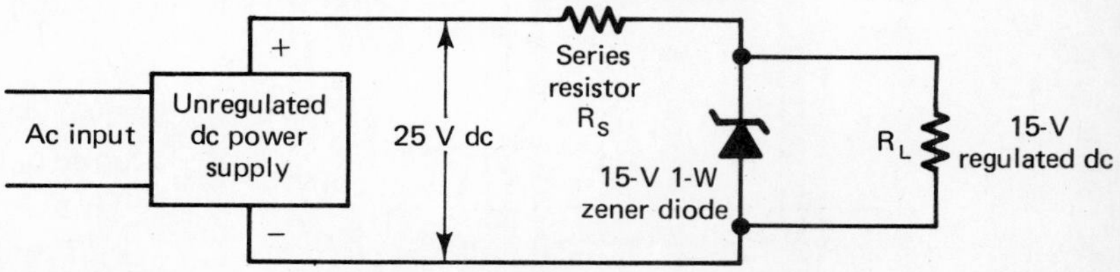

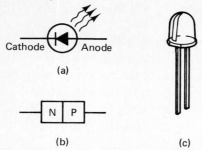

Figure 10-22. Light-emitting diode: (a) symbol; (b) crystal structure; (c) transparent lens.

are shown in Fig. 10-22(a) and (b). The lens of an LED [Fig. 10-22(c)] is transparent and focuses on the P-N junction.

When an LED is forward biased, electrons cross the junction and combine with holes. This action causes electrons to fall out of conduction and return to the valence band. The energy possessed by each free electron is then released. Some of this appears as heat energy and the rest of it is given off as light energy. Special materials such as gallium arsenide (GaAs) and gallium phosphide (GaP) cause this to take place when used in a P-N junction. These materials are purposely used because they emit visible light energy.

Light-emitting diodes are relatively small, ruggedly constructed, and have a very long life expectancy. They can be switched on and off very quickly. Construction techniques permit this type of device to be made in various shapes and patterns. A very common application is in numeric displays indicating the numbers 0 through 9. A representative LED seven-segment display is shown in Fig. 10-23.

Figure 10-23. Seven-segment LED display. Segment forward biasing produces the number 6.

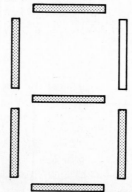

By selecting different combinations of segments, a desired number can be produced. Displays of this type are found in calculators, digital meters, clocks, and wristwatches.

Photovoltaic Cells

Photovoltaic cells are unique diode devices that generate a voltage when illuminated. These devices convert light energy directly into electrical energy. No outside source of electricity is needed to produce current. Figure 10-24 shows the schematic symbol and some representative photovoltaic cells.

Figure 10-25 shows the crystal structure of a junction photovoltaic cell. Basic construction is very similar to that of a junction diode. In operation with no light applied, the cell will not cause a current flow. Essentially, this means that the junction is not being properly biased. However, when light is applied, an interesting condition can be observed. Photocurrent (I_{PH}) will flow in the circuit because of the photovoltage (V_{PH}) that is developed across the junction. V_{PH} is due to the generation of new electron–hole pairs in the depletion region. Electrons move into the N material because of its normal positive ion concentration. Holes are likewise swept into the P material because of its negative-ion content. This action causes the N material to immediately take on a negative charge and the P material to take on a positive charge. The P-N junction therefore responds as a small voltage cell. Because of this, photovoltaic cells are often called solar cells or sun batteries.

Photovoltaic cells are generally classified as direct energy power converters. They will supply electrical power to a load device when light is applied. The operational efficiency of this device is based on the developed electrical power output divided by the light power input. Ideally, a high level of efficiency is desirable. The most efficient cells manufactured today are in the range 10 to 15%. For this reason, photovoltaic cells are used in applications where high levels of light energy are available.

Photodiodes

A photodiode is a P-N junction device that will conduct current when energized by light energy. This device is designed for operation in the reverse bias direction. An external voltage source is required

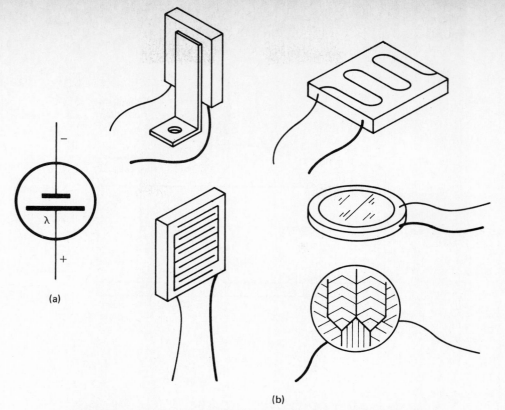

(a)

(b)

Figure 10-24. Photovoltaic cells: (a) symbol; (b) representative cells.

Figure 10-25. Photovoltaic cell operation.

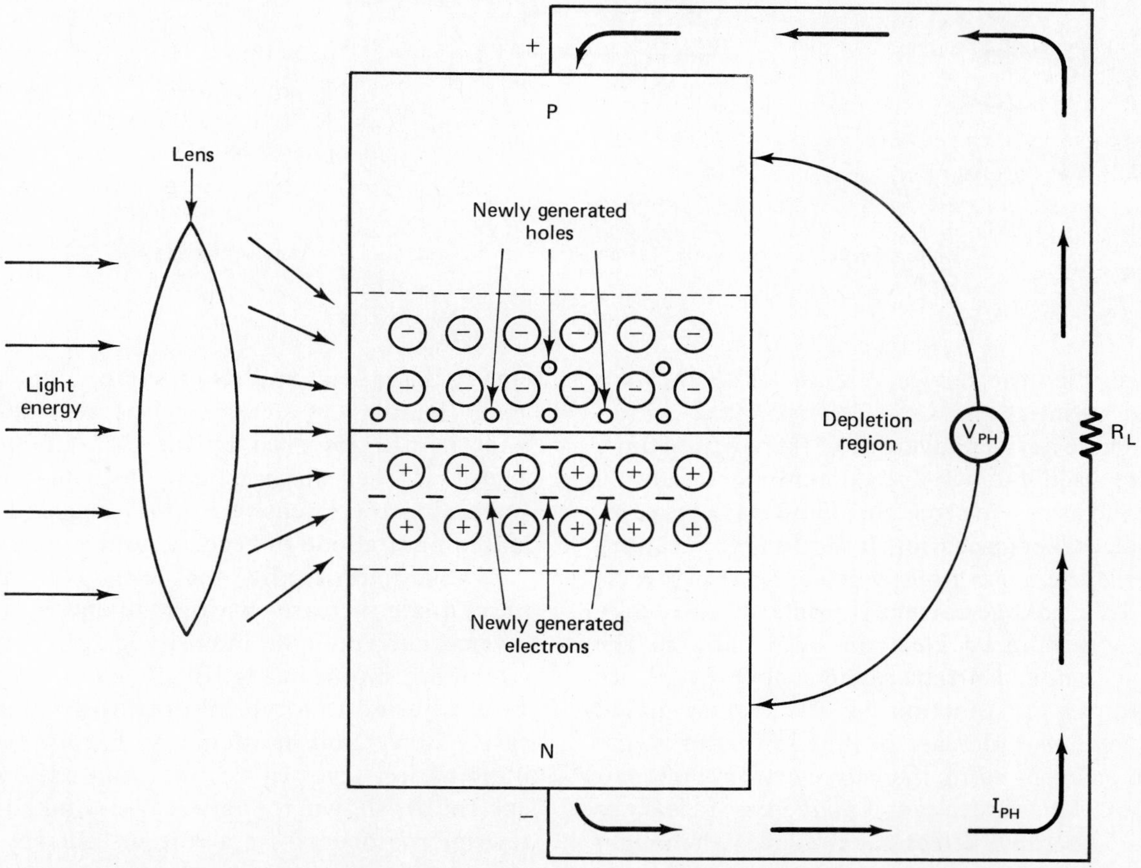

Lens

Light
energy

P

Newly generated
holes

Newly generated
electrons

N

Depletion
region

V_{PH}

R_L

I_{PH}

+

−

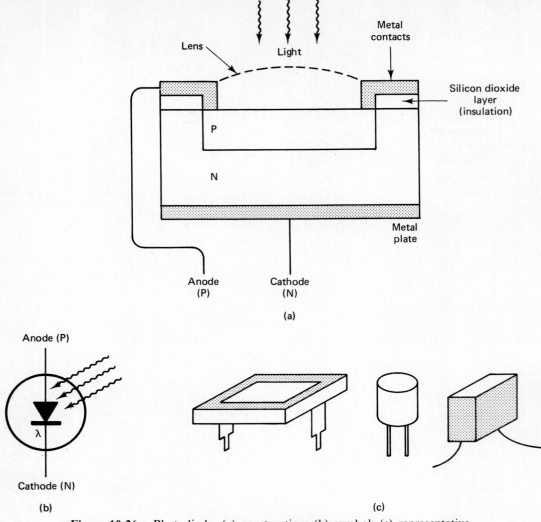

Figure 10-26. Photodiode: (a) construction; (b) symbol; (c) representative types.

to reverse bias the device. Figure 10-26 shows a schematic symbol and some representative photodiodes. These devices have a very fast response time. They are used in high-speed computer counting circuits and as movie projector sound track pickups.

In an earlier discussion, it was found that a very small reverse current flows when a diode is reverse biased. This leakage current is primarily caused by thermally generated electron–hole pairs in the depletion region. Current carriers of this type are swept across the junction by the reverse-biased electric field. An increase in junction temperature causes a corresponding increase electron–hole pair generation. This in turn causes an increase in leakage current. The same effect takes place when light

energy is applied to a P-N junction. Light travels in small bundles of energy called photons. When photons are absorbed by the P-N junction, their energy is used to free new electron–hole pairs. Increasing the intensity of light causes a reverse-biased photodiode to become more conductive.

Characteristically, the reverse current of a photodiode increases with the intensity of light. A reverse current–light intensity graph and its test circuit are shown in Fig. 10-27. The source voltage V_S is adjusted to a typical operating value, in this case -20 V. Note particularly that an increase in illumination level causes a linear rise in reverse current. As shown, the value of I_R is limited to only a few microamperes. As a rule, its current must be

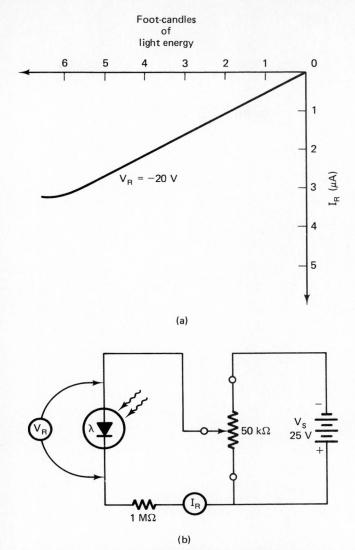

Foot-candles
of
light energy

(a)

(b)

Figure 10-27. Characteristic (a) reverse current light intensity graph and (b) its test circuit.

amplified before it can be made useful. Photodiodes are generally used to drive transistor amplifiers. Phototransistors are now available that have a photodiode and a transistor combined in a single device.

Varactor Diodes

In an ordinary junction diode, the depletion region is an area which separates the P and N material on each side of the junction. This area develops when the junction is initially formed. It represents a unique part of the diode that is essentially void of current carriers. In this regard, the depletion region serves as a dielectric medium or insulator.

When a diode has bias voltage applied, its depletion region will change in width. In a sense, this means that a diode responds as a voltage-variable capacitor. By definition, a capacitor is two or more conductors separated by an insulating material. The P and N materials, being semiconductors, are separated by a depletion region insulator. Devices designed to respond to the capacitance effect are called varactors, varicap diodes, or voltage-variable capacitors.

A varactor diode is a specially manufactured P-N junction with a variable concentration of impurities in its P-N materials. In a conventional diode doping impurities are usually distributed equally throughout the material. Varactors have a very light dose of impurities near the junction. Moving away from the junction the impurity level increases. This type of construction produces a much steeper voltage–capacitance relationship. Figure 10-28 shows a comparison of the capacitance between a conventional silicon diode and a varactor diode. It can be seen that an ordinary diode possesses only a small value of internal capacitance. In general this capacitance is too small to be of practical value.

Varactor diodes are normally operated in the reverse bias direction. With an increase in reverse biasing the depletion region increases its width. This means less resulting capacitance. A decrease in reverse bias voltage causes a corresponding increase in capacitance. In effect, the capacitance of a diode varies inversely with its bias voltage. This relationship, however, does not change linearly. Notice the nonlinear area between 0 and 0.2 V_R of the varactor diode of Fig. 10-28. Its normal range of operation is between zero and the reverse breakdown voltage.

A number of schematic symbols are used to represent the varactor diode. Figure 10-29 shows three rather popular methods of representation. The symbol on the left has a small capacitor inside the circle surrounding the diode. This symbol tends to be used more frequently than the other two. Industrial electronic circuits and specialized engineering schematics seem to use the other two symbols.

Varactor diodes are used primarily in resonant circuits where some level of tuning or frequency control is desired. Figure 10-30 shows a representative varactor tuned LC circuit. The resonant frequency of this circuit is determined by the inductor L and the series combination of C_D and C_1. C_1 normally has a larger capacitance value than C_D. C_1 is also used to prevent the dc bias voltage of C_D from

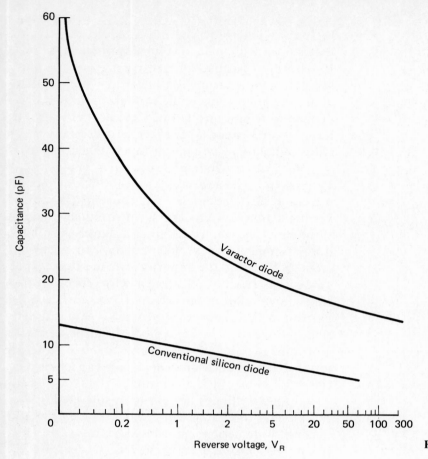

Figure 10-28. Diode capacitance values, $T_A = 25°\text{C}$.

Figure 10-29. Varactor diode symbols.

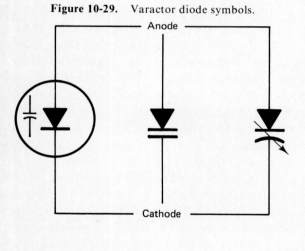

Figure 10-30. Varactor LC circuit.

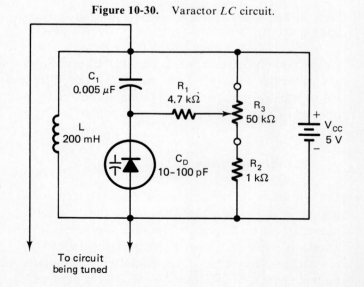

going into the inductor. R_1, R_2, and R_3 serve as the bias voltage control network for C_D. Adjusting R_2 will change the amount of reverse bias voltage to C_D. Moving the wiper arm of R_2 up causes more reverse bias voltage to C_D. This increases the capacitance of C_D. Moving the wiper arm down reduces the bias voltage, which causes a reduction in capacitance. Used in this manner the varactor becomes a tuning component for resonant frequency. Other applications of the varactor diode are automatic frequency control AFC in FM radio and TV receivers. Varactor tuning applications are widespread in new solid-state circuitry.

VACUUM TUBES

When certain metals are heated to a high enough temperature they will give off electrons. This principle is called thermionic emission. It is the operational basis of all vacuum tubes. Figure 10-31 shows an example of thermionic emission.

In the operation of a vacuum tube, heat is developed by an electrically energized metal conductor. This electrode is similar to the filament of an incandescent lamp. When current passes through the filament it generates heat. At a certain temperature electrons are emitted from its surface. This

action causes the electrons to form a cloud around the filament. An additional electrode, placed inside an evacuated globe, is called a plate or anode (see Fig. 10-32). When this electrode is made positive with respect to the filament, it will attract the emitted electrons. Electrons or current will flow between these electrodes when the tube is in operation. There is no direct connection between these two electrodes inside the globe. After reaching the plate, current will flow out of the tube and through the external circuit. This condition of operation is called the Edison effect. All vacuum tubes, regardless of their construction differences, respond to this effect.

The vacuum tube in Fig. 10-33 has a directly heated filament. This means that electrons are emitted directly from the surface of the filament. Tubes of this type are designed primarily for high-power applications. The filament is made of a rather

Figure 10-32. Edison effect.

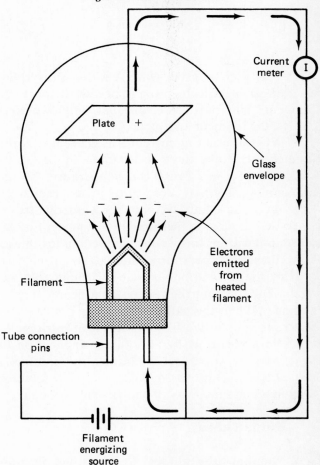

Figure 10-31. Principle of thermionic emission.

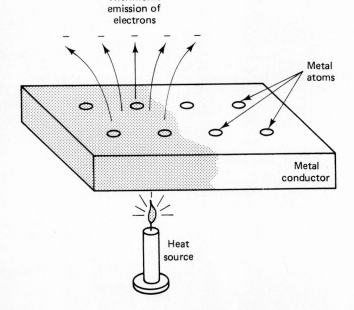

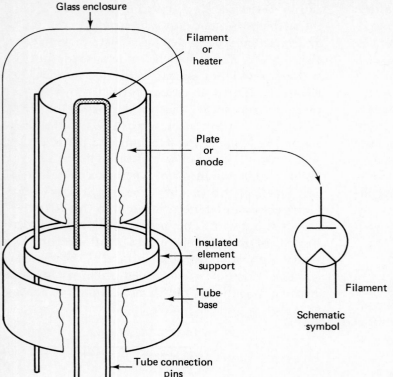

Glass enclosure

Filament
or
heater

Plate
or
anode

Insulated
element
support

Tube
base

Tube connection
pins

Filament

Schematic
symbol

Figure 10-33.   Directly heated vacuum tube.

large diameter tungsten wire. A rather substantial amount of electrical power is needed to heat the filament. This particular tube has only two electrodes, the filament and plate.

When alternating current is used to energize the filament of a tube there is usually a change in its structure. Ac will cause the temperature of the filament to fluctuate according to its frequency. Sixty-hertz ac is injected into the electron stream when the filament is energized by the power line. In most applications this would produce a noticable ac hum during its operation. To alter this problem, a tubular-shaped cathode is placed around the filament. The filament then serves as a heater for the cathode. When the cathode becomes hot enough, electrons are emitted from its surface. This type of tube emits electrons by an indirect application of heat. See the structural diagram of an indirectly heated tube of Fig. 10-34. This type of structure is used in a large majority of vacuum tubes today.

Diode Tubes

Vacuum-tube diodes are classified as unidirectional conduction devices. Electron flow or current occurs only in one direction. Initially, the

Figure 10-34.   Indirectly heated tube.

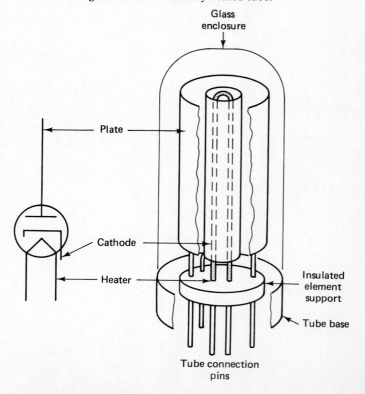

Glass
enclosure

Plate

Cathode

Heater

Insulated
element
support

Tube base

Tube connection
pins

filament or cathode emits electrons from its surface. Making the plate more positive than the cathode will cause it to attract electrons. Internally, electron flow occurs only between filament and cathode. Upon reaching the plate, current will then flow through an external circuit before returning to the cathode.

Two circuit conditions of a vacuum-tube diode are shown in Fig. 10-35. When the plate is made positive and the cathode negative, a tube is forward biased [see Fig. 10-35(a)]. This condition causes the tube to conduct plate current I_P. Reverse biasing occurs when the cathode is made positive and the anode negative [see Fig. 10-35(b)]. No current flows through a reverse-biased diode. Essentially, electrons are not emitted from the cold plate and they will not be attracted by the heated cathode. A normal diode has no current between its electrodes when it is reverse biased. The peak reverse voltage (PRV) of a vacuum tube diode is generally very high. This is due primarily to the space between electrodes.

When the heater of a diode tube is energized for a few seconds it is ready to be placed into operation. Different values of plate voltage V_P will cause a change in plate current I_P. With a constant filament voltage, changes in V_P and I_P can be plotted to make a characteristic curve for the diode. Figure 10-36 shows a representative characteristic curve of a vacuum-tube diode. Notice that an increase in V_P

causes a corresponding increase in I_P. At the beginning of the curve I_P does not increase very rapidly. This causes the curve to be nonlinear at the start. As the value of V_P increases, the curve begins to change into a straight line. This is called the linear region. The topmost part of the curve shows another, nonlinear region. This is called the saturation point. At this point a further increase in V_P does not cause an increase in I_P. Essentially, the plate is attracting all the electrons emitted by the cathode. The I_P therefore levels off when saturation occurs.

As a general rule, the V_P–I_P characteristics of a diode vacuum tube are only concerned with forward bias operation. The reverse characteristics are simply not shown. Ordinarily, no current flows in the reverse direction until the applied voltage is high enough to cause an arc between the electrodes. When this occurs there is permanent damage to the electrodes. The PRV rating usually indicates where this will occur.

When using a vacuum-tube diode there are four ratings that must be considered:

1. *Maximum plate current (I_{max})*: The maximum plate current that can flow through without causing electrode damage.
2. *Plate dissipation (P_D)*: Power loss due to heat developed by electrons striking the plate.

Figure 10-35. Operating conditions of a vacuum-tube diode: (a) forward bias; (b) reverse bias.

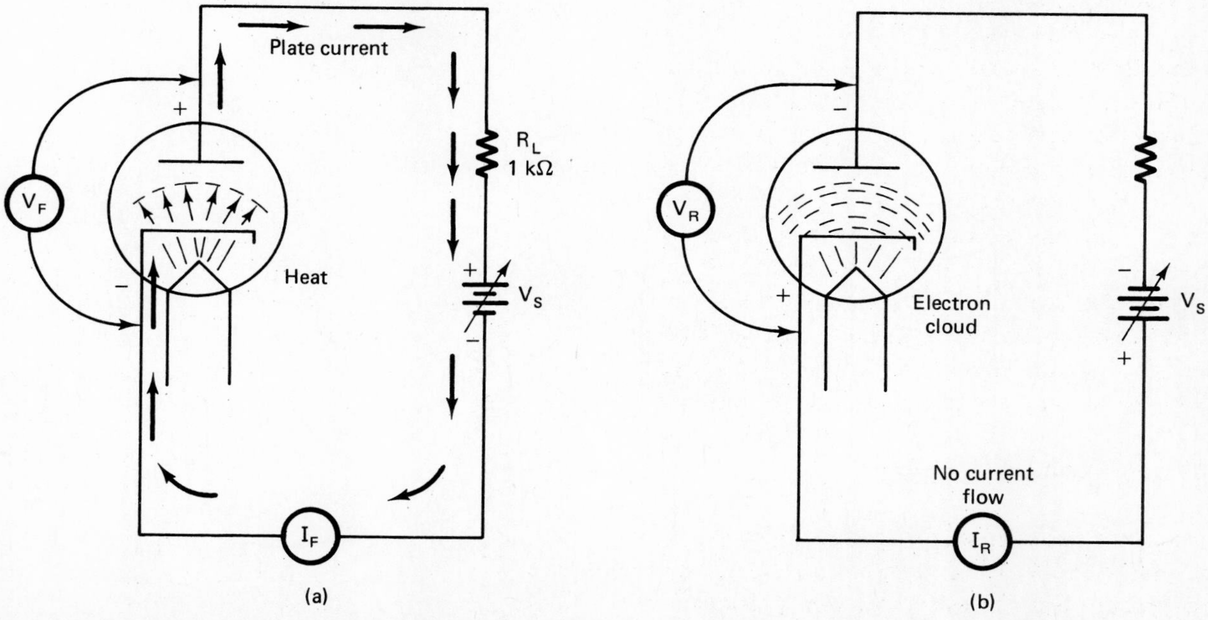

(a)

(b)

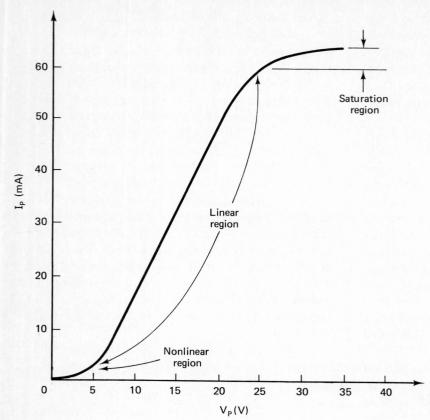

Figure 10-36. V_p–I_p characteristics of a vacuum-tube diode.

3. *Peak reverse voltage (PRV)*: The maximum reverse bias voltage that can be applied without causing electrode damage.

4. *Plate resistance (R_P)*: The internal resistance that current encounters while passing between electrodes. It is calculated by the formula

$$R_p = \frac{\Delta V_p}{\Delta I_p}$$

Some representative vacuum-tube diodes are shown in Fig. 10-37. Connection to the electrodes is made through the bottom pins or the top cap. The enclosure or envelope is usually made of glass or glass-lined metal. The size of the tube is dependent primarily on its current and voltage capability. In general, these devices are rather fragile and should be handled with care.

Figure 10-37. Representative vacuum-tube diodes.

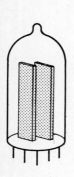
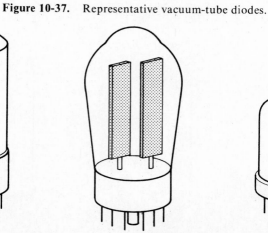
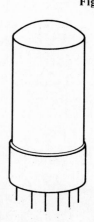

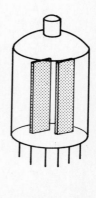

Review

1. When P and N materials are connected together, what causes the depletion zone to be formed?

2. What is barrier potential, and what causes it in a semiconductor diode?

3. Explain what is meant by forward biasing, and explain the influence that it has on the depletion zone and the barrier potential in a semiconductor diode.

4. Explain why conduction increases when a semiconductor diode is forward biased.

5. What causes leakage current in a semiconductor diode?

Student Activities

Diode Characteristics

1. Connect the diode characteristic test circuit of Fig. 10-10(b). Adjust the V_S control in 0.1-V increments of V_F while observing I_F.

2. Plot an I_F–V_F characteristic curve on graph paper.

3. Test the forward characteristics of a silicon diode and a germanium diode, if possible.

4. To see the reverse characteristic, simply reverse the polarity of the diode and alter the voltage.

Diode Testing

1. With an ohmmeter of a VTVM or a VOM, test the forward and reverse resistance of several diodes.

2. If possible, test a shorted and an open diode. Compare the resistance differences.

3. Identify the voltage polarity of the ohmmeter leads. Record the voltage and lead polarity. Identify the leads of several diodes.

Zener Diode Characteristics

1. Connect the zener diode characteristic test circuit of Fig. 10-19. Adjust V_S in 1 V increments while observing V_Z and I_Z. The diode should be from 2 to 15 V_Z at 1 W.

2. Reverse the polarity of the diode and test the forward conduction of the zener diode.

3. Plot an I_F–V_Z characteristic curve on graph paper.

Light-Emitting Diode Characteristics

1. Connect the LED circuit of Fig. 10-38. Starting at 0 V_S, increase V_F in 0.5-V increments until illumination occurs.

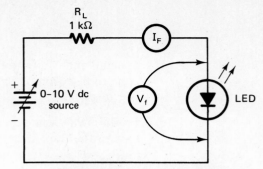

Figure 10-38. LED test circuit.

2. Record the V_F and I_F that occurs when the LED is lighting.

3. Reverse the polarity of the LED and repeat the procedure.

Photovoltaic Cell Characteristics

1. Connect the leads of a photovoltaic cell to a dc voltmeter.

2. Determine the output voltage of the cell at normal room light, near a window, in bright sunlight, and in total darkness.

3. Record the voltage values for each test condition.

Photodiode Characteristics

1. Connect the photodiode characteristic test circuit of Fig. 10-27. Adjust V_R to −20 V dc.

2. Apply varying levels of light intensity to the diode while observing I_R.

3. If the intensity of light does not cause a change in I_R, the diode is probably not connected in the reverse bias direction.

4. Record the I_R for three different levels of light from dark to bright.

Vacuum-Tube Diode Characteristics

1. Connect the forward conduction test circuit of Fig. 10-35. The V_F and I_F values are dependent on the tube selected. For a 6AL5 dual diode tube a V_F of 0 to 10 V will produce an I_F of 9 mA. The filament of this tube is heated by 6.3 V ac.

2. Plot an I_F–V_F curve for the tube on graph paper.

11

Electronic Power Supplies

All electronic systems require certain voltage and current values for operation. The energy source of the system is primarily responsible for this function. As a general rule, the source is more commonly called a power supply. In some systems, the power supply may be a simple battery or a single dry cell. Portable radios, tape recorders, and calculators are examples of systems that employ this type of supply. Television receivers, computer terminals, stereo amplifier systems, and electronic instruments usually derive their energy from the ac power line. This part of the system is called an electronic power supply. Electronic devices are used in the power supply to develop the required output energy.

In an electronic system the power supply serves as the primary source of electrical energy. As a rule,

this part of the system is responsible for changing energy of one form into something that is more useful. The block diagram of Fig. 11-1 shows the location of the power supply in an electronic system.

An electronic power supply is often considered to be a smaller part or subsystem within the overall system. It has its own energy source, path, control, load, and indicator. Figure 11-2 shows a block diagram of the electronic power supply as a subsystem.

The energy source of an electronic power supply is alternating current. In most systems this is supplied by the ac power line. Most small electronic systems use 120-V single-phase 60-Hz ac as their energy source. This energy is readily available in homes, buildings, and industrial facilities.

Figure 11-1. Electrical system parts.

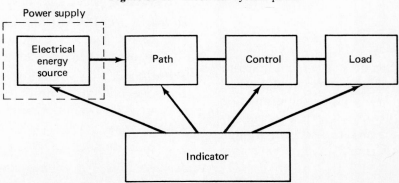

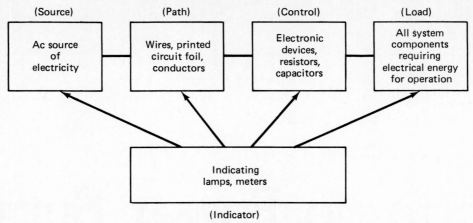

Figure 11-2. Power supply subsystem.

The path of electrical energy in a power supply is very similar to that of other systems. Copper wire and the metal foil of a printed circuit board serves as the conduction path. Conductors are generally insulated to prevent them from touching. Electrical energy flows to each system component through insulated conductors.

The control function of an electronic power supply is achieved by several components. Transformers, resistors, capacitors, inductors, and electronic devices are all used to achieve control. Each component has a unique role to play in the operation of the supply. The type of system, its operational capacity, and application have a great deal to do with the amount of control required.

The load of a power supply is somewhat unusual. A composite of all electrical and electronic components that are supplied energy serves as the load. These components do some type of work when energized. Heat, light, sound, and mechanical motion are typical examples of the work produced by load components. A specific resistor value will serve as the load of a power supply in this discussion.

IMPORTANT TERMS

In a discussion of electronic power supplies, one frequently encounters a number of new and unusual terms. These terms generally make the discussion more meaningful if they are reviewed before studying the chapter. Some of the terms are discussed in detail in the chapter, while others are listed for reference purposes.

Alternation. The positive or negative part of an ac sine wave.

Anode. The positive terminal or electrode of an electronic device such as a solid-state diode.

Average value. An ac voltage or current value based on the average of all instantaneous values for one alternation. $V_A = 0.637 \times V_{max}$.

Capacitance input filter. A filter circuit employing a capacitor as the first component of its input.

Cathode. A negative terminal of an electronic device such as a solid-state diode.

Center tap. An electrical connection point at the center of a wire coil or transformer.

Counter EMF. The voltage induced in a conductor through a magnetic field which opposes the original source voltage.

Forward bias. A voltage connection method that produces current conduction in a P-N junction.

Inductance. The property of an electric circuit that opposes a change in current flow.

Peak value. The maximum voltage or current value of an ac sine wave. $V_p = 1.414 \times$ rms.

Pi filter. A filter with an input capacitor connected to an inductor–capacitor filter, forming the shape of the Greek letter pi.

Pulsating dc. A voltage or current value that rises and falls at a rate or frequency with current flow always in the same direction.

Rectification. The process of changing ac into pulsating dc.

Reverse bias. A voltage connection method producing very little or no current through a P-N junction.

Schematic. A diagram of an electronic circuit made with symbols.

Time constant (RC). The amount of time required for the voltage of a capacitor in an *RC* circuit to increase 63.2% of the value or decrease to 36.7% of the value.

Turns ratio. The ratio of the number of turns in the primary winding to the number of turns in the secondary winding.

POWER SUPPLY FUNCTIONS

All electronic power supplies have a number of functions that must be performed in order to produce operation. Some of these functions are achieved by all power supplies. Others are somewhat optional and are dependent primarily on the parts being supplied. A block diagram of a general-purpose electronic power supply is shown in Fig. 11-3. Each block represents a specific power supply function.

Transformer

Electronic power supplies rarely operate today with ac obtained directly from the power line. A transformer is commonly used to step the line voltage up or down to a desired value. A schematic symbol and a simplification of a power supply transformer are shown in Fig. 11-4.

The coil on the left of the symbol is called the primary winding. The ac input is applied to this winding. The coil on the right side is the secondary winding. The output of the transformer is developed by the secondary winding. The parallel lines near the center of the symbol are representative of the core. In a transformer of this type, the core is usually made of laminated soft steel. Both windings are placed on the same core. Power supply transformers may have more than one primary and secondary winding on

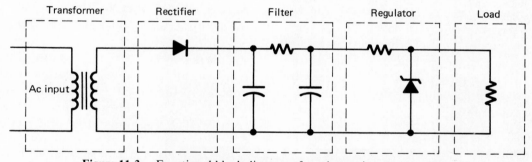

Figure 11-3. Functional block diagram of an electronic power supply.

Figure 11-4. (a) Power supply transformer; (b) schematic symbol.

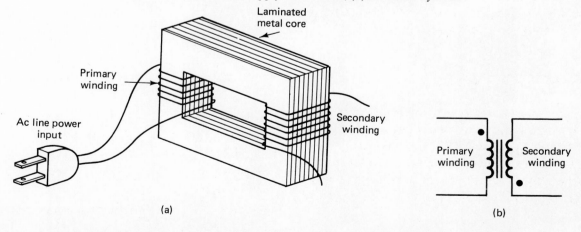

the same core. This permits it to accommodate different line voltages and to develop alternate output values.

The output voltage of a power supply transformer may have the same polarity as the input voltage or it may be reversed. This depends on the winding direction of the coils. A dot is often used on the symbol to indicate the polarity of a winding. The polarity of the input and output are the same where the dots are indicated.

The output voltage developed by a transformer is primarily dependent on the turns ratio of its windings. If the primary winding has 500 turns of wire, and the secondary has 1000 turns, the transformer has a 1:2 turns ratio. This type of transformer will step up the input voltage by a factor of 2. With 120 V applied from the power line, the output voltage will be approximately 240 V. It is interesting to note that the current capability of a transformer is the reverse of its turns ratio. A 1:2 turns ratio will step up the voltage and step down the current capability. One ampere of primary current is capable of producing only 0.5 A of secondary current. The resistance of the wire is largely responsible for this condition. A 1:2 turns ratio will have twice as much resistance in the secondary as it has in the primary winding. An increase in resistance will therefore cause a decrease current. This is all based on the wire size of the two windings being the same. Figure 11-5 shows some representative transformer types that are widely used today.

Power supplies for solid-state systems generally develop low voltage values. Transformers for this type of supply are designed to step down the line voltage. A turns ratio of 10:1 is very common. With 120 V applied to the input, the output will be approximately 12 V. The output current capabilities of this transformer are increased by a ratio of 1:10. This means that a step-down transformer has a low voltage output with a high current capacity. Supplies of this type are well suited for solid-state applications.

Rectification

The primary function of a power supply is to develop dc for the operation of electronic devices. Most power supplies are initially energized by ac. This energy must be changed into dc before it can be used. The process of changing ac into dc is called *rectification*. All electronic power supplies have a rectification function. Figure 11-6 shows a graphic representation of the rectification function. Notice that rectification can be either half-wave or full-wave. Half-wave rectification uses one alternation of the ac input, while full-wave uses both alternations.

HALF-WAVE RECTIFICATION. In half-wave rectification, only one alternation of the ac input appears in the output. The negative alternation of Fig. 11-7(a) has been eliminated by the diode rectifier. The resulting output of this circuit is called

Figure 11-5. Representative transformers. (Courtesy of TRW/UTC Transformers.)

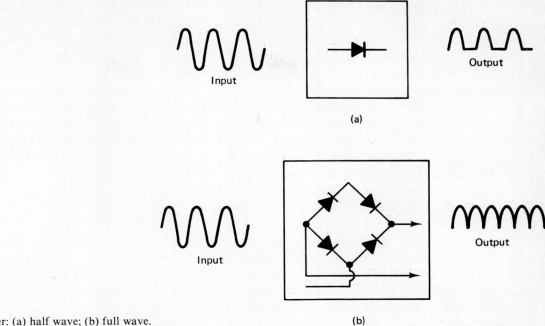

Figure 11-6. Rectifier: (a) half wave; (b) full wave.

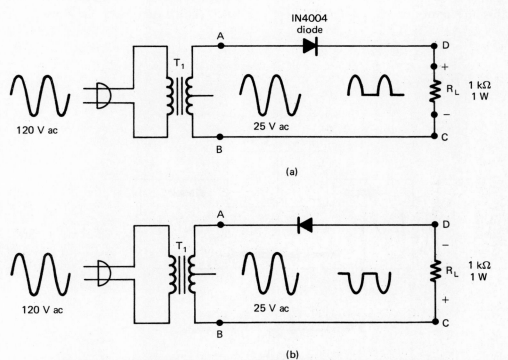

Figure 11-7. Half-wave power supplies: (a) positive output; (b) negative output.

pulsating dc. A half-wave rectifier can remove either the positive or negative alternation, depending on its polarity in the circuit. When the output has a positive alternation, the resulting dc is positive with respect to the common point or ground. A negative

output occurs when the positive alternation is removed. Figure 11-7(b) shows an example of a rectifier with a negative output.

The rectification function of a modern power supply is performed by solid-state diodes. This **type** of device has two electrodes, known as the anode and cathode. Operation is based on its ability to conduct easily in one direction but not in the other direction. As a result of this, only one alternation of the applied ac will appear in the output. This occurs only when the diode is forward biased. The anode must be positive and the cathode negative. Reverse biasing will not permit conduction through a diode.

A simplification of the rectification process is shown in Fig. 11-8. Part (a) shows how the diode will respond during the positive alternation. This alternation causes the diode to be forward biased. Note that the anode is positive and the cathode is negative. The resulting current flow is shown by arrows. The output of the rectifier appears across R_L. It is positive at the top of R_L and negative at the bottom.

Figure 11-8(b) shows how the diode will respond during the negative alternation. This alternation causes the diode to be reverse biased. The anode is now negative and the cathode is positive. No current will flow during this alternation. The output across R_L is zero for this alternation.

The dc output of a half-wave rectifier appears as a series of pulses across R_L. A dc voltmeter or ammeter would therefore indicate the average value of these pulses over a period of time. Figure 11-9(a) shows an example of the pulsating dc output. The average value of one pulse is 0.637 of the peak value. Since the second alternation of the output is zero, the composite average value must take into account the time of both alternations. The composite output is therefore $0.637 \div 2$ or 0.318 of the peak value. Figure 11-9(b) shows a graphic example of this value.

The equivalent dc output of a half-wave rectifier without filtering can be calculated when the value of the ac input is known. The peak value of one alternation is determined by first multiplying the rms value by 1.414. The composite dc output is then determined by multiplying the peak value by 0.318.

A simplified method of calculating the dc output of a half-wave rectifier is to combine different values. This means that $1.414 \times 0.637 \div 2$ equals 0.45. The composite dc output of a half-wave

Figure 11-8. Simplification of rectification: (a) positive alternation; (b) negative alternation.

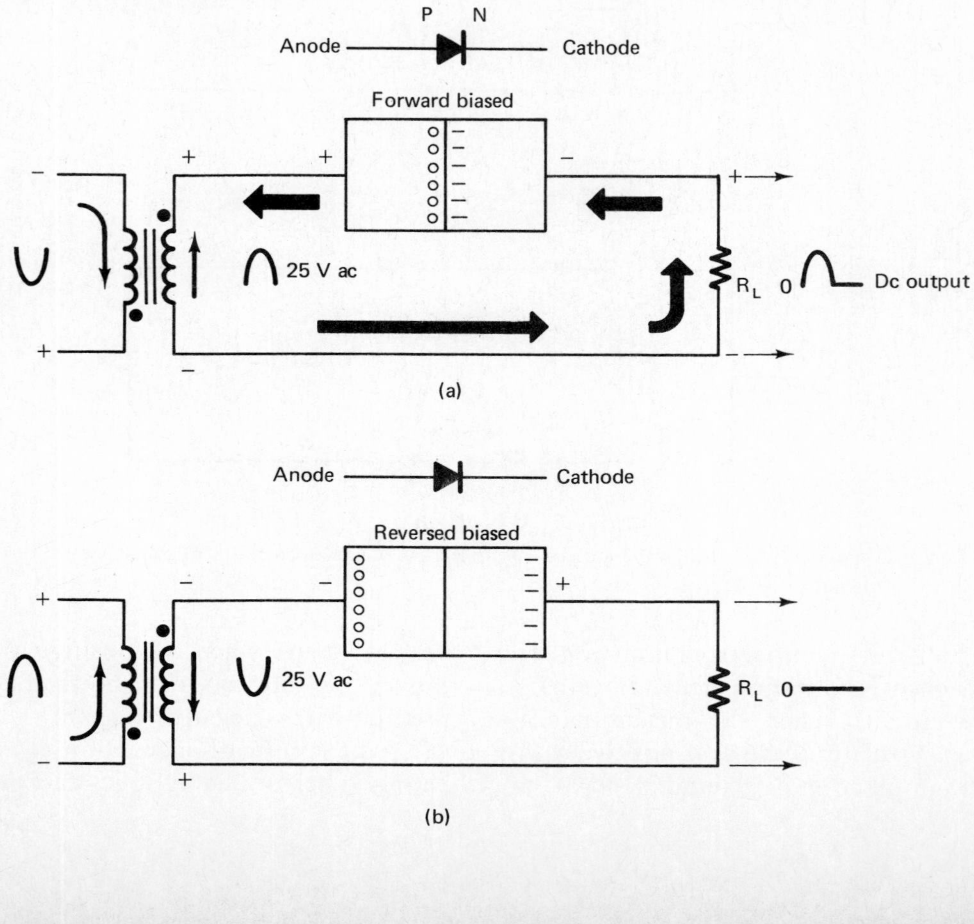

(a)

(b)

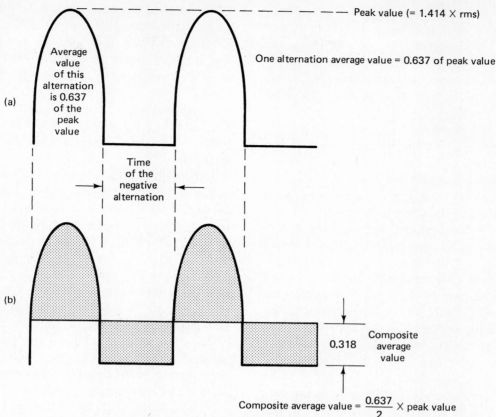

(a)

Peak value (= 1.414 × rms)

Average value of this alternation is 0.637 of the peak value

One alternation average value = 0.637 of peak value

Time of the negative alternation

(b)

0.318

Composite average value

Composite average value = $\dfrac{0.637}{2}$ × peak value

Figure 11-9. Half-wave rectifier output: (a) pulsating dc output; (b) composite output.

rectifier is therefore 45%, or 0.45 of the rms input voltage. In an actual circuit we must also take into account the voltage drop across the diode. When a silicon diode is used, this would be 0.6 V. The dc output of an unfiltered half-wave rectifier would therefore be 0.45 × rms − 0.6 V. Using this procedure, determine equivalent dc voltage of the half-wave rectifier of Fig. 11-8.

The frequencies of the output pulses of a half-wave rectifier occur at the same rate as the applied ac input. With 60 Hz input, the pulse rate or ripple frequency of a half-wave rectifier is 60 Hz. Only one pulse appears in the output for each complete sine-wave input. Potentially, this means that only 50% of the ac input is transformed into a usable dc output. Half-wave rectification is therefore only 50% efficient. Due to its low efficiency rating, half-wave rectification is not widely used today.

FULL-WAVE RECTIFICATION. A full-wave rectifier responds as two half-wave rectifiers that conduct on opposite alternations of the input. Both alternations of the input are changed into pulsating dc output. Full-wave rectification can be achieved by two diodes and a center-tapped transformer or by four diodes in a bridge circuit. Both types of rectifiers are widely used in solid-state power supplies today.

A schematic diagram of a two-diode full-wave rectifier is shown in Fig. 11-10. Note that the anodes of each diode are connected to opposite ends of the transformer secondary winding. The cathode of

Figure 11-10. Two-diode full-wave rectifier.

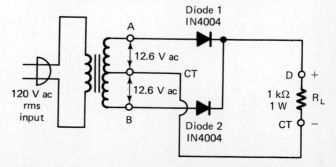

each diode is then connected together to form a common positive output. The load of the power supply (R_L) is connected between the common cathode point and the center-tap connection of the transformer. A complete path for current is formed by the transformer, two diodes, and the load resistor.

When ac is applied to the primary winding of the transformer, it steps the voltage down in the secondary winding. The *center tap* serves as the electrical neutral or center of the secondary winding. Half of the secondary voltage will appear between points CT and A and the other half between CT and B. These two voltage values are equal and will always be 180° out of phase with respect to point CT.

Assume now that the 60-Hz input is slowed down so that we can see how one complete cycle responds. In Fig. 11-11 note the polarity of the secondary winding voltage. For the first alternation point A is positive and point B is negative with respect to the center tap. The second alternation causes point A to be negative and point B to be positive with respect to CT. The center tap will therefore always be negative with respect to the positive end of the winding. This point then serves as the negative output of the power supply.

Conduction of a specific diode in a full-wave rectifier is based on the polarity of the applied

voltage during each alternation. In Fig. 11-11 this is made with respect to point CT. For the first alternation, point A is positive and B is negative. This polarity causes D_1 to be forward biased and D_2 to be reverse biased. Current flow for this alternation is shown by the solid arrows. Starting at CT, electrons will flow through a conductor to R_L and D_1, and return to point A. This current flow causes a pulse of dc to appear across R_L for the first alternation.

For the second alternation, point A becomes negative and point B is positive. This polarity will forward bias D_2 and reverse bias D_1. Current flow for this alternation is shown by the dashed arrows. Starting at point CT, electrons flow through a conductor to R_L, through D_2, and return to point B. This causes a pulse of dc to appear across R_L for the negative alternation.

It should be obvious at this point that current flow through R_L is in the same direction for each alternation of the input. This means that each alternation is transposed into a pulsating dc output. Full-wave rectification therefore changes the entire ac input into dc output.

The resulting dc output of a full-wave rectifier is 90%, or 0.90 of the ac voltage between the center tap and the outer ends of the transformer. This voltage value is determined by calculating the peak value of the rms input voltage, then multiplying it by

Figure 11-11.   Conduction of a full-wave rectifier.

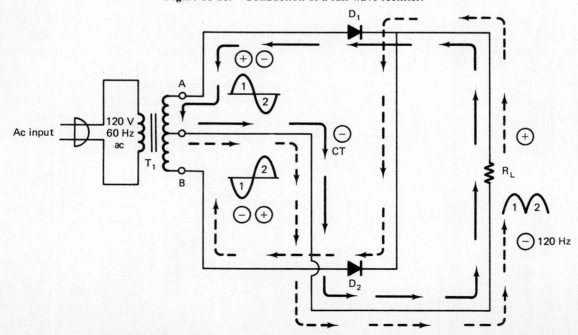

the average value. Since 1.414×0.637 equals 0.90, 90% of the rms value is the equivalent dc output. The output of a full-wave rectifier is 50% more efficient than that of an equivalent half-wave rectifier.

The actual dc output of a full-wave rectifier will be slightly less than 0.90 of the rms input just described. Each diode, for example, will reduce the dc voltage by 0.6 V. A more practical calculation would be dc output = rms $\times$ 0.90 − 0.6 V. In a low-voltage power supply a reduction of the dc by 0.6 V may be quite significant.

The ripple frequency of a full-wave rectifier is somewhat different from that of a half-wave rectifier. Each alternation of the input, for example, produces a pulse of output current. The ripple frequency will therefore be twice the alternation frequency. For a 60-Hz input the ripple frequency of a full-wave rectifier is 120 Hz. As a general rule, a 120-Hz ripple frequency is easier to filter than the 60 Hz of a half-wave rectifier.

FULL-WAVE BRIDGE RECTIFIERS. A bridge rectifier requires four diodes to achieve full-wave rectification. In this configuration, two diodes will conduct during the positive alternation and two will conduct during the negative alternation. A center-tapped transformer is not required to achieve full-wave rectification with a bridge circuit.

The component parts of a full-wave bridge rectifier are shown in Fig. 11-12. In this circuit, one side of the ac input is applied to the junction of diodes D_1 and D_2. The alternate side of the input is applied to diodes D_3 and D_4. The diodes at each input are reversed with respect to each other. Output

of the bridge occurs at the other two junctions. The cathodes of D_1 and D_3 serve as the positive output. Negative output appears at the anode junction of D_2 and D_4. The load of the power supply is connected across the common anode and common cathode connection points.

When ac is applied to the primary winding of a power supply transformer, it can be either stepped up or down, depending on the desired dc output. In Fig. 11-12, the 120-V input is stepped down to 25 V rms. In normal operation, one alternation will cause the top of the transformer to be positive and the bottom to be negative. The next alternation will cause the bottom to be positive and the top to be negative. Opposite ends of the secondary winding will always be 180° out of phase with each other.

Assume now that the ac input causes point A to be positive and B to be negative for one alternation. The schematic diagram of a bridge in Fig. 11-13 shows this condition of operation. With the indicated polarity, diode D_1 is forward biased and D_2 is reverse biased at the top junction. At the bottom junction D_4 is forward biased and D_3 is reverse biased. When this occurs, electrons will flow from point B through D_4, up through R_L, through D_1, and return to point A. Solid arrows are used to show this current path. The load resistor sees one pulse of dc across it for this alternation.

With the next alternation, point A of the schematic diagram becomes negative and point B becomes positive. When this occurs, the bottom diode junction becomes positive and the top junction goes negative. This condition forward biases diodes D_2 and D_3 while reverse biasing D_1 and D_4. The

Figure 11-12. Bridge rectifier components.

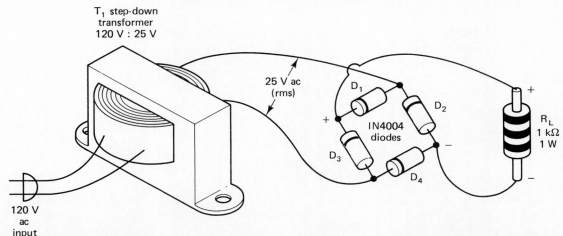

resulting current flow is indicated by the dashed arrows starting at point A. Electrons flow from A through D_2, up through R_L, through D_3 and return to the transformer at point B. Dashed arrows are used to show this current path. The load resistor also sees a pulse of dc across it for this alternation.

The output of the bridge rectifier has current flow through R_L in the same direction for each alternation of the input. Ac is therefore changed into a pulsating dc output by conduction through a bridge network of diodes. The dc output, in this case, has a ripple frequency of 120 Hz. Each alternation produces a resulting output pulse. With 60 Hz applied, the output is 2×60, or 120 Hz.

The dc output voltage of a bridge circuit is slightly less than 90% of the rms input. Each diode, for example, reduces the output by 0.6 V. With two diodes conducting during each alternation, the dc output voltage is reduced by 0.6×2, or 1.2 V. In the circuit of Fig. 11-13, the dc output would be $25 V \times 0.9 - 1.2 V$, or 21.3 V. The output of a bridge circuit is slightly less than that of an equivalent two-diode full-wave rectifier. The voltage drop of the second diode accounts for this difference.

The four diodes of a bridge rectifier can be obtained today in a single package. Figure 11-14 shows several different package types. As a general rule, there are two ac input terminals and two dc output terminals. These devices are generally rated according to their current-handling capability and peak reverse voltage rating. Typical current ratings are from 0.5 to 50 A in single-phase ac units. PRV ratings are 50, 200, 400, 800, and 1000 V. A bridge package usually takes less space than four single diodes. Nearly all bridge power supplies employ the single package assembly today.

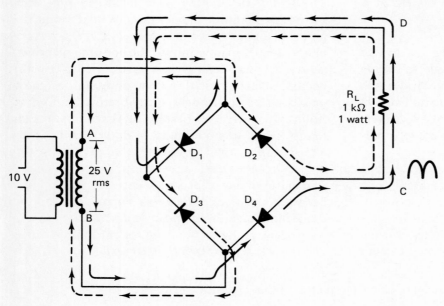

Figure 11-13. Bridge rectifier conduction.

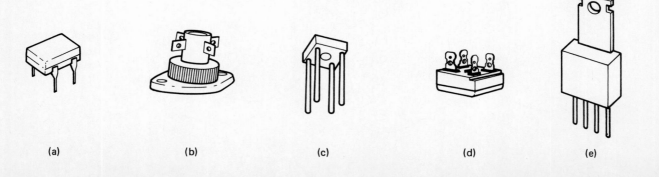

Figure 11-14. Packaged bridge rectifier assemblies: (a) dual-in-line package; (b) sink-mount package; (c) epoxy package; (d) tab-pack enclosure; (e) thermo-tab package.

(a)   (b)   (c)   (d)   (e)

DUAL POWER SUPPLIES. A rather new variation of the power supply is now being used in some systems. This supply has both negative and positive output with respect to ground. Dual or split power supplies, as they are known, have been developed as a voltage source for integrated circuits. The secondary winding of the input transformer is devided into two parts. The center tap or neutral serves as a common ground connection for the two outside windings. Each half of the winding has a complete full-wave rectifier.

A dual power supply with two full-wave rectifiers is shown in Fig. 11-15. Notice that the output of this supply is +10.7 and −10.7 V.

Diodes D_1 and D_2 are rectifiers for the positive supply. They are connected to opposite ends of transformer T_1. Diodes D_3 and D_4 are rectifiers for the negative supply. They are connected in a reverse direction to the opposite ends of the transformer. The positive and negative output is with respect to the center tap of the transformer.

For one alternation assume that the top of the transformer is positive and the bottom negative. Current flows out of CT and divides at the junction of R_{L_1} and R_{L_2}. Equal amounts of current flow in each direction. The top of R_{L_1} becomes positive with the bottom of R_{L_2} being negative. The path is made complete through D_1 to point A and D_4 to point B of the transformer. Solid arrows show the path of current flow for this alternation.

The next alternation makes the top of the transformer negative and the bottom positive. Current again flows out of CT and divides at the junction of R_{L_1} and R_{L_2}. Equal amounts of current flow in each direction. The top of R_{L_1} continues to be positive while the bottom of R_{L_2} is negative. One path is made complete through D_3 to point A and the other path is through D_2 to point B. The direction of current flow through R_{L_1} and R_{L_2} is the same as the first alternation. The resulting output voltage of the supply appears at the top and bottom of R_{L_1} and R_{L_2}. Dashed arrows show the direction of current flow for the second alternation.

Filtering

The output of a half- or full-wave rectifier is pulsating direct current. This type of output is generally not usable for most electronic circuits. A rather pure form of dc is usually required. The filter section of a power supply is designed to change pulsating dc into a rather pure form of dc. Filtering takes place between the output of the rectifier and the input to the load device. Power supplies discussed up to this point have not employed a filter circuit.

Figure 11-15. Dual power supply conduction.

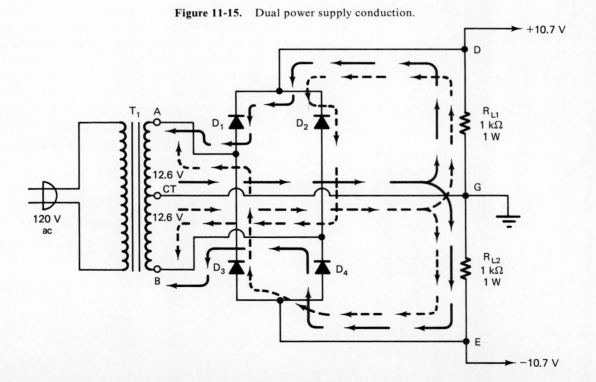

The pulsating dc output of a rectifier contains two components. One of these deals with the dc part of the output. This component is based on the combined average value of each pulse. The second part of the output refers to its ac component. Pulsating dc, for example, occurs at 60 Hz or 120 Hz, depending on the rectifier being employed. This part of the output has a definite ripple frequency. Ripple must be minimized before the output of a power supply can be used by most electronic devices.

Power supply filters fall into two general classes according to the type of component used in its input. If filtering is first achieved by a capacitor, it is classified as a capacitor or *C* input filter. When a coil of wire or inductor is used as the first component, it is classified as an inductive or *L*-input filter. *C* input filters develop a higher value of dc output voltage than does an *L* filter. The output voltage of a *C* filter usually drops in value when the load increases. *L* filters, by comparison, tend to keep the output voltage at a rather constant value. This is particularly important when large changes in the load occur. The output voltage of an *L* filter is however somewhat lower than that of a *C* filter. Figure 11-16 shows a graphic comparison of output voltage and load current for *C* and *L* filters.

C-INPUT FILTERS. The ac component of a power supply can be effectively reduced by a *C*-input filter. A single capacitor is simply placed across the

Figure 11-16. Load current versus output voltage comparisons for *L* and *C* filters.

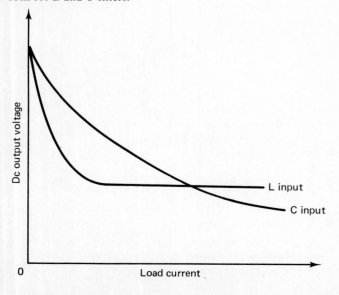

load resistor as shown in Fig. 11-17. For alternation 1 of the circuit [Fig. 11-17(a)] the diode is forward biased. Current flows according to the arrows of the diagram. *C* charges very quickly to the peak voltage value of the first pulse. At the same point in time, current is also supplied to R_L. The initial surge of current through a diode is usually quite large. This current is used to charge *C* and supply R_L at the same time. A large capacitor however, responds somewhat like low resistance when it is first being charged. Notice the amplitude of the I_d waveform during alternation 1.

When alternation 2 of the input occurs the diode is reverse biased. Figure 11-17(b) shows how the circuit responds for this alternation. Notice that there is no current flow from the source through the diode. The charge acquired by *C* during the first alternation now finds an easy discharge path through R_L. The resulting discharge current flow is indicated by the arrows between *C* and R_L. In effect, R_L is now being supplied current even when the diode is not conducting. The voltage across R_L is therefore maintained at a much higher value. See the V_{RL} waveform for alternation 2 in Fig. 11-17(c).

Discharge of *C* continues for the full time of alternation 2. Near the end of the alternation there is somewhat of a drop in the value of V_{RL}. This is due primarily to a depletion of capacitor charge current. At the end of this time, the next positive alternation occurs. The diode is again forward biased. The capacitor and R_L both receive current from the source at this time. With *C* still partially charged from the first alternation, less diode current is needed to recharge *C*. In Fig. 11-17(c), note the amplitude change in the second I_d pulse. The process from this point on is a repeat of alternations 1 and 2.

The effectiveness of a capacitor as a filter device is based on a number of factors. Three very important considerations are:

1. The size of the capacitor being used
2. The value of the load resistor R_L
3. The time duration of a given dc pulse

These three factors are related to one another by the formula

$$T = R \times C$$

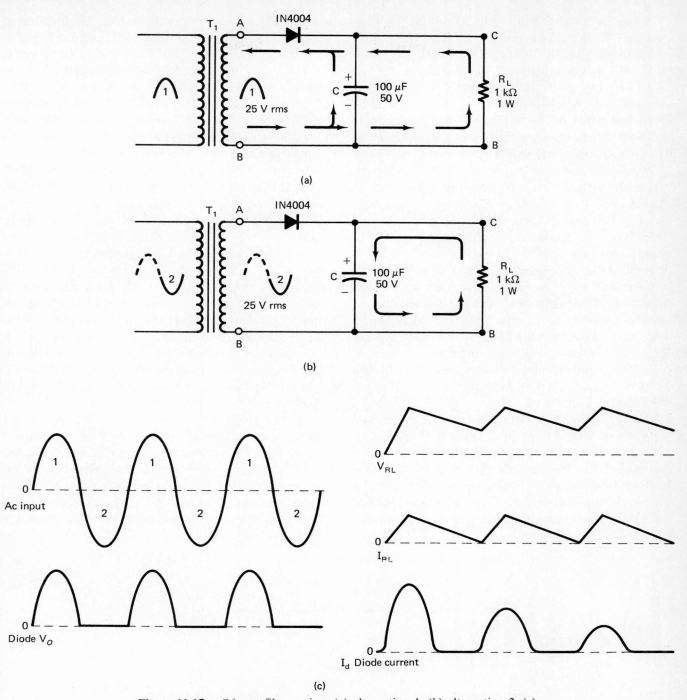

Figure 11-17. *C*-input filter action. (a) alternation 1; (b) alternation 2; (c) waveforms.

where *T* is the time in seconds, *R* the resistance in ohms, and *C* the capacitance in farads. The product *RC* is an expression of the filter circuit's time constant. *RC* is a measure of how rapidly the voltage and current of the filter respond to changes in input voltage. A capacitor will charge to 63.2% of the applied voltage in one time constant. A discharging capacitor will have a 63.2% drop of its original value in one time constant. It takes five time constants to fully charge or discharge a capacitor.

The filter capacitor of Fig. 11-17 charges very quickly during the first positive alternation. Essentially, there is very little resistance for the RC time constant during this period. The discharge of C is, however, through R_L. If R_L is small, C will discharge very quickly. A large value of R_L will cause C to discharge rather slowly. For good filtering action, C must not discharge very rapidly during the time of one alternation. When this occurs, there is very little change in the value of V_{RL}. A C-input filter works very well when the value of R_L is relatively large. If the value of R_L is small, as in a heavy load, more ripple will appear in the output.

A rather interesting comparison of filtering occurs between half-wave and full-wave rectifier power supplies. The time between reoccurring peaks is twice as long in a half-wave circuit as it is for full-wave. The capacitor of a half-wave circuit therefore has more time to discharge through R_L. The ripple of a half-wave filter will be much greater than that of a full-wave circuit. In general, it is easier to filter the output of a full-wave rectifier. Figure 11-18 compares half- and full-wave rectifier filtering with a single capacitor.

The dc output voltage of a filtered power supply is usually a great deal higher than that of an unfiltered power supply. In Fig. 11-18 the waveforms show an obvious difference between outputs. In the filtered outputs, the capacitor will charge to the peak value of the rectified output. The amount of discharge action that takes place is based on the resistance of R_L. For a light load or high resistance R_L, the filtered output will remain charged to the peak value. The peak value of the 10 V rms is 14.14 V. For the unfiltered full-wave rectifier the dc output would be approximately $0.9 \times$ rms, or 9.0 V. Comparing 9 V to 14 V shows a rather decided difference in the two outputs. For the half-wave rectifier the output difference is even greater. The unfiltered half-wave output is approximately 45% of 10 V rms, or 4.5 V. The filtered output is 14 V less 20% for the added ripple, or approximately 11.2 V. The filtered output of a half-wave rectifier is nearly 2.5 times more than that of the unfiltered output. These values are only rough generalizations of power supply output with a light load.

INDUCTANCE FILTERING. An inductor is a device that has the ability to store and release electrical energy. It does this by taking some of the applied current and changing it into a magnetic field. An increase in current through an inductor causes the magnetic field to expand. A decrease in current causes the field to collapse and release its stored energy.

The ability of an inductor to store and release energy can be used to achieve filtering. An increase in the current passing through an inductor will cause a corresponding increase in its magnetic field. Voltage induced into the inductor due to a change in its field will oppose a change in the current passing through it. A decrease in current flow causes a

Figure 11-18. (a) Half-wave and (b) full-wave filter comparisons.

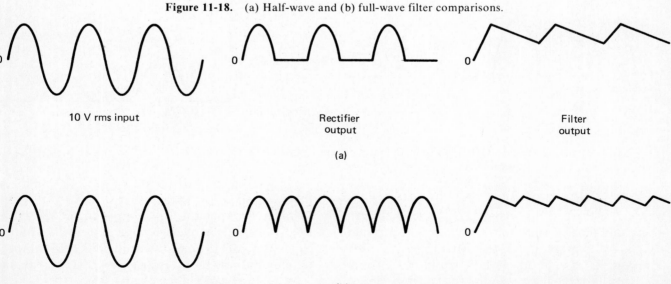

similar reaction. A drop in current will cause its magnetic field to collapse. This action also induces voltage into the inductor. The induced voltage in this case causes a continuation of the current flow. As a result of action, the added current tends to bring up its decreasing value. An inductor will oppose any change in its current flow. Inductive filtering is well suited for power supplies with a large load current.

An inductive input filter is shown in Fig. 11-19. The inductor is simply placed in series with the rectifier and the load. All of the current being supplied to the load must pass through the inductor (L). The filtering action of L will not let a pronounced change in current take place. This prevents the output voltage from reaching an extreme peak or valley. Inductive filtering in general does not produce as high an output voltage as does capacitive filtering. Inductors tend to maintain the current at an average value. A larger load current can be drawn from an inductive filter without causing change in its output voltage.

When an inductor is used as the primary filtering element it is commonly called a choke. The term "choke" refers to the ability of an inductor to reduce ripple voltage and current. In electronic power supplies, choke filters are rarely used as a single filtering element. A combination inductor–capacitor or LC filter is more widely used. This filter has a series inductor with a capacitor connected in parallel with the load. The inductor controls large changes in load current. The capacitor, which follows the inductor, is used to maintain the load voltage at a constant voltage value. The combined filtering action of the inductor and capacitor produces a rather pure value of load voltage. Figure 11-20 shows a representative LC filter and its output.

PI FILTERS. When a capacitor is placed in front of the inductor of an LC filter it is called a pi filter. In effect, this circuit becomes a CLC filter. The two capacitors are in parallel with R_L and the inductor is in series. Component placement of this circuit in a schematic diagram resembles the Greek capital letter pi (Π). This accounts for the name of the filter. See the schematic diagram of a pi filter in Fig. 11-21.

The operation of a pi filter can best be understood by considering L_1 and C_2 as an LC filter. This part of the circuit acts upon the output voltage developed by the capacitor-input filter C_1. C_1 charges

Figure 11-19.   Inductive filtering.

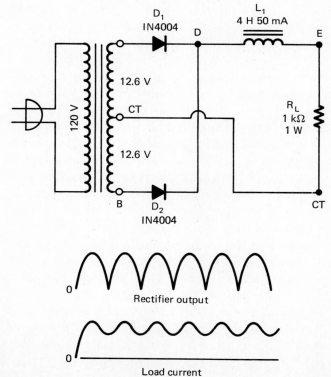

Figure 11-20. LC filter.

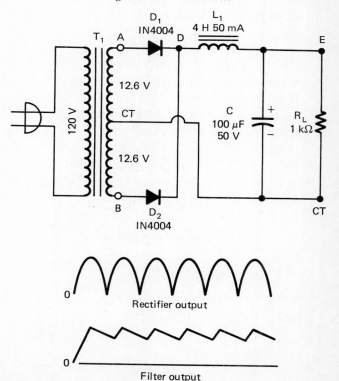

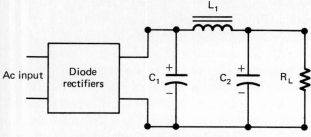

Figure 11-21. *CLC* (pi) filter.

to the peak value of the rectifier input. It has a ripple content that is very similar to that of *C*-input filter of Fig. 11-17. This voltage is then applied to C_2 through inductor L_1. C_1 charges C_2 through L_1. C_2 then holds its charge for a time interval determined by the time constant of C_2–R_L. As a result of this action, there is additional filtering by L_1 and C_2. The ripple content of this filter is much lower than that of a single *C*-input filter. There is, however, a slight reduction in the dc supply voltage to R_L due to the voltage drop across L_1.

A pi filter has a very low ripple content when used with a light load. An increase in load, however, tends to lower its output voltage. This condition tends to limit the number of applications of pi filters. Today we find them used to supply radio circuits, in stereo amplifiers, and in TV receivers. The input to this filter is nearly always supplied by a full-wave rectifier.

RC FILTERS. In applications where less filtering can be tolerated, an *RC* filter can be used in place of a pi filter. As shown in Fig. 11-22, the inductor is replaced with a resistor. An inductor is rather expensive, quite large physically, and weighs a great deal more than a resistor. The performance of the *RC* filter is not quite as good as that of a pi filter. There is usually a reduction in dc output voltage and increased ripple.

Figure 11-22. *RC* filter.

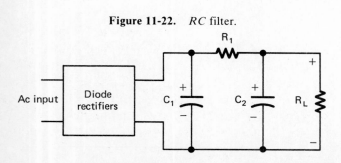

In operating, C_1 charges to the peak value of the rectifier input. A drop in rectifier input voltage causes C_1 to discharge through R_1 and R_L. The voltage drop across R_1 lowers the dc output to some extent. C_2 charges to the peak value of the R_L voltage. The dc output of the filter is dependent on the load current. High values of load current cause more voltage drop across R_1. This in turn lowers the dc output. Low values of load current have less voltage drop across R_1. The output voltage therefore increases with a light load. In practice, *RC* filters are used in power supplies that have 100 mA or less of load current. Its primary advantage is reduced cost.

Voltage Regulation

The dc output of an unregulated power supply has a tendency to change value under normal operating conditions. Changes in ac input voltage and variations in the load are primarily responsible for these fluctuations. In some power supply applications, voltage changes do not represent a serious problem. In many electronic circuits voltage changes may cause improper operation. When a stable dc voltage is required, power supplies must employ a voltage regulator. The block diagram of a power supply in Fig. 11-23 shows where the regulator is located.

A number of voltage regulator circuits have been developed for use in power supplies. One very common method of regulation employs the zener diode. Figure 11-24 shows this type of regulator located between the filter and the load. The zener diode is connected in parallel or shunt with R_L. This regulator requires only a zener diode D_z and a series resistor (R_S). Notice that D_z is placed across the filter circuit in the reverse bias direction. Connected in this way, the diode will go into conduction only when it reaches the zener breakdown voltage V_z. This voltage then remains constant for a large range of zener current I_z. Regulation is achieved by altering the conduction of I_z through the zener diode. The combined I_z and load current I_L must all pass through the series resistor. This current value then determines the amount of voltage drop across R_s. Variations in current through R_s are used to keep the output voltage at a constant value.

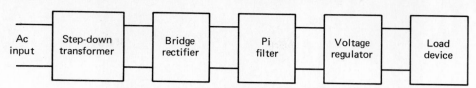

Figure 11-23. Block diagram of a power supply.

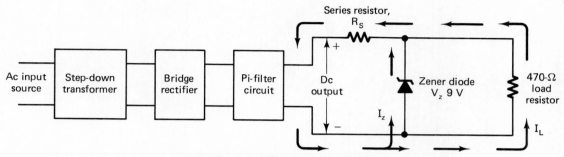

Figure 11-24. Zener diode regulator.

A schematic diagram of a 9-V regulated power supply is shown in Fig. 11-25. This circuit derives its input voltage from a 12.6-V transformer. Rectification is achieved by a self-contained bridge rectifier assembly. Filtering is accomplished by an *RC* filter. The series resistor of this circuit has two functions. It first couples capacitors C_1 and C_2 together in the filter circuit. Second, it serves as the series resistor for the regulator circuit. Diode D_z is a 9-V 1-W zener diode.

Operation of the regulated power supply is similar to that of the bridge circuit discussed earlier. Full-wave output from the rectifier is applied to C_1 of the filter circuit. C_1 then charges to the peak value of the rms input less the voltage drop across two silicon diodes (1.2 V). This represents a value of $12.6 \times 1.414 - 1.2$, or 16.6 V dc input. R_s therefore has a voltage drop of $16.6 - 9$ V, or 7.6 V. This represents a total current flow passing through R_s of $7.6 \text{ V}/100\Omega = 0.076$ A or 76 mA. With the bleeder resistor R_8 serving as a fixed load, there is $9 \text{ V}/10 \text{ k}\Omega = 0.9$ mA or 0.0009 A of load current. The difference in I_{RS} and I_L is therefore 0.076 A $- 0.0009$ A $= 0.0751$ A, or 75.1 mA. This current must all pass through the zener diode when the circuit is in operation. Nine volts dc will then appear at the two output terminals of the power supply. This represents the no-load (NL) condition of operation.

When the power supply is connected to an external load, more current is required. Ideally, this current should be available with 9 V of output. Assume now that the power supply is connected to a 270-Ω external load. The total resistance that the power supply sees at its output is 10 kΩ in parallel with 270 Ω. This represents a load resistance of 263 Ω. With 9 V applied, the total load current is $9 \text{ V}/263 = 0.0342$, or 34.2 mA. The total output current available for the power supply is that which passes through R_s. This was calculated to be 0.076 A or 76 mA. With the 270-Ω load, I_L will be 34.2 mA with an I_z of 41.8 mA. The output voltage therefore remains at 9 V with the increase in load.

With a slightly smaller external load, the power supply would go out of its regulation range. Should this occur, no I_z will flow and the value of I_L alone determines the current passing through R_s. An I_{RS} value in excess of 76 mA would cause a greater voltage drop across R_s. This, in effect, would cause the output voltage to be less than 9 V. All voltage regulator circuits of this type have a maximum load limitation. Exceeding this limit will cause the output voltage to go out of regulation.

If the resistance of the external load is increased in value, it will cause a reduction in load current. This condition causes a reverse in the operation of the regulator. With a larger R_L there is less I_L. As a result of this, the zener diode must conduct more heavily. The maximum current-handling rating of D_z would determine this condition. An infinite load would, for example, demand no I_L. D_z would

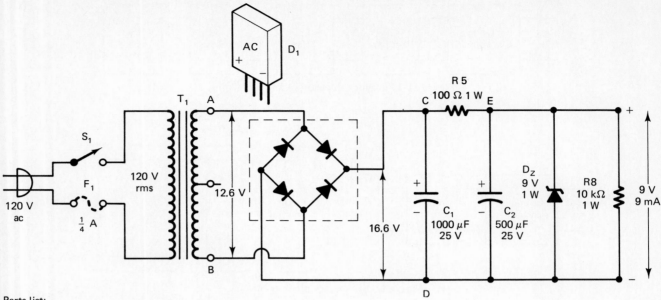

Parts list:

T_1 : 12.6-V 2-A Stancor P-8130, or Triad F-44X

D_1 : silicon bridge rectifier 2 A 50 V; General Instrument Co., KBF-005 or Motorola MDA-200

C_1 : 1000-μF 25-V electrolytic capacitor

C_2 : 500-μF 25-V electrolytic capacitor

R_5 : 100-Ω 1-W resistor

R_8 : 10-kΩ $\frac{1}{2}$-W resistor

D_Z : 9-V 1-W zener diode IN5346A or HEP Z-2513

S_1 : SPST toggle switch

F_1 : Fuse, $\frac{1}{4}$ A 250 V

Misc. : metal chassis, PC board, line cord, fuse holder, grommets, solder, cabinet

Figure 11-25. 9-V regulated power supply.

therefore conduct the full current passing through R_s. This value was calculated to be 76 mA for our circuit. For a 1-W 9-V zener diode the maximum current rating is 1 W/9 V = 0.111 A, or 111 mA. In this case, the diode is capable of handling the maximum possible I_L that will occur for an infinite load. In the actual circuit the bleeder resistor R_8 demands 0.9 mA of current when the load is infinite. The current through I_D will therefore not be in excess of 75.1 mA, which is well below its maximum rating. In effect, the regulating range of this power supply is from a infinite value to approximately 120 Ω. The output voltage will remain at 9 V over this entire range of load values.

A regulator must also be responsive to changes in input voltage. If the input voltage, for example,

were to increase by 10%, it would cause the supply to see 13.86 V instead of 12.6 V. The peak value of 13.86 V would then be 19.6 V − 1.2 V, or 18.4 V instead of 16.6 V. The voltage drop across R_s would therefore increase from 7.6 V to 9.4 V. This, in turn, would cause an increase in total current through R_S of 94 mA. The current flow of the zener diode would increase to 94.0 − 0.9 = 93.1 mA with no external load. Since the zener is capable of handling up to 111 mA, it could respond to this change to maintain the output voltage at 9 V.

The response of a power supply to a decrease in input voltage must also be taken into account. A 10% decrease in input voltage would cause the supply to see only 11.34 V instead of 12.6 V. The peak value of 11.34 V is 16 V. Capacitor C_1 would then charge to 16.0 − 1.2 V, or 14.8 V. The 1.2 V is the voltage drop across two silicon diodes in the bridge. The voltage drop across R_S will now be 5.8 V. With this reduced voltage, the total current will drop to 5.8 V/100 Ω = 0.058 A, or 58 mA. This value can certainly be handled by the zener diode. The maximum low-resistance value of the load will increase somewhat. Under this condition the load may drop to only 160 Ω before going out of regulation.

Review

1. What function does the power supply of a system perform?

2. Explain the fundamental differences between half-wave and full-wave rectification.

3. Why does a rectifying diode conduct in only one direction?

4. What is the dc output of an unfiltered half-wave rectifier with 25 V rms of input?

5. Why is the ripple frequency of a half-wave rectifier 60 Hz and of a full-wave rectifier 120 Hz?

6. Why is full-wave rectification more efficient than half-wave rectification?

7. Explain how full-wave rectification is achieved with two diodes.

8. Why are four diodes needed in a full-wave bridge rectifier?

9. Describe the primary differences between a two-diode full-wave rectifier and a dual or split power supply.

10. If 18 V rms is applied to the input of a bridge rectifier, what is the equivalent dc output?

11. Compare the voltage and current characteristics of a *C*-input filter and an *L* filter.

12. If 25 V rms is applied to a full-wave rectifier with a *C*-input filter, what value of dc would appear across *C*?

13. What are the primary functions of a power supply?

14. How long would it take a 100-mF capacitor to discharge to 63.2% of its peak value when connected to a 1-kΩ resistor?

15. If 10 V dc is applied to the *RC* circuit of question 14, what would be the capacitor voltage in five time constants?

16. How does a zener diode achieve voltage regulation?

Student Activities

Half-Wave Rectifiers

1. Construct the half-wave rectifier of Fig. 11-7(a). Nearly any value of low-voltage transformer may be used for T_1.

2. Calculate the dc output voltage for the circuit.

3. With a dc voltmeter, measure and record the dc output at points D–C. How do the measured and calculated values compare?

4. With an oscilloscope, observe the waveform at points A–B and D–C. Make a sketch of the observed waveforms.

5. Turn off the electrical power and reverse the diode as in Fig. 11-7(b).

6. Turn on the power and repeat steps 3 and 4.

Full-Wave Rectifiers

1. Construct the two-diode full-wave rectifier of Fig. 11-10. Nearly any value of low-voltage center-tapped transformer may be used for T_1.

2. Calculate the dc output voltage for the circuit.

3. With a dc voltmeter, measure and record the dc output voltage at points CT and D. How do the measured and calculated values compare?

4. With an oscilloscope, observe the ac input at points A–CT and B–CT. Use line sync for the oscilloscope. Make a sketch of the observed waveforms. How do they compare?

5. Connect the oscilloscope at points CT and D. Make a sketch of the observed waveform.

Bridge Rectifiers

1. Construct the bridge rectifier of Fig. 11-13. All four diodes are 1N4004 or the equivalent.

2. Calculate the dc output voltage for the circuit.

3. With a dc voltmeter, measure and record the dc output voltage. How does it compare with the calculated value?

4. With an oscilloscope, observe the ac input at points A–B and the dc output at points C–D. Make a sketch of the observed waveforms.

Dual Power Supply

1. Construct the dual full-wave power supply of Fig. 11-15. All diodes are 1N4004 or equivalent.

2. Calculate the dc output voltage for the top section and the bottom section.

3. With a dc voltmeter, measure and record the dc output voltage at points D–G, E–G, and D–E.

4. With an oscilloscope, observe the ac input at points A–CT, B–CT, D–G, E–G, and D–E. Make a sketch of the observed waveforms.

C-Input Filtering

1. Construct the half-wave rectifier with a C-input filter in Fig. 11-17.

2. Calculate the dc voltage value appearing across C_1. With a dc voltmeter, measure and record the dc voltage.

3. With an oscilloscope, observe the waveforms at points A–B and C–B. Make a sketch of the observed waveforms.

4. Turn off the circuit and change the value of R_L to 10 kΩ, then 470 Ω 1 W. Repeat step 3 for each value of R_L. How does the value of R_L influence the dc ripple?

L-Input Filter

1. Construct the full-wave power supply with an *L*-input filter in Fig. 11-19.
2. Calculate the dc voltage output of the rectifier section. With a dc voltmeter measure and record the dc voltage at CT–D, CT–E, and D–E.
3. With an oscilloscope, observe the waveforms at points A–CT, B–CT, D–CT, and E–CT. Make a sketch of the observed waveforms.
4. Turn off the circuit and change the value of R_L to 470 Ω. Turn on the circuit and observe the waveform at D–CT and E–CT. What influence does increased load have on the output ripple?

LC Filters

1. Construct the full-wave power supply with an *L–C* filter in Fig. 11-20.
2. With a dc voltmeter measure and record the dc voltage at test points CT–D, CT–E, and D–E.
3. With an oscilloscope, observe the waveforms at points A–CT, B–CT, D–CT, and E–CT. Make a sketch of the observed waveforms.
4. Momentarily turn off the circuit and change R_L to a 470-Ω 1-W resistor. Repeat step 3. How does an increased value of R_L influence the ripple?

Zener Diode Regulators

1. Construct the 9-V regulated power supply of Fig. 11-25.
2. With a dc voltmeter measure and record the dc voltage at points C–D, C–E, and E–D.
3. Turn off the circuit and change R_L to a 1-kΩ 1-W resistor. Turn on the circuit and repeat step 2.
4. Repeat step 3 for a 470-Ω 1-W resistor.
5. Repeat step 3 for a 220-Ω 1-W resistor.
6. How does the voltage output of the power supply respond to different values of R_L?

Power Supply Project

1. Construct a 9-V regulated power supply using the circuit of Fig. 11-25.
2. A representative printed-circuit-board layout is shown in Fig. 11-26. The circuit board may require some modification according to the part selection. Make a prototype layout of the PC board on a piece of card stock. Use the actual dimensions of the parts being used.
3. Layout the PC board using tape and PC resist dots.
4. Etch the board.
5. Drill holes at the designated locations.
6. Solder components in place.

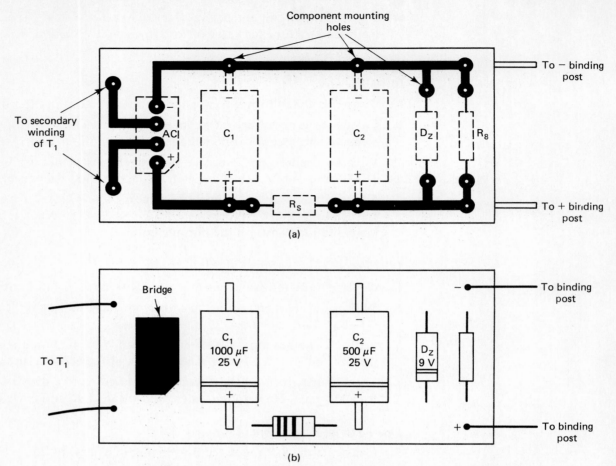

Figure 11-26. Printed-circuit-board layout of regulated power supply: (a) bottom view; (b) top view.

7. Wire the external circuitry.

8. Assemble the external circuitry and PC board.

9. Test the operation of the circuit with a voltmeter and oscilloscope. Record the representative voltages observed at points *A–B*, *C–D*, *E–D*, and *C–E*.

10. Assemble the entire unit in a cabinet and test it again for operation.

12

Transistors

Invention of the transistor in 1948 signaled the start of a whole new era in electronics. Circuits could be built exceedingly small, operate without heating power, amplify signals with a low-voltage source, and be extremely rugged. Integrated circuits, microprocessors, pocket calculators, home computers, and video games have all been made possible through this development. The impact that this device has had on the electronics field has still not been fully realized.

Transistors are three-element devices made of semiconductor materials. These devices have one or more P-N junctions in their construction. Bipolar transistors have two P-N junctions. Field-effect transistors (FETs) have one junction in their construction. FETs are classified as unipolar devices. This type of device responds to one type of current carrier passing through a channel. Unijunction transistors are similar in construction to the FET. These three transistors play an important role in the field of electronics.

IMPORTANT TERMS

In the study of transistors there are several terms that are used for the first time. In general, these terms have an important meaning in the discussion. Review each term very carefully. Then proceed with the chapter.

Active region. An area of transistor operation between cutoff and saturation.

Base. A thin layer of semiconductor material between the emitter and collector of a bipolar transistor.

Beta. A designation of transistor current gain determined by I_C/I_B.

Bipolar. A type of transistor with two P-N junctions. Current carriers have two polarities, including holes and electrons.

Channel. The controlled conduction path of a field-effect transistor.

Collector. A section of a bipolar transistor that collects majority current carriers.

Current carriers. Electrons or holes that support current flow in solid-state devices.

Cutoff. A condition or region of transistor operation where current carriers cease to flow through the device.

Depletion region. The area of a P-N junction where the number of current carriers has diminished.

Drain. The output terminal of a field-effect transistor.

Emitter. A semiconductor section that is responsible for the release of majority current carriers.

Enhancement. A conduction function where current carriers are pulled from the substrate into the channel of a MOSFET.

Gain. A ratio of voltage or current output to voltage or current input.

Gate. The control element of a field-effect transistor.

Induced channel. The channel of an E-MOSFET that is developed by gate voltage.

Interbase resistance. The resistance between B_1 and B_2 of a unijunction transistor.

Negative resistance. A UJT characteristic where an increase in emitter current causes a decrease in emitter voltage.

Peak voltage point (P_V). A characteristic point of the unijunction transistor where the emitter voltage rises to a peak value.

Saturation region. A condition of operation where a device is conducting to its full capacity.

Source. The common or energy input lead of a field-effect transistor.

Substrate. A piece of N or P semiconductor material which serves as a foundation for other parts of a MOSFET.

Unipolar. A semiconductor device that has only one P-N junction and one type of current carrier. FETs and UJTs are unipolar devices.

Valley voltage (V_V). A characteristic voltage point of the UJT where the voltage value drops to its lowest level.

BIPOLAR TRANSISTORS

When the word "transistor" is used, it usually has some kind of letter designations to identify its type. NPN and PNP are two types of bipolar transistors. The letters are used to denote the polarity of the semiconductor material used in its construction. Material polarity is determined by the type of impurity added to silicon or germanium when it is manufactured.

A PNP transistor has a thin layer of N material placed between two pieces of P material. The schematic symbol of the PNP transistor and its crystal structure are shown in Fig. 12-1. Leads attached to each piece of material are indentified as the emitter, base, and collector. In the symbol, when the arrowhead of the emitter lead "*Points iN*" toward the base it indicates that the device is of the PNP type. Lead identification is extremely important when working with transistors.

An NPN transistor has a thin layer of P material placed between two pieces of N material. The schematic symbol, crystal structure, and lead designations of the transistor are shown in Fig. 12-2. The schematic symbol of this device is the same as the PNP except the pointing direction of the arrowhead. An NPN transistor has the arrow "*Not*

Figure 12-1. PNP transistor symbol and crystal structure.

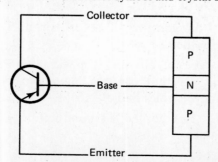

Figure 12-2. NPN transistor symbol and crystal structure.

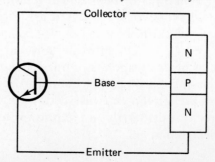

Pointing" toward the base. The polarity of the source voltage applied to each element is determined by the material used in its construction.

PNP and NPN transistors are considered to be bipolar devices. The term "bipolar" refers to conduction by holes and electrons. Electrons are the majority current carriers of the N material. In the P material, holes are the majority current carriers. Current conduction in the transistor is based on the movement of both current carriers. The term *bipolar* denotes this condition of conduction.

TRANSISTOR BIASING

For a transistor to function properly, it must have electrical energy applied to all its electrodes. In semiconductor circuits the source voltage is called bias voltage or simply bias. Bipolar transistors must have both junctions biased in order to function. In an earlier chapter we discussed diode biasing. Forward biasing of a P-N junction takes place when the polarity of the source voltage is positive to the P material and negative to the N material. This condition causes a current flow. The depletion zone of the device is reduced and majority current carriers are driven toward the junction. One junction of a transistor must be forward biased when it operates.

Reverse biasing of a P-N junction is achieved when the polarity of the source is negative to the P material and positive to the N material. This condition does not ordinarily permit current flow. The width of the depletion zone is increased by this action and minority current carriers move toward the junction. In normal circuit operation reverse bias current is extremely small. One junction of a transistor must be reverse biased in order for the device to be operational.

NPN Transistor Biasing

Consider now the biasing of the NPN transistor in Fig. 12-3. In this diagram an external voltage source has been applied only to the emitter–base regions of the transistor. In an actual circuit, this source is called the emitter–base voltage or V_{BE}. Note that the polarity of V_{BE} causes the emitter–base junction to be forward biased. This condition causes majority current carriers to be forced together at the emitter–base junction. The resulting current flow in this case would be quite large. In our example, note that the same amount of current flows into the emitter and out of the base. Emitter current (I_E) and base current (I_B) are used to denote these values. Under normal circuit conditions a V_{BE} source would not be used independently.

Figure 12-4 shows an external voltage source connected across the base–collector junction of an NPN transistor. The base–collector voltage or V_{CB} is used to reverse bias the base–collector junction. In this case there is no indication of base current I_B or collector current I_C. In an actual reverse-biased junction there could be a very minute amount of current. This would be supported primarily by minority current carriers. As a general rule, this is called leakage current. In a silcon transistor leakage current is usually considered to be negligible. In an actual circuit V_{CB} is not normally applied to a transistor without the V_{BE} voltage source.

Figure 12-3. Emitter–base biasing.

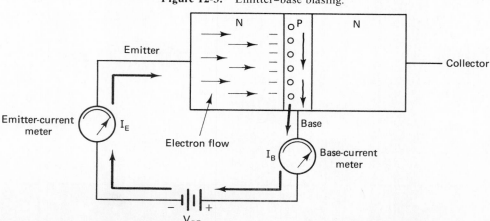

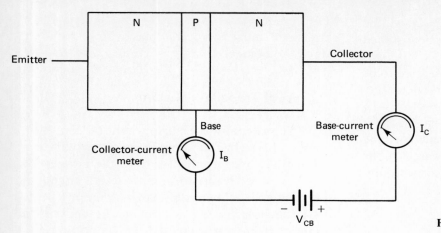

Figure 12-4. Base–collector biasing.

For a transistor to function properly the emitter–base junction must be forward biased and the base–collector junction reverse biased. Both junctions must have bias voltage applied at the same time. In some circuits this voltage may be achieved by separate V_{BE} and V_{CB} sources. Other circuits may utilize a single battery with specially connected bias resistors. In either case, the transistor responds quite differently when all its terminals are biased.

Consider now the action of a properly biased NPN transistor. Figure 12-5 shows separate V_{BE} and V_{CB} sources connected to the transistor. V_{BE} provides forward bias for the emitter–base junction while V_{CB} reverse biases the collector junction. Connected in this manner, the two junctions do not respond as independent diodes.

Figure 12-6 shows how current carriers pass through a properly biased NPN transistor. The emitter–base junction being forward biased causes a large amount of I_E to move into the emitter–base junction. Upon arriving at the junction, a large number of the electrons do not effectively combine with holes in the base. The base is usually made very thin (0.0025 cm or 0.001 in.) and it is lightly doped. This means that the majority current carriers of the emitter exceed the majority carriers of the base. Most of the electrons that cross the junction do not combine with holes. They are, however, immediately influenced by the positive V_{CB} voltage applied to the collector. A very high percentage of the original emitter current enters the collector. Typically, 95 to 99% of I_E flows into the collector junction. It then

Figure 12-5. NPN transistor biasing.

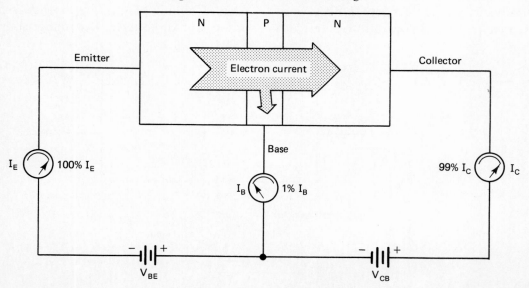

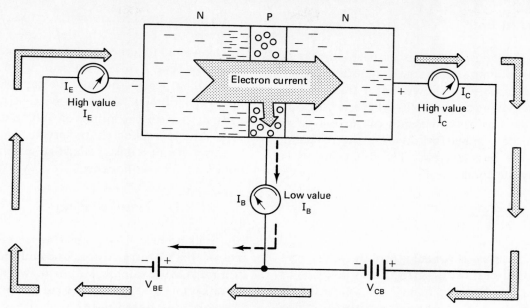

Figure 12-6. Current carriers passing through an NPN transistor.

becomes collector current. After passing through the collector region, I_C returns to V_{CB}, V_{BE}, and ultimately to I_E. The current flow inside the transistor is indicated by a large arrow. Outside the transistor current flow is indicated by small arrows.

The difference between the amount of I_E and I_C of Fig. 12-6 is base current. Essentially, base current is due to the combining of a small number of electrons and holes in the emitter–base junction. In an typical circuit, I_B is approximately 1 to 5% of I_E. With the base region being very narrow, it cannot support a large number of current carriers. A small amount of base current is needed, however, to make the transistor operational. Note the direction of I_B and its flow path in the diagram.

The relationship of I_E, I_B, and I_C in a transistor can be expressed by the equation

emitter current = base current + collector current

or

$$I_E = I_B + I_C$$

This shows that the emitter current is equal to the sum of I_B and I_C. The largest current flow in a transistor takes place in the emitter. I_C is slightly less than I_E. This means that the current flow through a reverse-biased collector–base junction is nearly equal to that of the forward-biased emitter–base junction. The difference in I_E and I_C is I_B.

If the base of a transistor were not made extremely thin, it would not respond as just described. The thin base region makes it possible for large amounts of I_E to pass through the base and into the collector region. A thicker base would cause more I_E to combine with holes in the base. This would reduce the operational effectiveness of the transistor.

The amount of base current that flows in a transistor is very small but extremely important. Suppose, for example, that the base lead of the transistor in Fig. 12-6 is momentarily disconnected. With this element open it should be obvious that there will be no base current. Closer examination will also show that V_{BE} and V_{CB} are now connected in series. This means that their voltages are added together. As a result of this action, the collector becomes more positive and the emitter more negative. One would immediately think that this condition would cause the transistor to conduct very heavily. It does not, however, permit any conduction at all. In effect, this means that base current has a direct influence on emitter and collector current. The base–emitter junction of a transistor must be forward biased in order for it to produce collector current.

Transistor Beta

A small change in transistor base current is capable of producing a very large change in collector current. This, in effect, shows that a transistor is capable of achieving current gain. The term "beta" is commonly used as an expression of current gain. Beta can be determined by dividing the collector current by the base current. This relationship is expressed by the equation

$$\text{beta} = \frac{\text{collector current}}{\text{base current}} \quad \text{or} \quad \beta = \frac{I_C}{I_B}$$

For most transistors beta is usually quite large. In a previous circuit, we described I_C as being 95 to 99% of I_E, with I_B being from 1 to 5%. For this example, the beta of a transistor with this ratio would be

$$\beta = \frac{95\%}{5\%} = 19 \quad \text{or} \quad \beta = \frac{99\%}{1\%} = 99$$

In both examples note that beta is expressed as a pure number. The percentage signs in the numerator and denominator of the equation cancel each other. If actual current values in milliamperes or amperes were used, they would also cancel. Beta is simply expressed as a whole-number value.

Transistors are manufactured today with a very wide range of beta capabilities. Power transistors, which respond to large current values, usually have a rather low beta. Typical values are in the range of 20. Small-signal transistors may be capable of beta in the range of 400. A good average beta for all transistors is in the range of 100. Beta is a very important transistor characteristic. It is not very meaningful unless it is applied to a specific transistor and its circuit components.

PNP Transistor Biasing

Biasing of a PNP transistor is very similar to that of the NPN transistor. The emitter–base junction must, for example, still be forward biased and the collector–base reverse biased. The polarity of the bias voltage is, however, reversed for each transistor. The majority current carriers of a PNP transistor are holes instead of electrons as in the NPN device. Except for these two differences, operation is primarily the same.

Figure 12-7 shows how current carriers pass through a properly biased PNP transistor. The emitter–base junction being forward biased causes a large number of holes to move through the junction. Upon arriving in the base region, a very small number of these holes combine with electrons. This is representative of the base current. Ninety-five to

Figure 12-7. Current carriers of a PNP transistor.

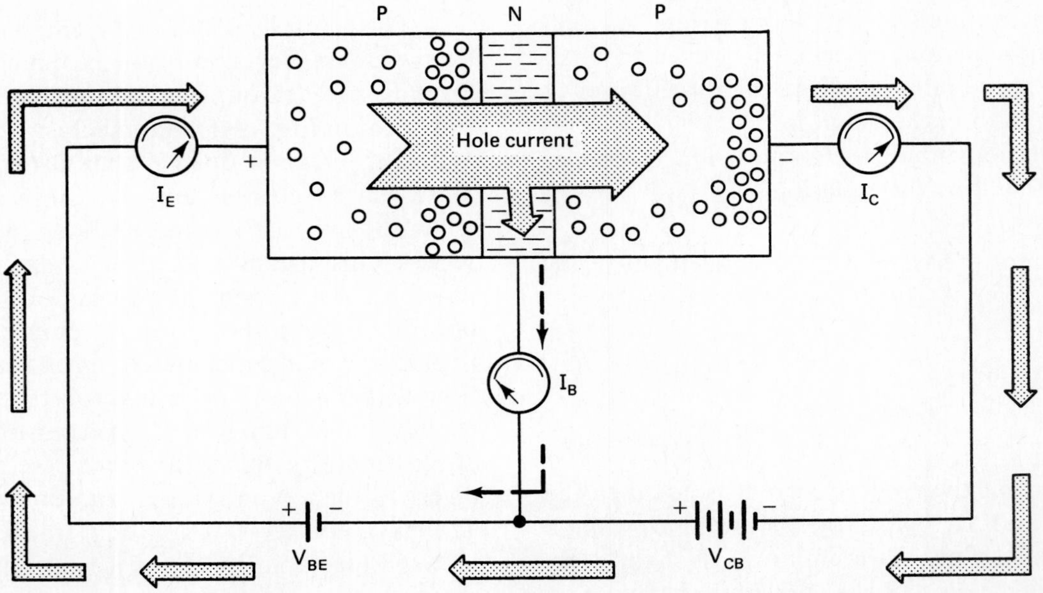

99% of the holes move through the base region and enter the collector junction. These holes do not find electrons to combine with in the thin base region. They are, however, immediately influenced by the negative V_{CB} voltage of the collector. This action causes hole current to flow through the reverse-biased base–collector junction. The end result of this is I_C passing through the collector region. It then leaves the collector and flows into V_{CB}, V_{BE}, and returns to the emitter. Hole current flow is indicated by the large arrow inside the transistor. Outside the transistor, current flow is achieved by electrons. Small arrows in the diagram show current outside the transistor.

It is rather interesting to compare the transistors of Figs. 12-6 and 12-7. Specific attention should be directed to the polarity of V_{BE} and V_{CB}. This shows the primary difference in device biasing. Note also the type of majority current flow through each transistor. This is represented by the large arrow of each diagram. It is different for each device. This means that the two transistor types are not directly interchangeable. Substituting one for the other would necessitate a complete reversal of bias voltage.

PNP transistors are not as widely used today as is the NPN type. Electron current flow of the NPN device has much better mobility than the hole flow of a PNP type. This means that electrons have a tendency to move more quickly through the crystal material than do holes. Because of this characteristic, NPN transistors tend to respond better at high frequencies. In general, this means that the NPN device has a wider range of applications. Manufacturers usually have a larger selection of these transistors in their lines. With a better selection of transistors available, circuit designers find it easier to select devices of a desired characteristic. As a result of this, the NPN device is much more popular than its PNP counterpart.

TRANSISTOR CHARACTERISTICS

When a transistor is manufactured it has characteristic data that distinguish it from other devices. These data are extremely important in predicting how a transistor will perform in a circuit. One very common way of showing this information is with the use of graphs. A collector family of characteristic curves is widely used to show transistor data.

A collector family of characteristic curves for an NPN transistor is shown in Fig. 12-8. The vertical part of this graph shows different collector current values in milliamperes. The horizontal part shows the collector–emitter voltage. Individual lines of the graph represent different values of base current. The zero base-current line is normally omitted from the graph. This condition shows when the transistor is not conducting.

A family of collector curves tells a great deal about the operation of a transistor. Take for example, point A on the 60-μA base-current line. If the transistor has 60 μA of I_B and a V_{CE} of 6 V, there is a collector current of 3 mA. This is determined by projecting a line to the left of the intersection of 60 μA and 6 V. Any combination of I_C, I_B, and V_{CE} can be quickly determined from the display of these curves.

If the values of I_C and I_B can be determined from a family of curves, they can be used to predict other values. The emitter current (I_E) of a transistor can be determined from the values of I_C and I_B. Remember that $I_E = I_B + I_C$. For point A, the I_E is 60 μA + 3 mA, or 3.06 mA.

The current gain or beta of a transistor can also be determined from a family of collector curves. Since beta is I_C/I_B, it is possible to determine this condition of operation. For point A on the curves, I_C is 3 mA and I_B is 60 μA. The current gain of a transistor operating at this point would be determined by the equation

$$\text{beta} = \frac{I_C}{I_B} = \frac{3 \text{ mA}}{60 \text{ } \mu\text{A}} = \frac{0.003}{0.00006} = 50$$

This condition is called the direct-current beta. It shows how the transistor will respond when dc voltage is applied. Dc beta is an important consideration when a transistor is operated as a direct current amplifier.

In many applications a transistor must amplify ac signals. Dc voltage values are applied to the device to make it operational. An ac signal voltage applied to the base would then cause a change in the value of I_B. In Fig. 12-8, assume that the input voltage causes I_B to change between points B and C.

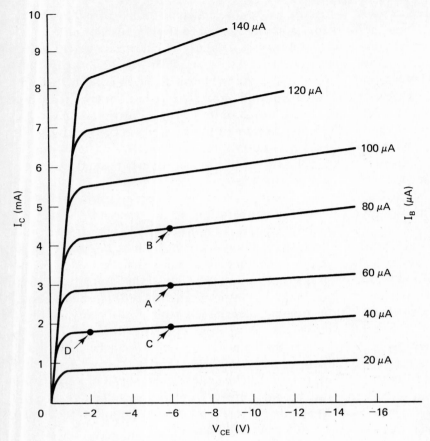

Figure 12-8. Collector family of characteristic curves for an NPN transistor.

Point *A* is considered to be the operating point. The change in I_B is from 80 μA to 40 μA. This is normally called ΔI_B. The Greek letter delta (Δ) is used to denote a changing value in I_B. A corresponding change in I_C is determined by projecting a line to the left each I_B value. ΔI_C is therefore 4.5 mA to 2 mA.

The ac beta of a transistor is determined by dividing ΔI_C by ΔI_B. For points *B–C* this is

$$\text{ac beta} = \frac{\Delta I_C}{\Delta I_B} = \frac{4.5 \text{ mA to 2 mA}}{80 \ \mu\text{A to 40} \ \mu\text{A}} = \frac{2.5 \text{ mA}}{40 \ \mu\text{A}}$$

$$= \frac{0.0025}{0.00004} = 62.5$$

The ac beta in this case is somewhat different from that of the dc beta. As a general rule, ac beta is larger than dc beta. To amplify an ac voltage, a transistor must operate over a range of different values. This usually accounts for the difference in beta values.

The same family of collector curves can also be used to show the effectiveness of a transistor in controlling collector current. Note points *D* and *C*

on the 40 μA curve. A change in V_{CE} from 2 to 6 V occurs between these two points. Projecting points *D* and *C* to the left indicates a change in I_C from 1.8 to 2 mA. A 4-V change in V_{CE}, therefore, causes only a 0.2-mA change in I_C. Using the ac beta for comparison, a 40-μA change in I_B causes a ΔI_C of 2.5 mA. This indicates that I_C is more effectively controlled by I_B changes than by V_{CE} changes.

Operation Regions

A collector family of characteristic curves is also used to show the desirable operating regions of a transistor. When there is operation in the center area it is called the active region. This area is located anywhere between the two shaded areas of Fig. 12-9. Amplification is achieved when the transistor operates in this region.

A second possible region of operation is called cutoff. On the family of curves, this is where $I_C = 0$ mA. When a transistor is cut off no I_B and I_C flows. Any current flow that occurs in this region is due to leakage current. As a general rule, a transistor

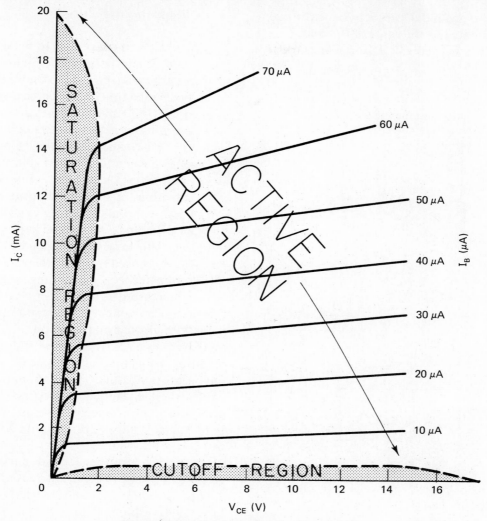

Figure 12-9. Operating regions of a transistor.

operating in the cutoff region responds as a circuit with an open switch. When a transistor is cut off there is infinite resistance between the emitter and collector.

The shaded area on the left side of the family of curves is called the saturation region. This area of operation is where maximum I_C flows. A transistor operating in this region responds as a closed switch. A transistor is considered to be fully conductive in the saturation region. The resistance between emitter and collector is extremely small when a transistor is saturated.

Figure 12-10 shows how a transistor responds in the three regions of operation. In part (a) the transistor is operating in the active area. The value of V_C is somewhat less than that of the source voltage

V_{CC}. Collector current through the transistor causes a voltage drop across R_C. The base current of a transistor operating in this region would be some moderate value between zero and saturation. Transistor amplifiers usually operate in this area.

Figure 12-10(b) shows how a transistor responds when it is in the cutoff region. With no current flow through the transistor, there is no voltage drop across R_C. V_C of this circuit is therefore the same as the source voltage. A transistor is cut off when no I_B occurs.

Figure 12-10(c) shows how a transistor operates in the saturation region. In this condition of operation the transistor responds as a closed switch. V_C is approximately equal to 0 V. Heavy I_C values cause nearly all V_{CC} to appear across R_C. In a sense, V_C is considered to be at ground. Current passing through the transistor is limited by the value of V_{CC} and R_C. A large value of I_B is needed to cause a transistor to saturate.

Figure 12-10. Transistor operation by regions: (a) active region; (b) cutoff region; (c) saturation region.

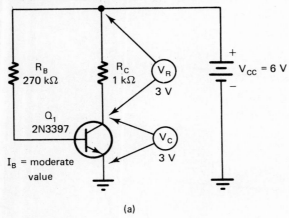

(a)

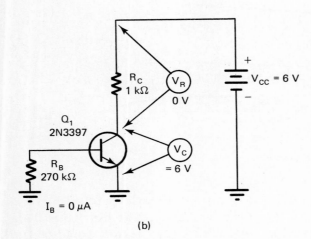

(b)

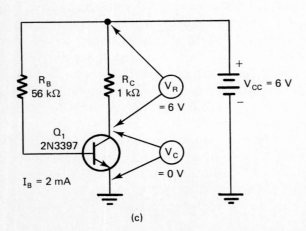

(c)

Characteristic Curves

Special circuits are used to obtain the data for a family of characteristic curves. Figure 12-11 shows a test circuit that can be used to find the data points of a family of collector curves. Three meters are used to monitor this information. Base current and collector current are observed by meters connected in series with the base and collector. Collector–emitter voltage is measured with a voltmeter connected across these two terminals. V_{CE} is adjusted to different values by a variable power supply.

The data points for a single curve are developed by first adjusting I_B to a constant value. A representative value for a small-signal transistor is $10\,\mu$A. Resistor R_B is used to adjust the value of I_B. The V_{CE} is then adjusted through its range starting at 0 V. As V_{CE} is increased in 0.1-V steps up to 1 V, the I_C increases very quickly. Corresponding I_C and V_{CE} values are plotted on the graph. V_{CE} should then be increased in 1-V steps while monitoring I_C. The resulting collector current usually levels off to a fairly constant value. The corresponding I_C and V_{CE} values are then added to the first data points. All points are then connected together by a continuous line. The completed curve would be labeled $10\,\mu$A of I_B.

To develop the second curve, V_{CE} is first returned to zero. I_B is then increased to a new value. A common value for a small-signal transistor would be $20\,\mu$A. The second curve is then plotted by recording the I_C values for each V_{CE} value.

To obtain a complete family of curves, the process would be repeated for several other I_B values. A representative family of curves may have 8 to 10 different I_B values. The step values of V_{CE} and the increments of I_B and I_C will change with different transistors. Small-signal transistors may use 10-μA I_B steps with I_C values ranging from 0 to 20 mA. Large-signal transistors may use 1-mA I_B steps with I_C values going to 100 mA. Each transistor may require different values of I_B, I_C, and V_{CE} to obtain a suitable curve.

TRANSISTOR PACKAGING

Once a transistor is constructed, the entire crystal assembly must be placed in an enclosure or package. The unit is then sealed to protect it from dust,

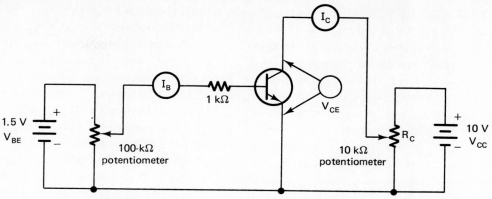

Figure 12-11. Transistor characteristic curve circuit.

moisture, or outside contaminates. Electrical connections are made to each element through leads attached to the package. When metal packages are used, the housing also serves as a heat sink. Heat is carried away from the crystal by the metal package. A common practice in this kind of unit is to attach the collector to the metal case. This helps the collector to dissipate heat. Packages of this type should not be permitted to touch other circuit components when it is in operation.

Ceramic and epoxy packaged transistors are also available. As a rule, this type of packaging is somewhat cheaper to produce. These devices are usually not as rugged as the metal-packaged devices. Most small-signal transistors are now being housed in epoxy packages. The outside case of this device is insulated from all transistor elements. It is specifically designed for printed-circuit-board construction and close placement with other circuit components.

Several of the more popular transistor package styles are shown in Fig. 12-12. Part (a) shows devices designed for small-signal applications. The letter designation "TO" stands for transistor outline. These same packages are also used to house other solid-state devices. Part (b) shows some of the packages used to house large-signal or power transistors. Heavy metal cases of this type are used to dissipate heat generated by the transistor. The collector is usually attached to the metal case. Part (c) shows a few of the common epoxy packages. The letter designation "SP" stands for semiconductor package. This type of package is used to house a number of other semiconductor devices.

With the wide range of package styles available today, it is difficult to have a standard method of

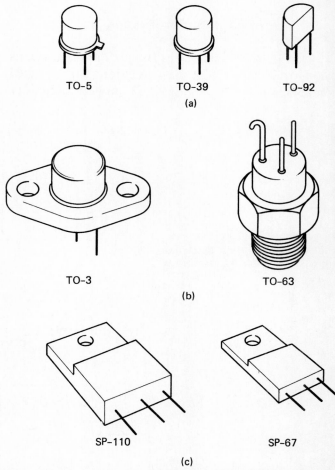

Figure 12-12. Transistor packages: (a) small-signal devices; (b) large-signal devices; (c) epoxy devices.

identifying transistor leads. Each package generally has a unique lead location arrangement. It is usually a good idea to consult the manufacturer's specifications for proper lead identification.

BIPOLAR TRANSISTOR TESTING

Transistor testing is a rather common procedure that must be performed periodically when working with semiconductor devices. Lead identification, gain, open leads, shorted conditions, and leakage are some of the tests commonly performed on a transistor. Curve tracers, gain testers, and sound-producing instruments can effectively be used for these tests. These instruments, as a general rule, are not always available. An ohmmeter, however, can be used to perform many of these tests. An ohmmeter is one of the functions of a volt-ohm-milliammeter (VOM) or vacuum-tube voltmeter (VTVM).

Transistor Junction Testing

A good transistor has two P-N junctions in its construction. Both of these junctions can be tested with an ohmmeter. Figure 12-13 shows the junctions of PNP and NPN transistors. Each junction will respond as a diode when it is tested. Forward biasing occurs when the material polarity of the junction matches the voltage polarity of the ohmmeter. When the ohmmeter polarity is reversed, the same junction will be reverse biased. Forward biasing is indicated by a low resistance reading, while reverse bias will show an extremely high resistance value. Both junctions of a transistor will respond to this test in the same way if the transistor is good.

A faulty P-N junction will not show a difference in its forward and reverse resistance connections. Low resistance in both directions indicates that a junction is shorted. Excessive current generally causes this type of failure in a transistor. Infinite or exceedingly high resistance in both directions indicates an open condition. An open junction is normally caused by a broken internal connection. This may be the result of a current overload or excessive shock. Open and shorted transistor failures, as a

Figure 12-13. Junction polarity of (a) PNP and (b) NPN transistors.

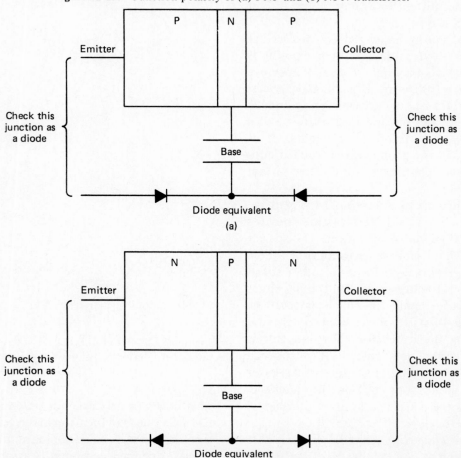

general rule, occur suddenly and cannot be corrected.

A rather quick test of both junctions at the same time can be made with the ohmmeter. Connect one ohmmeter lead to the emitter and the other to the collector. Ohmmeter lead polarity in this case is not important. This connection should cause an extremely high resistance reading. Then reverse the two ohmmeter leads. High resistance should also occur in this direction. A good silicon transistor will show an infinite resistance in either direction. Any measurable resistance in either direction indicates leakage. Germanium transistors will show some leakage in this test. Unless the leakage is excessive, it can be tolerated in a germanium transistor. A perfect transistor would have no indication of emitter–collector leakage.

Lead Identification

An ohmmeter can also be used to identify the leads of a transistor. The polarity of the ohmmeter voltage source must be known in order for this test to be meaningful. Straight polarity ohmmeters have the black or common lead negative and the red or probe lead positive. Reverse polarity ohmmeters are connected in the opposite direction. The polarity of an ohmmeter can be tested with a separate dc volt meter if it is unknown. The ohmmeter used in this explanation has straight polarity.

To identify transistor leads, inspect the lead location of the device under test. Pick out the center lead of the transistor. Assume that it is the base lead. Connect the negative lead of the ohmmeter to it as shown in Fig. 12-14. Then alternately touch the positive ohmmeter lead to the two outside transistor leads. If a low resistance indication occurs for each lead, the center lead is actually the base. The test also indicates that the transistor is the PNP type. If the resistance is high between the center lead and the two outside leads, reverse the meter polarity. The positive lead should now be connected to the assumed base. Once again alternately switch the negative ohmmeter lead between the two outside transistor leads. If a low resistance reading is obtained, the transistor is an NPN type.

If the center lead does not produce low resistance in either of the two conditions, it is not the base. Then select one of the outside leads as the assumed base. Try the same procedure again with

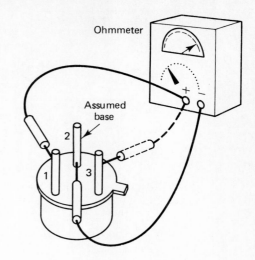

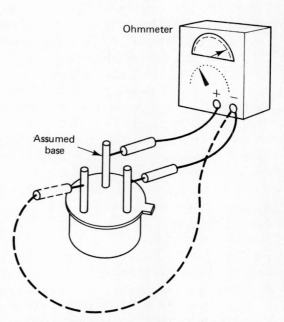

Figure 12-14.   Ohmmeter transistor testing.

the newly assumed base. If this does not produce results, try the other outside lead. One of the three leads must respond as the base if the transistor is good. If a response cannot be obtained, the transistor must be open or shorted.

Thus far we have identified the base lead and determined the polarity of the transistor to be NPN or PNP. It is now possible to test the gain and to identify the remaining two leads. A 100,000-Ω resistor and an ohmmeter are needed for this part of the test. The resistor is used to provide base current from the ohmmeter. If a power transistor is being tested, use the $R \times 1$ meter range and a 1000-Ω base resistor.

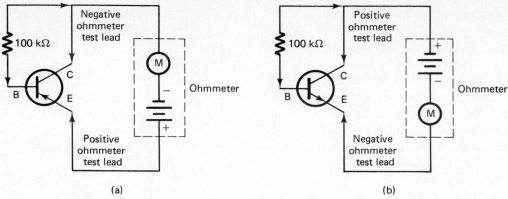

Figure 12-15.   Transistor gain test: (a) PNP test circuit; (b) NPN test circuit.

The process of testing transistor gain should be more meaningful if we know how it works. Figure 12-15 shows how PNP and NPN transistors will respond to the ohmmeter test. The energy source of the ohmmeter is used to supply the bias voltage to each of the transistor elements. For the PNP transistor of part (a) the positive ohmmeter lead is connected to the emitter and the negative lead to the collector. When a resistor is connected between the collector and the base, it causes base current. With the emitter and base forward biased and the collector reverse biased, the transistor becomes low resistant. The ohmmeter will respond to this condition by showing a change in resistance.

The ohmmeter test of an NPN transistor is shown in Fig. 12-15(b). For this type of transistor the emitter is connected to the negative ohmmeter lead and the collector to the positive. When a resistor is connected between the collector–base it will cause base current to flow. With the emitter–base forward biased and base–collector reverse biased, the transistor becomes low resistant. An indication of this type on the ohmmeter shows that the transistor has gain and correct lead selection.

The procedure when checking an NPN transistor is shown in Fig. 12-16. In this case, the ohmmeter is connected to the two outside leads. We have previously identified the center lead as the base. If another lead were found to be the base, it would be used in place of the center lead. Note that one end of the resistor is connected to the base with the other lead open. We have in this case assumed that the negative ohmmeter lead is connected to the emitter and the positive lead to the collector. If the assumed

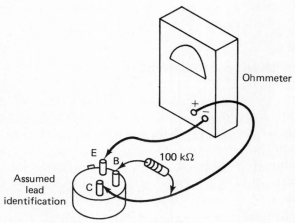

Figure 12-16.   NPN gain test.

leads are correct, the emitter will be forward biased and the collector reverse biased. Touching the open end of the base resistor to the positive lead will cause a base current flow. If it does, the ohmmeter will indicate a low resistance. In effect, the emitter–base junction is forward biased and the base–collector junction reverse biased. If no base current flows, the assumed emitter–collector leads are reversed. Simply reverse the ohmmeter leads and again touch the resistor lead to the positive ohmmeter lead. If the transistor is good and the assumed leads are correct, the ohmmeter will show a low resistance. If it does not, the transistor has low gain.

Figure 12-17 shows the procedure for testing a PNP transistor. In this case, the ohmmeter is connected to the two outside leads. We have identified the center lead as the base. The base resistor is now connected to this lead. For a PNP transistor,

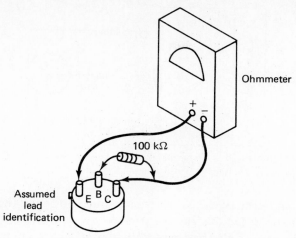

Figure 12-17. PNP gain test.

the positive ohmmeter lead goes to the emitter and the negative lead to the collector. If the assumed leads are correct, touching the open end of the base resistor to the negative lead will cause base current to flow. If it does, the ohmmeter will show a low resistance reading. This indicates that the emitter–base is forward biased and the base–collector reverse biased as in the circuit. If no current flows, the assumed emitter–collector leads are reversed. Reverse the two ohmmeter leads and again touch the base resistor to the negative ohmmeter lead. A good transistor with correct lead identification will indicate low resistance. A transistor with low gain will not cause much of a change in resistance.

UNIPOLAR TRANSISTORS

A transistor that only has one P-N junction in its construction is called a *unipolar* device. Unlike the bipolar transistor, a unipolar transistor conducts current through a single piece of semiconductor material. The current carriers of this device have only one polarity. The term unipolar is used to describe this characteristic.

Field-effect transistors and unijunction transistors are unipolar devices. These transistors are very similar in construction but differ a great deal in operation and function. Field-effect transistors are normally used to achieve amplification. Unijunction transistors are used as wave-shaping devices. Both devices play an important role in solid-state electronics.

JUNCTION FIELD-EFFECT TRANSISTORS

The junction field-effect transistor (JFET) is a three-element electronic device. Its operation is based on the conduction of current carriers through a single piece of semiconductor material. This piece of material is called the *channel*. An additional piece of semiconductor material is diffused into the channel. This element is called the *gate*. The entire unit is built on a third piece of semiconductor known as the *substrate*. The assembled device is housed in a package. The physical appearance of a JFET in its housing is very similar to that of a bipolar transistor.

N-Channel JFETs

Figure 12-18 shows the crystal structure, element names, and schematic symbol of an N-channel JFET. Its construction has a thin channel of N material formed on the P *substrate*. The P material of the gate is then formed on top of the N channel. Lead wires are attached to each end of the *channel* and the gate. No connection is made to the substrate of this device.

The schematic symbol of an N-channel JFET is somewhat representative of its construction. The bar part of the symbol refers to the channel. The drain (*D*) and source (*S*) are attached to the channel. The gate (*G*) has an arrow. This shows that it forms a P-N junction. The arrowhead of an N-channel symbol "*Points iN*" toward the channel. This indicates that it is a P-N junction. The arrowhead or gate is P material and the channel N material. This part of the device responds as a junction diode.

The operation of a JFET is somewhat unusual when compared with that of a bipolar transistor.

Figure 12-18. N-channel junction field-effect transistor.

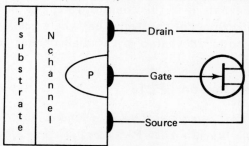

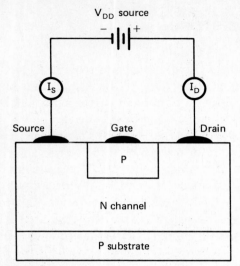

Figure 12-19.   Source–drain conduction of a JFET.

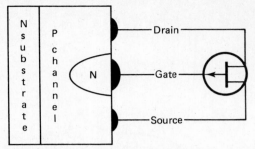

Figure 12-20.   P-channel junction field-effect transistor.

Figure 12-19 shows the source and drain leads of a JFET connected to a dc voltage source. In this case, maximum current will flow through the channel. The I_S and I_D meters will show the same amount of current. The value of V_{DD} and the internal resistance of the channel determines the amount of channel current flow. Typical source–drain resistance values of a JFET are several hundred ohms. Essentially this means that full conduction will take place in the channel even when the gate is open. A JFET is considered to be a normally-on device.

Current carriers passing through the channel of a JFET are controlled by the amount of bias voltage applied to the gate. In normal circuit operation, the gate is reverse biased with respect to the source. Reverse biasing of the gate–source of any P-N junction will increase the size of its depletion region. In effect, this will restrict or deplete the number of majority carriers that can pass through the channel. This means that drain current (I_D) is controlled by the value of V_{GS}. If V_{GS} becomes great enough, no I_D will be permitted to flow through the channel. The voltage that causes this condition is called the cutoff voltage. I_D can be controlled anywhere between full conduction and cutoff by a small change in gate voltage.

P-Channel JFETs

The crystal structure, element names, and schematic symbol of a P-channel JFET are shown in Fig. 12-20. Construction of this device has a thin channel of P material formed on an N substrate. The

N material of the gate is then formed on top of the P channel. Lead wires are attached to each end of the channel and to the gate. Other construction details are the same as those of the N-channel device.

The schematic symbol of a JFET is different only in the gate element. In a P-channel device the arrow is "*Not Pointing*" toward the channel. This means that the gate is N material and the channel is P material. In conventional operation, the gate is made positive with respect to the source. Varying values of reverse bias gate voltage will change the size of the P-N junction depletion zone. Current flow through the channel can be altered between cutoff and full conduction. P-channel and N-channel JFETs cannot be used in a circuit without a modification in the source voltage polarity.

JFET Characteristic Curves

The JFET is a very unique device compared with other solid-state components. A small change in gate voltage will, for example, cause a substantial change in drain current. The JFET is therefore classified as a voltage-sensitive device. By comparison, bipolar transistors are classified as current-sensitive devices. A JFET has a rather unusual set of characteristics compared with other solid-state devices.

A family of JFET characteristic curves is shown in Fig. 12-21. The horizontal part of the graph shows the voltage appearing across the source–drain as V_{DS}. The vertical axis shows the drain current (I_D) in milliamperes. Individual curves of the graph show different values of gate voltage (V_{GS}). The cutoff voltage of this device is approximately -7 V. The control range of the gate is from 0 to -6 V.

A drain family of characteristic curves tells a great deal about the operation of a JFET. Refer to point A on the -3 V_{GS} curve. If the device has -3 V

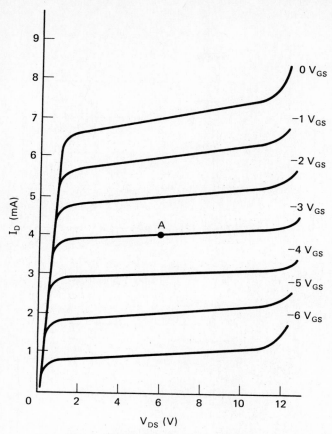

Figure 12-21. Family of drain curves.

DEVELOPING JFET CHARACTERISTIC CURVES.

A special circuit is used to develop the data for a drain family of characteristic curves. Figure 12-22 shows a circuit that is used to find the data points of a characteristic curve for an N-channel JFET. Three meters are used to monitor this data. Gate voltage is monitored by a V_{GS} meter connected between the gate-source leads. Drain current is measured by a milliampere meter connected in series with the drain. V_{DS} is measured with a voltmeter connected across the source and drain. V_{GS} and V_{DS} are adjusted to different values while monitoring I_D. Two variable dc power supplies are used in the test circuit.

The data points of a single curve are developed by first adjusting V_{GS} to 0 V. V_{DS} is then adjusted through its range starting at 0 V. Normally, V_{DS} is increased in 0.1-V steps up to 1 V. I_D increases very quickly during this operational time. V_{DS} is then increased in 1-V steps recording the I_D values. Corresponding I_D and V_{GS} values are then plotted on a graph as the 0 V_{GS} curve.

To develop the second curve, V_{DS} must be returned to zero. V_{GS} is then adjusted to a new value. 0.5 V would be a suitable value for most JFETs. V_{DS} is again adjusted through its range while monitoring I_D. Data for the second curve are then recorded on the graph.

To obtain a complete family of drain curves, the process would be repeated for several other V_{GS} values. A typical family of curves may have 8 to 10 different values. The step values of V_{DS}, V_{GS}, and I_D will obviously change with different devices. Full conduction is usually determined first. This gives an approximation of representative I_D and V_{DS} values.

applied to its gate and a V_{DS} of 6 V, there is approximately 4 mA of I_D flowing through the channel. This is determined by projecting a line to the left of the intersection of -3 V_{GS} and 6 V_{DS}. Any combination of I_D, V_{GS}, and V_{DS} can be determined from the family of curves.

Figure 12-22. Drain curve test circuit for N-channel JFET.

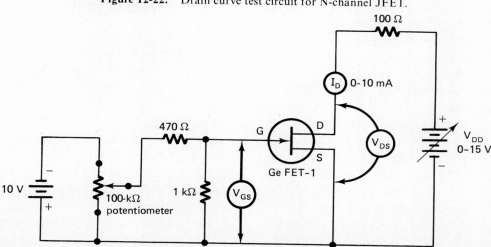

Normally, I_D will level off to a rather constant value when V_{DS} is increased. V_{DS} can be increased in value to a point where I_D starts a slight increase. Generally, this indicates the beginning of the breakdown region. JFETs are usually destroyed if conduction occurs in this area of operation. Maximum V_{DS} values for a specific device are available from the manufacturer. It is a good practice to avoid operation in or near the breakdown region of the device.

MOS FIELD-EFFECT TRANSISTORS

Metal-oxide semiconductor field-effect transistors (MOSFETs) are a rather new addition to the unipolar transistor family. A distinguishing feature of this device is its gate construction. The gate is, for example, completely insulated from the channel. Voltage applied to the gate will cause it to develop an electrostatic charge. No current is permitted to flow in the gate area of the device. The gate is simply an area of the device coated with metal. Silicon dioxide is used as an insulating material between the gate and the channel. Construction of a MOSFET is rather easy to accomplish.

The channel of a MOSFET has a number of differences in its construction. One type of MOSFET responds to the depletion of current carriers in its channel. Depletion-type MOSFETs have an inter-connecting channel built on a common substrate. Direct connection is made between the source and drain of this device. Channel construction is very similar to that of the JFET. A second type of MOSFET is called an enhancement type of device. E-MOSFETs do not have an interconnecting channel. Current carriers pulled from the substrate form an induced channel. Gate voltage controls the size of the induced channel. Channel development is a direct function of the gate voltage. E- and D-type MOSFETs both have unique operating characteristics.

Enhancement-type MOSFET construction, element names, and schematic symbols are shown in Fig. 12-23. Note in the crystal structure that no channel appears between the source and drain. The gate is also designed to cover the entire span between the source and drain. The gate is actually a thin layer of metal oxide. Conductivity of this device is controlled by the voltage polarity of the gate. A P-N junction is not formed by the gate, source, and drain.

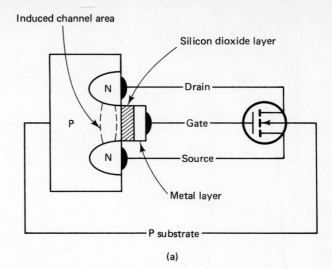

(a)

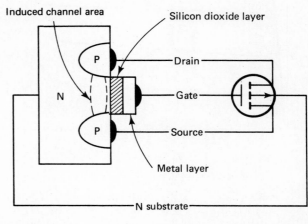

(b)

Figure 12-23. Enhancement-mode MOSFET: (a) N channel; (b) P channel.

An E-MOSFET is considered to be a normally-off device. This means that without an applied gate voltage, there is no conduction of I_D. When the gate of an N-channel device is made positive, electrons are pulled from the substrate into the source–drain region. This action causes the induced channel to be developed. With the channel complete, conduction occurs between the source and drain. In effect, drain current is aided or enhanced by the gate voltage. Without V_G, the channel is not properly developed and no I_D will flow.

A family of characteristic curves for an E-type MOSFET is shown in Fig. 12-24. This particular set of curves is representative of an N-channel device. A P-channel device would have the polarity of the gate voltage reversed. Note that an increase of gate voltage causes a corresponding increase in I_D. The

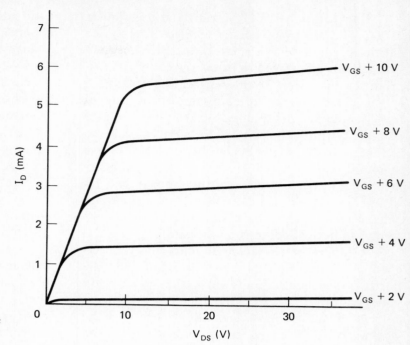

Figure 12-24. Family of drain curves for an E-type MOSFET (N-channel type).

input impedance of this device is extremely high regardless of the crystal material used or the polarity of the gate voltage.

Depletion-type MOSFET construction, element names, and schematic symbols are shown in Fig. 12-25. This N-channel device has a thin channel of N material formed on a P substrate. Source and gate connections are made directly to the channel. A thin layer of silicon dioxide (SiO_2) insulation covers the channel. The gate is a metal-plated area formed on the silicon dioxide layer. The entire unit is then built on a substrate of P material. The arrow of the schematic symbol refers to the material of the substrate. When it "*Points iN*" this shows that the substrate is P material and the channel is N.

The construction of P- and N-channel devices is essentially the same. The crystal material of the channel and substrate is the only difference. Current carriers are holes in the P-channel device, while electrons are carriers in the N-channel unit. The schematic symbol differs only in the direction of the substrate arrow. It does "*Not Point*" toward the substrate in a P-channel device. This means that the substrate is N material and the channel is P material.

Figure 12-25. D-type MOSFET: (a) N channel; (b) P channel.

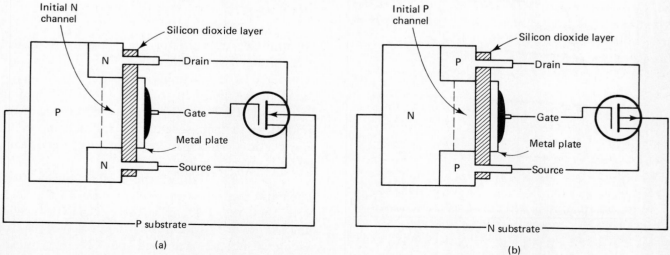

(a)

(b)

The operation of a D-MOSFET is very similar to that of a JFET. With voltage applied to the source and drain, current carriers pass through the channel. The gate does not need to be energized in order to produce conduction. In this regard, a D-MOSFET is considered to be a normally-on device. Control of I_D is, however, determined by the polarity of gate voltage. When the polarity of V_G is the same as the channel, current carrier depletion occurs. A reverse in polarity causes I_D to increase due to the enhancement of the current carriers. In a sense, the D-MOSFET should be classified as a depletion–enhancement device. Its conduction normally responds to both conditions of operation.

A family of characteristic curves for an N-channel D-MOSFET is shown in Fig. 12-26. The horizontal part of the graph shows the source–drain voltage as V_{DS}. The vertical axis shows the drain current in milliamperes. Individual curves of the graph show different values of gate voltage. Note that zero V_G is near the center of the curve. This means that the gate can swing above or below zero. For an N-channel device, negative gate voltage reduces I_D. This voltage, in effect, pulls holes from the substrate into the initial channel area. Electrons normally in the channel move in to fill the holes. This

action depletes the number of electrons in the initial channel. An increase in negative gate voltage causes a corresponding decrease in I_D.

When V_G swings positive, it causes the number of current carriers in the initial channel to be increased. A positive gate voltage attracts electrons from the P substrate. This action increases the width of the initial channel. As a result, more I_D flows. Making V_G more positive increases I_D. This means that the channel is aided or enhanced by a positive gate voltage. A D-MOSFET therefore responds to both the depletion and enhancement of its current carriers. Selection of an operating point is easy to accomplish when this device is used as an amplifier.

In a P-channel D-MOSFET, operation is very similar to that of the N-channel device. In the P-channel device, holes are the current carriers. Current conduction is normally on when V_G is zero. A positive gate voltage will reduce I_D. Electrons are drawn out of the substrate to fill holes in the channel. This causes a reduction in channel current. When the gate voltage swings negative the channel current carriers are enhanced. Holes pulled into the initial channel cause an increase in current carriers. Current carriers are enhanced or depleted according to the polarity of V_G.

P-channel D-MOSFETs are not as readily available as the N-channel type of device. Fabrication of the P-channel device is somewhat more difficult to achieve. As a general rule, the N-channel unit is somewhat more costly to produce. Applications of D-MOSFETs tend to use the N-channel device more than the P-channel device.

UNIJUNCTION TRANSISTORS

The unijunction transistor (UJT) is often described as a voltage-controlled diode. This device is not considered to be an amplifying device. It is, however, classified as a unipolar transistor. The UJT and an N-channel JFET are often confused because of the similarities in their schematic symbols and crystal construction. The operation and function of each device is entirely different.

Figure 12-27 shows the crystal construction, schematic symbol, and element names of a UJT. Note that the UJT is a three-terminal, single-junction device. The crystal is an N-type bar of silicon with contacts at each end. The end connec-

Figure 12-26. Family of drain curves for a D-type N-channel MOSFET.

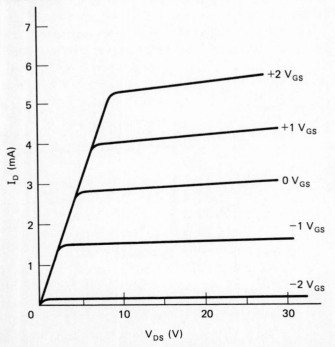

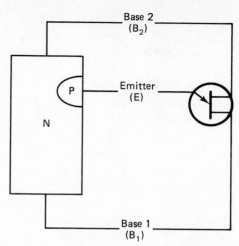

Figure 12-27. Unijunction transistor.

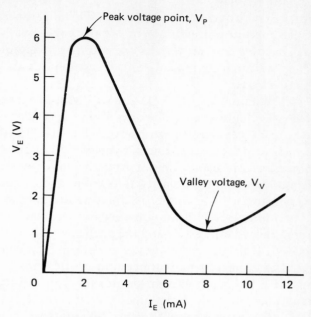

Figure 12-28. Characteristic curve for a UJT.

tions are called base 1 (B_1) and base 2 (B_2). A small heavily doped P region is alloyed to one side of the silicon bar. This serves as the emitter (E). A P-N junction is formed by the emitter and the silicon bar. The arrow of the symbol "*Points in N*," which means that the emitter is P material and the silicon bar is N material. The arrow of the symbol is slanted, which distinguishes it from the N-channel JFET.

An interbase resistance (R_{BB}) exists between B_1 and B_2. Typically, R_{BB} is between 4 and 10 kΩ. This value can be easily measured with an ohmmeter. The resistance of the silicon bar is represented by R_{BB}. This resistance can be divided into two values. R_{B1} is between the emitter and B_1. R_{B2} is between the emitter and B_2. Normally, R_{B2} is somewhat less than R_{B1}. The emitter is usually closer to B_2 than B_1. When a UJT is made operational, the value of R_{B1} will change with different emitter voltages.

In circuit applications B_1 is usually placed at circuit ground or the source voltage negative. The emitter then serves as the input to the device. B_2 provides circuit output. A change in E–B_1 voltage will cause a change R_{BB}. The output current of the device will increase when E–B_1 turns on. A UJT is normally used as a trigger device. It is used to control some other solid-state device. No amplification is achieved by a UJT.

A characteristic curve for a typical UJT is shown in Fig. 12-28. The vertical part of this graph shows the emitter voltage as V_E. An increase in V_E causes the curve to rise vertically. The horizontal part of the graph shows the emitter current (I_E). Note that the curve has *peak voltage* (V_P) and *valley voltage* (V_V) points. An increase in I_E causes V_E to rise until it reaches the peak point. A further increase in I_E will cause the emitter voltage to drop to the valley point. This condition is called the negative resistance region. A device that has a negative resistance characteristic is capable of regeneration or oscillation.

In operation, when the emitter voltage reaches the peak voltage point, E–B_1 becomes forward biased. This condition draws holes and electrons to the P-N junction. Holes are injected into the B_1 region. This action causes B_1 to be more conductive. As a result, the E–B_1 region become low resistant. A sudden drop in R_{B1} will cause a corresponding increase in B_1–B_2 current. This change in current can be used to trigger or turn on other devices. The trigger voltage of a UJT is a predictable value. Knowing this value will permit the device to be used as a control element.

UNIPOLAR TRANSISTOR TESTING

Unipolar transistors, like bipolar transistors, periodically require testing. The condition of the device can be tested with a curve tracer oscilloscope, sound-producing instruments, or in-circuit/out-of-circuit testers. This equipment is normally used to evaluate a device and to identify leads. Test procedures are rather easy to accomplish and a reliable indicator of device condition. If equipment of this type is available, become familiar with its operation by review of the operational manual. Use it to test devices and identify leads.

The ohmmeter of a VOM or VTVM can also be used to evaluate unipolar devices. This instrument is more readily available than specialized testers. Lead identification, shorted devices, open conditions, and gain can be evaluated. Before using the ohmmeter it is important that the polarity of its voltage source be known. In the following procedure, we are describing an ohmmeter with straight polarity. The black or common lead is negative and the probe or red lead is positive. If the meter polarity is unknown, it may be tested by connecting a voltmeter across its leads.

JFET testing with an ohmmeter is relatively easy to accomplish. Remember that the JFET is a single bar of silicon with one P-N junction. To identify the leads, first select any two of the three leads. Connect the ohmmeter to these leads. Note the resistance. If it is 100 to 1000 Ω the two leads are the source and drain. Reverse the ohmmeter and connect it to the same leads. If the two leads are actually the source and drain, the same resistance value will be indicated. The third lead is gate. If the two selected leads show a different resistance in each direction, one must be the gate. Select the third lead and one of the previous leads. Repeat the procedure. A good JFET must show the same resistance in each direction between two leads. This represents the source–drain channel connections.

The gate of a JFET responds as a diode. It will be low resistant in one direction and high resistant in the reverse direction. If the device shows low resistance when the positive probe is connected to the gate and negative to the source or drain, the JFET is an N-channel device. A P-channel device will show low resistance when the gate is made negative and the source or drain positive. The resistance ratio in the forward and reverse direction must be at least 1:1000 for a good device. Figure 12-29 indicates some representative resistance values for good JFETs.

A test of JFET amplification is indicated by connecting the ohmmeter between source and drain. Polarity of the meter is not significant. Then touch the gate with your finger. This action should cause a decrease in resistance. A more significant indication will take place when a 200-kΩ resistor is used.

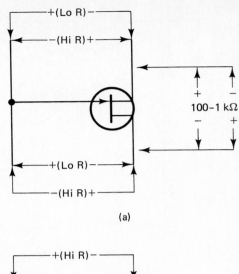

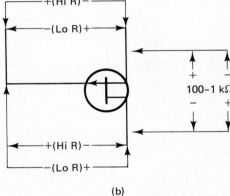

(a)

(b)

Figure 12-29.   Resistance values of a JFET: (a) N-channel; (b) P-channel.

Touching the resistor between the gate and either the source or drain will cause reduced resistance. If very little or no resistance change occurs during this test, it indicates an open or faulty gate.

D-MOSFET testing is very similar to the JFET. The gate of the D-MOSFET is insulated from the channel. It will show infinite resistance between each lead regardless of the polarity of the ohmmeter. Figure 12-30 shows some representative values for good D-MOSFETs. Three lead devices have the substrate and source internally connected. Four lead devices have an independent substrate lead.

A four-lead device can be identified as an N-channel or a P-channel device. A P-channel device will show low resistance when the positive lead is connected to the substrate and the negative lead is connected to either the source or drain. An N-channel device has low resistance when the negative lead is connected to the substrate and the positive lead to either the source or drain.

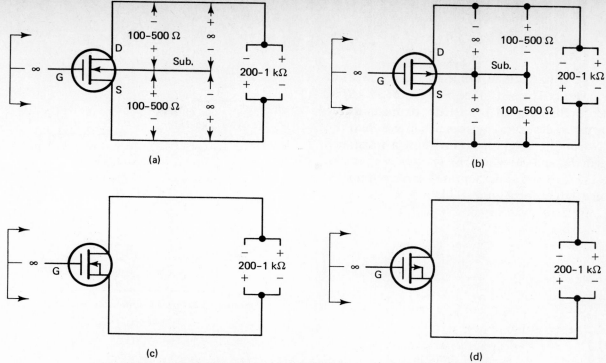

Figure 12-30. D-MOSFET resistances: (a) N-channel; (b) P-channel (c) N-channel; (d) P-channel.

E-MOSFET testing with an ohmmeter is not very meaningful. The structure of this device does not effectively have an interaction between different leads. The gate, source, and drain of a good device will show infinite resistance between these leads. A four-lead device will respond as a diode between the substrate and the source or drain. Polarity of the ohmmeter can be used to identify the substrate and the channel. A three-load device responds as a diode between the source and drain. Figure 12-31 shows some representative ohmmeter resistances for good E-MOSFETs.

Figure 12-31. Resistances of E-MOSFET: (a) N-channel; (b) P-channel; (c) N-channel; (d) P-channel.

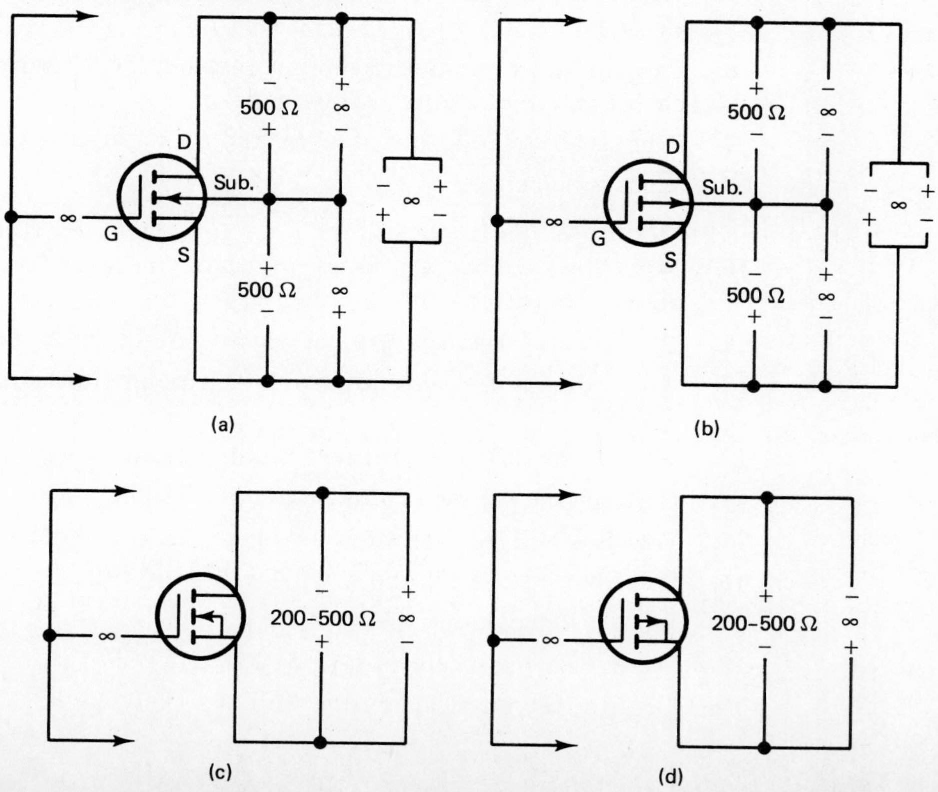

UJT testing with an ohmmeter is identical to that of a JFET. The resistance between B_1 and B_2 is, however, in the range of 1 to 10 kΩ. In most devices the emitter B_2 resistance is somewhat less than the E–B_1. Normally, a UJT will not show a resistance change in B_1 and B_2 when the emitter is touched. Figure 12-32 shows some representative ohmmeter resistance values for a good UJT.

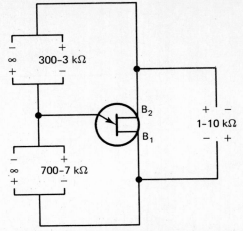

Figure 12-32.   UJT ohmmeter resistance values.

Review

1. Explain what is meant by the term "bipolar transistor."

2. Why is it important that the base of a transistor be thin and lightly doped?

3. Why must the emitter–base junction and base–collector junction of a bipolar transistor both be biased at the same time to produce operation?

4. Why does current flow through the reverse-biased base–collector junction of a bipolar transistor?

5. Why can PNP and NPN transistors not be used interchangeably in a circuit?

6. Explain how current carriers flow through the forward-biased emitter–base junction of a bipolar transistor.

7. What is the relationship of I_B, I_C, and I_E in a bipolar transistor?

8. What is dc beta?

9. What is ac beta?

10. What does a collector family of characteristic curves show about a bipolar transistor?

11. How does a transistor respond when operating in the saturation region?

12. How do I_C and V_{CE} respond when a bipolar transistor is in the cutoff region?

13. What is the difference between a unipolar and a bipolar transistor?

14. Describe the physical construction of a JFET.

15. Why is a JFET described as a voltage-controlled device?

16. How do a MOSFET and a JFET compare?

17. What is a depletion-type MOSFET?

18. What is an enhancement-type MOSFET?

19. Describe the physical construction of a UJT.

20. Why does the interbase resistance of a UJT change when the emitter becomes forward biased?

21. What is the negative resistance region of a UJT characteristic?

Student Activities

Bipolar Transistor Type Identification

1. With an ohmmeter, identify the type NPN or PNP of several bipolar transistors.

2. Record the number of the device and indicate its type according to your findings.

Bipolar Transistor Lead Identification

1. Select a known NPN transistor type.

2. Make a sketch of the lead placement.

3. Identify the leads of the device.

4. Add the labels *E*, *B*, and *C* to the sketch.

5. Repeat the procedure of steps 1 to 4 for a PNP device.

6. If the manufacturer's data are available, compare your lead identification with the known information.

7. What type of packaging is used for each device?

Bipolar Gain Test

1. Using the known lead data of a specific transistor, determine the gain using an ohmmeter.

2. Use a 100-kΩ resistor for the base connection.

3. Record the resistance change observed for each device.

JFET Testing

1. With an ohmmeter, determine the type of device being tested. Use both N- and P-channel devices.

2. Identify the leads.

3. Make a sketch showing the lead placement.

4. Test the device for gain.

5. Note any faulty devices.

MOSFET Testing

1. With an ohmmeter, determine the type of MOSFET device being tested. Indicate E, D, N, or P channel.

2. Identify the leads. Note if it is a three- or a four-lead device.

3. Make a sketch of the lead location.

Bipolar Transistor Characteristic Curves

1. Construct the I_C–V_{CE} data circuit of Fig. 12-11.

2. Collect data for three or four individual I_B curves.

3. On a sheet of graph paper plot curves using the data collected.

Transistor Operation Regions

1. Construct the circuit of Fig. 12-10(a).

2. Use the small-signal transistor indicated or an equivalent device.

3. Turn on the source voltage V_{CC}.

4. Measure and record V_C and V_R.

5. Turn off V_{CC} and modify the circuit as in Fig. 12-10(b).

6. Turn on V_{CC}. Measure and record V_C and V_R.

7. Turn off V_{CC} and modify the circuit to conform with Fig. 12-10(c).

8. Turn on V_{CC}. Measure and record V_C and V_R.

9. Note the operational differences of the transistor in its active, cutoff, and saturation regions.

JFET Characteristic Curves

1. Construct the drain curve test circuit of Fig. 12-22.

2. Adjust V_{GS} to zero, then alter V_{DD} from 0 to 15 V while recording I_D values for each 0.5- or 1-V increment.

3. Adjust V_{GS} to 0.5 V, 1 V and 1.5 V values while recording I_D values for the same V_{DD} values.

4. On a sheet of graph paper plot an I_D family of curves.

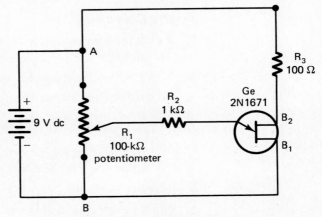

Figure 12-33.   UJT test circuit.

UJT Characteristics

1. Construct the UJT characteristic test circuit of Fig. 12-33.

2. Turn on the power supply and adjust it to 9 V dc as measured at point $A-B$.

3. Connect the voltmeter across E and B_1 of the UJT.

4. Adjust the 100-kΩ potentiometer to an indication of 0 V.

5. Gradually adjust the potentiometer while observing V_E-B_1. Note where the peak voltage (V_P) and the valley voltage (V_V) are located.

6. Turn off the power supply and disconnect the circuit.

13

Amplification

Electronic systems must perform a variety of basic functions in order to accomplish a particular operation. An understanding of these functions is an essential part of system operational theory. In this chapter we investigate the amplification function. Amplification is achieved by devices that produce a change in signal amplitude. Transistors, integrated circuits, and vacuum tubes are typical amplifying devices.

In the first part of this chapter, we look at amplification in general. No particular amplifying device is considered. Second, we discuss how a particular type of device achieves amplification. Circuit operation, biasing methods, circuit configurations, classes of amplification, and operational conditions are considered. As a general rule, this chapter is directed toward small-signal amplification. The following chapter deals with large-signal or power amplification.

IMPORTANT TERMS

The study of amplification involves a number of new and somewhat unusual terms. In general, these terms are extremely important in the understanding of amplifiers. Review each term very carefully before proceeding with the text of the chapter. Refer back to a specific term when it is encountered. Mastery of these terms will be a definite aid in the understanding of this chapter.

Amplification. The ability to control a large force by a smaller force. There is voltage, current, and power amplification.

Audio amplifiers. Amplifiers that respond to the human range of hearing or frequencies from 15 to 15 kHz.

Beta-dependent biasing. A method of biasing that

responds in some way to the amplifying capabilities of the device.

Bypass capacitor. A capacitor that provides an alternative path around a component or to ground.

Circuit configuration. A method of connecting a circuit, such as common-emitter, common-source, and so on.

Control grid. An element of a vacuum tube that achieves control.

Digital integrated circuit. An IC that responds to two-state (on–off) data.

Dual-in-line package (DIP). A packaging method for integrated circuits.

Dynamic characteristics or operation. Electronic device operational characteristics that show how the device responds to ac or changing voltage values.

Feedback network. A resistor combination that returns some of the output signal to the input of an IC.

Impedance. A form of opposition to alternating current measured in ohms. Electronic devices have input and output impedance values.

Linear amplifiers. Amplifying circuits that increase the amplitude of a signal that is a duplicate of the input. These devices operate in the straight-line part of the characteristic curves.

Load line. A line drawn on a family of characteristic curves that shows how the device will respond in a circuit with a specific load resistor value.

Microphone. A device that changes sound energy into electrical energy.

Noninverting input. An input lead for an integrated circuit that does not shift the phase of a signal.

Open loop. An operation amplifier connection method that causes full or maximum amplification.

Plate. The element of a vacuum tube that collects electrons emitted from the cathode.

Power dissipation. An electronic device characteristic that indicates the ability of the device to give off heat.

Q point. An operating point for an electronic device that indicates its dc or static operation with no ac signal applied.

Radio frequency (RF). An alternating current (ac) frequency in excess of the human range of hearing which starts at 30,000 Hz.

Static state. A dc operating condition of an electronic device with operating energy but no signal applied.

Thermal stability. The condition of an electronic device that indicates its ability to remain at an operating point without variation due to temperature.

Triode. A three-element vacuum tube containing a cathode, grid, and plate.

Voltage follower. An amplifying condition where the input and output voltages are of the same value across different impedance.

AMPLIFICATION PRINCIPLES

Amplification is achieved by an electronic device and its associated components. As a general rule, the amplifying device is placed in a circuit. The components of the circuit usually have about as much influence on amplification as the device itself. The circuit and the device must be supplied electrical energy for it to function. Typically, amplifiers are energized by direct current. This may be derived from a battery or a rectified ac power supply. The amplifier then processes a signal of some type when it is placed in operation. The signal may be either ac or dc, depending on the application.

Reproduction and Amplification

The signal to be processed by an amplifier is first applied to the input part of the circuit. After being processed the signal appears in the output circuitry. The output signal may be reproduced in its exact form, amplified, or both amplified and reproduced. The value and type of input signal, operating source energy, device characteristics, and circuit components all have some influence on the output signal.

Figure 13-1 illustrates the process of reproduction, amplification, and the combined amplification–

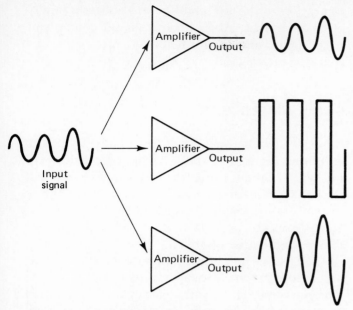

Figure 13-1. Amplification and reproduction: (a) reproduction; (b) amplification; (c) combined amplification and reproduction.

reproduction functions of an amplifier. In part (a), the amplifier performs only the reproduction function. Note that the input and output signals have the same size and shape. Part (b) shows only the amplification function. In this case, the input signal is increased in amplitude. The output signal is amplified but does not necessarily resemble the input signal. In many applications amplification and reproduction must both be achieved at the same time. Part (c) shows this function of an amplifier. Depending on the application, an amplifier must be capable of developing any of these three output signals.

Voltage Amplification

A system designed to develop an output voltage that is greater than its input voltage is called a voltage amplifier. A voltage amplifier has voltage gain. This function is defined as a ratio of the output signal voltage to the input signal voltage. In equation form voltage amplification is expressed as

$$\text{amplification voltage} = \frac{\text{output voltage}}{\text{input voltage}}$$

or

$$A_V = \frac{V_O}{V_{in}} \quad \text{or} \quad \frac{\Delta v_o}{\Delta v_{in}}$$

The uppercase letters in V_O and V_{in} denote dc voltage values. Ac voltages are usually expressed by lowercase letters. The Greek capital letter delta (Δ) preceding a voltage indicates a changing value. Voltage amplifiers are capable of a wide range of amplification values.

Current Amplification

An amplifying system designed to develop output current that is greater than the input current is called a current amplifier. A current amplifier has current gain. This type of amplification is defined as a ratio of output current to input current. In equation form, it is expressed as:

$$\text{current amplification} = \frac{\text{output current}}{\text{input voltage}}$$

or

$$A_i = \frac{I_O}{I_{in}} \quad \text{or} \quad \frac{\Delta i_o}{\Delta i_{in}}$$

Current gain takes into account the beta of a transistor and all of the associated components that make the device operational.

Power Amplification

Power amplification is a ratio of the developed output signal power to the input signal power. Note that this refers only to the power gain of the signal. All amplifiers consume a certain amount of power from the energy source during operation. This is usually not included in an expression of power gain. An equation of power gain shows that

$$\text{power amplification} = \frac{\text{output signal power}}{\text{input signal power}}$$

or

$$A_p = \frac{P_O}{P_{in}}$$

Power signal gain can also be expressed as the product of voltage gain and current gain. In this regard, the equation is

power amplification = voltage amplification

$$\times \text{ current amplification}$$

or

$$A_p = A_V \times A_i$$

It should be apparent from the A_p equation that A_V and A_i do not both need to be large to have power gain. Typical power amplifiers may have a large current gain and a voltage gain of less than 1. Power gain must take into account both signal voltage and current gain in order to be meaningful.

BIPOLAR TRANSISTOR AMPLIFIERS

The primary function of an amplifier is to provide voltage gain, current gain, or power gain. The resulting gain depends a great deal on the device being used and its circuit components. Normally, the circuit is energized by dc voltage. An ac signal is then applied to the input of the amplifier. After passing through the transistor an amplified version of the signal appears in the output. Operating conditions of the device and the circuit determine the level of amplification to be achieved.

In order for a bipolar transistor to respond as an amplifier, the emitter–base junction must be forward biased and the collector–base junction reverse biased. Specific operating voltage values must then be selected that will permit amplification. If both reproduction and amplification are to be achieved, the device must operate in the center of its active region. Remember that this is between the

saturation and cutoff regions of the collector family of characteristic curves. Proper circuit component selection permits a transistor to operate in this characterisitc region.

BASIC AMPLIFIERS

Suppose now that we look at the circuitry of a basic amplifier. Figure 13-2 shows a schematic diagram of this type of circuit. Notice that a schematic symbol of the NPN transistor is used in this diagram. Schematic symbols are normally used in all circuit diagrams. Keep in mind the crystal structure of the device represented by the symbol.

The basic amplifier has a number of parts that are needed to make it operational. V_{CC} is the dc energy source. The negative side of V_{CC} is connected to ground. The emitter is also connected to ground. This type of circuit configuration is called a grounded-emitter or common-emitter amplifier. The emitter, one side of the input, and one side of the output are all commonly connected together. In an actual circuit these points are usually connected to ground.

Resistor R_B of the basic amplifier is connected to the positive side of V_{CC}. This connection makes the base positive with respect to the emitter. The emitter–base junction is forward biased by R_B. Resistor R_L connects the positive side of V_{CC} to the collector. This connection reverse biases the collector. Through the connection of R_B and R_L the transistor is properly biased for operation. A transistor connected in this manner is considered to be in its *static* or *dc operating state*. It has the necessary dc energy applied for it to be operational. No signal is applied to the input for amplification.

Let us now consider the operation of the basic amplifier in its static state. With the given values of Fig. 13-2 base current (I_B) can be calculated. I_B, in

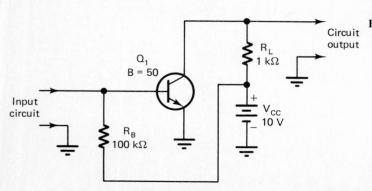

Figure 13-2. Basic bipolar amplifier.

this case, is limited by the value R_B and the base–emitter junction resistance. When the emitter–base junction is forward biased, its resistance becomes very small. The value of R_B is therefore the primary limiting factor of I_B. I_B can be determined by an application of Ohm's law. The equation is

$$\text{base current} = \frac{\text{source voltage}}{\text{base resistance}} \quad \text{or} \quad I_B = \frac{V_{CC}}{R_B}$$

For our basic transistor amplifier, this is determined to be

$$I_B = \frac{V_{CC}}{R_B} = \frac{10 \text{ V}}{100,000 \ \Omega} \quad \text{or} \quad \frac{10 \text{ V}}{100 \times 10^3 \ \Omega}$$
$$= 100 \times 10^{-6} \text{ or } 100 \ \mu A$$

The value of base current, in this case, is extremely small. Remember that only a small amount of I_B is needed to produce I_C.

The beta of the transistor used in Fig. 13-2 has a given value. With beta and the calculated value of I_B, it is possible to determine the collector current of the circuit. Beta is the current gain of a common-emitter amplifier. In Chapter 12 beta was described by the formula

$$\text{beta} = \frac{\text{collector current}}{\text{base current}} \quad \text{or} \quad \beta = \frac{I_C}{I_B}$$

In our transistor amplifier I_B has been determined and beta has a given value of 50. By transposing the beta formula, I_C can be determined by the equation

$$\text{collector current} = \text{beta} \times \text{base current}$$

or

$$I_C = \beta \times I_B$$

For the amplifier circuit, collector current is determined to be

$$I_C = \beta \times I_B = 50 \times 100 \times 10^{-6}$$
$$= 5 \times 10^{-3} \text{ or } 5 \text{ mA}$$

This means that 5 mA of I_C will flow through R_L when an I_B of 100 μA flows.

In any electric circuit, we know that current flow through a resistor will cause a corresponding voltage drop. In a basic transistor amplifier, I_C will cause a voltage drop across R_C. In the preceding step I_C was calculated to be 5 mA. By using Ohm's law again, the voltage drop across R_L can be determined. This is figured by the equation

$$R_L \text{ voltage} = \text{collector current}$$
$$\times \text{ collector resistance}$$

or

$$V_{RL} = I_C \times R_L$$

For the amplifier circuit, the voltage drop across R_L is

$$V_{RL} = I_C \times R_L$$
$$= 5 \times 10^{-3} \times 1 \times 10^3$$
$$= 5 \text{ V}$$

This means that half or 5 V or V_{CC} will appear across R_L. With a V_{CC} value of 10V, the other 5V will appear across the collector–emitter of the transistor. A V_{CE} voltage of 5 V from a V_{CC} of 10 V means that the transistor is operating near the center of its active region. Ideally, the transistor should respond as a linear amplifier. When a suitable signal is applied to the input, it should amplify and reproduce the signal in the output.

The operational voltage and current values of the basic transistor amplifier in its static state are summarized in Fig. 13-3. Note that an I_B of 100 μA will cause an I_C of 5 mA. With a beta of 50, the

Figure 13-3. Static operating condition of a basic amplifier.

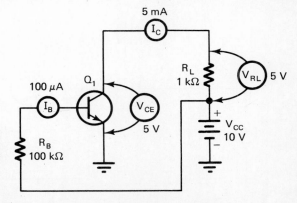

amplifier has the capability of rather substantial amount of output current. The collector current passing through R_L causes V_{CC} to be divided. This establishes operation in the approximate center of the collector family of characteristic curves. Ideally, this static condition should cause the amplifier to respond as a linear device when a signal is applied.

Signal Amplification

In order for the basic transistor circuit of Fig. 13-3 to respond as an amplifier, it must have a signal applied. The signal may be either voltage or current. An applied signal causes the transistor to change from its static state to a dynamic condition. Dynamic conditions involve changing values. All ac amplifiers respond in the dynamic state. The output of this should develop an ac signal.

Figure 13-4 shows a basic transistor amplifier with an ac signal applied. A capacitor is used, in this case, to couple the ac signal source to the amplifier. Remember that ac passes easily through a capacitor while dc is blocked. As a result of this, the ac signal is injected into the base–emitter junction. Dc does not flow back into the signal source. The ac signal is therefore added to the dc operating voltage. The emitter–base voltage is a dc value that changes at an ac rate.

Consider now how the applied ac signal alters the emitter–base junction voltage. Figure 13-5 shows some representative voltage and current values that appear at the emitter–base junction of the transistor. They can be measured with a voltmeter or observed with an oscilloscope. Part (a) shows the dc operating

voltage and base current. This occurs when the amplifier is in its steady or static state. Part (b) shows the ac signal that is applied to the emitter–base junction. Note that the amplitude change of this signal is a very small value. This also shows how the resulting base current and voltage change with ac applied. The ac signal is essentially riding on the dc voltage and current.

Refer now to the schematic diagram of the basic amplifier in Fig. 13-6. Note in particular the waveform inserts that appear in the diagram. These show how the current and voltage values respond when an ac signal is applied.

The ac signal applied to input of our basic amplifier rises in value during the positive alternation and falls during the negative alternation. Initially, this causes an increase and decrease in the value of V_{BE}. This voltage has a dc level with an ac signal riding on it. The indicated I_B is developed as a result of this voltage. The changing value of I_B causes a corresponding change in I_C. Note that I_B and I_C both appear to be the same. There is however, a very noticable difference in values. I_B is in microamperes while I_C is in milliamperes. The resulting I_C passing through R_L causes a corresponding voltage drop (V_{RL}) across R_L. V_{CE} appearing across the transistor is the reverse of V_{RL}. These signals are both ac values riding upon a dc level. The output, V_O, is changed to an ac value. Capacitor C_2 blocks the dc component and passes only the ac signal.

It is interesting to note in Fig. 13-6 that input and output signals of the amplifier are reversed. When the ac input signal rises in value it causes the output to fall in value. The negative alternation of

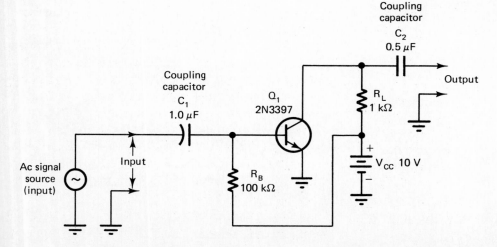

Figure 13-4. Amplifier with ac signal applied.

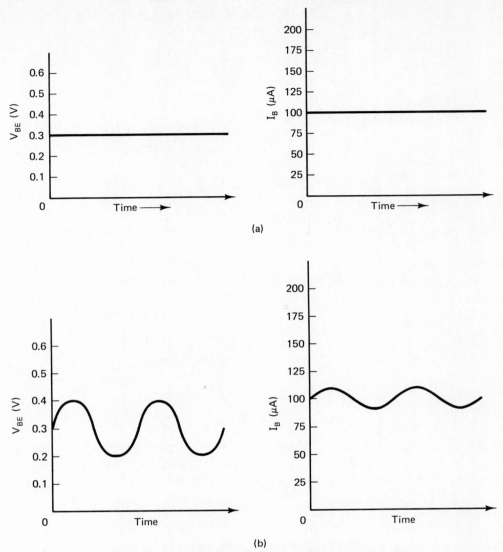

Figure 13-5. I_B and V_{BE} conditions: (a) static; (b) dynamic.

the input causes the output to rise in value. This condition of operation is called phase inversion. Phase inversion is a distinguishing characteristic of the common-emitter amplifier.

Amplifier Bias

If a transistor is to operate as a *linear amplifier* it must be properly biased and have a suitable operating point. Its steady state of operation depends a great deal on base current, collector voltage, and collector current. Establishing a desired operating point requires proper selection of bias resistors and a load resistor to provide proper input current and collector voltage.

Stability of operation is a very important consideration. If the operating point of a transistor is permitted to shift with temperature changes, unwanted distortion may be introduced. In addition to distortion, operating point changes may also cause damage to a transistor. Excessive collector current, for example, may cause the device to develop too much heat.

The method of biasing a transistor has a great deal to do with its thermal stability. Several different methods of achieving bias are in common use today. One method of biasing is considered to be beta dependent. Bias voltages are largely dependent on transistor beta. A problem with this method of biasing is transistor response. Transistor beta is

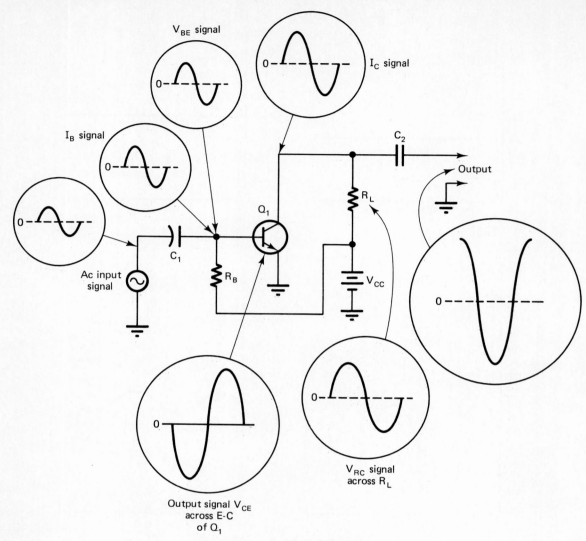

Figure 13-6. Ac amplifier operation.

rarely ever the same for a specific device. Biasing set up for one transistor will not necessarily be the same for another transistor.

Biasing that is independent of beta is very important. This type of biasing responds to fixed voltage values. Beta does not alter these voltage values. As a general rule, this form of biasing is more reliable. The input and output of a transistor is very stable and the results are very predictable.

BETA-DEPENDENT BIASING. Four common methods of beta-dependent biasing are shown in Fig. 13-7. The steady-state or static conditions of operation are a function of the transistor's beta.

These methods are rather easy to achieve with a minimum of parts. They are not very widely used today.

Circuit (a) is a very simple form of biasing for a common-emitter amplifier. This method of biasing was used for our basic transistor amplifier. The base is simply connected to V_{cc} through R_B. This causes the emitter–base junction to be forward biased. The resulting collector current is beta times the value of I_B. The value of I_B is V_{cc}/R_B. As a general rule, biasing of this type is very sensitive to changes in temperature. The resulting output of the circuit is rather difficult to predict. This method of biasing is often called fixed biasing. It is primarily used because of its simplicity.

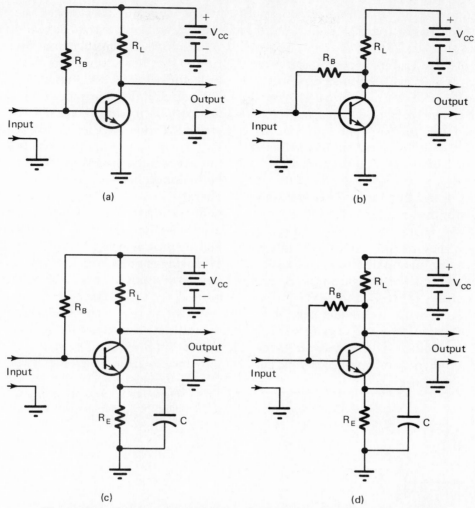

Figure 13-7. Methods of beta-dependent biasing: (a) fixed biasing; (b) self-biasing; (c) fixed-emitter biasing; (d) self-emitter biasing.

Circuit (b) of Fig. 13-7 was developed to compensate for the temperature sensitivity of circuit (a). Bias current is used to counteract changes in temperature. In a sense, this method of biasing has negative feedback. R_B is connected to the collector rather than V_{CC}. The voltage available for base biasing is the leftover after voltage drop across the load resistor. If the temperature rises, it causes an increase in I_B and I_C. With more I_C, there is more voltage drop across R_L. Reduced V_C voltage will cause a corresponding drop in I_B. This, in turn, brings I_C back to normal. The opposite reaction will occur when transistor temperature becomes less. This method of biasing is called *self-biasing*.

Circuit (c) is an example of emitter biasing. Thermal stability is improved with this type of construction. I_B is again determined by the value of R_B and V_{CC}. An additional resistor is placed in series with the emitter of this circuit. Emitter current passing through R_E produces emitter voltage (V_E). This voltage opposes the base voltage developed by R_B. Proper values R_B and R_E are selected so that I_B and I_E will flow under ordinary operating conditions. If a change in temperature should occur, V_E will increase in value. This action will oppose the base bias. As a result, collector current will drop to its normal value. The capacitor connected across R_E is called an emitter bypass capacitor. It provides an ac path for signal voltages around R_E. With C in the circuit, an average dc level is maintained at the emitter. Without C in the circuit, amplifier gain would be reduced. The value of C is dependent on the frequencies being amplified. At the lowest possible frequency being amplified, the capacitive reactance (X_C) must be 10 times smaller than the resistance of R_E.

Circuit (d) of Fig. 13-7 is a combination of circuits (b) and (c). It is often called *self-emitter bias*. In the same regard, circuit (c) could be called *fixed-emitter bias*. As a general rule, emitter biasing is not very effective when used independently. Circuit (d) has good thermal stability. The output has reduced gain because of the base resistor connection.

INDEPENDENT BETA BIASING.

Two methods of biasing a transistor that is independent of beta are shown in Fig. 13-8. These circuits are extremely important because they do not change operation with beta. As a general rule, these circuits have very reliable operating characteristics. The output is very predictable and stability is excellent.

Circuit (a) is described as the divider method of biasing. It is widely used today. The base voltage (V_B) is developed by a voltage-divider network made up of R_B and R_1. This network makes the circuit

independent of beta changes. Voltage at the base, emitter, and collector all depend on external circuit values. By proper selection of components, the emitter–base junction is forward biased, with the collector being reverse biased. Normally, these bias voltages are all referenced to ground. The base, in this case, is made slightly positive with respect to ground. This voltage is somewhat critical. A voltage that is too positive, for example, will drive the transistor into saturation. With proper selection of bias voltage, however, the transistor can be made to operate in any part of the active region. The temperature stability of the circuit is excellent. With proper R_E bypass capacitor selection, this method of biasing produces very high gain. The divider method of biasing is often a universal biasing circuit. It can be used to bias all transistor amplifier circuit configurations.

Circuit (b) is very similar in construction and operation to circuit (a). One less resistor is used. The power supply requires two voltage values with reference to ground. A split power supply is used as

Figure 13-8. Independent beta biasing: (a) divider bias; (b) divider bias with split supply.

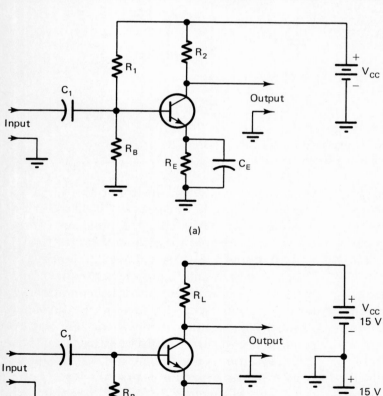

(a)

(b)

an energy source for this circuit. Note the indication of $+V_{CC}$ and $-V_{CC}$. R_B is connected to ground. In this case, the value of R_B would determine the value of I_B with only half of total supply voltage. The values of R_L and R_E are usually larger to accommodate the increased supply voltage. If R_E is properly bypassed, the gain of this circuit is very high. Thermal stability is excellent.

Load-Line Analysis

Earlier in the chapter we looked at the operation of an amplifier with respect to its beta. Current and voltage were calculated and operation was related to these values. This method of analysis is very important. Amplifier operation can also be determined graphically. This method employs a collector family of characteristic curves. A load line is developed for the graph. With the load-line method of analysis, it is possible to predict how the circuit will respond graphically. A great deal about the operation of an amplifier can be observed through this method.

The load-line method of circuit analysis is widely used today. Engineers use this method in designing new circuits. The operation of a specific circuit can also be visualized by this method. In circuit design, a specific transistor is selected for an amplifier. Source voltage, load resistance, and input signal levels may be given values in the design of the circuit. The transistor is made to fit the limitation of the circuit.

For our application of the load line, assume that a circuit is to be analyzed. Figure 13-9(a) shows the circuit being analyzed. A collector family of characteristic curves for the transistor is shown in Fig. 13-9(b). Note that the power dissipation rating of the transistor is included in the diagram.

Figure 13-9. (a) Circuit and (b) characteristic curves of a circuit to be analyzed.

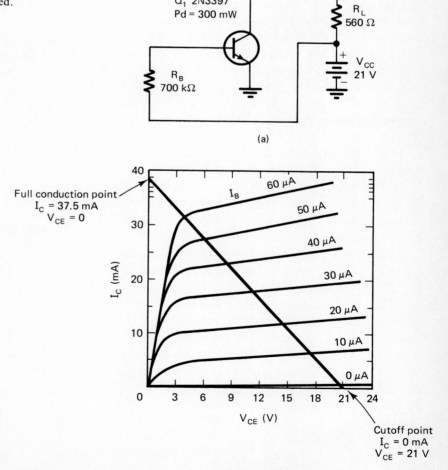

POWER DISSIPATION CURVE. A common practice in load-line analysis is to first develop a power dissipation curve. This gives some indication of the maximum operating limits of the transistor. *Power dissipation* (*Pd*) refers to maximum heat that can be given off by the base–collector junction. Usually, this value is rated at 25° C. *Pd* is the product of I_C and V_{CE}. In our circuit the *Pd* rating for the transistor is 300 mW.

To develop a *Pd* curve each value of V_{CE} is used with the *Pd* rating to determine an I_C value. The formula is

$$\text{collector current} = \frac{\text{power dissipation}}{\text{collector–emitter voltage}}$$

or

$$I_C = \frac{Pd}{V_{CE}}$$

Using this formula, calculate the I_C value for each of the V_{CE} values of the family of curves. Using the V_{CE} value and the corresponding calculated I_C value, note the location on the curve. These points connected together are representative of the 300-mW power dissipation curve. In practice, the load line must be located to the left of the established *Pd* curve. Satisfactory operation without excessive heat generation can be assured in that area of the curve.

STATIC LOAD-LINE ANALYSIS. The load line of a transistor amplifier represents two extreme conditions of operation. One of these is in the cutoff region. When the transistor is cut off there is no I_C flowing through the device. V_{CE} equals the source voltage with zero I_C. The second load-line point is in the saturation region. This point assumes full conduction of I_C. Ideally, when a transistor is fully conductive, $V_{CE} = 0$ and $I_C = V_{CC}/R_L$

The two load-line construction points for the analysis circuit are shown on the curves of Fig. 13-9. The cutoff point is located at the zero I_C, 21 V_{CE} point. The value of V_{CC} determines this point. At cutoff, $V_{CE} = V_{CC}$. The saturation points is located at the 37.5 mA I_C and 0 V_{CE} point. The I_C value is calculated using V_{CC} and the value of R_L. The formula is

$$I_C = \frac{V_{CC}}{R_L} = \frac{21 \text{ V}}{560 \text{ }\Omega}$$

$$= 0.0375 \text{ A or } 37.5 \text{ mA}$$

These two points are connected together with a straight line.

Figure 13-10 shows a family of characteristic curves with a load line for the circuit of Fig. 13-9. Development of the load line makes it possible to determine the operating conditions of the amplifier. For linear amplification, the operating point should be located near the center of the load line. In our circuit, an operating point of 30 μA is used. In the circuit diagram the value of R_B determines I_B. It is calculated by the equation $I_B = V_{CC}/R_B$. The value is

$$I_B = \frac{V_{CC}}{R_B} = \frac{21 \text{ V}}{700 \text{ k}\Omega}$$

$$= 0.00003 \text{ A or } 30 \text{ }\mu\text{A}$$

The operating point for this value is located at the intersection of the load line and the 30-μA I_B curve. It is indicated as point Q. Knowing this much about an amplifier shows how it will respond in its steady or static state. The Q point shows how the amplifier will respond without a signal applied.

Operation of our amplifier in its static state is displayed by the family of curves in Fig. 13-9. Projecting a line from the Q point to the I_C scale shows the resulting collector current. In this case the I_C is 17.5 mA. The dc beta for the transistor at this point is determined by the formula $\beta = I_C/I_B$. This value is

$$\beta = \frac{I_C = 17.5 \text{ mA}}{I_B = 30 \text{ }\mu\text{A}} \quad \text{or} \quad \frac{0.0175 \text{ A}}{0.000030 \text{ A}}$$

$$= 583.3$$

Figure 13-10. Family of characteristic curves for Fig. 13-9.

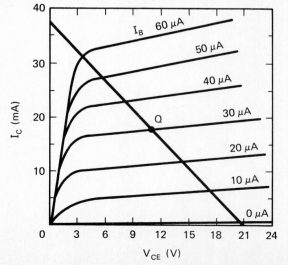

The resulting V_{CE} that will occur for the amplifier can also be determined graphically. Projecting a line directly down from the Q point will show the value of V_{CE}. In our circuit the V_{CE} is approximately 11 V. This means that 10 V will appear across R_L when the transistor is in its static state.

DYNAMIC LOAD-LINE ANALYSIS. Dynamic load-line analysis shows how an amplifier will respond to an ac signal. In this case the collector curves and circuit of Fig. 13-11 are used. A load line and Q point for the circuit have been developed on the curves. This establishes the static operation of the amplifier. Note the values of V_{CE} and I_C in the static state.

Assume now that a 0.1-V peak-to-peak ac signal is applied to the input of the amplifier. In this case the signal will cause a 20-μA peak-to-peak change in I_B. During the positive alternation of the input, I_B will change from 30 μA to 40 μA. This is shown as point P on the load line. During the negative alternation, I_B will drop from 30 μA to 20 μA. This is indicated as point N on the load line. In effect, this means that 0.1 V p-p causes I_B to change 20 μA p-p. The I_B signal extends to the right of the load line. Its value is shown as ΔI_B.

To show how a change in I_B influences I_C, lines are projected to the left of the load line. Note the projection of lines P, Q, and N toward the I_C values. The changing value of I_C is indicated as ΔI_C. An increase and decrease in I_B causes a corresponding increase and decrease in I_C. This shows that I_B and I_C are in phase.

Figure 13-11. Dynamic load line: (a) circuit; (b) characteristic curves.

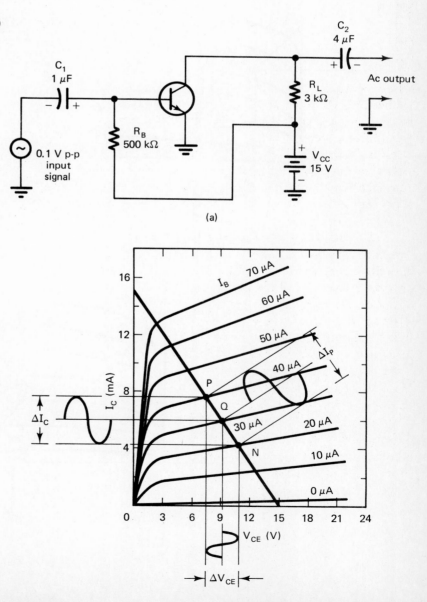

The ac beta of the amplifier can be determined by ΔI_C and ΔI_B. First determine the peak-to-peak I_C and I_B values. Then divide ΔI_C by ΔI_B. Determine the beta of the amplifier circuit. Using the same procedure, determine the dc beta of the transistor at point Q. How do the ac and dc beta values of this circuit compare?

Projecting points P, Q, and N downward from the load line shows how V_{CE} changes with I_B. The

value of V_{CE} is indicated as ΔV_{CE}. Note that an increase in I_B causes a decrease in the value of V_{CE}. A decrease in I_B causes V_{CE} to increase. This shows that I_B and V_{CE} are 180° out of phase. The difference in V_{CE} at any point appears across R_L.

The ac voltage gain of the amplifier can be determined from the dynamic load line. Remember that 0.1 V p-p input caused a change of 20 μA p-p in the I_B signal. The ac voltage gain can be determined

Figure 13-12. Amplifier operation: (a) linear operation; (b) operation near saturation; (c) operation near cutoff.

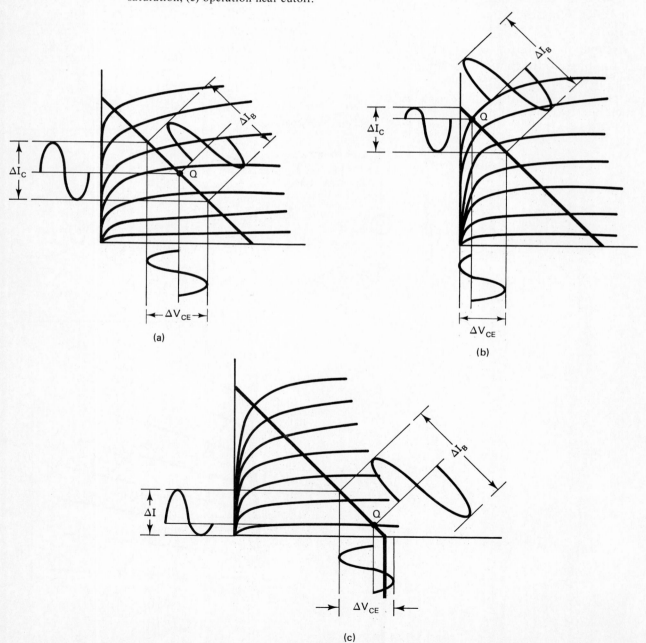

by dividing ΔV_{CE} by ΔV_B. The ΔV_B value is 0.1 V p-p. Using the ΔV_{CE} value from the graph and ΔV_B, determine the ac voltage gain of the amplifier circuit.

Linear and Nonlinear Operation

The V_{CE} or output of an amplifier should be a duplicate of the input with some gain. When a sine wave is applied, the output should develop a sine wave. When an amplifier operates in this manner it is considered to be *linear*. For this to be achieved, the amplifier must operate in or near the center or linear area of the collector curves. As a rule, nonlinearity occurs near the saturation and cutoff regions. If the operating point is adjusted near these regions it usually causes the output to be distorted. Normally, this is called nonlinear distortion.

Figure 13-12 shows how an amplifier will respond at three different operating points. Part (a) shows linear operating. The input and output are

duplicates in this case. Part (b) shows operation in the saturation region. Note that the top of the input wave distorts the negative alternation of the output voltage. Part (c) shows operation near the cutoff region. The lower part of the input wave causes distortion of the positive alternation. As a general rule, nonlinear distortion can be tolerated in some applications. In other applications nonlinear distortion is very obvious.

An input signal that is too large can also produce distortion. Figure 13-13 shows how an amplifier will respond to a large signal. Note that the input signal swings into the saturation region during the positive alternation and cutoff region during the negative alternation. This condition causes distortion of both alternations of the output. Normally, this condition is described as overdriving. In a radio receiver or stereo amplifier overdriving causes speech or music to sound bad. This usually occurs when the volume control is turned too high. The audio or

Figure 13-13. Overdriven amplifier.

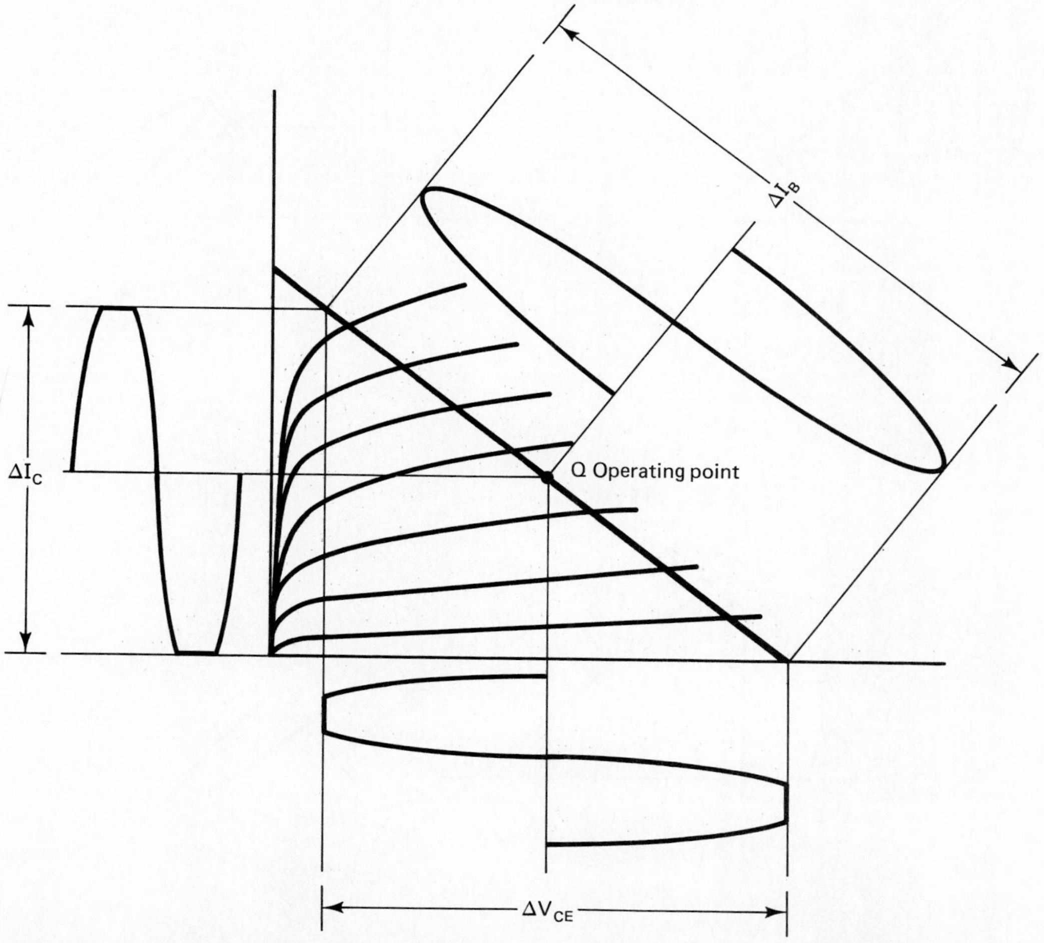

sound signal has its peaks clipped. The volume level may be higher, but the quality of reproduction is usually very poor when this occurs. It is interesting to note that the operating point is near the center of the active region. Overdriving can occur even with a properly selected operating point.

CLASSES OF AMPLIFICATION

Transistor amplifiers are frequently classified according to their bias operating point. This means of classification describes the shape of the output wave. Three general groups of amplifiers are class A, class B, and class C. Figure 13-14 shows a graphical display of these three amplifier classes.

Class A amplifiers generally have linear operation. The bias operating point is set near the center of the active region. With a sine wave applied to the input, the output is a complete sine wave. Figure 13-14(a) shows the input–output waveforms of a class A amplifier. This type of amplifier is used when a true reproduction of the input signal is required.

Figure 13-14(b) shows the input–output waveforms of a class B amplifier. The operating point of this amplifier is adjusted near the cutoff point. With a sine wave applied to the input, only one alternation of the signal reproduced. When using class B amplifiers it is possible to get a large change in output for one alternation. Two class B amplifiers working together can each amplify one alternation of a sine wave. When the wave is restored, a

Figure 13-14.   Classes of amplification: (a) class A; (b) class B; (c) class C.

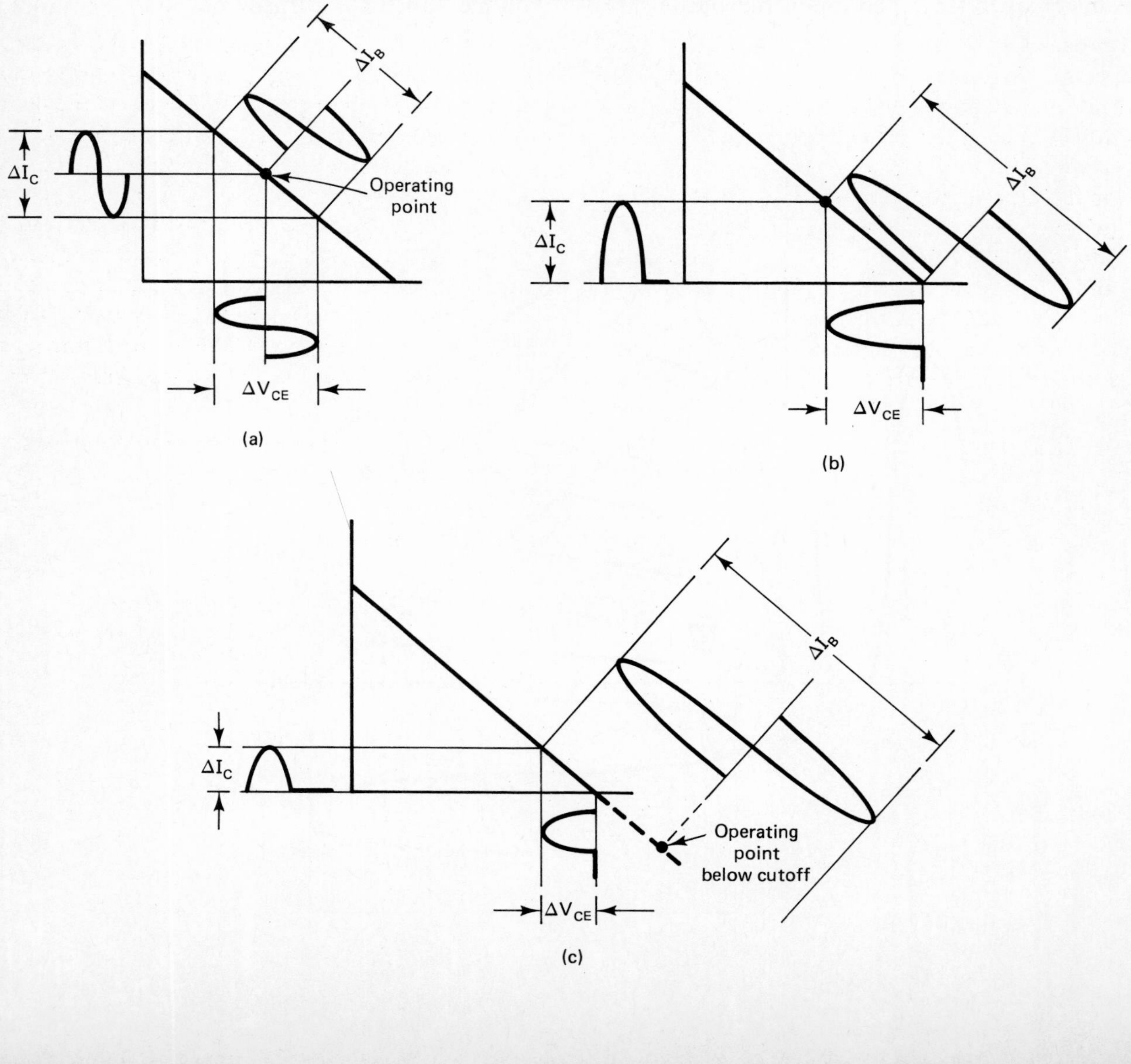

complete sine wave is developed. Class B amplifiers are commonly used in push-pull audio output circuits. These amplifiers are usually concerned with power amplification.

A class C amplifier is shown in Fig. 13-14(c). In this amplifier the bias operating point is below cutoff. With a sine wave applied to the input, the output is less than half of one alternation. *Radio-frequency* circuits are the primary application of class C amplifiers today. The operational efficiency of this amplifier is quite high. It consumes energy only for a small portion of the applied sine wave.

TRANSISTOR CIRCUIT CONFIGURATIONS

The elements of a transistor can be connected together in one of three different circuit configurations. These are usually described as common-emitter, common-base, and common-collector circuits. Of the three transistor leads, one is connected to the input and one to the output. The third lead is commonly connected to both the input and output. The common lead is generally used as a reference point for the circuit. It is usually connected to the circuit ground or common point. This has brought about the terms "grounded emitter," "grounded base," and "grounded collector." The terms "common" and "ground" mean the same thing. In some circuit configurations the emitter, base, or collector may be connected directly to ground. When this

occurs, the lead is at both dc and ac ground potential. When the lead goes to ground through a battery or resistor that is *bypassed* by a capacitor it is at an ac ground only.

Common-Emitter Amplifier

The common-emitter amplifier is a very important transistor circuit. A very high percentage of all amplifiers in use today are of the common-emitter type. The input signal of this amplifier is applied to the base and the output is taken from the collector. The emitter is the common or grounded element.

Figure 13-15 shows a circuit diagram of the common-emitter amplifier. This circuit is very similar to the basic amplifier used in the first part of the chapter. In effect, we have used this circuit to acquaint you with amplifiers in general. It is presented here to make a comparison with the other circuit configurations.

The signal being amplified by the common-emitter amplifier is applied to the emitter–base junction. This signal is superimposed on the dc bias of the emitter–base. The base current then varies at an ac rate. This action causes a corresponding change in collector current. The output voltage developed across the collector–emitter is inverted 180°. The current gain of the circuit is determined by beta. Typical beta values are in the range of 50. Voltage gain of the circuit ranges from 250 to 500. Power gain is in the range of 20,000. Input impedance is moderately high, typical values being 100 Ω.

Figure 13-15. Common-emitter amplifier.

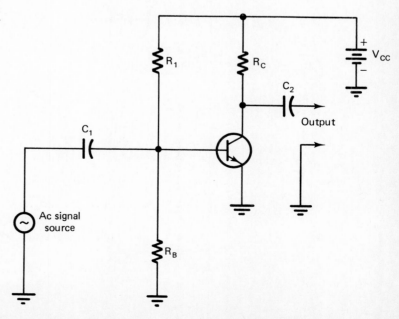

Output impedance is moderate, typical values being 2000 Ω. In general, common-emitter amplifiers are used in small-signal applications or as voltage amplifiers.

Common-Base Amplifiers

A common-base amplifier is shown in Fig. 13-16. This type of amplifier has the emitter–base junction forward biased and the collector–base junction reverse biased. In this circuit configuration, the emitter is the input. An applied input signal changes the circuit value of I_E. The output signal is developed across R_L by changes in collector current. For each value change in I_E there is a corresponding change in I_C.

In a common-base amplifier the current gain is called alpha. Alpha is determined by the formula I_C/I_E. In a common-base amplifier the gain is always less than a value of 1. Remember that $I_E = I_B + I_C$. The I_C will therefore always be slightly less than I_E by the value of I_B. Typical values of alpha are 0.98 to 0.99.

In Fig. 13-16 V_{EE} forward biases the emitter–base while V_{CC} reverse biases the collector–base. Resistor R_E is an emitter-current-limiting resistor and R_L is the load resistor. Note that the transistor is a PNP type.

When an ac signal is applied to the input, it is added to the dc operating value of V_E. The positive alternation therefore adds to the forward bias voltage of V_E. This condition causes an increase in I_E. A corresponding increase in I_C also takes place. With more I_C through R_L there is an increase in its voltage drop. The collector therefore becomes less negative or swings positive. In effect, the positive alternation of the input produces a positive alternation in the output. The input and output of this amplifier are in phase. See the waveform inserts in the diagram. The negative alternation causes the same reaction, the only difference being a reverse in polarity.

A common-base amplifier has a number of rather unique characteristics that should be considered. Its current gain is called alpha. The current gain is always less than 1. Typical values are 0.98 to 0.99. Voltage gain is usually very high. Typical values range from 100 to 2500 depending on the value of R_L. Power gain is the same as the voltage gain. The input impedance of the circuit is quite low. Values of 10 to 200 Ω are very common. The output impedance of the amplifier is somewhat moderate. Values range from 10 to 40 kΩ. This amplifier does not invert the applied signal.

Common-base amplifiers are used primarily to match a low-impedance input device to a circuit. This type of circuit configuration is also used in

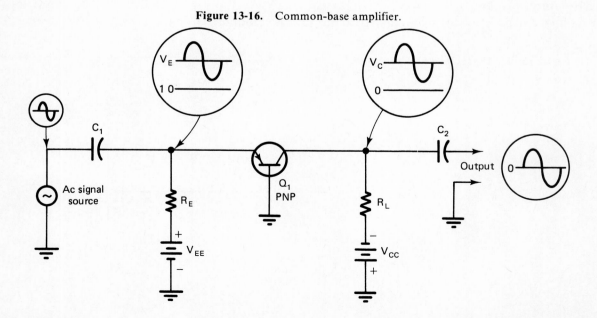

Figure 13-16.   Common-base amplifier.

radio-frequency amplifier applications. As a general rule, the common-base amplifier is not used very commonly today.

Common-Collector Amplifiers

A common-collector amplifier is shown in Fig. 13-17. In this circuit the base serves as the signal input point. The input of this amplifier is primarily the same as the common-emitter circuit. The collector is connected to ground through V_{CC}. Note that the input, output, and collector are all commonly connected. The unique part of this circuit is the output. It is developed across the load resistor (R_L) in the emitter. There is no resistor in the collector circuit.

When an ac signal is applied to the input it adds to the base current. The positive alternation increases the value of I_B above its static operating point. An increase in I_B causes a corresponding increase in I_E

and I_C. With an increase in I_E, there is more voltage drop across R_L. The top side of R_L, the emitter, becomes more positive. A positive input alternation therefore causes a positive output alternation across R_L. Essentially, this means that the input and output are in phase. The negative alternation reduces I_B, I_E, and I_C. With less I_E through R_L, the output swings negative. The output is again in phase for this alternation.

The common-collector amplifier is capable of current gain. A small change in I_B causes a large change in I_E. This current gain does not have a descriptive term such as alpha or beta. In general, the current gain is assumed to be greater than 1. Values approximate those of the common-emitter amplifier.

Other factors of importance are voltage gain, power gain, input impedance, and output impedance. Voltage gain is less than 1. Power gain is

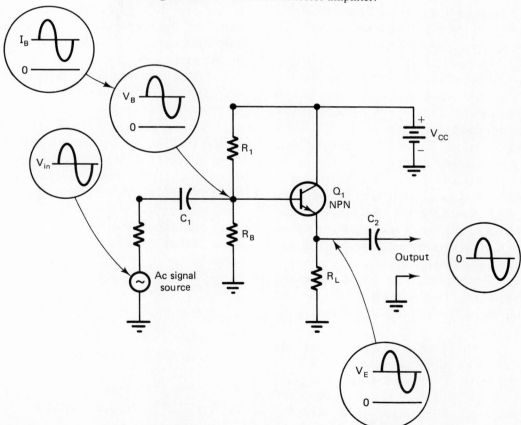

Figure 13-17. Common-collector amplifier.

moderately high. Typical values are in the range of 50. Input impedance is in the range of 50 kΩ. The output impedance is quite low because of the location of R_L. Typical values are 50 Ω.

Common-collector amplifiers are used primarily as impedance-matching devices. They can match a high-impedance device to a low-impedance load. Practical applications include preamplifiers operated from a high-impedance microphone or phonograph pickup. Common-collector amplifiers are also used as driver transistors for the last stage of amplification. In this application power transistors generally require large amounts of input current to deliver maximum power to the load device. Since the emitter output follows the base, this type of amplifier is often called an emitter follower.

FIELD-EFFECT TRANSISTOR AMPLIFIERS

The primary function of an amplifier is to reproduce the applied signal and provide some level of amplification. Unipolar transistors can be used to achieve this function. The junction field-effect transistor (JFET) and metal-oxide semiconductor field-effect transistor (MOSFET) are examples of unipolar amplifying devices. Amplification depends on the device being used and its circuit components. The composite circuit is normally energized by DC voltage. An ac signal is then applied to the input of the amplifier. After passing through the device, an amplified version of the signal appears in the output. Operating conditions of the device and the circuit combine to determine the level of amplification to be achieved.

For an FET to respond as an amplifier, it must be energized with dc voltage. Specific voltage values and polarities must be applied to each type of device. Forward and reverse biasing are not as meaningful in FET circuits. Remember that current conduction is through a channel. The value of voltage and its polarity determines the amount of control being achieved.

An illustration of the supply voltages and polarities needed for different types of FETs is shown in Fig. 13-18. JFETs must have the gate–source junction reverse biased to achieve control. The gate voltage of a enchancement MOSFET is opposite that of the source. Depletion MOSFETs have some flexibility in the polarity of the gate voltage. Some of these devices have a zero value of V_G and swing positive or negative during operation.

Basic FET Amplifier Operation

We will now take a look at the circuitry of a basic FET amplifier. In this example an N-channel JFET is used. A P-channel device could also have been selected. The primary difference in operation is the polarity of the source voltage and the type of current carrier in the channel.

Figure 13-19 shows an N-channel JFET amplifier connected in the common-source configuration. The source of this device is common to both the input signal and the output. This circuit is very similar to the common-emitter amplifier. V_{DD} serves as the dc energy source for the source and drain. V_{GS} reverse biases the gate with respect to the source. The value of V_{GS} establishes the static operating point of the circuit. R_g is normally extremely high. No gate current flows through R_g. The input signal is applied to the gate through capacitor C. Selecting a large value of R_g keeps the input impedance of the amplifier high. If R_g is 1 MΩ, the signal source sees an impedance of 1 MΩ.

Let us now consider the operation of the JFET amplifier of Fig. 13-19 in its static state. The family of drain curves of Fig. 13-20 will be used in this explanation. A load line for the circuit must be established. Remember that the two extreme conditions of operation are needed for the load line. In a JFET, these extremes are full conduction and cutoff. At the cutoff point, no current passes through the channel. The V_{DD} voltage will all appear across V_{DS}. Full conduction occurs when a maximum value of I_D flows through R_L. This value is determined by the formula

$$\text{drain current} = \frac{\text{source voltage}}{\text{load resistance}} = I_D = \frac{V_{DD}}{R_L}$$

For the amplifier circuit, the maximum value of I_D is

$$I_D = \frac{25 \text{ V}}{10 \text{ K}\Omega} = \frac{25 \text{ V}}{10 \times 10^3}$$

$$= 0.0025 \text{ A} \quad \text{or} \quad 2.5 \text{ mA}$$

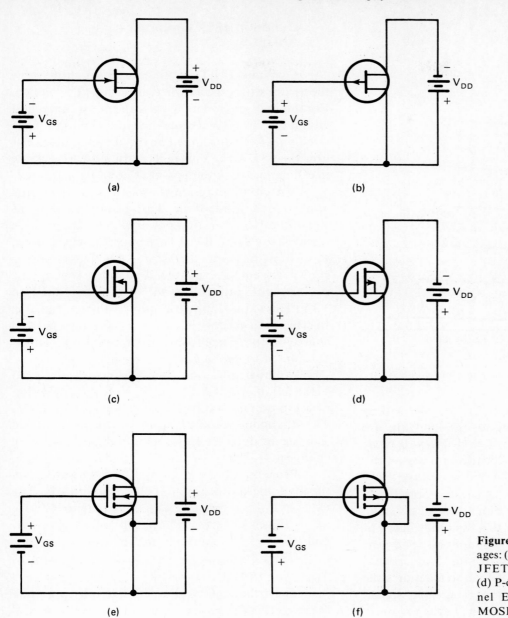

(a)

(b)

(c)

(d)

(e)

(f)

Figure 13-18. FET operational voltages: (a) N-channel JFET; (b) P-channel JFET; (c) N-channel D-MOSFET; (d) P-channel D-MOSFET; (e) N-channel E-MOSFET; (f) P-channel E-MOSFET.

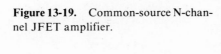

Figure 13-19. Common-source N-channel JFET amplifier.

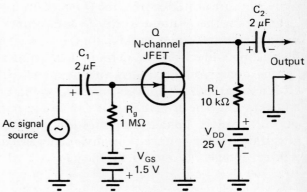

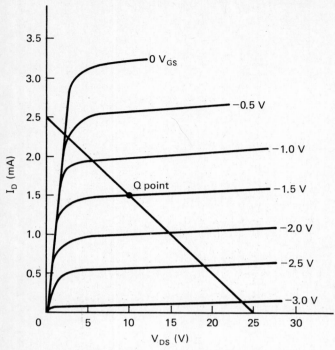

Figure 13-20. Family of drain curves.

With this conduction through R_L, the V_{DS} will be zero. The two locating points of the load line are $V_{DS} = 25$ V with 0 mA of I_D and $V_{DS} = 0$ V at 2.5 mA I_D. Check these two location points on the load line. A straight line connects the points.

With the load line developed, it is possible to see how the JFET will respond in its static state. Static operation occurs without a signal applied. For linear operation, the amplifier should respond near the center of the active region. In our basic circuit, V_{GS} is −1.5 V. Point Q on the load line denotes the operating point. With an appropriate input signal, the amplifier should respond in the linear region.

To see how the JFET responds in its static state, lines are projected from the *Q point*. Projecting a line from the Q point to the I_D scale shows the resulting drain current. In this case, the I_D is 1.5 mA. Projecting a line from the Q point down shows the drain–source voltage. V_{DS} is approximately 10 V. This means that 15 V appears across R_L at this point of operation. The dc voltage gain can be determined with this value. Voltage amplification (A_V) equals the drain–source voltage (V_{DS}) divided by the gate–source voltage. For our circuit the gain is

$$A_V = \frac{V_{DS}}{V_{GS}} = \frac{10 \text{ V}}{1.5 \text{ V}}$$
$$= 6.667$$

Dynamic JFET Amplifier Analysis

Dynamic JFET amplifier analysis shows how the device will respond when an ac signal is applied. For this explanation we will use the drain curves and circuit of Fig. 13-21. A load line and Q point for the circuit has been developed on the curves. This shows how the JFET responds in its static state. Check the load-line location points and Q point for accuracy.

The circuit diagram shows a 1-V p-p ac signal applied to the input of the JFET amplifier. With this signal the V_{GS} operating point will change in value from −2.0 V to −3.0 V. For the positive alternation, V_{GS} will swing from −2.5 V to −2.0 V. This change is shown as point P on the load line. During the negative alternation, V_{GS} will drop from −2.5 V to −3 V. This value is shown as point N on the load line. In effect, this means that a 1-V p-p input signal will cause V_{GS} to change from −2.0 V to −3.0 V. This is considered to be the ΔV_{GS} value.

To show how ΔV_{GS} alters I_D, lines are projected to the left of points P, Q, and N. Note the projection of these points toward the I_D values in Fig. 13-21. The changing value of I_D is shown as ΔI_D. An increase or decrease in V_{GS} causes a corresponding change in I_D. This shows that V_{GS} and I_D are in phase.

Projecting lines P, Q, and N downward from the load line shows how V_{DS} changes with V_{GS}. The value of V_{DS} is shown as ΔV_{DS}. Note that an increase in V_{GS} causes a decrease in V_{DS}. This tells us that V_{GS} and V_{DS} are 180° out of phase. The value difference in V_{DS} and V_{DD} appears across the load resistor as V_{RL}.

The voltage gain of a JFET amplifier can be readily observed by load-line data. A ΔV_{GS} of 1 V p-p causes a ΔV_{DS} of approximately 10 V p-p. This shows an obvious gain of 18. Using the formula,

$$A_V = \frac{\Delta V_{DS}}{\Delta V_{GS}} = \frac{10 \text{ V p-p}}{1 \text{ V p-p}}$$
$$= 10$$

MOSFET Circuit Operation

D-type MOSFET and E-type MOSFET operation is very similar to that of the JFET. There is an obvious difference in the required operational voltage values and polarity. A load-line analysis of either amplifier type shows the same basic principle of

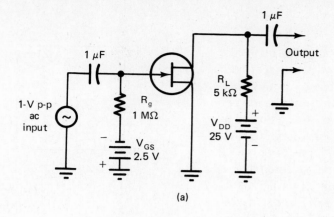

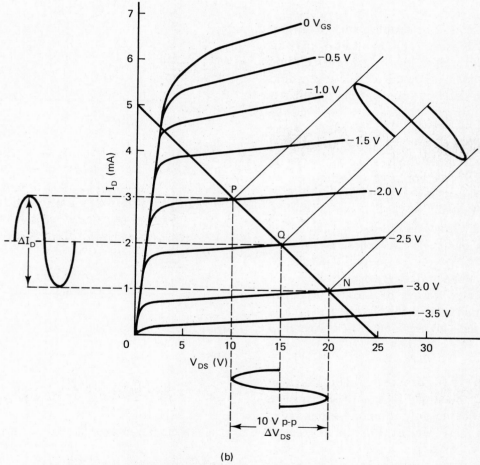

Figure 13-21. JFET amplifier: (a) N-channel amplifier; (b) drain family of curves.

operation. The gate is highly resistant or infinite for all devices. Current flow in the channel is controlled by the voltage value and polarity of the gate signal. The only difference of any concern is with the D-type MOSFET. It is frequently biased at the 0 V_{GS} point.

An ac input signal would cause V_{GS} to vary above and below the zero value. It is important to remember that all FETs are voltage-sensitive devices. They produce only voltage amplication. No current flows in the gate circuit.

FET Biasing Methods

For FET operation biasing refers to the necessary dc voltage needed by the gate with respect to the source lead. Essentially, biasing refers to the development of a suitable V_{GS} value. By selection of a proper bias voltage, it becomes possible to force a device to operate at a chosen Q point. The V_{DD} voltage applied to the source–drain is generally not considered to be bias voltage. The biasing method selected for a specific circuit has a great deal to do with type of device involved. Enchancement MOSFETs and depletion MOSFETs require different biasing procedures.

FIXED BIASING.   The simplest way to bias an FET is with fixed biasing. This method of biasing can be used equally well be E- or D-MOSFETs or JFETs. In fixed biasing, the correct voltage value and polarity is supplied by a small battery or cell. An electronic power supply could also be used to perform the same operation. Figure 13-22 shows fixed biasing applied to three different N-channel FET types. If P-channel devices were used, the polarity of V_{CC} and V_{DD} would be reversed. V_{CC} is used as the fixed bias source for the three FET types.

If an enhancement–depletion type of MOSFET is to be fixed biased, the circuit of Fig. 13-23 would be used. In this type of device the operating point is usually $V_{GS} = 0$ V. No voltage source is needed to establish the operating point. R_g is normally of a high value. Typically, values of 1 to 10 MΩ are used. R_g must be included in the circuit. It serves as a return for the gate to the common connection point.

VOLTAGE-DIVIDER BIASING.   The voltage-divider method of biasing FETs is rather easy to achieve. It is applicable only to enhancement-type MOSFETs. With these devices, V_{GS} and V_{DD} are of the same polarity. As a result of this condition, a single voltage source can be used as a supply. The V_{GS} voltage is only a fraction of the value of V_{DD}. The divider resistors are used to develop the necessary voltage values. Figure 13-24 shows the divider method of biasing for N- and P-channel E-MOSFETs.

The values of R_1 and R_g must be properly chosen to establish a desired operating point. By the

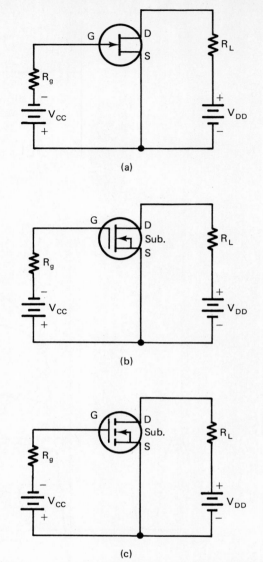

Figure 13-22.   Methods of fixed biasing: (a) N-channel JFET; (b) N-channel D-type MOSFET; (c) N-channel E-type MOSFET.

Figure 13-23.   Fixed bias of an N-channel E-D-type MOSFET.

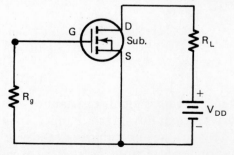

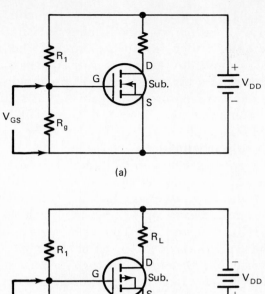

(a)

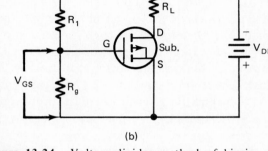

(b)

Figure 13-24. Voltage-divider method of biasing for E-MOSFETs: (a) N-channel; (b) P-channel.

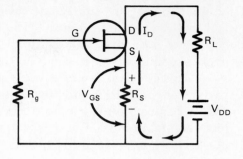

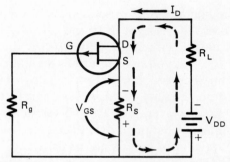

Figure 13-25. Self-biasing of JFETs: (a) N-channel; (b) P-channel.

voltage-divider equation, we find that the bias voltage is

gate–source voltage = R_g voltage

$$= \text{source voltage} \left(\frac{R_g}{R_1 + R_g} \right)$$

or

$$V_{GS} = V_{Rg} = V_{DD} \left(\frac{R_g}{R_1 + R_g} \right)$$

SELF-BIASING OF FETs. Self-biasing is commonly called source biasing. Self-biasing is rather easy to recognize. The source current of an FET is used to develop the bias voltage. The voltage developed is achieved by a resistor (R_S) connected in series with source. Current flow through the source–drain causes a voltage drop across R_S. The polarity of the developed voltage is based on the direction of current flow through R_S. Figure 13-25 shows biasing for N- and P-channel JFETs. D-type MOSFETs can also be biased by the same method.

For the N-channel circuit, current flow through R_S causes the source to be slightly positive with respect to the gate. With R_g connected to the lower side of R_S, this makes R_g negative and the source positive. The value of R_S and drain current (I_D) determines the bias operating point of the circuit. Note the polarity difference in the P-channel circuit.

FET Circuit Configurations

FET amplifiers, like their bipolar counterparts, can be connected in three different circuit configurations. These are described as common-source, common-gate, and common-drain circuits. One lead is connected to the input with a second lead connected to the output. The third lead is commonly connected to both input and output. This lead is used as a circuit reference point. It is often connected to the circuit ground. This has brought about the terms "grounded source," "grounded gate," and "grounded drain." The terms "common" and "ground" refer to the same type of connection.

COMMON-SOURCE AMPLIFIERS. The common-source amplifier is the most widely used FET circuit configuration. This circuit is similar in many respects to the common-emitter bipolar amplifier. The input signal is applied to the gate–source and the output signal is taken from the drain–source. The source lead is common to both input and output.

A practical common-source amplifier is shown in Fig. 13-26. This circuit is essentially the same as the one used in our basic JFET amplifier. It is shown here so that a comparison can be made with the other circuit configurations. The device can be a JFET, a D-MOSFET, or an E-MOSFET. Circuit characteristics are primarily the same for all three devices.

The signal being processed by the common-source amplifier is applied to the gate–source. Self biasing of the circuit is achieved by the source resistor R_2. This voltage establishes the static operating point. The incoming signal voltage is superimposed on the gate voltage. This causes the gate voltage to vary at an ac rate. This action causes a corresponding change in drain current. The output voltage developed across the source and drain is inverted 180°. Voltage gain A_V is V_{DS}/V_{GS}. Typical A_V values are 5 to 10. The input impedance is extremely high for nearly any signal source. One to several megohms is common. The output impedance (Z_O) is moderately high. Typical values are in the range of 2 to 10 kΩ. Z_O is dependent primarily on the value of R_L.

A common-source amplifier has a very good ratio between input and output impedance. Circuits of this type are extremely valuable as impedance matching devices. Primarily, common-source amplifiers are used exclusively as voltage amplifiers. They respond well in radio-frequency signal applications.

COMMON-GATE AMPLIFIERS. The common-gate FET amplifier is similar in many respects to the common-base bipolar transistor circuit. The input signal is applied to the source–gate and the output appears across the drain–gate. The common-gate amplifier is capable of voltage gain but has a current gain capablility that is less than 1. JFETs, E-MOSFETs, and D-MOSFETs may all be used as common-gate amplifiers.

A practical common-gate amplifier is shown in Fig. 13-27. This circuit employs an N-channel JFET. The operating voltages are the same as those of the common-source circuit. Self-biasing is achieved by the source resistor R_1. This voltage is used to establish the static operating point. An input signal is applied to R_1 through capacitor C_1. A variation in signal voltage causes a corresponding change in source voltage. Making the source more positive has the same effect on I_D as making the gate more negative. The positive alternation of the input signal will make the source more positive. This, in turn, reduces drain current. With less I_D, there is a smaller voltage drop across the load resistor R_2. The drain or output voltage therefore swings positive. The negative alternation of the input reduces the source voltage by an equal amount. This is the same as making the gate less negative. As a result, I_D is increased. The voltage drop across R_2 similarly increases. This, in turn, causes the drain or output to be less positive or negative going. The input signal is therefore in phase with the output signal.

The common-gate amplifier has a number of rather unusual characteristics. Its voltage gain is

Figure 13-26.   Common-source JFET amplifier.

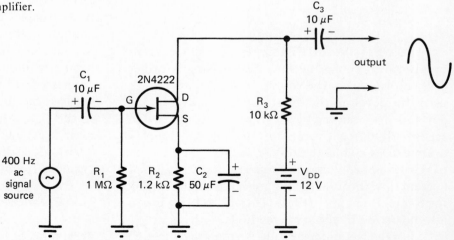

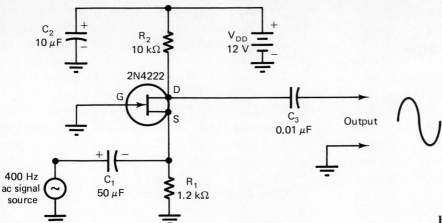

Figure 13-27. Common-gate JFET amplifier.

somewhat less than that of the common-source amplifier. Representative values are 2 to 5. The common-gate circuit has very low input impedance (Z_{in}). The output impedance (Z_O) is rather moderate. Typical Z_{in} values are 200 to 1500 Ω with Z_O being 5 to 15 kΩ. This type of circuit configuration is often used to amplify radio-frequency signals. Amplification levels are very stable for *RF*, without feedback between the input and output.

COMMON-DRAIN AMPLIFIERS. A common-drain amplifier has the input signal applied to the gate and the output signal removed from the source. The drain is commonly connected to one side of the input and output. Common-drain amplifiers are also called source followers. This circuit has similar characteristics to those of the common-collector bipolar amplifier.

Figure 13-28 shows a practical common-drain amplifier using in N-channel JFET. The input of this amplifier is primarily the same as that of the common-source amplifier. The input impedance is therefore very high. Z_{in} is determined largely by the value of R_1. The operating point of our amplifier is determined by the value of R_2. Essentially, this circuit has the same operating point as our other circuit configurations. Resistor R_3 has been switched from the drain to the source in this circuit. Resistors R_2 and R_3 are combined to form the load resistance. The output impedance is based on this value.

When an ac signal is applied to the input of the amplifier it changes the gate voltage. The dc operating point is established by the value of source resistor R_2. The positive alternation of an input signal makes the gate less negative. This causes an N-channel device to be more conductive. With more current

Figure 13-28. Common-drain JFET amplifier.

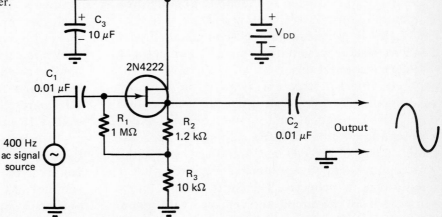

through R_3 and R_2, the source swings positive. The negative alternation of the input then makes the gate more negative. This action causes the channel to be less conductive. A smaller current therefore causes the source to swing negative. In effect, this means that the input and output voltage values of the amplifier are in step with each other. These signals are in phase in a common-drain amplifier.

Common-collector amplifiers are primarily used today as impedance-matching devices. They are used to match a high-impedance device to a low-impedance load. This type of circuit is capable of handling a high input signal level without causing distortion. The input impedance places a minimum load on the signal source. Common-collector amplifiers are used frequently today to match high-impedance devices such as *microphones* and phonograph pickups to the input of an *audio amplifier*.

INTEGRATED-CIRCUIT AMPLIFIERS

Within the last few years integrated circuits (ICs) have taken over practically all small-signal linear amplifier applications. Inside each small IC package are hundreds and even thousands of individual components. Transistors, resistors, diodes, and occasionally capacitors are built into *IC chips*. Bipolar transistors, JFETs, and MOSFETs are key devices used in IC construction. Each chip is designed to perform a specific function or to respond as a complete system.

A unique feature of the integrated circuit is the simplicity of its use. These devices require an input signal, an output connection, and an energy source. When a signal is applied to the input, an amplified version of it appears at the output. The circuitry of the device is practically all self-contained. Figure 13-29 shows a representative chip and a diagram of its internal connections.

The internal structure of an IC is generally quite small and very complex. The physical size of a chip varies a great deal for different types of ICs. For an operational amplifier, the chip size is approximately 0.003 in. $\times$ 0.003 in. Components within the chip cannot be effectively altered or repaired when it fails to operate properly. The entire chip is simply removed from the circuit and replaced with a new

one. Testing is generally achieved by observing signals at the input and output terminals. Gain is determined by observing signal levels at the input and output.

There are two basic types of ICs today: *linear* and *digital*. Linear ICs usually perform some type of amplifying function. *Operational amplifiers* are typical linear amplifiers. These amplifiers are discussed in this chapter. Digital ICs are discussed in a later chapter.

Operational Amplifiers

A large number of the linear ICs in use today are called operational amplifiers or op-amps. These devices have high dc and ac gain capabilities with stable operating characteristics. Op-amps are used as signal amplifiers in waveform-generation, waveshaping, and impedance-matching applications. The triangle-shaped amplifier symbol is commonly used to identify an op-amp. The point of the triangle is the output. The flat side with two leads is the input. A plus sign on the input lead denotes noninverting input. The negative sign indicates inverting input. The chip and all of its components are located inside the triangle symbol. Connection points and external components are connected outside the triangle symbol.

The internal structure of an op-amp IC is rather complex. Hundreds of discrete components are formed together on a single *substrate*. Electrical isolation among components is achieved by polarization of the material. *Dielectric barriers* separate components electrically and physically. Metal plating is used to connect individual parts to lead terminals. The entire assembly is then placed in an enclosure or package. Figure 13-30 shows three representative IC package styles. The TO-5 case shown in part (a) is similar to the one used to house bipolar devices and FETs. The *flat-pack* [part (b)] has horizontal leads. The *dual-in-line package* or DIP [part (c)] comes in a variety of different package sizes. DIP packages are very popular today.

The op-amp is essentially a high-gain amplifier. It is designed to operate over a wide range of voltages. Its internal structure is small and quite complex. As a general rule, components connected outside the op-amp determine its operating capabil-

Operational Amplifiers

LM741/LM741C operational amplifier
general description

The LM741 and LM741C are general purpose operational amplifiers which feature improved performance over industry standards like the LM709. They are direct, plug-in replacements for the 709C, LM201, MC1439 and 748 in most applications.

The offset voltage and offset current are guaranteed over the entire common mode range. The amplifiers also offer many features which make their application nearly foolproof: overload protection on the input and output, no latch-up when the common mode range is exceeded, as well as freedom from oscillations.

The LM741C is identical to the LM741 except that the LM741C has its performance guaranteed over a 0°C to 70°C temperature range, instead of –55°C to 125°C.

schematic and connection diagrams

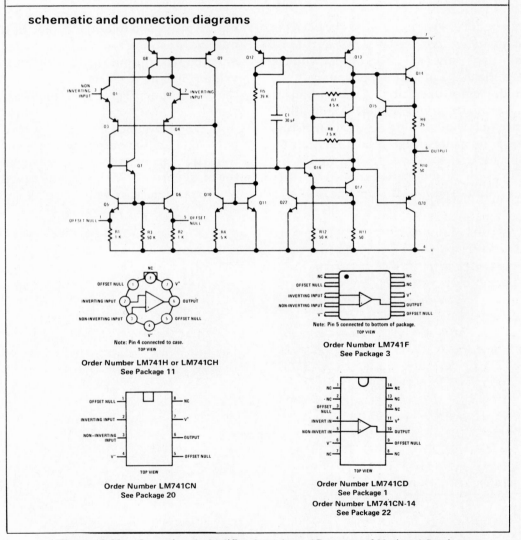

Figure 13-29. Operational amplifier data sheet. (Courtesy of National Semiconductor Corp.)

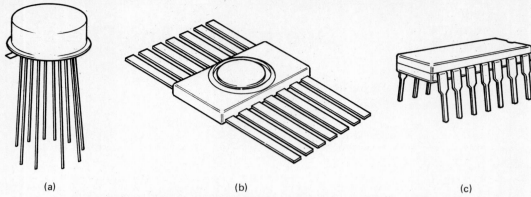

Figure 13-30. Integrated-circuit packages: (a) 10-terminal round; (b) flat pack; (c) 14-terminal DIP.

ities. Gain can be 100,000 or more when the device is operated in an *open-loop* circuit. As a *voltage follower*, gain may only be 1. The output voltage cannot exceed the value of the supply voltage.

The fundamental operation of an op-amp is based on a *differential amplifier* circuit. A simplification of this circuit is shown in Fig. 13-31. Transistors Q_1 and Q_2 respond as the differential amplifier. Q_3 serves as constant current source for the amplifier.

Normally, with no input signal applied to either Q_1 or Q_2, no output appears. If identical values of

input are applied to both Q_1 and Q_2, there is still no output. In a sense, each transistor conducts the same amount. The output signal is therefore of the same value or balanced. There is no potential difference in the two outputs. This means that there is no output signal.

When different signals are applied to the input of Q_1, and Q_2, the output becomes unbalanced. The transistor receiving the greatest signal will conduct the most current. With current regulated by Q_3, the alternate transistor will conduct less current. A

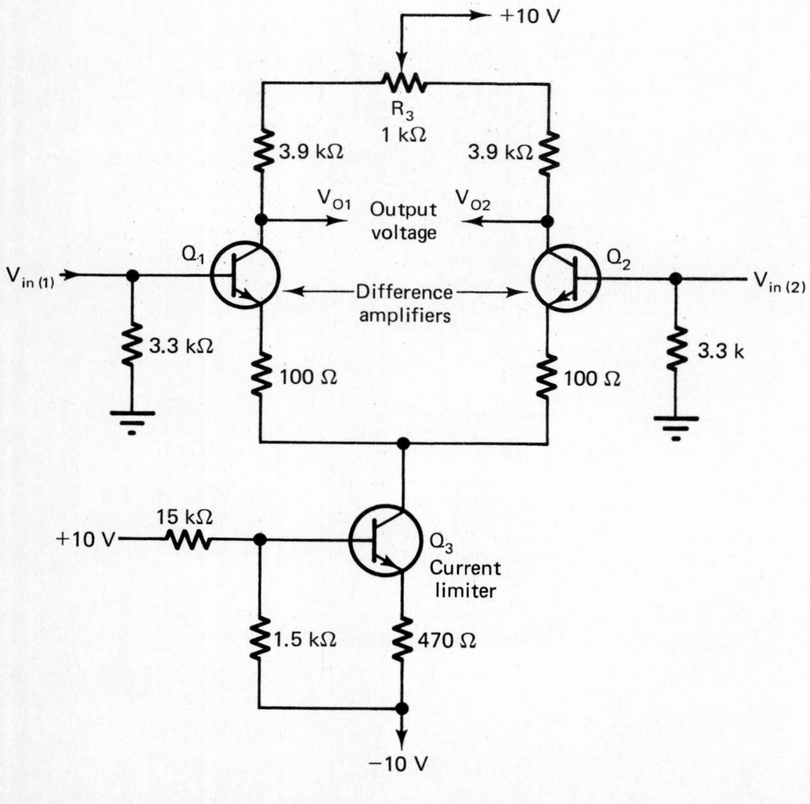

Figure 13-31. Transistor difference amplifier.

pronounced difference in the two outputs will occur. The resulting output is indicative of the conduction difference in Q_1 and Q_2.

If a signal is applied only to the *inverting input* of a difference amplifier, the output will be amplified and inverted. A signal applied only to the *noninverting* input will be amplified without inversion. For the circuit to respond in this manner, the input not being used must be grounded. The output will be the difference between input and ground.

Op-amps have a number of other circuits in their construction. A high-gain amplifier, for example, follows the differential amplifier. This amplifier brings up the signal level so that very small input signals can be amplified. An emitter-follower stage comes after the high-gain amplifier. The emitter follower provides the op-amp with a low-impedance output. This permits it to be connected to a variety of different output devices. The combination of all this makes the circuitry of an op-amp rather complex. As a general rule, the internal construction is not very important to a person using the device. Operation is usually controlled by the selection of external components.

OPEN-LOOP OPERATION. Figure 13-32 shows a circuit diagram of an op-amp designed for open-loop operation. In this circuit configuration, gain is controlled by the strength of the input signal. When the input signal voltage is zero, the output voltage is zero. If an input signal is applied, the output voltage will rise accordingly. Voltage gains of 100,000 or

Figure 13-32. Op-amp connected for open-loop gain.

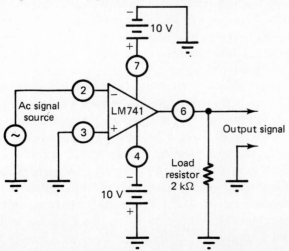

more are typical. If the output voltage rises to the value of the source voltage, the op-amp is *saturated*. An input signal of only a few millivolts is needed to cause the op-amp to reach saturation.

Applications of the op-amp in an open-loop configuration are not very common. Very few circuits require a gain factor of 100,000. The input signal level for a circuit of this type must be extremely small. A simple calculation of voltage amplification demonstrates this limitation. The formula for voltage amplification is

$$\text{voltage amplification} = \frac{\text{voltage output}}{\text{voltage input}}$$

or

$$A_V = \frac{V_O}{V_{\text{in}}}$$

For the circuit in Fig. 13-32, the A_V formula is transposed to determine the input voltage. The formula now becomes

$$V_{\text{in}} = \frac{V_O}{A_V}$$

We will assume, in this case, that an A_V value of 100,000 is achieved by the op-amp. The output voltage (V_O) cannot exceed the value of the source voltage. In this circuit V_{CC} is 10 V. The input voltage value is

$$V_{\text{in}} = \frac{V_O}{A_V} = \frac{10}{100,000} = 0.0001 \text{ V} \quad \text{or} \quad 0.1 \text{ mV}$$

Saturation occurs when the input voltage reaches 0.1 mV. An ac signal of 0.1 mV p-p would also cause distortion of the output. This shows that the op-amp in the open-loop configuration is very easy to distort. Closed-loop operation is generally used to alter this problem.

CLOSED-LOOP OPERATION. When an op-amp is used in a closed-loop configuration the voltage gain can be adjusted to nearly any desired value. This is achieved by a feedback resistor network connected between the output and the input. Figure 13-33 shows two simplified op-amps connected to a

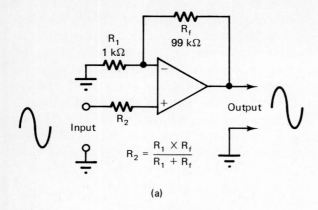

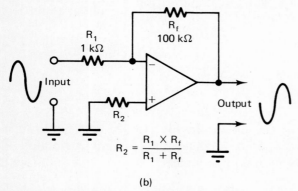

Figure 13-33. Closed-loop op-amp simplification: (a) noninverting amplifier; (b) inverting amplifier.

feedback network. Part (a) shows the op-amp connected in a noninverting circuit. The output will be the same polarity as the input. Ideally, the two inputs are balanced when resistors R_1 and R_2 are equal. In an actual circuit, R_2 is a calculated value based on the parallel arrangement of R_1 and the feedback resistor R_F. Amplifier gain is based on the formula

voltage amplification

$$= \frac{\text{input resistance} + \text{feedback resistance}}{\text{input resistance}}$$

or

$$A_V = \frac{R_1 + R_f}{R_1}$$

For the circuit in part (a), the gain will be

$$A_V = \frac{R_1 + R_f}{R_1} = \frac{1 \text{ k}\Omega + 99 \text{ k}\Omega}{1 \text{ k}\Omega} = 100$$

If R_f is changed to 9.9 kΩ and R_1 to 100Ω, the gain would be reduced. How would the value differ?

Figure 13-33(b) shows the inverting circuit configuration. The input signal is applied to the negative terminal. The feedback network is composed of R_1 and R_F. R_2 returns the noninverting input to ground. The value of R_2 is determined by the parallel value of R_1 and R_F. R_2 is generally smaller than R_1. The voltage gain of this amplifier is determined by the formula

$$\text{voltage amplification} = \frac{\text{feedback resistance}}{\text{input resistance}}$$

or

$$A_V = \frac{R_f}{R_1}$$

For the inverting op-amp of part (b) the gain will be

$$A_V = \frac{R_f}{R_1} = \frac{100 \text{ k}\Omega}{1 \text{ k}\Omega} = 100$$

PRACTICAL OP-AMP CIRCUITS. Figure 13-34 shows a functioning op-amp circuit that is connected for noninverting amplification. The numbers near each lead are the pin numbers. The LM741 used is housed in an eight-pin dual-in-line package. Notice that operating voltage is supplied from a split or divided power supply. Common leads on each side of the supply are connected to ground. The ground is also connected commonly to a number of circuit points. The output of this circuit cannot exceed 10 V dc or 20 V p-p ac. The gain achieved by this circuit closely follows the calculated values of the feedback network. When the circuit is made operational, gain and signal phase can be observed at the input and output with an oscilloscope. Input and output voltages can also be measured with a voltmeter.

A practical inverting op-amp circuit is shown in Fig. 13-35. This amplifier has the same gain capabilities as the noninverting circuit. Note the connection differences on the input circuit. The operational waveforms, signal phase, and voltage values can be observed on an oscilloscope and measured with a voltmeter. Op-amp circuits of this type are widely used today.

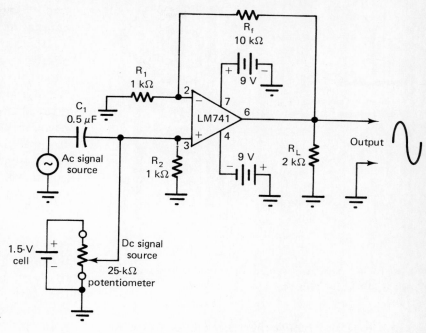

Figure 13-34.   Noninverting op-amp circuit.

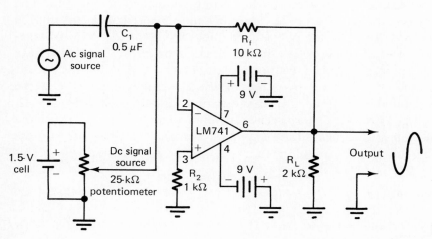

Figure 13-35.   Inverting op-amp circuit.

VACUUM TUBES

A vacuum tube is an amplifying device that is still being used in some electronic circuits. As a general rule, this device is found in equipment that has been in operation for some time. Replacements are available for nearly all tubes in operation today.

The simplest vacuum-tube amplifier in operation today is a *triode* or three-electrode device. It contains a heater-cathode, *control grid*, and *plate*. A schematic diagram of the tube, its element names, and a cutaway drawing of the elements is shown in

Fig. 13-36. Note that the heater-cathode element responds as a single unit. The grid is a coil-like structure placed around the cathode. The cathode emits or gives off electrons when heated by the filament. The plate or anode surrounds the other elements. The entire structure is then placed in a glass enclosure. Air is removed from the enclosure and the unit is sealed. A partial vacuum exists inside the enclosure.

Vacuum-tube operation is based on a principle called *thermionic emission*. This principle states that electrons are released or emitted from the surface of

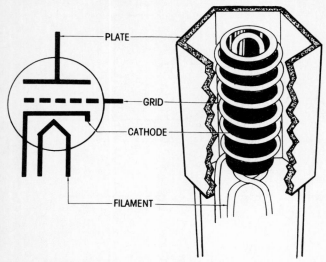

Figure 13-36. Schematic symbol and cutaway view of a triode vacuum tube.

certain materials when they are heated. In effect, electrons are forced out of their orbits by heat energy. Most metals have an emitting capability to some extent. The cathode of a vacuum tube is the emitting electrode. The filament heats the cathode so that it will emit electrons. The filament or heater is an essential part of the vacuum tube. An independent source of energy is needed to heat the filament.

After electrons have been emitted from the cathode, they must pass through the grid. This electrode serves as the primary control element. When a negative voltage is applied to the control

grid, electrons from the cathode are repelled. The amount of repelling action that takes place is based on the value of the negative voltage. A strong negative voltage may cause the electron flow to be cut off entirely. Smaller values of negative voltage let varying amounts of electron flow take place. The greatest conduction occurs when the grid is not negative. A fixed value of grid voltage is needed to establish the static or dc operating point for the tube. An ac signal applied to the grid, at the operating point, will cause varying values of grid voltage to occur. An outside ac signal can be superimposed upon the electron flow through this process.

The plate or anode of a vacuum tube is responsible for collecting the electrons emitted from the cathode. The plate is a metal structure connected to the positive side of the dc voltage source. In effect, the plate is positively charged with respect to the cathode. It attracts the electrons that are emitted from the cathode. The structure of the tube has electrons traveling through a space between cathode and plate while being controlled by the grid. The flow of electrons in a tube is called plate current (I_p).

A vacuum tube is classified as a voltage-sensitive device. Small changes in grid voltage are used to control plate current. I_p initially starts at the negative side of the dc energy source. See the schematic diagram of a vacuum tube in Fig. 13-37. Electrons then travel through the space between the cathode and plate. Upon reaching the plate, it flows out of the tube and into the positive side of V_{BB}. This

Figure 13-37. Triode amplifier.

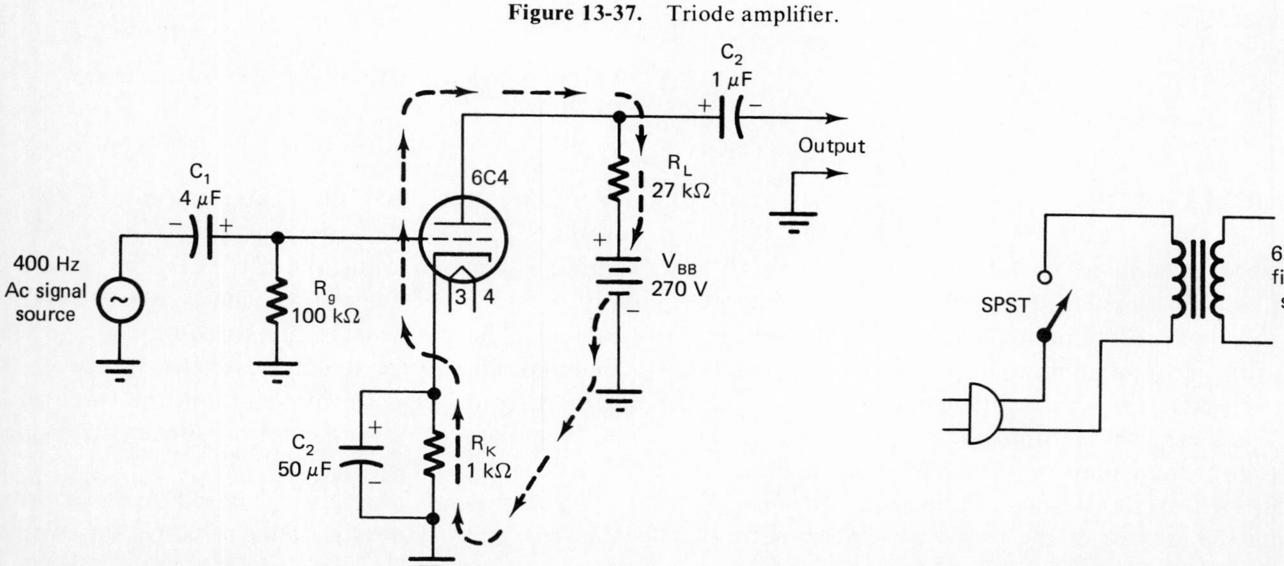

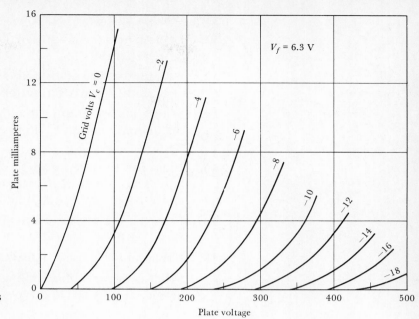

Figure 13-38. Plate family of characteristic curves for a triode.

makes a complete path for I_p. A change in I_p at any point along this path will cause the same response throughout the circuit.

A typical family of plate characteristic curves is shown in Fig. 13-38. This display plots plate current and plate voltage (V_p) for different values of grid voltage V_g. A small change in grid voltage will cause a rather large change in I_p. Individual curve lines represent different grid voltage values. A curve of this type is developed by selecting a fixed value of grid voltage. The V_p is then changed, starting at zero, while observing the value change in I_p. Operation of the circuit is determined by establishing a load line. Changes in V_g can then be related to I_p and V_p values.

Review

1. What is the difference between amplification and reproduction?

2. If the output voltage of an amplifier is 5 V p-p and the input voltage is 0.025 V p-p, what is the voltage gain?

3. If the output current of an amplifier is 8 mA when the input current is 100 μA, what is the current gain?

4. How much base current is produced by a transistor amplifier when a 120-kΩ resistor is connected to 9 V?

5. What is the beta of a transistor amplifier having an I_C or 6 mA, an I_B of 120 μA, and an I_E of 6.12 mA?

6. Explain how an ac signal is amplified by an NPN bipolar transistor.

7. What is beta-dependent biasing?

8. Why is independent biasing better than beta-dependent biasing?

9. Explain what is meant by the static state of a bipolar amplifier.

10. What are the two extreme conditions of operation of a bipolar transistor used to develop a load line?

11. What is meant by the term "dynamic characteristics of an amplifier"?

12. Explain how the input signal of a bipolar amplifier causes the output to be inverted 180°.

13. Explain the type of distortion that will occur when a bipolar amplifier is near cutoff or near saturation.

14. With a sine wave applied to the input of a bipolar amplifier, where must the operating point be adjusted to achieve class A, B, and C amplification?

15. What are the characteristic differences in common-emitter, common-base, and common-collector amplifiers with respect to A_V, current gain, phase, input impedance, and output impedance?

16. Describe the physical differences between a JFET, D-MOSFET, and E-MOSFET.

17. Explain how a JFET achieves amplification.

18. What is the difference in biasing an N-channel JFET compared with a bipolar-transistor NPN type?

19. What are the distinguishing features of fixed biasing, self-biasing, and divider biasing for a JFET?

20. Why are op-amps rarely used in the open-loop gain configuration?

21. If the feedback resistor for a noninverting circuit is 78 kΩ and the input resistor is 2 kΩ, what is the A_V?

22. An inverting op-amp has a 10-kΩ R_f and a 2-kΩ R_{in}. What is the potential A_V?

23. Explain the principle of thermionic emission associated with a vacuum tube.

Student Activities

Amplifier-Static Operation

1. Construct the amplifier of Fig. 13-3.

2. Turn on the energy source.

3. Measure and record the static operating conditions for I_B, I_C, V_{CE}, and V_{RL}. These values will differ a great deal among transistors.

Amplifier-Dynamic Operation

1. Modify the circuit of Fig. 13-3 to conform with the ac amplifier of Fig. 13-4.

2. With an oscilloscope connected to the output, adjust the ac input signal level to produce the greatest output with a minimum of distortion. Record the peak-to-peak output signal level.

3. Move the oscilloscope leads to the signal input points. Measure and record the signal level.

4. Calculate the ac voltage gain.

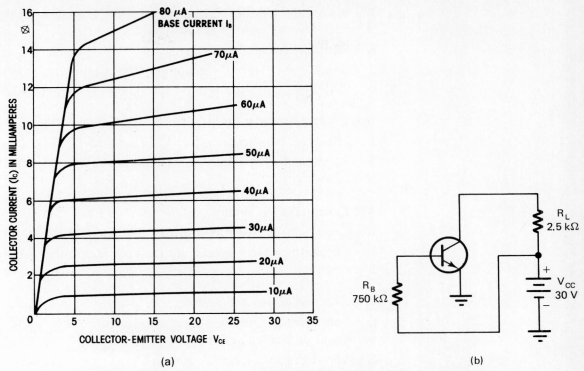

Figure 13-39. Transistor load-line analysis exercise: (a) I_C–V_{CE} curves; (b) transistor circuit.

Load-Line Analysis of an Amplifier

1. Using the collector family of characteristic curves in Fig. 13-39(a), develop a dc load line for the circuit in Fig. 13-39(b).

2. The two locating points are _____ and _____ .

3. With a straightedge, connect the locating points with a lead pencil.

4. Determine the Q point for the circuit. Locate this point on the load line.

5. Determine the static operation points for V_{CE} and I_C.

6. If a 0.5-V p-p signal is applied, it causes 20 μA of I_B.

7. Locate these points on the load line. By the projection of lines, determine the ΔI_C, ΔV_{CE}, and ΔI_B.

8. Calculate the dc beta, ac beta, and voltage gain.

JFET Amplifier—Common-Source

1. Construct the JFET amplifier of Fig. 13-26.

2. Apply energy to the circuit.

3. Measure and record the static operating conditions without a signal applied.

4. Apply the signal source to the amplifier.

5. Connect an oscilloscope to the output of the amplifier.

6. Adjust the signal source output to produce the greatest signal with a minimum of distortion. Note the peak-to-peak voltage signal level.

7. Move the oscilloscope leads to the amplifier input across R_1. Record the peak-to-peak voltage input signal level.

8. Calculate the voltage gain.

9. Connect the external sync input of the oscilloscope to the output. Observe the phase of the input and output signals. Describe the phase relationship.

JFET Amplifier—Common-Gate

1. Construct the common-gate JFET amplifier of Fig. 13-27.

2. Repeat steps 2 through 9 of the common-source amplifier for the common-gate amplifier.

JFET Amplifier—Common-Drain

1. Construct the common-drain JFET amplifier of Fig. 13-28.

2. Repeat steps 2 to 9 of the common-source amplifier for the common-drain amplifier.

Op-Amp Circuit—Noninverting

1. Construct the noninverting op-amp circuit of Fig. 13-34. Be certain to use a split or divider power supply.

2. Adjust the dc voltage input to 0.1 V.

3. Measure and record the dc voltage across R_L.

4. Calculate the dc voltage gain. Compare measured and calculated values.

5. Disconnect the dc signal source.

6. Turn on the ac signal source and adjust it to 0.1 V p-p.

7. With an oscilloscope or voltmeter, measure and record the peak-to-peak output voltage across R_L.

8. Calculate A_V. Compare calculated and measured values.

9. Observe the phase of the input and output signals.

Op-Amp Circuit—Inverting

1. Construct the inverting op-amp circuit of Fig. 13-35.

2. Repeat steps 2 to 9 of the noninverting circuit for the inverting amplifier.

Triode Amplifier

1. Construct the triode amplifier of Fig. 13-37.

2. Turn on the ac signal source, filament source, and dc power source. Be very careful in working with the circuit with the power applied. Do not touch any part of the circuit when the power is applied.

3. Connect the oscilloscope across the output.

4. Adjust the signal source to produce the greatest output with a minimum of distortion. Measure and record the peak-to-peak output voltage.

5. Measure and record the input voltage across Rg.

6. Measure the dc voltage between ground and cathode, grid and plate.

14

Amplifying Systems

Amplifying systems are widely used in the electronics field today. This particular type of system may be only one part or section of a rather large and complex system. A television receiver is a good example of this application. Several individual amplifying systems are included in a TV receiver. A phonograph, by comparison, has only one amplifying system. Its primary function is to process a sound signal and build it to a level that will drive a speaker. The function of an amplifier is basically the same regardless of its application.

In this chapter we investigate how amplifying devices are used in an operating system. Transistors, integrated circuits, and vacuum tubes can all be used for this type of operation. A number of specific circuit functions are also discussed. These include gain, coupling, load devices, output circuits, transducers, and signal processing. The amplifying devices and their circuit components play an important role in the operation of a system.

IMPORTANT TERMS

A number of new, and in some cases unusual terms will be encountered in this chapter. As a rule, these terms play an important part in the understanding of amplifier systems. Review each item very carefully before proceeding with the chapter. Refer back to a specific item when its meaning is not fully understood.

Amplitude modulation (AM). A method of transmitter modulation achieved by varying the strength of the RF carrier at an audio signal rate.

Audio frequency (AF). A range of frequency from 15 Hz to 15 kHz, the human range of hearing.

Bel (B). A measurement unit of gain that is equivalent to a 10:1 ratio of power levels.

Cascade. A method of amplifier connection in

which the output of one stage is connected to the input of the next amplifier.

Characteristic. The magnitude range of a number of logarithms.

Complementary transistors. PNP and NPN transistors with identical characteristics.

Coupling. A process of connecting active components.

Crossover distortion. The distortion of an output signal at the point where one transistor stops conducting and another conducts, in class B amplifiers.

Crystal microphone. Input transducer that changes sound into electrical energy by the piezoelectric effect.

Darlington amplifier. Two transistor amplifiers cascaded together.

Decibel. One-tenth of a bel; used to express gain or loss.

Dynamic microphone. An input transducer that changes sound into electrical energy by moving a coil through a magnetic field.

Impedance. An opposition to ac measured in ohms.

Impedance ratio. The ac resistance ratio between the input and output of a transformer.

Logarithm. A mathematical expression dealing with exponents showing the power to which a value called the base must be raised in order to produce a given number.

Mantissa. Decimal part of a logarithm.

Midrange. A speaker designed to respond to audio frequency in the middle part of the AF range.

Peak to peak (p-p). An ac value measured from the peak positive to the peak negative alternations.

Piezoelectric effect. The property of a crystal material that produces voltage by changes in shape or pressure.

Push-pull. An amplifier circuit configuration using two active devices with the input signal being of equal amplitude and 180° out of phase.

Read head. A signal pickup transducer that generates a voltage in a tape recorder.

Root mean square (RMS). Effective value of ac or $0.707 \times$ peak value.

Speaker. A transducer that converts electric current and/or voltage into sound waves. Also called a loudspeaker.

Stage of amplification. A transistor or vacuum tube and all the components needed to achieve amplification.

Stereophonic (stereo). Signals developed from two sources that give the listener the impression of sound coming into both ears from different points.

Stylus. A needle-like point that rides in a phonograph record groove.

Transducer. A device that changes energy from one form to another. A transducer can be on the input or output.

Tweeter. A speaker designed to reproduce only high frequencies.

Voice coil. The moving coil of a speaker.

Voltage divider. A combination of resistors that produce different or multiple voltage values.

Woofer. A speaker designed to reproduce low audio frequencies with high quality.

AMPLIFYING SYSTEM FUNCTIONS

Regardless of its application, an amplifying system has a number of primary functions that must be performed. An understanding of these functions is an extremely important part of operational theory. A block diagram of an amplifying system is shown in Fig. 14-1. The triangular-shaped items of the diagram show where amplification is performed. A *stage* of amplification is represented by each triangle. Three amplifiers are included in this particular system. A stage of amplification consists of an active device and all of its associated components. Small-signal amplifiers are used in the first three stages of this system. The amplifier on the right side of the diagram is an output stage. A rather large signal is needed to control the output amplifier. An output stage is generally called a large-signal amplifier or a power amplifier. In effect, this amplifier is used to control a rather large amount of current and voltage. Remember that power is the product of current and voltage.

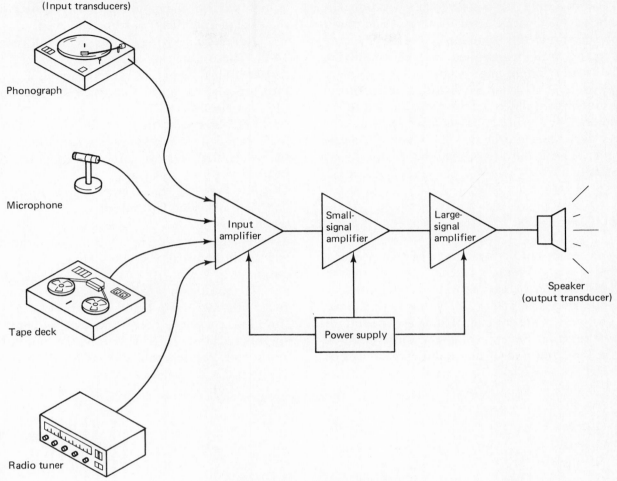

(Input transducers)

Phonograph

Microphone

Tape deck

Radio tuner

Input amplifier

Small-signal amplifier

Large-signal amplifier

Power supply

Speaker
(output transducer)

Figure 14-1.   Amplifying system.

In an amplifying system, a signal must be developed and applied to the input. The source of this signal varies a great deal with different systems. In a *stereophonic* (stereo) phonograph, the signal is generated by a phonograph cartridge. Variations in the groove of a record are changed into electrical energy. This signal is then applied to the input of the system. The amplitude level of the signal is increased to a suitable value by the amplifying devices. A variety of different input signal sources may be applied to the input of an amplifying system. In other systems the signal may also be developed by the input. *Transducers* are responsible for this function. An input transducer changes energy of one form into energy of a different type. Microphones, phonograph pickup cartridges, and tape-recorder heads are typical input transducers. Input signals may also be received through the air. Antenna coils and networks serve as the input transducer for this type of system. An antenna changes electromagnetic waves into radio-frequency (RF) voltage signals. The signal is then processed through the remainder of the system.

Signals processed by an amplifying system are ultimately applied to an output transducer. This type of transducer changes electrical energy into another form of energy. In a sound system the *speaker* is an output transducer. It changes electrical energy into sound energy. Work is performed by the speaker when it achieves this function. Lamps, motors, relays, transformers, and inductors are frequently considered to be transducers. An output transducer is also considered to be the load of a system.

For an amplifying system to be operational, it must be supplied with electrical energy. A dc power supply performs this function. A relatively pure form of dc must be supplied to each amplifying device. In most amplifying systems, ac is the primary energy source. Ac is changed into dc, filtered, and in some systems regulated before being applied to the amplifiers. The reproduction quality of the amplifier is very dependent on the dc supply voltage. Some portable stereo amplifiers may be energized by batteries.

AMPLIFIER GAIN

The gain of an amplifier system can be expressed in a variety of ways. Voltage, current, power, and, in some systems, decibels are expressed as gain. Nearly all input amplifier stages are voltage amplifiers. These amplifiers are designed to increase the voltage level of the signal. Several voltage amplifiers may be used in the front end of an amplifier system. The

voltage value of the input signal usually determines the level of amplification being achieved.

A three-stage voltage amplifier is shown in Fig. 14-2. These three amplifiers are connected in *cascade*. This refers to a series of amplifiers where the output of one stage is connected to the input of the next amplifier. The voltage gain of each stage could be observed with an oscilloscope. The waveform would show representative signal levels. Note the voltage-level change in the signal and the amplification factor of each stage.

The first stage has a voltage gain of 5 V. With 0.25 V p-p input, the output is 1.25 V p-p. The second stage also has a voltage gain of 5. With 1.25 V p-p input, the output is 6.25 V p-p. The output stage has a gain factor of 4. With 6.25 V p-p input, the output is 25 V p-p.

The total gain of the amplifier is $5 \times 5 \times 4$, or 100. With 0.25 V p-p input the output is 25 V p-p. Note that output is the product of the individual amplifier gains. It is not just the addition of $5 + 5 + 4$. Voltage gain (A_V) is an expression of

Figure 14-2.   Three-stage voltage amplifier.

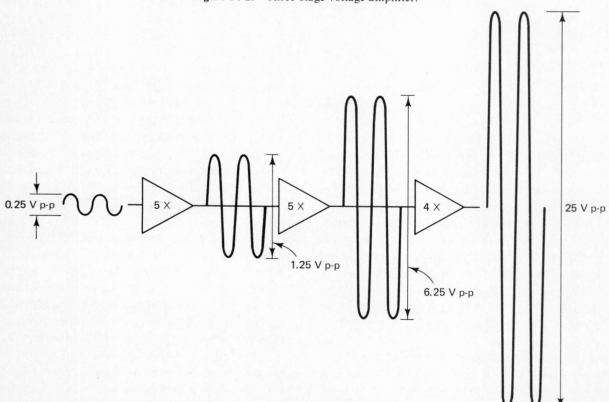

output voltage (V_O) divided by input voltage (V_{in}). For the amplifier system, A_V is expressed by the formula

$$A_V = \frac{V_O}{V_{in}} = \frac{25 \text{ V p-p}}{0.25 \text{ V p-p}} = 100$$

Note that the units of voltage cancel each other in the problem. Voltage gain is therefore expressed as a pure number. It is not good practice to say that the voltage gain of an amplifier is 100 V p-p. The voltage gain is best described as a value such as 100.

Power gain is generally used to describe the operation of the last stage of amplification. Power amplification (A_p) is equal to power output (P_O) divided by power input (P_{in}). If the last stage of amplification in Fig. 14-2 were a power amplifier, its gain would be expressed in watts rather than volts. In this case, the gain would be

$$A_p = \frac{P_O}{P_{in}} = \frac{25 \text{ W}}{6.25 \text{ W}} = 4$$

Note that the power units of this problem also cancel each other. Power amplification is expressed as a pure number such as 4.

DECIBELS

The human ear does not respond to sound levels in the same manner as an amplifying system. An amplifier, for example, has a linear rise in signal level. An input signal level of 1 V could produce an output of 10 V. The voltage amplification would be 10:1 or 10. The human ear, however does not respond in a linear manner. It is, essentially, a nonlinear device. As a result of this, sound amplifying systems are usually evaluated on a *logarithmic* scale. This is an indication of how our ears will actually respond to specific signal levels. Gain expressed in logarithms is much more meaningful than linear gain relationships.

The logarithm of a given number is the power to which another number, called the base, must be raised to equal the given number. Basically, a logarithm has the same meaning as an exponent. A common logarithm is expressed in powers of 10.

This is illustrated by the following:

$$10^3 = 1000$$
$$10^2 = 100$$
$$10^1 = 10$$
$$10^0 = 1$$

This means that the logarithm of any number between 9999 and 1000 would have a *characteristic* value of 3. The characteristic is an expression of the magnitude range of the number. Numbers between 999 and 100 would have a characteristic of 2. Numbers between 99 and 10 would have a characteristic of 1. Between 9.0 and 1.0 the characteristic would be 0. Number values less than 1.0 would be expressed as a negative characteristic. It is generally not customary to use negative characteristic values.

When a number is not an even multiple of 10, it must have a decimal expression. The decimal part of the logarithm is called the *mantissa*. A number such as 4000 would be expressed as 3.6021. The characteristic is 3 because 4000 is between 9999 and 1000. The mantissa of 4000 is 0.6021. The mantissa is found from a table of natural logarithms (see Appendix 8). To find this value, simply search down the vertical number column (N) to find the number 40. Moving horizontally to the "0" column the mantissa for 400 is 6021. The log of 4000 is therefore 3.6021. The characteristic is 3 and the mantissa is 0.6021.

The mantissa is always the same for a given sequence of numbers regardless of the location of the decimal point. For example, the mantissa is the same for 1630, 163.0, 16.3, 1.63, 0.163, and so on. The only difference in these values would be the characteristic. The mantissa for 1630 is 2122. The log of the five values would be 3.2122, 2.2122, 1.2122, 0.2122, and −1.2122. Practice changing several number values into logarithms using the natural logarithm table.

If an electronic calculator is available with logarithms, the conversion process is very easy. Simply place the number into the calculator. Then press the log button. For example, the log of the number 1590 is 3.2013971. Try this operation on a calculator. Use a variety of numbers, such as 2503., 603.8, 14.32, 8.134, and 0.2461.

Consider now the gain of a sound system with several stages of amplification. Gain is best expressed as a ratio of two signal levels. Specifically, gain is expressed as the output level divided by the input level. This is determined by the expression

$$\text{power amplification} = \log \frac{P_{\text{out}}}{P_{\text{in}}} = \text{bels}$$

For an amplifier with 0.1 W of input and 100 W of output the A_p would be

$$A_p = \log \frac{100 \text{ W}}{0.1 \text{ W}} = \log 1000 = 3 \text{ bels}$$

The fundamental unit of sound level gain is the *bel* (B). As noted in this calculation, the bel represents a rather large ratio in sound level. A *decibel* (dB) is a more practical measure of sound level. A decibel is one-tenth of a bel. The bel is named for Alexander Graham Bell, the inventor of the telephone.

The gain of a single stage of amplification within a system can be determined with decibels. A single amplifier stage could have an input of 10 mW and an output of 150 mW. The power gain would be determined by the formula

$$\text{power amplification (dB)} = 10 \log \frac{P_{\text{out}}}{P_{\text{in}}}$$

$$= 10 \log \frac{150 \text{ mW}}{10 \text{ mW}}$$

$$\text{dB} = 10 \log 15$$

$$= 10 \times 1.1761$$

$$= 11.761$$

The voltage gain of an amplifier can also be expressed in dB values. In order to do this the power-level expression must be adapted to accommodate voltage values. The dB voltage gain formula is

voltage amplification (dB)

$$= 20 \times \log \frac{\text{voltage output}}{\text{voltage input}}$$

$$= 20 \log \frac{V_O}{V_{\text{in}}}$$

Notice that the logarithm of V_O/V_{in} is multiplied by 20 in this equation. Power is expressed as V^2/R. Power gain using voltage and resistance values

would therefore be expressed as

$$\text{dB} = \frac{V_O{}^2 \times R_{\text{out}}}{V_{\text{in}}{}^2 \times R_{\text{in}}}$$

If the value of R_{in} and R_{out} are equal, the equation would be simplified to be

$$\text{dB} = \frac{V_O{}^2}{V_{\text{in}}{}^2}$$

The squared voltage values can be expressed as two times the log of the voltage value. Decibel voltage gain therefore becomes

$$\text{dB} = 2 \times 10 \log \frac{V_O}{V_{\text{in}}} \quad \text{or} \quad 20 \log \frac{V_O}{V_{\text{in}}}$$

It is important to remember that dB voltage gain assumes the value of R_{in} and R_{out} to be equal. Decibel gain is primarily an expression of power levels. Voltage gain is therefore only an adaptation of the power-level expression.

To demonstrate the use of the dB voltage gain equation, let us apply it to the first amplifier of Fig. 14-2. Note that the input voltage is 0.25 V p-p and the output voltage is 1.25 V p-p. The voltage gain in dB is

$$\text{dB} = 20 \log \frac{V_O}{V_{\text{in}}}$$

$$= 20 \log \frac{1.25 \text{ V p-p}}{0.25 \text{ V p-p}}$$

$$= 20 \log 5$$

$$= 20 \times 0.69897$$

$$= 13.9794 \text{ or } 14$$

Using the procedure just described, determine the gain of the second amplifier stage. Determine the gain of the output amplifier stage. The total voltage gain of the three amplifiers would be the sum of the individual dB voltage values. To check the accuracy of your calculations, determine the total voltage gain of amplifier using 0.25 as V_{in} and V_O as 25. This should compare with the addition of individual-stage dB values.

AMPLIFIER COUPLING

In many systems one stage of amplification will not provide the desired level of signal output. Two or more stages of amplification are therefore connected together to increase the overall gain. These stages must be properly coupled for the system to be operational. *Coupling* refers to a method of connecting amplifiers together. The signal being processed is transferred between stages by the coupling component.

Several methods of interstage amplifier coupling are in use today. Three very common methods include, capacitive-coupling, direct coupling, and transformer coupling. Each type of coupling has specific applications. In most amplifying devices, the output voltage is greater than the input voltage. If the output of one stage is connected directly to the input of the next stage, this voltage difference could cause a problem. Signal distortion and component damage might take place. Proper coupling procedures reduce this type of problem.

Capacitive Coupling

Capacitive coupling is particularly useful when amplifier systems are designed to pass ac signals. A capacitor will pass an ac signal and block dc voltages. The capacitor selected must have low capacitive reactance (X_C) at its lowest operating frequency. This is done to ensure amplification over a wide range of frequency. In general, the capacitive reactance of a good capacitor is extremely high at low frequency. At zero frequency or dc, the X_C is nearly infinite. Capacitive coupling has some difficulty in passing low-frequency ac signals. Large capacitance values are selected when good low-frequency response is desired.

A two-stage amplifier employing capacitor coupling is shown in Fig. 14-3. Transistor Q_1 is the input amplifier. Its collector voltage is 8.5 V. At the base of the second amplifier stage (Q_2), the voltage is approximately 1.7 V. The voltage difference across C_1 is 8.5 V − 1.7 V, or 6.8 V. The capacitor isolates these two operating voltages. It must have a dc working voltage value that will withstand this difference in potential.

The value of a coupling capacitor is a very important circuit consideration. In low-frequency amplifying systems, large values of electrolytic capacitors are normally used. These capacitors respond well to low-frequency ac. In high-frequency amplifier applications, small capacitor values are very common. Selection of a specific coupling capacitor for an amplifier is dependent on the frequency being processed.

Figure 14-3. Two-stage capacitor-coupled amplifier.

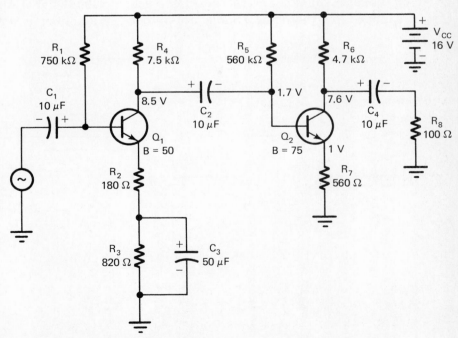

Direct Coupling

In direct coupling the output of one amplifier is connected directly to the input of the next stage. In a circuit of this type a connecting wire or conductor couples the two stages together. Circuit design must take into account device voltage values. This type of coupling is an outgrowth of the divider method of biasing. A transistor acts as one resistor in a divider network. The output voltage of one amplifier is the same as the input voltage of the second amplifier. When the circuit is designed to take this into account, the two can be connected together without isolation.

Figure 14-4 shows a two-stage direct-coupled transistor amplifier. Notice that the signal passes directly from the collector into the base. The base current (I_B) of transistor Q_2 is developed without a base resistor. Any I_B needed for Q_2 also passes through the load resistor R_3. The collector voltage V_C of Q_1 remains fairly constant when the two transistors are connected. Note also that the emitter bias voltage of Q_2 is quite large (7.2 V). The base–emitter voltage (V_{BE}) of Q_2 is the voltage difference between V_B and V_E. This is 7.8 V_B − 7.2 V_E, or 0.6 V_{BE}. This value of V_{BE} forward biases the base–emitter junction of Q_2. In direct-coupled amplifiers each stage has a different operating point based on a common voltage source. Several direct-coupled stages supplied by the same source are rather

difficult to achieve. As a general rule, only two stages are coupled together by this method in an amplifying system.

Direct-coupled amplifiers are very sensitive to changes in temperature. The beta of a transistor will, for example, change rather significantly with temperature. An increase in temperature causes an increase in beta and leakage current. This tends to shift the operating point of a transistor. All stages that follow will amplify according to the operating point shift. Changes in the operating point can cause nonlinear distortion.

When two transistors are directly coupled it is often called a *Darlington amplifier*. The transistors are usually called a Darlington pair. Two transistors connected in this manner are also manufactured in a single case. This type of unit has three leads. It is generally called a Darlington transistor. The gain produced by this device is the product of the two transistor beta values. If each transistor has a beta of 100, the total current gain is 100 × 100, or 10,000. Figure 14-5 shows a Darlington transistor amplifier circuit.

Darlington amplifiers are used in a system where high current gain and high input impedance are needed. Only a small input signal is needed to control the gain of a Darlington amplifier. In effect, this means that the amplifier does not load down the input signal source. The output impedance of this amplifier is quite low. It is developed across the

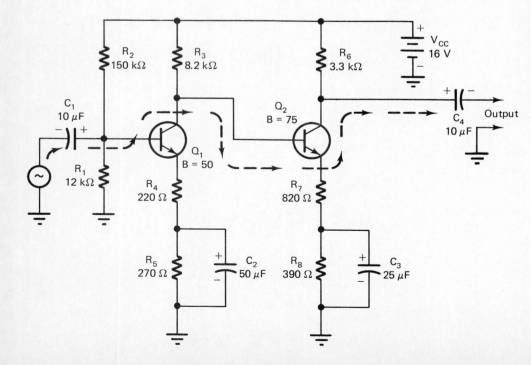

Figure 14-4. Two-stage direct-coupled amplifier.

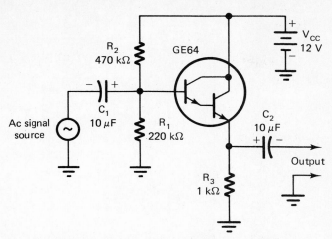

Figure 14-5. Darlington transistor amplifier.

Figure 14-6. Coupling transformers. (Courtesy of TRW/UTC Transformers.)

emitter resistor R_3. A Darlington amplifier has an emitter-follower output. Several different transistor combinations are used in Darlington amplifiers today.

Transformer Coupling

In transformer coupling, the output of one amplifier is connected to the input of the next amplifier by mutual inductance. Depending on the frequency being amplified, a coupling transformer may use a metal core or an air core. The output of one stage is connected to the primary winding and the input of the next stage is connected to secondary

winding. The number of primary and secondary turns determines the *impedance* ratio of the respective windings. The input and output impedance of an amplifier stage can be easily matched with a transformer. Ac signals pass easily through the transformer windings. Dc voltages are isolated by the two windings. Figure 14-6 shows some representative coupling transformers.

Figure 14-7 shows a two-stage transformer coupled radio-frequency amplifier. This particular

Figure 14-7. Two-stage transformer-coupled RF amplifier.

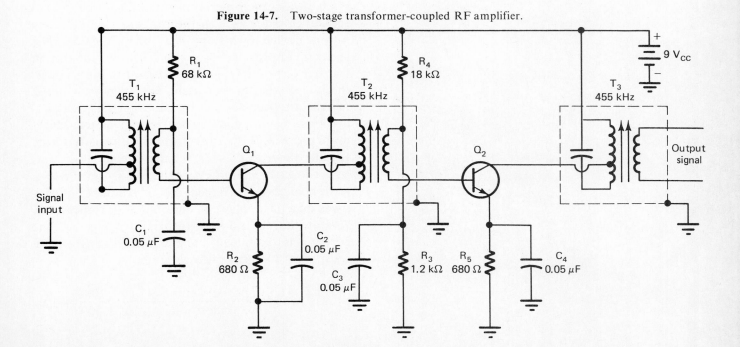

circuit is used in an *amplitude-modulated* (AM) radio receiver. Circuits of this type operate very well when they are built on a printed circuit board. The operational frequency of this amplifier is 455 kHz.

The operation of a transformer coupled circuit is very similar to that of a capacitive–coupled amplifier. Biasing for each transistor element is achieved by resistance and transformer impedance. The primary impedance of T_2 serves as the load resistor of Q_1. The secondary winding of the transformer and R_4 serves as the input impedance for Q_2. The low output impedance of Q_1 is matched to the high input impedance of Q_2 by the transformer. The primary tap connection is used to assure proper load impedance for Q_1. The impedance of each winding is dependent on the primary and secondary turns ratio. This particular transformer-coupled amplifier is tuned to pass a specific frequency. Selection of this frequency is achieved by coil inductance (L) and capacitance (C). Tuned transformer-coupled amplifiers are widely used in radio receivers and TV circuits. The dashed line surrounding each transformer indicates that it is housed in a metal can. This is done purposely to isolate the transformers from one another. Several transformers of this type are shown in Fig. 14-8. The large transformer is used in vacuum-tube circuits. The smaller units are found in transistor radios.

Transformers are also used to couple an amplifier to a load device. Figure 14-9 shows a transistor

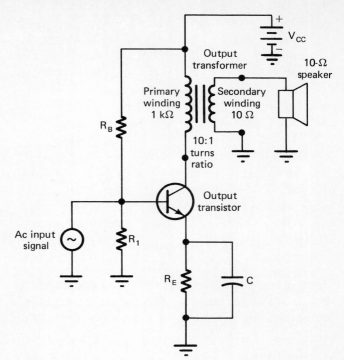

Figure 14-9. Transformer-coupled transistor circuit.

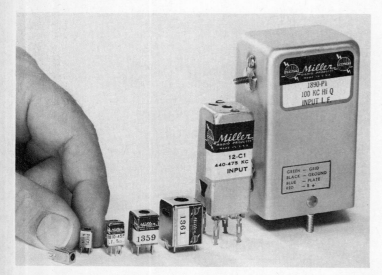

Figure 14-8. IF transformers. (Courtesy of J. W. Miller Div./ Bell Industries.)

circuit with a coupling transformer. In this application the coupling device is called an output transformer. The primary transformer winding serves as a collector load for the transistor. The secondary winding couples transistor output to the load device. The transformer serves as an impedance-matching device. Its primary impedance matches the collector load. In a common-emitter amplifier, typical load resistance values are in the range of 1000 Ω. The output device in this application is a speaker. The speaker impedance is 10 Ω. To get maximum transfer of power from the transistor to the speaker, the impedances must match. The primary winding matches the transistor output impedance. The secondary winding matches the speaker impedance. Maximum power is transferred from the transistor to the speaker through the transformer impedance.

Closer examination of the transformer tells a great deal about its impedance. The number of wire turns on a particular coil determines its impedance. Assume that the output transformer of Fig. 14-9 has 1000 turns on the primary and 100 turns on the secondary. Its turns ratio would be

$$\text{turns ratio} = \frac{\text{primary turns } (N_P)}{\text{secondary turns } (N_S)} = \frac{1000}{100} = 10$$

The impedance ratio of a transformer is the square of its turns ratio. For the output transformer the impedance ratio is 10^2 or 100 Ω. This means the impedance of the collector will be 100 times more than the impedance of the load device. If the load (speaker) of our circuit has an impedance of 10 Ω, the input impedance will therefore be 100×10, or 1000 Ω.

Transformer coupling has a number of problems. In sound system applications, the impedance increases with frequency. As a result of this, the high-frequency response of a sound signal is rather poor. Better-quality sound systems do not use transformer coupling. The physical size of a transformer must also be quite large when high-power signals are being amplified. Because of this, low- and medium-power amplifying circuits only use transformer coupling. In addition to this, good-quality transformers are rather expensive. These disadvantages tend to limit the number of applications of transformer coupling today.

POWER AMPLIFIERS

The last stage of an amplifying system is nearly always a power amplifier. This particular type of amplifier is generally a low-impedance device. It controls a great deal of current and voltage. Power is the product of current and voltage. Normally, a power amplifier dissipates a great deal more heat than does a comparable signal amplifier. In transistor systems, when 1 W or more of power is handled by a device, it is considered to be a power amplifier. Power amplifiers are generally designed to operate

over the entire active region of the device. With careful selection of the device and circuit components it is possible to achieve power gain with a minimum of distortion.

Single-Ended Power Amplifiers

An amplifier system that has only one active device that develops output power is called a single-ended amplifier. This type of amplifier is very similar in many respects to a small-signal amplifier. It is normally operated as a class A amplifier. An ac signal applied to the input will be fully reproduced in the output. The operating point of the amplifier is near the center of its active region. As a general rule, this type of amplifier operates with a minimum of distortion. The operational efficiency of this amplifier is, however, quite low. Efficiency levels are in the range of 30%. Efficiency refers to a ratio of developed output power to dc supply power. The equation for efficiency is

$$\text{efficiency percentage} = \frac{\text{ac power output}}{\text{applied dc operating power}}$$
$$= \frac{P_O}{P_{\text{in}}}$$

For a class A amplifier it takes 9 W of dc power to develop 3 W of output signal when an amplifier is 33% efficient. Because of their low efficiency rating, single-ended amplifiers are generally used only to develop low-power output signals.

Figure 14-10 shows a single-ended *audio* output amplifier. The term "audio" refers to the range of human hearing. *Audio frequency* (AF) refers to

Figure 14-10. Single-ended audio amplifier.

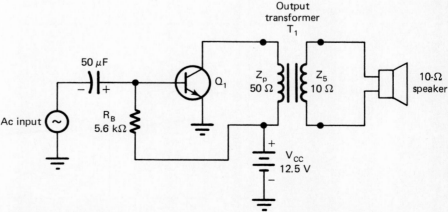

frequencies that are from 15 Hz to 15 kHz. This particular amplifier will respond to signals somewhere within this range.

A family of characteristic curves for the transistor used in the single-ended power amplifier is shown in Fig. 14-11. Note that the power dissipation

curve of this amplifer is 5 W. The developed load lines for the amplifier are well below the indicated P_D curve. This means that the transistor can operate effectively without becoming overheated.

Investigation of the family of collector curves shows that two load lines are plotted. The static or dc

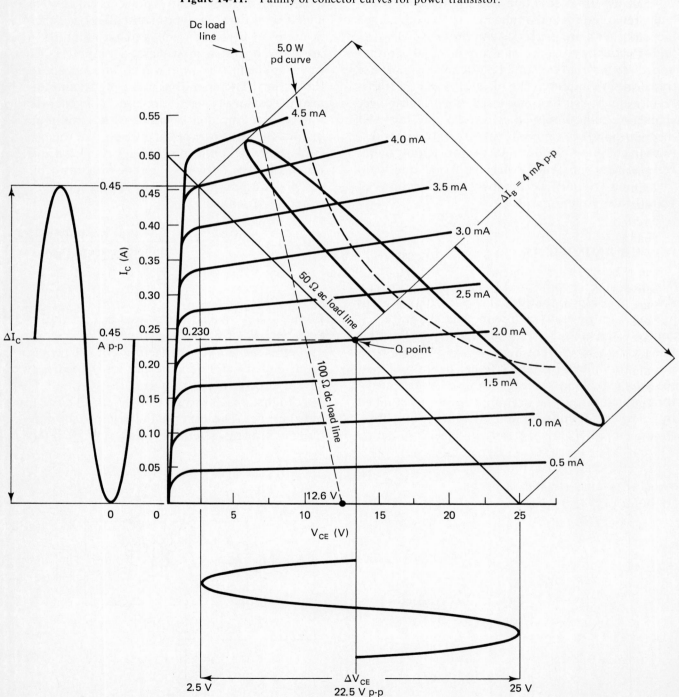

Figure 14-11. Family of collector curves for power transistor.

load line is based on the resistance of the transformer primary. For a primary impedance of 50 Ω, the resistance would be approximately 10 Ω. The dc load line is therefore based on 12.5 V_{CC} and 1.25 A of I_C. This means that the dc load line will extend almost vertically from the 12.5 V_{CC} point on the curve. The ac load line is determined with the primary winding impedance (50 Ω) and twice the value of V_{CC} (25 V). The ac output signal will therefore be twice the value of V_{CC}. The operational points are 25 V_{CC} and 500 mA of I_C. Note the location of these operating points on the ac load line.

When an ac signal is applied to the input it can cause a swing 12.5 V above and below V_{CC}. The transformer, in this case, accounts for the voltage difference. A transformer is an inductor. When the electromagnetic field of an inductor collapses, a voltage is generated. This voltage is added to the source voltage. In an ac load line, V_{CE} will therefore swing to twice the value of the source voltage. This usually occurs only in transformer-coupled amplifier circuits. Note that the operating point (Q point) is near the center of the ac load line.

With the load line of the amplifier established, the base resistance value must be determined. For our circuit the emitter voltage is zero. The emitter–base junction of the transistor, being silicon, will have a 0.7 V across it when it is in conduction. The base resistor would therefore have 12.5 V − 0.7, or 11.8 V across it at the Q point. By Ohm's law, the value of R_B is 11.8 V/2 mA, or 5900 Ω. A standard resistor value of 5.6 kΩ would be selected for R_B. Notice the location of this resistor in the circuit.

With the available data that we have, it is possible to look at the operational efficiency of the single-ended amplifier. The dc power supplied to the circuit for operation is

applied dc operating power = dc source voltage

$\times$ collector current

or

$$P_{in} = V_{CC} \times I_C$$
$$= 12.6 \text{ V} \times 0.230 \text{ A}$$
$$= 2.898 \text{ W}$$

The developed ac output power (P_O) is a peak-to-peak value. The V_{CE} is 2.5 to 25 V p-p. The I_C is 0 to 0.45 A or 0.45 A p-p.

In order to make a comparison in power efficiency, the peak-to-peak values must be changed to *rms values*. An rms ac value and a dc value do the same effective work. Conversion of peak-to-peak voltage to rms is done by the expression

$$V_{CE}(\text{rms}) = \frac{V_{p\text{-}p}}{2} \times 0.707$$
$$= \frac{22.5 \text{ V p-p}}{2} \times 0.707 = 7.953 \text{ V}$$

For the rms, I_C value the conversion is

$$I_C(\text{rms}) = \frac{I_{p\text{-}p}}{2} \times 0.707$$
$$= \frac{0.45 \text{ A}}{2} \times 0.707 = 0.1591 \text{ A}$$

The developed ac output power (p_o) is determined by the formula

$$\text{ac power output } (p_o) = V_{CE}(\text{rms}) \times I_C(\text{rms})$$
$$= 7.953 \text{ V} \times 0.1591 \text{ A}$$
$$= 1.2652 \text{ W}$$

The operational efficiency of the single-ended amplifier of Fig. 14-10 is determined to be

$$\% \text{ efficiency} = \frac{P_O}{P_{in}} = \frac{1.2652 \text{ W}}{2.898 \text{ W}} = 0.44 \text{ or } 44\%$$

This shows that an amplifier of this type is only 44% efficient. It takes 2.898 W of dc supply power to develop 1.2652 W of ac output power. This low operational efficiency does not make the single-ended amplifier very suitable for high-power applications. As a general rule, amplifiers of this type are used only where the output is 10 W or less. The power wasted by a single-ended amplifier appears in the form of heat. Obviously, it would be very desirable to get more sound output and less heat during operation.

Push-Pull Amplifiers

Power amplifiers can also be connected in a *push-pull* circuit configuration. Two amplifying devices are needed to achieve this type of output.

The operational efficiency of this circuit is much higher than that of a single-ended amplifier. Each transistor handles only one alternation of the sine-wave input signal. Class B amplifier operation is used to accomplish this. A class B amplifier has sine-wave input and half sine-wave output. The bias operating point is at cutoff. This means that no power will be consumed unless a signal is applied. The operational efficiency of a push-pull amplifier is approximately 80%.

A simplified transformer-coupled push-pull transistor amplifier is shown in Fig. 14-12. The circuit is somewhat like two single-ended amplifiers placed back to back. The input and output transformers both have a common connection. The V_{CC} supply voltage is connected to this common point. For operation, the base voltage must rise slightly above zero to produce conduction. The turn-on voltage is usually about 0.6 V. Base voltage is developed by the incoming signal between the center tap and outside ends of the transformer. The transistor will not conduct until the base voltage rises above 0.6 V. The secondary winding resistance of the input transformer (T_1) is several thousand ohms. The divided transformer winding means that each transistor is alternately fed an input signal. The signals will be the same value but 180° out of phase.

A class B push-pull amplifier will produce an output signal only when an input signal is applied. For the first alternation of the input, Q_1 will swing positive and Q_2 negative. Only Q_1 will go into conduction with this input. Note the resulting I_C waveform. At the same time, Q_2 is at cutoff. The negative alternation drives it further into cutoff. The resulting I_C from Q_1 flows into the top part of primary winding of T_2. This causes a corresponding change in the electromagnetic field. Voltage is induced in the secondary winding through this action. Note the polarity of the secondary voltage for this alternation.

For the next alternation of the input signal, the process is reversed. The base of Q_1 swings negative. It becomes nonconductive. Q_2 swings positive at the same point in time. Q_2 goes into conduction. Note the indicated I_C of Q_2 for this alternation. The I_C of Q_2 flows into the lower side of T_2. It, in turn, causes the electromagnetic field to change. Voltage is again induced into the secondary winding. Note the polarity of the secondary voltage for this alternation. It is reversed because the primary current is in an opposite direction to that of the first alternation.

Figure 14-13 shows a combined collector family of characteristic curves for Q_1 and Q_2. These transistors are operated as class B amplifiers. With each transistor operating at cutoff (point Q), only one alternation of output will occur. The entire active region of each transistor can be used in this case to amplify the applied alternation. The combined alternations make a noticable increase in output power.

Using the data of Fig. 14-13, we will analyze the power output and operational efficiency of the push-pull amplifier. The combined change in I_C for both Q_1 and Q_2 is 0.25 A + 0.25 A, or 0.5 A p-p. The combined change in collector voltage is 12 V + 12 V or 24 V p-p. The effective ac current and voltage values are

$$V_{CC} \text{ (rms)} = \frac{V_{\text{p-p}}}{2} \times 0.707$$

$$= \frac{24 \text{ V p-p}}{2} \times 0.707 = 8.484 \text{ V}$$

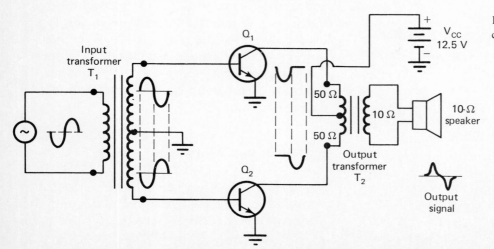

Figure 14-12. Transformer-coupled class B push-pull amplifier.

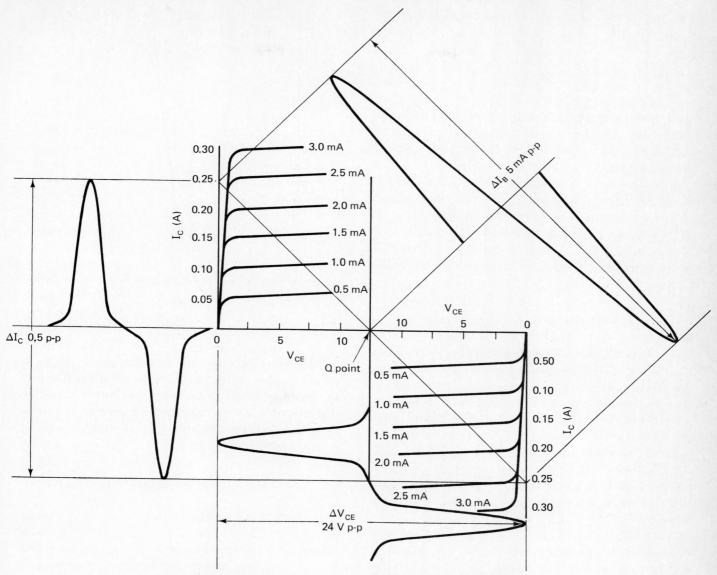

Figure 14-13. Combined collector curves for push-pull operation.

$$I_C\,(\text{rms}) = \frac{I_{\text{p-p}}}{2} \times 0.707$$

$$= \frac{0.5\ \text{A}}{2} \times 0.707 = 0.17675\ \text{A}$$

The effective power output is

ac power output $(P_O) = V_{CC}(\text{rms}) \times I_C(\text{rms})$

$$= 8.484 \times 0.17675\ \text{A}$$

$$= 1.4995\ \text{or}\ 1.5\ \text{W}$$

The operating dc power of our push-pull amplifier must be determined in order to evaluate its operating efficiency. The dc supply voltage V_{CC} is 12.5 V. The I_C changes from 0 to 0.25 A in each transistor. The average value of the I_C is

applied I_C (average) $= \Delta I_C \times 0.637$

$$= 0.25\ \text{A} \times 0.637$$

$$= 0.15925\ \text{A}$$

The average dc operating power (P_{in}) is equal to the dc voltage times the current flow. This value is

$P_{\text{in}} = $ dc input V $\times$ dc input current

$$= 12.5\ \text{V} \times 0.15925\ \text{A}$$

$$= 1.9906\ \text{W}$$

The operational efficiency of our push-pull amplifier is therefore

$$\% \text{ efficiency} = \frac{P_O}{P_{\text{in}}} = \frac{1.5 \text{ W}}{1.9906 \text{ W}} = 0.753 = 75.3\%$$

This means that it takes nearly 2 W of dc input power to develop 1.5 W of ac output power. This type of operation is very good. The operational efficiency of a push-pull amplifier is much better than that of a comparable class A power amplifier. Its efficiency does not effectively change with the value of the signal-level input. The power output of a push-pull amplifier can be nearly four times the output of a single-ended class A power amplifier. Push-pull output circuitry is used primarily to develop high-power signal levels.

Crossover Distortion

Class B push-pull amplification has a very good operating efficiency and does a good job in reproducing high-power signals. It does, however, have a distortion problem when conduction changes back and forth between each transistor. This distortion is at the midpoint of the signal. The term "crossover distortion" is used to describe this condition. Figure 14-13 shows crossover distortion near the center of the I_C and V_{CE} waveforms. In sound-amplifying systems the human ear is able to detect crossover distortion.

Crossover distortion is due to the nonlinear characteristic of a transistor when it first goes into conduction. The emitter–base junction of a transistor responds as a diode. A silicon diode does not go into conduction immediately when it is forward biased. It takes approximately 0.6 V to produce conduction. This means that no current will flow between 0 and 0.6 V of base–emitter voltage (V_{BE}). Figure 14-14 shows the input characteristic of a silicon transistor. Notice the nonlinear area between 0 and 0.6 V.

When a transistor is cut off there is no conduction of base current or collector current. When it is forward biased conduction does not start immediately. There is a slight delay until 0.6 V_{BE} is reached. In a push-pull amplifier circuit, this initial delay causes distortion of the input signal. Distortion occurs when the input signal crosses from the conducting transistor to the off transistor. One

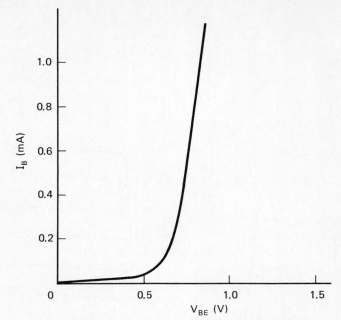

Figure 14-14. Input characteristic of a silicon transistor.

transistor is turning off and the other starts to conduct. The trailing edge of the off transistor and the leading part of the on transistor both cause some distortion of the waveform. As a general rule, this distortion is very noticable in small input signals. It is less noticable in large signals.

Class AB Push-Pull Amplifiers

Crossover distortion can be reduced by changing the operating point of a transistor. If each transistor is forward biased by 0.6 V, there will not be any noticable crossover distortion. A transistor biased slightly above cutoff is considered to be a class AB amplifier. Operation of this type prevents the base–emitter voltage from reaching the nonlinear part of the curve. Figure 14-15 shows the operating point of a class AB amplifier on an input curve and its location on a load line. The load line also shows operating points for class A and class B operation.

The operational efficiency of a class AB amplifier is not quite as good as that of a class B amplifier. There must be some conduction when no signal is applied. The operating point of a class AB amplifier is usually held to an extremely low conduction level. Its operational efficiency is much better than that of a class A amplifier. In a sense, AB operation is a design compromise to minimize distortion at a reasonable efficiency level.

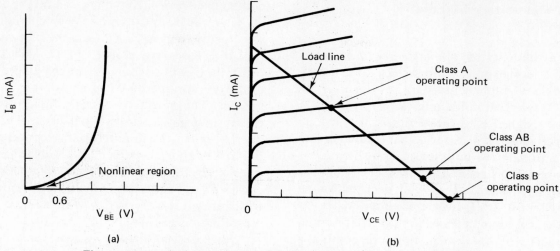

Figure 14-15. Transistor characteristics and operating points: (a) input characteristic for a silicon transistor; (b) bias operating points on a load line.

A class AB push-pull amplifier is shown in Fig. 14-16. The circuit is very similar to that of the class B amplifier of Fig. 14-12. A divider network composed of R_{B1} and R_{B2} is used to bias transistors Q_1 and Q_2. The secondary winding resistance of T_1 is rather low compared to the values of R_{B1} and R_{B2}. Resistors R_{E1} and R_{E2} are used to establish emitter bias. The circuit actually has divider bias and emitter base combined. Emitter biasing is generally used to provide thermal stability for the two transistors. An amplifier connected in this manner is used for high-power audio systems. Low-power versions of this circuit are widely used in small transistor radios.

Figure 14-16. Transformer-coupled class AB P-P amplifier.

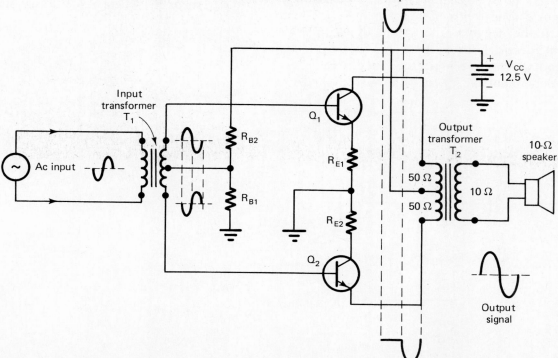

Complementary-Symmetry Amplifiers

The availability of *complementary transistors* has made the design of transformerless power amplifiers possible. Complementary transistors are PNP and NPN types with similar characteristics. Power amplifiers designed with these transistors have an operating efficiency and linearity that is equal to that of a conventional push-pull circuit. Transformerless amplifiers have the high- and low-frequency response of the output extended. Component cost and weight are also reduced with this type of circuitry.

An elementary complementary-symmetry amplifier is shown in Fig. 14-17. This circuit responds as a simple voltage divider. Transistors Q_1 and Q_2 serve as variable resistors with the load resistor (R_L) connected to their common center point. When the positive alternation of the input signal occurs, it is applied to both Q_1 and Q_2. This alternation will cause Q_1 to be conductive and Q_2 to be cut off. Conduction of Q_1 causes it to be low resistant. This connects the load resistor to the $+ V_{CC}$ supply. Capacitor C_2 charges in the direction indicated by the solid-line arrows. Note the direction of the resulting current through the load resistor. The polarity of voltage developed across R_L is the same as the input alternation. This means that the input and output signals are in phase. This is typical of a common-collector or emitter-follower amplifier.

When the negative alternation of the input occurs, it goes to both Q_1 and Q_2. The negative alternation will reverse bias Q_1. This causes it to become nonconductive. Q_2 will, however, become conductive. A positive voltage going to the base of an NPN transistor causes it to become low resistant. The positive charge on C_2 from the first alternation forward biases the emitter of Q_2. With Q_2 low resistant, C_2 will discharge through it to ground. The discharge current of C_2 is shown by the dashed arrows. The resulting current flow through R_L causes the negative alternation to occur. The negative alternation of the input therefore causes the same polarity to appear across R_L. Q_2 is also connected as an emitter follower. In effect, conduction of Q_1 causes C_2 to charge through R_L and conduction of Q_2 causes it to discharge. The applied input is reproduced across R_L.

Biasing for Q_1 and Q_2 is achieved by a voltage-divider network consisting of R_1, R_2, and R_3. Each transistor is biased to approximately 0.6 V above cutoff. This biasing must make the emitter junction of Q_1 and Q_2 one-half of the value of V_{CC} when no signal is applied. The voltage drop across R_2 should be 1.2 V if the two transistors are silicon. The voltage drop across R_2 is quite small compared to that across R_1 and R_3. The voltage across R_1 and R_3 must be of an equal value to set the emitter junction at $V_{CC}/2$. To assure proper current output, the value of R_1 and R_3 is based on transistor beta and R_L. This is determined by the formula

$$R_1 \text{ or } R_3 = \text{beta} \times \text{load resistance}$$

or

$$R_1 = R_3 = \beta \times R_L$$

Figure 14-17. Complementary-symmetry amplifier.

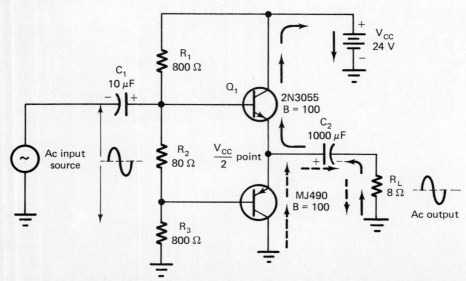

In Fig. 14-17, R_1 and R_3 would be 100×8, or $800\,\Omega$. The current flow through this network is

$$\text{network current } (I_n) = \frac{\text{supply voltage } (V_{CC})}{R_1 + R_3}$$

$$I_n = \frac{V_{CC}}{R_1 + R_3} = \frac{24\text{ V}}{800 + 800} = \frac{24\text{ V}}{1600} = 15\text{ mA}$$

The value of R_2 is then determined by the Ohm's law expression:

$$R_2 = \frac{V_2}{I_n} = \frac{1.2\text{ V}}{15\text{ mA}} = 80\,\Omega$$

V_2 is the voltage value used when Q_1 and Q_2 are silicon transistors. The base–emitter voltage V_{BE} is 0.6 V for each transistor. V_2 is therefore 0.6 V_{BE} plus 0.6 V_{BE} or 1.2 V. The value of V_2 would be 0.4 V if germanium transistors were used in the circuit.

INTEGRATED-CIRCUIT AMPLIFYING SYSTEMS

A number of manufacturers market amplifying systems in integrated-circuit packages. A wide range of different ICs are available today. Some of these are designed to perform one specific system function. This includes such things as preamplifiers, linear signal amplifiers, and power amplifiers. Other ICs may perform as a complete system. Electrical power for operation and only a limited number of external components are needed to complete the system. Low-power audio amplifiers and stereo amplifiers are examples of these devices. Only two of a wide range of IC applications are presented here.

Integrated-circuit power amplifiers are frequently used in sound amplifying systems today. An example of such a circuit is shown in Fig. 14-18. An internal circuit diagram of the IC is shown in part (a). A pin-out diagram of the unit is shown in part (b). This particular chip has a differential amplifier input. Darlington pairs are used to increase each input resistance. Inverting and noninverting inputs are both available. As a general rule, when a signal is applied to one input, the other is grounded. This IC will only respond to a difference in input signal levels. It has a fixed gain of 20 and variable gain capabilities up to 200. Without any external components connected to pins 1 and 8, the device has a

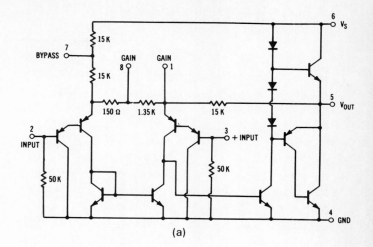

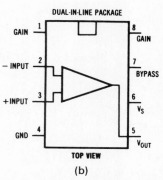

Figure 14-18. Low-voltage audio power amplifier IC: (a) internal circuit diagram; (b) pin diagram. (Courtesy of National Semiconductor Corp.)

gain of 20. A 10-mF capacitor connected between these two pins causes a gain of 200. The output of the unit is a Darlington-pair power amplifier. The supply voltage is automatically set to $V_{CC}/2$ at the output transistor common point.

An application of the IC power amplifier is shown in Fig. 14-19. Pins 1 and 8 have a capacitor-resistor combination. This will cause the gain to be less than the maximum value of 200. A decoupling capacitor is also connected between pin 7 and ground. R_1 serves as a volume control for the circuit. The amplitude level of the input signal is controlled by this potentiometer. The power output of the IC is 500 mW into a 16-Ω load resistor. Applications include portable AM-FM radio amplifiers, portable tape player amplifiers, and TV sound systems.

A dual 6-W audio amplifier is shown in Fig. 14-20. This particular IC offers high-quality performance for stereo phonographs, tape players, recorders, and AM-FM stereo receivers. The internal circuitry of this IC has 52 transistors, 48

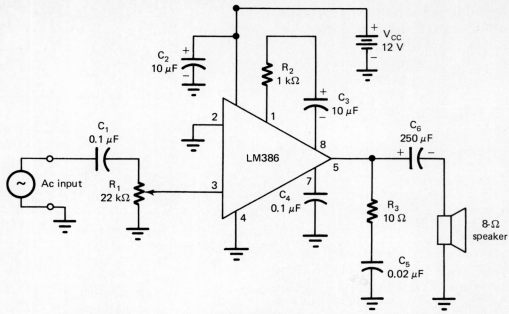

Figure 14-19. IC power amplifier. (Courtesy of National Semiconductor Corp.)

Figure 14-20. LM 379 dual 6-W audio amplifier. (Courtesy of National Semiconductor Corp.)

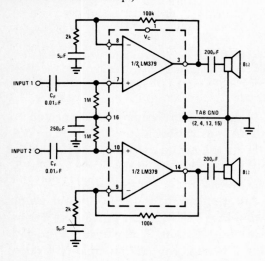

Dual-In-Line Package

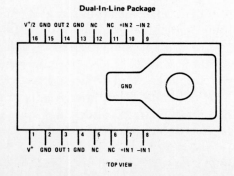

resistors, a zener diode, and two silicon diodes. Two triangle amplifier symbols are used to represent the internal components of this device. This particular amplifier is designed to operate with a minimum of external components. It contains internal bias regulation circuitry for each amplifier section. Overload protection consists of internal current limiting and thermal shutdown circuits.

The LM 379 has gain capabilities of 90 dB per amplifier. The supply voltage can be from 10 to 35 V. Approximately 0.5 A of current is needed for each amplifier. This particular IC must be mounted to a heat sink. A hole is provided in the housing for a bolted connection to the heat sink. A number of companies manufacture amplifying system ICs today. As a general rule, a person should review the technical data developed by these manufacturers before attempting to use a device.

SPEAKERS

The output of an amplifying system is always applied to a *transducer*. A transducer is a device that changes energy of one type into an entirely different form of energy. Work is performed by a transducer when energy conversion occurs. The load device of a system is a transducer. Functionally, the load may be resistive, inductive, and in some cases capacitive.

Most of our applications have used resistive loads. Resistors generally change electrical energy into heat energy. In an audio amplifying system the load device is a speaker. It changes electical energy into sound energy. Variations in current cause the mechanical movement of a stiff paper "cone." Movement of the cone causes alternate compression and decompression of air molecules. This causes sound waves to be set into motion. The human ear responds to these waves.

The operation of a speaker is based on the interaction of two magnetic fields. One field is usually developed by permanent magnet. The second field is electromagnetic. Permanent-magnet (PM) speakers are of this type. The electromagnetic part of the speaker is generally called a *voice coil*. The voice coil is attached to the cone and suspended around the permanent magnet. See the cross-sectional view of a speaker in Fig. 14-21.

When current flows through the voice coil, it produces an electromagnetic field. The polarity of the field (north or south) depends on the direction of current flow. If ac flows in the voice coil, the field varies in both strength and polarity. The power amplifier of a sound system supplies ac to the voice coil. The changing field reacts with the permanent magnetic field. This causes the voice coil to move. With the voice coil attached to the cone, the cone also moves. This action causes air molecules to be set into motion. Sound waves are emitted from the speaker cone.

The frequency of the applied ac signal determines how slow or fast the cone of a speaker responds. Operational frequency is based on the rate or speed change of the electromagnetic field. The loudness of the developed sound wave is based on the moving distance of the cone. This depends on the amount of current supplied to the voice coil by the power amplifier. The primary function of an amplifying system is to develop electrical power to drive a speaker.

When selecting a speaker for a specific application one must take into account a number of considerations. Generally, it takes a large speaker to properly develop low-frequency sounds. Large volumes of air must be set into motion for low-frequency reproduction. Small speakers cannot effectively move enough to produce low tones. Small speakers respond better to high-frequency tones. High-frequency reproduction requires rapid development of air pressure. Small cones can move very rapidly. Large speakers with a big cone and voice coil cannot react quickly enough to produce high-frequency tones. A speaker obviously cannot be large and small at the same time.

In a high-fidelity sound system, at least two speakers are needed to reproduce a typical audio signal. Small speakers are commonly called *tweeters*. The cone of this speaker is generally made of a rather stiff material. Some units employ thin metal cones. Large speakers are called *woofers*. The cone of this speaker is usually quite flexible. Some systems may also employ an intermediate range or *midrange* speaker. Speakers of this type are designed to respond efficiently to frequencies in the center of the audio range. A great majority of the sound being reproduced falls in the range. These speakers are frequently housed in a wooden enclosure.

Figure 14-21. Cross-sectional view of a speaker.

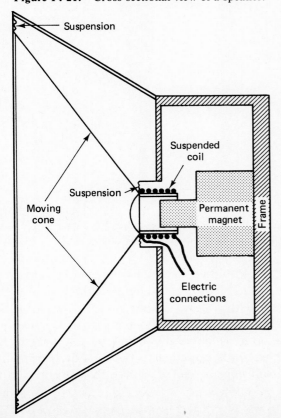

INPUT TRANSDUCERS

The input transducer of an amplifying system is primarily responsible for the development of the signal that is being processed by the system. In audio systems, a transducer is used to change sound energy into electrical signal energy. Systems that respond to frequencies above the human range of hearing are called radio-frequency or RF systems. Electromagnetic waves must be changed into electrical signal energy in this type of system. Radio and television receivers respond to RF signals. The type of input transducer being utilized is dependent on the signal being processed. RF systems pick up signals that travel through the air or through an interconnecting cable network. AF systems develop a signal where a system is being operated. Microphones, tape heads, and phonograph pickup cartridges serve as input transducers.

Microphones

The primary function of a microphone is to change variations in air pressure into electrical energy. These variations in air pressure are called sound waves. The pressure changes of a sound wave are easily converted into voltage current or resistance. Modern microphones generally change sound energy into a changing voltage signal. *Crystal microphones* and *dynamic microphones* are widely used today.

Crystal microphones generate voltage through a mechanical stress placed on a piece of crystal material. This action is called the *piezoelectric effect.* Rochelle salt crystals produce this effect. Synthetic ceramic materials respond in the same way. Both materials respond to changes in pressure by producing a voltage. Sound waves striking the crystal cause it to bend or squeeze together. Metal plates attached to opposite sides of the crystal develop a potential difference in charge (voltage). The output voltage is then routed away from the crystal by connecting wires. Figure 14-22 shows the construction of a simplified crystal microphone. Part (a) shows a sketch of the internal structure. Part (b) shows an assembled cartridge.

Dynamic microphones are considered to be mechanical generators of electrical energy. The mechanical energy of a sound wave causes a small wire coil to move through a magnetic field. This action causes electrons to be set into motion in the coil. A difference in potential charge or voltage is induced in the coil. The resulting output of a dynamic microphone is ac voltage.

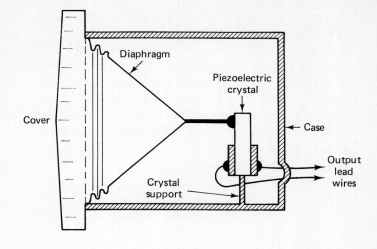

(a)

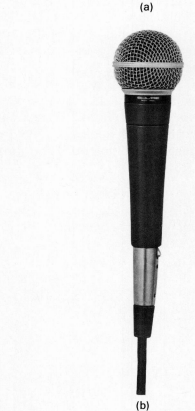

(b)

Figure 14-22. Crystal microphone: (a) construction; (b) assembled. [(b) Courtesy of Shure Brothers Inc.]

A simplified dynamic microphone is shown in Fig. 14-23. Note that a lightweight coil of wire is attached to a cone-shaped piece of material. The cone piece is made of flexible plastic or thin metal. The cone is generally called a diaphragm. The diaphragm is attached to the outside frame of the microphone. Its center is free to move in and out

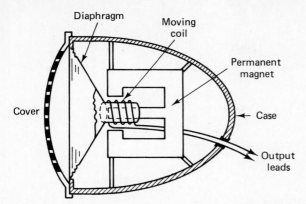

Figure 14-23. Structure of a dynamic microphone.

when sound waves are applied. Sound waves striking the diaphragm cause it and the coil to move through a permanent magnetic field. Through this action sound energy is changed into voltage. Flexible wires attached to the moving coil transport signal voltage out of the microphone. This voltage is applied to the input of a sound system.

Magnetic Tape Input

Magnetic tape is a very common input for many amplifying systems. A long strip of plastic tape is coated with iron oxide. Individual molecules of this material can be easily magnetized. A magnetic tape can therefore store information in the form of minute charged areas.

Figure 14-24 shows how a sound signal is placed on a magnetic tape. This is considered to be the recording function. Essentially, a sound signal is applied to the electromagnetic head. Variations in the applied signal will cause a corresponding change in the electromagnetic field. Tape passing under the air gap of the head will cause oxide molecules to arrange themselves in a specific order. In a sense, a changing magnetic field is transferred to the tape. The recorded signal will remain stored on the tape for a long period of time. Amplifying systems are used to place information on a tape. This is achieved by pushing the record button of a tape machine and speaking into a microphone. The signal is then processed by the amplifier. The output of the system supplies signal current to the electromagnetic recording head. A motor-drive mechanism is needed to move the tape at a selected recording speed. The tape head responds as a load device when the system responds as a recording device.

The recovery of taped information is achieved by a *read head*. This part of the system is very much like a dynamic microphone. However, the electromagnetic coil is stationary in this type of device. Voltage is induced in the coil by a moving magnetic field. As the magnetic tape moves under the read head, voltage is induced in the coil. Changes in the field will cause variations in voltage values. Figure 14-25 shows a simplification of the tape reading function. The tape head responds as an input transducer when the system responds as a playback device.

The head of a tape machine is usually designed to perform both record and playback functions. A record–playback head is shown in Fig. 14-26. The air gap of a tape head is 6 to 9 mils (0.006 to 0.009 in.)

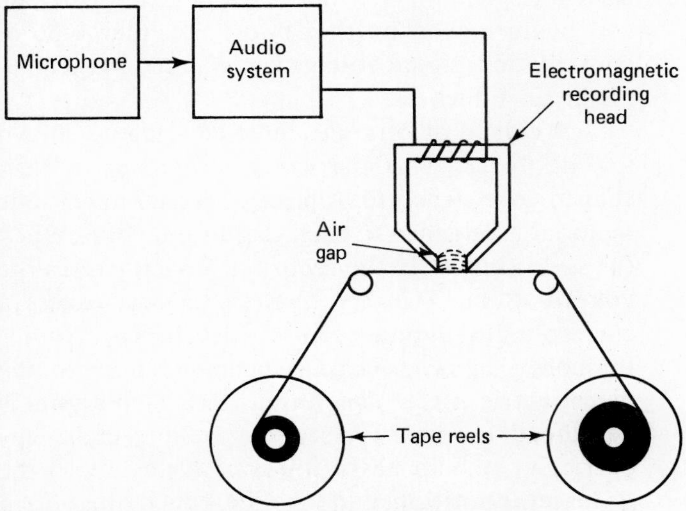

Figure 14-24. Tape recording function.

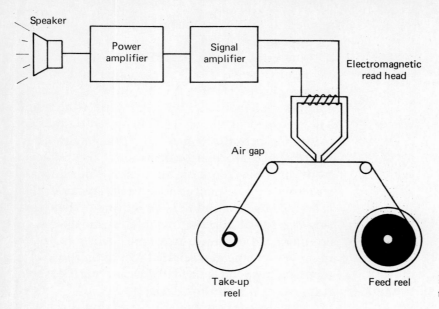

Figure 14-25. Tape reading or playback function.

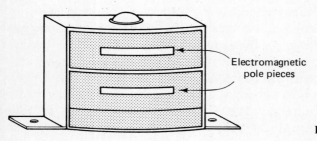

Figure 14-26. Record–playback tape head.

for audio tape recorders. This gap is provided by a thin layer of nonmagnetic material between the poles of an electromagnetic coil. The head must be kept clean and free of foreign material in order to function properly.

Phonograph Pickup Cartridges

The phonograph pickup cartridge is a very important amplifying system input transducer. This device develops an audio signal from the grooves of a record. Phono cartridges are either of the electromagnetic or piezoelectric crystal type. The output of a phono pickup is much greater than that of a microphone. Typical voltage levels are 5 to 10 mV for magnetic devices and 200 to 1000 mV for crystal types. The resulting frequency output can be 15 to 16,000 Hz.

The construction of an electromagnetic phono pickup is shown in Fig. 14-27. This particular cartridge is designed for stereophonic reproduction.

The grooves of a record are cut with two audio signals on opposite sides of a 45° angle. As the *stylus* moves the yoke, voltage is induced in each coil. The yoke is a small permanent magnet core. In other cartridges, the magnet may be stationary with the coil being moved by the stylus. In either type of construction, signal voltage is generated by electromagnetic induction.

A crystal or ceramic stereo cartridge is shown in Fig. 14-28. This particular design has a "Y"-shaped yoke structure. A piece of crystal or ceramic material is attached to each side of the yoke. When the stylus moves in the record groove it causes the yoke to twist. This force on the crystal causes a corresponding voltage to be developed. Audio-frequency signal voltage is generated due to the piezoelectric effect. The output from each crystal is developed by wire connections at one end. This particular pickup has a flip-over stylus. With the stylus lever positioned to the right side, it operates in fine-groove records. Positioned to the left it operates in wide-groove records.

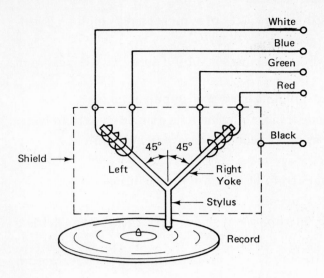

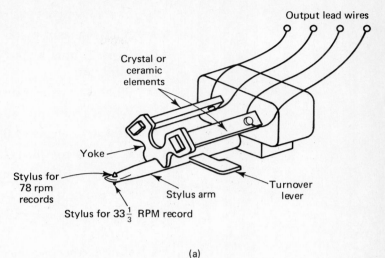

(a)

(b)

Figure 14-27. Electromagnetic cartridges.

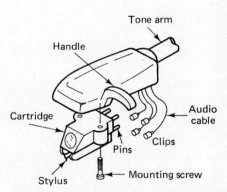

Figure 14-28. Stereo cartridge: (a) crystal or ceramic cartridge structure; (b) assembled phonograph cartridge. [(b) Courtesy of Shure Brothers Inc.]

Review

1. Explain what is included in an amplifying system.

2. If the input of an amplifier is 0.1 V and the output voltage is 10 V, what is the dB gain?

3. If an amplifying system has 0.02 W of input power and develops 20 W of output power, what is the dB gain?

4. If an amplifying system has 0.025 V input and 2.25 V output, how much dB gain is achieved?

5. Why is the value of the capacitance important in capacitance coupling?

6. What is a Darlington amplifier?

7. What is meant by impedance matching in transformer coupling?

8. Why is impedancing matching important in transformer coupling?

9. Explain why the operational efficiency of a single-ended amplifier is low.

10. What is meant by the term "audio frequency"?

11. Explain how a sine-wave input causes output in a class B push-pull amplifier.

12. What is crossover distortion in a class B push-pull amplifier?

13. Why is a push-pull amplifier more efficient than a single-ended amplifier?

14. How does a class AB amplifier reduce crossover distortion?

15. What is a complementary-symmetry amplifier?

16. Why is a speaker considered to be an output transducer?

17. Explain how a speaker functions.

18. Explain how a dynamic microphone changes sound energy into electrical energy.

19. How is the piezoelectric effect used in a microphone to generate an electric signal?

20. What is the primary function of a tape head?

21. How does a dynamic phonograph cartridge change a record groove vibration into electrical signal energy?

Student Activities

Capacitor-Coupled Amplifier

1. Construct the two-stage capacitor-coupled amplifier of Fig. 14-3. The transistors should be of the silicon NPN type with at least 250-mW power dissipation. A type 2N3397 works fairly well.

2. Measure the dc bias voltages at emitter, base, and collector of each transistor.

3. Apply an ac signal of 400 Hz to the input. Adjust the signal level to produce the greatest output with a minimum of distortion. Observe this on an oscilloscope.

4. Determine the voltage gain of each stage and the combined system gain.

Direct-Coupled Amplifier

1. Construct the two-stage direct-coupled amplifier of Fig. 14-4. The transistors should be of the silicon NPN type with at least 250 mW power dissipation. A type 2N3397 could be used.

2. Measure and record the dc emitter, base, and collector voltage of each transistor.

3. Apply an ac signal of 1 kHz to the input. With an oscilloscope observe the output and input waveforms. Adjust the signal level to produce the greatest output with a minimum of distortion.

4. Determine the ac gain using the measured input and output voltages. Determine the gain of each transistor and the overall system gain.

Darlington Transistor Amplifier

1. Construct the Darlington amplifier of Fig. 14-5. A GE-62 amplifier can be used.

2. Apply a small signal to the input. Something less than 1 mV should be used.

3. With an oscilloscope, observe the output signal. Adjust the signal level to produce the greatest output with a minimum of distortion.

4. Measure the input and output signal levels. Calculate the ac voltage gain. Calculate the dB gain.

5. What is the phase relationship of the input–output signals?

Signal-Ended Power Amplifier

1. Construct the single-ended power amplifier of Fig. 14-10. Use a 2N3053 transistor or a suitable equivalent. A 10:1-ratio filament transformer could be used for T_1 if a suitable audio output transformer is not available.

2. Measure and record the emitter, base, and collector bias voltages.

3. Apply an ac signal to the input. Adjust the output across the speaker to the greatest signal level with a minimum of distortion.

4. If a crystal microphone or phonograph input is available, connect it to the amplifier input.

5. Test the operation of the amplifier.

6. Observe the input and output signals with an oscilloscope.

7. Determine the signal voltage gain and dB gain.

Push-Pull Amplifier

1. Construct the push-pull amplifier of Fig. 14-16. Transistors Q_1 and Q_2 are of the NPN silicon type. A 2N3053 or its equivalent can be used. R_{E1} and R_{E2} are 1-Ω resistors. R_{B1} and R_{B2} are selected to produce +0.6 V at the base of Q_1 and Q_2. Experiment with different values. An R_{B1} of 500 Ω and R_{B2} of 15 Ω are some representative values. T_2 should have a 2:1 impedance ratio.

2. After proper bias voltage selection for Q_1 and Q_2, test circuit operation. Measure and record the voltage of the emitter, base, and collector for both transistors.

3. Apply an ac signal to the input of T_1. Adjust the signal level to produce an undistorted output across the secondary winding of T_2. Proper base bias voltage is essential for undistorted output.

4. If undistorted output is achieved, observe the input signals to Q_1 and Q_2. Note the signal level and phase of each input. Observe the output signal at each collector. Note the amplitude and phase of each signal. Does this amplifier have crossover distortion?

5. Calculate the voltage gain in dB. Assume that the input and output impedances are equal.

IC Power Amplifier

1. Construct the IC power amplifier of Fig. 14-19.

2. Connect a 400-Hz ac signal to the input. Use an audio signal generator.

3. With an oscilloscope, observe the output signal across the speaker.

4. Adjust R_1 through its range. Does the output become distorted?

5. Measure the input signal at pin 3. Then measure the output signal at pin 5. Calculate the voltage gain in dB for low-, medium-, and high-output signal levels.

6. Test the frequency response of the amplifier. What is the low-frequency limit where the signal output drops 30% in level?

Complementary-Symmetry Amplifier

1. Connect the direct-coupled complementary-symmetry amplifier of Fig. 14-29.

Figure 14-29. Direct-coupled complementary-symmetry amplifying system.

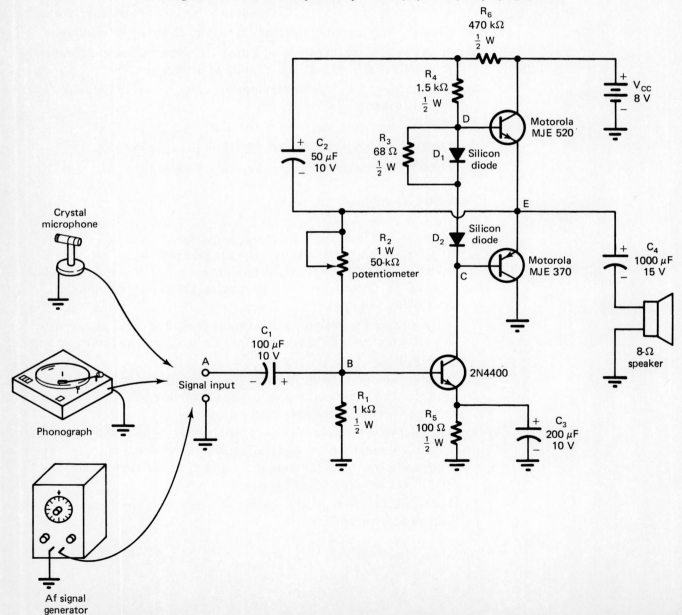

2. Adjust R_2 to $V_{CC}/2$ at the transistor common point.

3. Connect an audio signal into the amplifier. Test the circuit for operation.

4. With an oscilloscope, trace the signal path through the circuit starting at test point A and finishing at point F. Note any change in signal level or shape.

5. Test the operation of the circuit with a phonograph or crystal microphone input.

6. Measure the emitter, base, and collector voltages appearing at each transistor.

7. Measure the voltage across D_1 and D_2.

15

Oscillators

Many electronic systems employ circuits that convert the dc energy of a power supply into a useful form of ac. Oscillators, generators, and electronic clocks are typical circuits. In a radio receiver, dc is converted into high-frequency ac to achieve signal tuning. Television receivers also have an oscillator in their tuner. In addition to the tuner, oscillators are used to produce horizontal and vertical sweep signals. These signals control the electron beam of the picture tube. In the same regard, calculators and computers employ an electronic clock circuit. Timing pulses are produced by this circuit. Every radio-frequency transmitter employs an oscillator. This part of the system generates the signal that is radiated into space. Oscillators are extremely important in the electronic field.

An oscillator generally uses an amplifying device to aid in the generation of the ac output signal. Transistors, vacuum tubes, and ICs can all be used for this function. Transistors and ICs are widely used today. Vacuum tubes are usually found in older circuit applications. In this chapter we concentrate on solid-state oscillator circuits. Some vacuum-tube circuits will be presented for comparison purposes. As a general rule, vacuum tubes can be used interchangeably for transistors without a great deal of circuit modification.

IMPORTANT TERMS

In a discussion of oscillator circuits, one frequently encounters a number of new and unusual terms. These terms usually play a rather important part of the presentation. The chapter is much more meaningful if there is some understanding of the terms. Study the following definitions before proceeding with the chapter.

Abstable multivibrator. A free-running generator that develops a continuous square-wave output.

Blocking oscillator. An oscillator circuit that drives an active device into cutoff or blocks its conduction for a certain period of time.

Capacitor divider. A network of series-connected capacitors, each having a voltage drop of ac.

Comparator. An amplifier that has a reference signal input and a test signal input. When the test signal differs in value from the reference signal, a state change in the output occurs.

Continuous wave (CW). Uninterrupted sine waves, usually of the radio-frequency type, that are radiated into space by a transmitter.

Damped wave. A wave in which successive oscillations decrease in amplitude.

Differentiator network. A resistor–capacitor combination where the output appears across the resistor. An applied square wave will produce a positive and negative spiked output.

Electromagnetic field. The space around an inductor that changes due to current flow through a coil. The field expands and collapses with applied ac.

Electrostatic field. The space or area around a charged body where there is some reaction to the developed charge.

Feedback. Transferring voltage from the output of a circuit back to its input.

Ferrite core. An inductor core made of molded iron particles.

Flip-flop. A multivibrator circuit having two stable states that can be switched back and forth.

Free running. An oscillator circuit that develops a continuous waveform that is not stabilized, such as an astable multivibrator.

Monostable. A multivibrator with one stable state. It changes to the other state momentarily and then returns to its stable state.

Nonsinusodial. Waveforms that are not sine waves, such as square, sawtooth, and pulsing waves.

Radio-frequency choke (RFC). An inductor or coil that offers high impedance to radio-frequency ac.

Regenerative feedback. Feedback from the output to the input that is in phase so that it is additive.

Relaxation oscillator. A nonsinusodial oscillator that has a resting or nonconductive period during its operation.

Resonant frequency. The frequency at which a tuned circuit oscillates.

Saturation. An active device operational region where the output current levels off to a constant value.

Stability. The ability of an oscillator to stay at a given frequency or given condition of operation without variation.

Symmetrical. A condition of balance where parts, shapes, and sizes of two or more items are the same.

Synchronized. A condition of operation that occurs at the same time. The vertical and horizontal sweep of an oscilloscope are often synchronized or placed in sync.

Tank circuit. A parallel resonant *LC* circuit.

Threshold. The beginning or entering point of an operating condition. A terminal connection of the LM555 IC timer.

Time constant (*RC*). The period required for voltage of a capacitor to increase 63.2% of the maximum value or decrease to 36.7% of the maximum value.

Toggle. A switching condition that changes back and forth.

Triggered. A control technique that causes a device to change its operational state, such as a triggered monostable multivibrator.

Vertical blocking oscillator. A TV circuit that generates the vertical sweep signal for deflection of the cathode-ray tube or picture tube.

OSCILLATOR TYPES

There are several possible ways of describing an oscillator. This can be done according to generated frequency, operational stability, power output, signal waveforms, and components. In this chapter it is convenient to divide oscillators according to their method of operation. This includes feedback oscillator and relaxation oscillators. Each group has a number of distinguishing features.

In *feedback oscillators* a portion of the output power is returned to the input circuit. This type of oscillator usually employs a tuned *LC* circuit. The operating frequency of the oscillator is established by this circuit. Most sine-wave oscillators are of this type. The frequency range is from a few hertz to millions of hertz. Applications of the feedback oscillator are widely used in radio and television receiver tuning circuits and in transmitters. Feedback oscillators respond very well to radio-frequency (RF) generator applications.

Relaxation oscillators respond to an electronic device that goes into conduction for a certain time, then turns off for a period of time. This condition of operation repeats itself on a continuous basis. This oscillator usually responds to the charge and discharge of an *RC* or an *RL* network. Oscillators of this type usually generate square- or triangular-shaped waves. The active device of the oscillator is "triggered" into conduction by a change in voltage. Applications include the vertical and horizontal sweep generators of a television receiver and computer clock circuits. Relaxation oscillators respond extremely well to low-frequency applications.

FEEDBACK OSCILLATORS

Feedback is a process by which a portion of the output signal of a circuit is returned to the input. This function is an important characteristic of any oscillator circuit. Feedback may be accomplished by inductance, capacitance, or resistance coupling. A variety of different circuit techniques have been developed today. As a general rule, each technique has certain characteristics that distinguishes it from others. This accounts for the large number of different circuit variations.

If you have ever been to a public gathering where a sound system was being used, you have probably had an opportunity to experience feedback. If a microphone is placed too close to a speaker, feedback can occur. Sound from the speaker is picked up by the microphone and returned to the amplifier. It then comes out of the speaker and is returned again to the microphone. The process continues until a hum or high-pitched howl is produced. Figure 15-1 shows an example of sound system feedback. This condition is considered to be mechanical feedback. In sound system operation, feedback is an undesirable feature. In an oscillator circuit operational feedback is a necessity.

Oscillator Fundamentals

An oscillator is an electronic circuit that generates a continuously repetitive output signal. The output signal may be alternating current or some

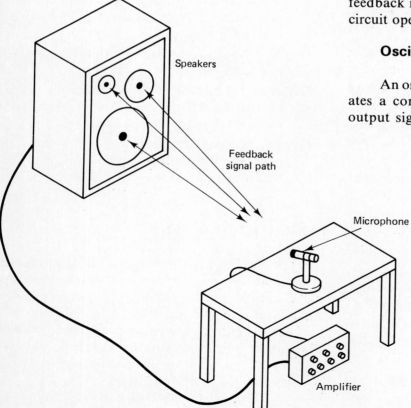

Speakers

Feedback
signal path

Microphone

Amplifier

Figure 15-1. Sound system feedback.

form of changing dc. The unique feature of an oscillator is its generation function. It does not receive an external input signal. It develops the resulting output signal from the dc operating power provided by the power supply. An oscillator in a sense is an electronic power converter. It changes dc operating power into ac or some useful form of changing dc. A block diagram of a feedback oscillator is shown in Fig. 15-2.

An oscillator has an amplifying device, a feedback network, frequency-determining components, and a dc power source. The amplifier is used primarily to build up the output signal to a usable level. Any device that has signal gain capabilities can be used for this function. The feedback network can be inductive, capacitive, or resistive. It is responsible for returning a portion of the amplifier's output back to the input. The feedback signal must be of the correct phase and value in order to cause oscillations to occur. In-phase or regenerative feedback is essential in an oscillator. An inductor–capacitor or *LC* network determines the frequency of the oscillator. The charge and discharge action of this network establishes the oscillating voltage. This signal is then applied to the input of the amplifier. In a sense, the *LC* network is energized by the feedback signal. This energy is needed to overcome the internal resistance of the *LC* network. With suitable feedback, a continuous ac signal can be generated. The output of a good oscillator must be uniform. It should not vary in frequency or amplitude.

LC Circuit Operation

The operating frequency of a feedback oscillator is usually determined by an inductance–capacitance network. An *LC* network is sometimes called a *tank circuit*. The tank, in this case, has a storage capability. It will store an ac voltage that occurs at its *resonant* frequency. This voltage can be stored in the tank for a short period of time. In an oscillator, the tank circuit is responsible for the frequency of the ac voltage that is being generated.

In order to see how ac is produced from dc, let us take a look at an *LC* tank circuit. Figure 15-3(a)

Figure 15-3. *LC* tank circuit being charged: (a) basic circuit; (b) charging action; (c) charged capacitor.

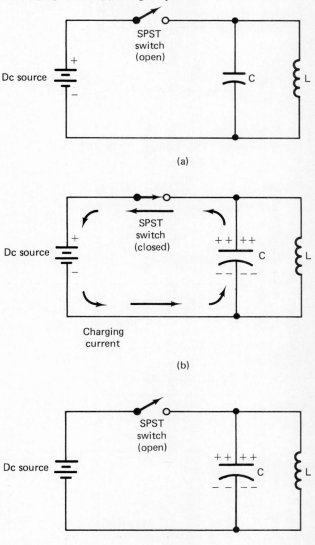

(a)

(b)

(c)

Figure 15-2. Fundamental oscillator parts.

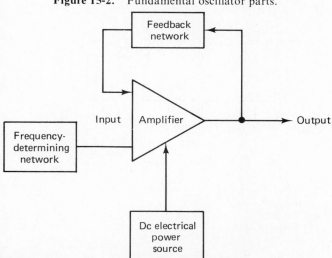

shows a simple tank circuit connected to a battery. In Fig. 15-3(b), the switch is closed momentarily. This action causes the capacitor to charge to the value of the battery voltage. Note the direction of the charging current. After a short time the switch is opened. Figure 15-3(c) shows the accumulated charge voltage on the capacitor.

We will now see how a tank circuit develops a sine-wave voltage. Assume now that the capacitor of Fig. 15-4(a) has been charged to a desired voltage value by momentarily connecting it to a battery. Figure 15-4(b) shows the capacitor discharging through the inductor. Discharge current flowing through *L* causes an electromagnetic field to expand

Figure 15-4. Charge–discharge action of an *LC* circuit.

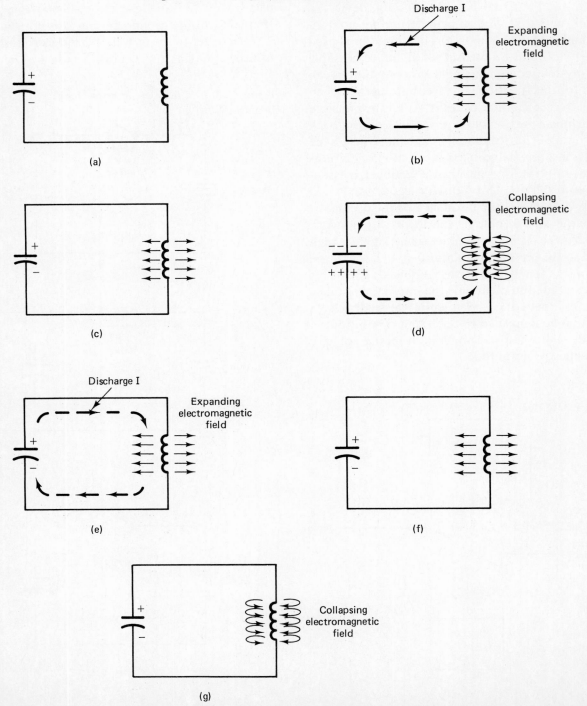

around the inductor. Figure 15-4(c) shows the capacitor charge depleted. The field around the inductor then collapses. This causes a continuation of current flow for a short time. Figure 15-4(d) shows the capacitor being charged by the induced voltage of the collapsing field. The capacitor once again begins to discharge through *L*. In Fig. 15-4(e), note that the direction of the discharge current is reversed. The electromagnetic field again expands around *L*. Its polarity is reversed. Figure 15-4(f) shows the capacitor with its charge depleted. The field around *L* collapses at this time. This causes a continuation of current flow. Figure 15-4(g) shows *C* being charged by the induced voltage of the collapsing field. This causes *C* to be charged to the same polarity as the beginning step. The process is then repeated. The charge and discharge of *C* through *L* causes an ac voltage to occur in the circuit.

The frequency of the ac voltage generated by a tank circuit is based on the value of *L* and *C*. This is called the resonant frequency of a tank circuit. The

formula for resonant frequency is

$$\text{frequency } (f) = \frac{1}{2\pi\sqrt{LC}}$$

where *f* is in hertz, *L* is the inductance in henries, and *C* is the capacitance in farads. This formula is extremely important because it applies to all *LC* sine-wave oscillators. Resonance occurs when the inductive reactance (X_L) equals the capacitive reactance (X_C). A tank circuit will oscillate at this frequency for a rather long period of time without a great deal of circuit opposition.

At resonant frequency an *LC* tank always has some circuit resistance. This resistance tends to oppose the ac that circulates through the circuit. The ac voltage will therefore decrease in amplitude after a few cycles of operation. Figure 15-5(a) shows the resulting wave of a tank circuit. Note the gradual decrease in signal amplitude. This is called *damped sine wave*. A tank circuit produces a wave of this type.

Figure 15-5. Wave types: (a) damped oscillatory wave; (b) continuous wave.

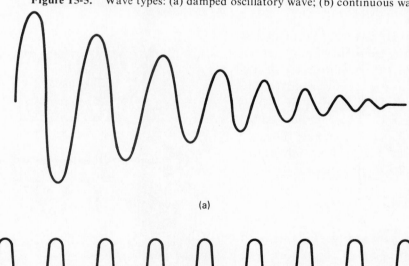

(a)

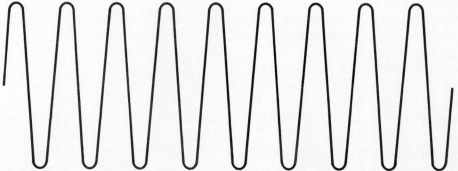

(b)

A tank circuit, as we have seen, generates an ac signal when it is energized with dc. The generated wave gradually diminishes in amplitude. After a few cycles of operation, the lost energy is transformed into heat. When the lost energy is great enough, the signal stops oscillating.

The oscillations of a tank circuit could be made continuous if energy were periodically added to the circuit. The added energy would restore that lost by the circuit through heat. The added energy must be in step with the circulating tank energy. If the two are additive, the circulating signal will have a continuous amplitude. Figure 15-5(b) shows the continuous wave (CW) of an energized tank circuit.

Our original tank circuit was energized by momentarily connecting a dc source to the capacitor. This action caused the capacitor to become charged by manual switching action. To keep a tank circuit oscillating by manual switch operation is physically impossible. Switching of this type is normally achieved by an electronic device. Transistors and vacuum tubes are commonly used to perform this operation in an oscillator circuit. The output of the active device is applied to the tank circuit in proper step with the circulating energy. When this is achieved the oscillator will generate a CW signal of a fixed frequency.

The inductance or coil of a tank circuit varies a great deal depending on the frequency being generated. Most applications of *LC* oscillators are in the radio-frequency range. Some representative RF oscillator coils are shown in Fig. 15-6. Coil inductance is usually a variable. Inductance is changed by moving the position of a *ferrite core* inside the coil. Adjustment of the inductance will permit some change in the frequency of the tank circuit.

Armstrong Oscillator

The Armstrong oscillator of Fig. 15-7 is used to show the basic operation of an *LC* oscillator. The characteristic curve of the transistor and its ac load line are also used to explain its operation. Note that conventional bias voltages are applied to the transistor. The emitter–base junction is forward biased and the collector is reverse biased. Emitter biasing is achieved by R_3. R_1 and R_2 serve as a divider network for the base voltage.

When power is first applied to the transistor, R_1 and R_2 establish the operating point (Q) near the

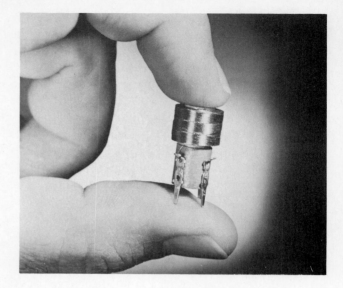

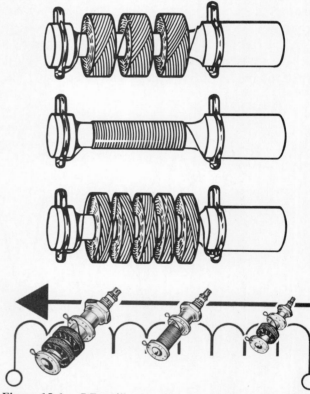

Figure 15-6. *RF* oscillator coils. (Courtesy of J. W. Miller Div./Bell Industries.)

center of load line. The output of the transistor, at the collector, should ideally be 0 V ac. Due to the initial surge of current at turn on time, some noise will immediately appear at the collector. This is usually of a very minute value. For our discussion, let us assume that a −1-mV signal appears at the

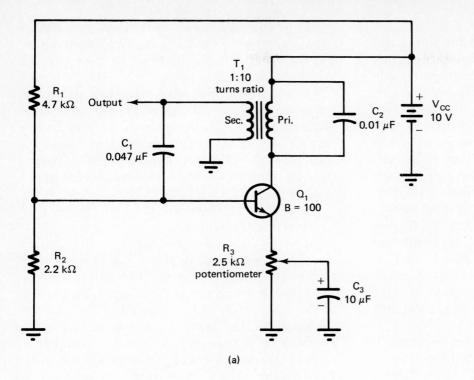

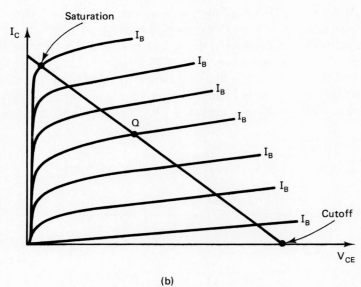

(b)

Figure 15-7. Armstrong oscillator: (a) circuit; (b) characteristic curves.

collector. The transformer will invert this voltage and step it down by a factor of 10. T_1 of the circuit has a 10:1 turns ratio. A +0.1-mV signal is therefore applied to C_1 in the base circuit.

It should be noted at this time that the transistor has a beta of 100. With +0.1 V applied to the base, Q_1 will develop a −10-mV output signal at the collector. The polarity change from + to − in the output signal is due to the inverting characteristic of a common-emitter amplifier. The output signal voltage is again stepped down by the transformer and applied to the base of Q_1. A collector signal of −10 mV will now cause a +1.0-mV base voltage. With transistor gain, the collector voltage will immediately rise to −100 mV. The process will continue, thus producing collector voltages of −1.0 V and finally −10 V. At this point, transistor operation will be driven up the load line until saturation is reached.

Note this point on the load line. The collector voltage of Q_1 will not change beyond this point.

With no change in V_C across the primary winding of T_1, the secondary voltage drops immediately to zero. The base voltage returns immediately to the Q point. This decrease or negative-going base voltage (from saturation to point Q) causes V_C to be positive going. Through transformer action this appears as a negative-going base voltage. This action continues to drive the transistor past point Q. The process continues until the cutoff point is reached. The transformer then stops supplying input voltage to the base. Transistor operation swings immediately in the opposite direction. R_1 and R_2 cause the base voltage to again rise to point Q. The process is then repeated. Q_1 goes to saturation, to point Q, to cutoff, and returns again to point Q. Operation is continuous. An ac voltage appears across the secondary of T_1.

The frequency of the Armstrong oscillator is based on the value of C_1 and S. S is the inductance of the transformer secondary winding. The frequency of the generated wave is determined by the LC resonant frequency formula. C_1 and S forms a tank circuit with the emitter–base junction of Q_1 and R_3 included.

The output of the Armstrong oscillator of Fig. 15-7 can be changed by altering the value of R_3. Gain is highest when the variable arm of R_3 is at the top of its range. This adjustment may cause a great deal of signal distortion. In some cases, the output voltage may appear as a square wave instead of a sine wave. When the position of the variable arm is moved downward, the output wave becomes a sine wave. Adjusting R_3 to the bottom of its range may cause oscillations to stop. The gain of Q_1 may not be capable of developing enough feedback for regeneration at this point. Normally 20 to 30% of output must be returned to the base to assure continuous oscillation.

Hartley Oscillator

The Hartley oscillator of Fig. 15-8 is used rather extensively in AM and FM radio receivers. The resonant frequency of the circuit is determined by the values of T_1 and C_1. Capacitor C_2 is used to couple ac to the base of Q_1. Biasing for Q_1 is provided by R_2 and R_1. Capacitor C_4 couples ac variations in collector voltage to the lower side of T_1. The *radio-frequency choke* coil (L_1) prevents ac from going into the power supply. L_1 also serves as the load for the circuit. Q_1 is an NPN transistor connected in a common-emitter circuit configuration.

When dc power is applied to the circuit current flows from the negative side of the source through R_1 to the emitter. The collector and base are both

Figure 15-8. Hartley oscillator.

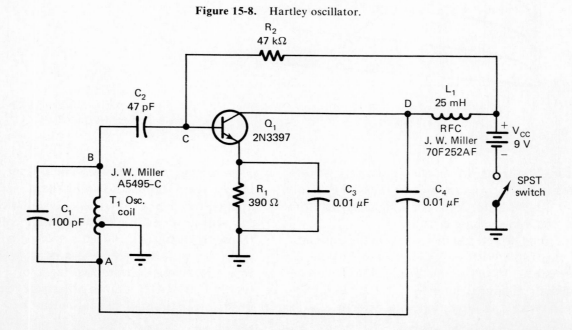

connected to the positive side of V_{CC}. This forward biases the emitter–base junction and reverse biases the collector. I_E, I_B, and I_C flow initially through Q_1. With I_C flowing through L_1, the collector voltage drops in value. This negative-going voltage is applied to the lower side of T_1 by capacitor C_4. This causes current flow in the lower coil. An electromagnetic field expands around the coil. This in turn cuts across the upper part of the coil. By transformer action, the top of the upper coil swings positive. Capacitor C_1 is charged by this voltage. This same voltage is also added to the forward bias voltage of Q_1 through C_2. Q_1 is eventually driven into saturation. When saturation occurs, there is no change in the value of V_C. The field around the lower part of T_1 collapses immediately. This causes a change in the polarity of the voltage at the top of T_1. The top plate of C_1 now becomes negative and the bottom positive.

The accumulated charge on C_1 will begin to discharge immediately through T_1 by normal tank circuit action. This negative voltage at the top of C_1 causes the base of Q_1 to swing negative. Q_1 is driven to cutoff. This in turn causes the V_C of Q_1 to rise very quickly. A positive-going voltage is then transferred to the lower part of T_1 by C_4. This adds to the voltage of C_1. C_1 continues to discharge. The value change in V_C eventually stops and no voltage is fed back through C_4. C_1 is fully discharged by this time. The field around the lower side of L_1 then collapses. C_1

charges again, with the bottom side being positive and the top negative. Q_1 again becomes conductive. The process is repeated on a continuous basis. The tank circuit produces a continuous wave. The circuit losses of the tank are restored by regenerative feedback.

A distinguishing feature of the Hartley oscillator is its tapped coil. A number of circuit variations are possible. The coil may be placed in series with the collector. Collector current flows through the coil in normal operation. A variation of this type is called a series-fed Hartley oscillator. The circuit of Fig. 15-8 is a shunt-fed Hartley oscillator. I_C does not flow through T_1. The coil is connected in parallel or shunt with the dc voltage source. Only ac flows through the lower part of T_1. Shunt-fed Hartley oscillators tend to produce more stable output.

Colpitts Oscillator

A Colpitts oscillator is very similar to the shunt-fed Hartley oscillator. The primary difference is in the tank circuit structure. A Colpitts oscillator uses two capacitors instead of a divided coil. Feedback is developed by an *electrostatic field* across the capacitor divider network. Frequency is determined by two capacitors in series and the inductor.

Figure 15-9 shows a schematic of the Colpitts oscillator. Bias voltage for the base is provided by

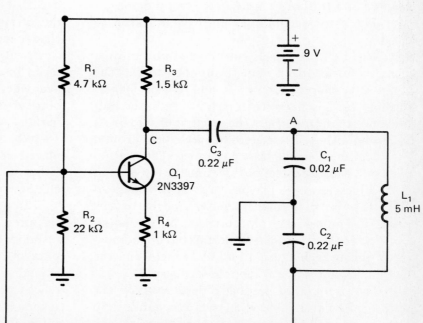

Figure 15-9. Colpitts oscillator.

resistors R_1 and R_2. The emitter is biased by R_4. The collector is reverse biased by connection to the positive side of V_{CC} through R_3. This resistor also serves as the collector load. The transistor is connected in a common-emitter circuit configuration.

When dc power is applied to the circuit, current flows from the negative side of V_{CC} through R_4, Q_1, and R_3. I_C flowing through R_3 causes a drop in the positive value of V_C. This negative-going voltage is applied to the top plate of C_1 through C_3. With C_1 and C_2 connected in series, the bottom plate of C_2 takes on a positive charge. This adds to the positive voltage of the base, which in turn causes an increase in I_B. The conduction of Q_1 increases. The process continues until Q_1 is saturated.

When Q_1 becomes saturated there is no further increase in I_C. The changing value of V_C also stops. There is no feedback to the top side of C_1. The composite charge across C_1 and C_2 discharges through L_1. Discharge current flow through L_1 is from the top plate of C_1 to the bottom of C_2. The positive charge on C_2 soon diminishes. The electromagnetic field around L_1 collapses. The discharge current continues for a short time. The bottom plate of C_2 now becomes negatively charged and the top plate of C_1 becomes positive. The negative charge voltage of C_2 reduces the forward bias voltage of Q_1. I_C decreases in value. V_C begins to increase in value. This change in voltage is again fed back to the top plate of C_1 through C_3. C_1 becomes more positively charged and the bottom of C_2 becomes more negative. The process continues until Q_1 is driven to cutoff.

When Q_1 reaches its cutoff point no I_C flows. There is no feedback voltage supplied to C_1. The composite charge across C_1 and C_2 discharges through L_1. Discharge current flows from the bottom of C_2 to the top of C_1. The negative charge on C_2 soon diminishes. The electromagnetic field around L_1 collapses. Current flow continues. The bottom plate of C_2 now becomes positive and the top plate of C_1 becomes negative. The positive charge voltage of C_2 pulls Q_1 out of the cutoff region. I_C begins to flow again. The process then repeats itself from this point. Feedback energy is added to the tank circuit momentarily during each alternation. As a general rule the Colpitts circuit is a very reliable oscillator.

The amount of feedback developed by the Colpitts oscillator is based on the capacitance ratio of C_1 and C_2. The value of C_1 in our circuit is much smaller than C_2. The capacitive reactance (X_C) of C_1 will therefore be much greater for C_1 than for C_2. The voltage across C_1 will be much greater than that across C_2. By making the value of C_2 smaller, the feedback voltage can be increased. As a general rule, large amounts of feedback may cause the generated wave to be distorted. Small values of feedback will not permit the circuit to oscillate. In practice, 10 to 50% of the collector voltage is returned to the tank circuit as feedback energy.

Crystal Oscillator

When extremely high frequency stability is desired, crystal oscillators are used. The crystal of an oscillator is a finely ground wafer of quartz or Rochelle salt. It has the property to change electrical energy into vibrations of mechanical energy. It can also change mechanical vibrations into electrical energy. These two conditions of operation are called the piezoelectrical effect.

The crystal of an oscillator is placed between two metal plates. Contact is made to each surface of the crystal by these plates. The entire assembly is then mounted in a holding case. Connection to each plate is made through terminal posts. When a crystal is placed in a circuit, it is plugged into a socket.

A crystal by itself behaves like a series resonant circuit. It has inductance (L), capacitance (C), and resistance (R). Figure 15-10(a) shows an equivalent circuit. The L function is determined by the mass of the crystal. Capacitance deals with its ability to change mechanically. Resistance corresponds to the electrical equivalent of mechanical friction. The series resonant equivalent circuit changes a great deal when the crystal is placed in a holder. Capacitance due to the metal support plates is added in parallel with the crystal equivalent circuit. Figure 15-10(b) shows the equivalent circuit of a crystal placed in a holder. In a sense, a crystal can have either a series resonant or a parallel resonant characteristic.

In an oscillator, the placement of a crystal in the circuit has a great deal to do with how it will respond. If placed in the tank circuit, it responds as a parallel resonant device. In some applications the crystal serves as a tank circuit. When a crystal is placed in the feedback path it responds as a series resonant device. It essentially responds as a sharp filter. It only permits feedback of the desired fre-

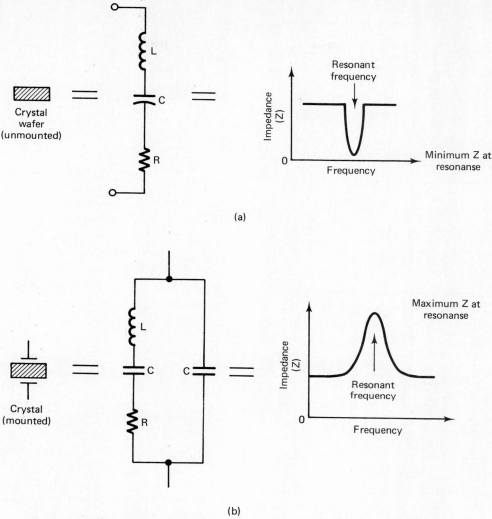

Figure 15-10. Crystal equivalent circuits: (a) series resonant; (b) parallel resonant.

quency. Hartley and Colpitts can be modified to accommodate a crystal of this type. The operating stability of this type of oscillator is much better than the circuit without a crystal. Figure 15-11 shows crystal-controlled versions of the Hartley and Colpitts. Operation is essentially the same.

Pierce Oscillator

The Pierce oscillator of Fig. 15-12 employs a crystal as its tank circuit. In this circuit the crystal responds as a parallel resonant circuit. In a sense, the Pierce oscillator is a modification of the basic Colpitts oscillator. The crystal is used in place of the tank circuit inductor. A specific crystal is selected for

the desired frequency to be generated. The parallel resonant frequency of the crystal is slightly higher than its equivalent series resonant frequency.

Operation of the Pierce oscillator is based on feedback from the collector to the base through C_1 and C_2. These two capacitors provide a combined 180° phase shift. The output of the common-emitter amplifier is therefore inverted to achieve in-phase or regenerative feedback. The value ratio of C_1 and C_2 determines the level of feedback voltage. From 10 to 50% of the output must be fed back to energize the crystal. When it is properly energized, the resonant frequency response of the crystal is extremely sharp. It will vibrate only over a narrow range of frequency. The output at this frequency is very stable. The

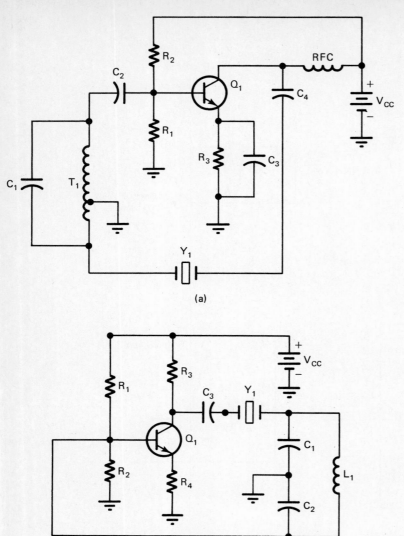

(a)

(b)

Figure 15-11. Crystal-controlled oscillators: (a) Hartley; (b) Colpitts.

Figure 15-12. Pierce oscillator.

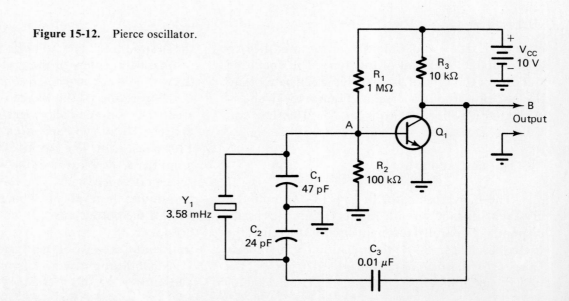

nonconduction of the active device regulates the charge and discharge of the *RC* network. Circuits connected in this manner are classified as relaxation oscillators. When the circuit device is conductive, it is active. When it is not conductive, it is relaxed. The active device of a relaxation oscillator switches states between conduction and nonconduction. This regulates the charge and discharge rate of the capacitor. A sawtooth waveform appears across the capacitor of this type of oscillator.

A unijunction transistor (UJT) is used in the relaxation oscillator of Fig. 15-16. The *RC* network is composed of R_1 and C_1. The junction of the network is connected to the emitter (*E*) of the UJT. The UJT will not go into conduction until a certain voltage value is reached. When conduction occurs the emitter–base 1 junction becomes low resistant. This provides a low-resistant discharge path for *C*. Current flows through R_3 only when the UJT is conducting. R_3 refers to the resistance of the speaker in this circuit.

Assume now that power is applied to the UJT oscillator circuit. The UJT is in a nonconductive state initially. The emitter–B_1 junction is reverse biased. C_1 begins to accumulate charge voltage after a short period of time. Time equals $R \times C$ in this case. The capacitor eventually charges to a voltage value that will cause the E–B_1 junction to become conductive. When this point is reached the E–B_1 junction becomes low resistant. C_1 discharges immediately through the low-resistant E–B_1 junction. This action removes the forward bias voltage from the emitter. The UJT immediately becomes noncon-

ductive. C_1 begins to charge again through R_1. The process is then repeated on a continuous basis.

UJT oscillators are found in applications that require a signal with a slow rise time and a rapid fall time. The E–B_1 junction of the UJT has this type of output. Between B_1 and circuit ground the UJT produces a spiked pulse. This type of output is frequently used in timing circuits and counting applications. The time that it takes for the waveform to repeat itself is called the *pulse repetition rate* (prr). This term is very similar to the hertz designation for frequency. As a general rule, UJT oscillators are very stable and are quite accurate when used with time constants of one or less.

Astable Multivibrator

Multivibrators are a very important classification of the relaxation oscillator. This type of circuit employs an *RC* network in its physical makeup. A rectangular-shaped wave is developed by the output. Astable multivibrators are commonly used in television receivers to control the electron beam deflection of the picture tube. Computers use this type of generator to develop timing pulses.

Multivibrators are considered to be either *triggered* devices or *free running*. A triggered multivibrator requires an input signal or timing pulse to be made operational. The output of this multivibrator is controlled or *synchronized* by an input signal. The electron beam deflection oscillators of a television receiver are triggered into operation. When the receiver is tuned to an operationing channel, its oscillators are synchronized by the incoming signal. When it is tuned to an unused channel the oscillators are free running. Free-running oscillators are self-starting. They operate continuously as long as electrical power is supplied. The shape and frequency of the waveform is determined by component selection. An astable multivibrator is a free-running oscillator.

A multivibrator is composed of two amplifiers that are cross-coupled. The output of amplifier 1 goes to the input of amplifier 2. The output of amplifier 2 is coupled back to the input of amplifier 1. Since each amplifier inverts the polarity of the input signal, the combined effect is a positive feedback signal. With positive feedback an oscillator is regenerative and produces continuous output.

Figure 15-17 shows a representative multivibrator using bipolar transistors. These amplifiers are

Figure 15-16. UJT oscillator.

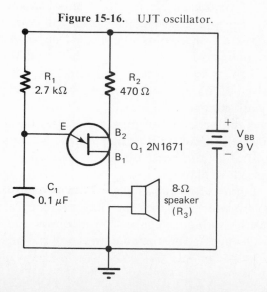

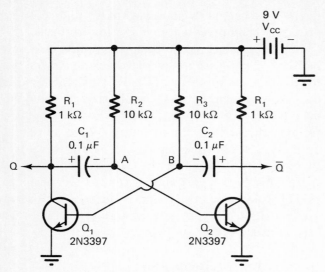

Figure 15-17. Astable multivibrator.

connected in a common-emitter circuit configuration. R_2 and R_3 provide forward bias voltage for the base of each transistor. Capacitor C_1 couples the collector of transistors Q_1 to the base of Q_2. Capacitor C_2 couples the collector of Q_2 to the base of Q_1. Because of the cross-coupling, one transistor will be conductive and one will be cut off. After a short period of time the two transistors will change states. The conducting transistor will be cut off and the off transistor will become conductive. The circuit changes back and forth between these two states. The output of the circuit will be a rectangular-shaped wave. An output signal can be obtained from the collector of either transistor. As a rule the output is labeled Q or $\overline{Q}$. This denotes that outputs are of an opposite polarity.

When power is first applied to the multivibrator of Fig. 15-17, one transistor will go into conduction first. A slight difference in component tolerances usually accounts for this condition. For explanation purposes, assume that Q_1 conducts first. When Q_1 becomes conductive, there is a voltage drop across R_1. V_C becomes less than V_{CC}. This causes a negative-going voltage to be applied to C_1. The positive base voltage of Q_1 is reduced by this voltage. The conduction of Q_2 decreases. The collector voltage of Q_2 begins to rise to the value of V_{CC}. A positive-going voltage is applied to C_2. This voltage is added to the base voltage of Q_1. Q_1 becomes more conductive. The process continues until Q_1 becomes saturated and Q_2 is cut off.

When the output voltages of each transistor becomes stabilized, there is no feedback voltage. Q_2 is again forward biased by R_2. Conduction of Q_2 causes a drop in V_C. This negative-going voltage is coupled to the base of Q_1 through C_2. Q_1 becomes less conductive. The V_C of Q_1 begins to rise toward the value of V_{CC}. This is coupled to the base of Q_2 by C_1. The process continues until Q_2 is saturated and Q_1 is cut off. The output voltages then become stabilized. The process is then repeated.

The oscillation frequency of a multivibrator is determined by the time constant of R_2–C_1 and R_3–C_2. The values of R_2 and R_3 are usually selected to cause each transistor to reach saturation. C_1 and C_2 are then chosen to develop the desired operating frequency. If C_1 equals C_2 and R_2 equals R_3, the output is symmetrical. This means that each transistor will be on and off for an equal amount of time. The output frequency of a symmetrical multivibrator is determined by the formula

$$\text{frequency } (f) = \frac{1}{1.4RC}$$

If the resistor and capacitor values are unequal, the output will not be symmetrical. One transistor could be on for a long period with the alternate transistor being on for only a short period. The output of a nonsymmetrical multivibration is described as a rectangular wave.

Monostable Multivibrator

A *monostable multivibrator* has one stable state of operation. It is often called a one-shot multivibrator. One trigger pulse will cause the oscillator to change its operational state. After a short period of time, however, the oscillator will return to its original starting state. The RC time constant of this circuit determines the time period of the state change. A monostable multivibrator will always return to its original state. No operational change will occur until a trigger pulse is applied. A monostable multivibrator is considered to be a triggered oscillator.

Figure 15-18 shows a schematic of a monstable multivibrator. This circuit has two operational states. Its stable state is based on conduction of Q_2 with Q_1 cut off. The unstable state occurs when Q_1 is

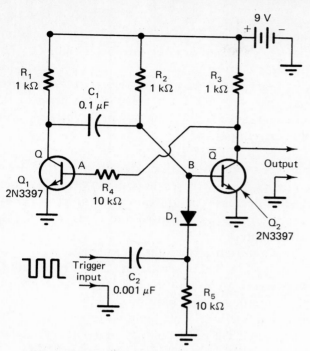

Figure 15-18. Monostable multivibrator.

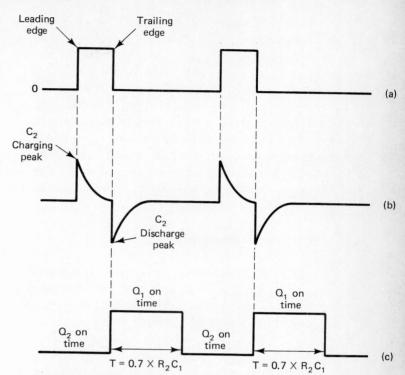

Figure 15-19. Monostable multivibrator waveforms: (a) trigger input waveform; (b) differentiator output waveform; (c) monostable multivibrator output waveform.

conductive and Q_2 is cut off. The circuit relaxes in its stable state when no trigger pulse is supplied. The unstable state is initiated by a trigger pulse. When a trigger pulse arrives at the input, the circuit changes from its stable state to the unstable state. After the time of $0.7 \times R_2 C_1$ the circuit goes back to its stable state. No circuit change will occur until another trigger is applied to the input.

Consider now the operation of a monostable multivibrator when power is first applied. No trigger pulse is applied initially. Q_2 is forward biased by a divider network consisting of R_2, D_1, and R_5. The value of R_2 is selected to cause Q_2 to reach saturation. Resistors R_1 and R_3 reverse bias each collector. With the base of Q_2 forward biased, it is driven into saturation immediately. The collector voltage of Q_2 drops to a very small value. This voltage coupled through R_4 is applied to the base of Q_1. V_B is not great enough to cause conduction of Q_1. The circuit will therefore remain in this conduction state as long as power is applied. This represents the stable state of the circuit.

To initiate a state change in a monstable multivibrator a trigger pulse must be applied to its input. Figure 15-19 shows a representative trigger pulse, a wave-shaped pulse, and the resulting output

of the multivibrator. C_2 and R_5 of the input circuit form a *differentiator* network. The leading edge of the applied trigger pulse will cause a large current flow through R_5. After C_2 begins to charge, the current through R begins to drop. When the trailing edge of the pulse arrives, the voltage applied to C drops to zero. With no source voltage applied to C_2, the capacitor will discharge through R_5. An opposite polarity pulse will therefore occur at the trailing edge of the input pulse. The input pulse is therefore changed into a positive and a negative spike that appears across R_5. D_1 will conduct only during the time of the negative spike. This will feed a negative spike to the base of Q_2. With a square trigger pulse applied to the input a single negative spike pulse is applied to the base of Q_2. This will initiate a state change in the multivibrator.

When the base of Q_2 receives a negative spike, it is driven into cutoff. This causes the collector voltage of Q_2 to rise very quickly to the value of $+V_{CC}$. This in turn causes the base of Q_1 to become positive. Q_1 therefore becomes conductive and Q_2 is driven into cutoff. When Q_1 conducts, the emitter–collector junction becomes very low resistant. Charging current will flow through Q_1, C_1, and R_2. The bottom of R_2 will immediately become negative due to the charging current of C_1. This will drive the base of Q_2 negative. Q_2 remains in its cutoff state. The process will continue until C becomes charged. The charging current through R_2 then begins to slow down and the top of R_2 eventually becomes positive. Q_2 immediately goes into conduction. This in turn drives Q_1 to cutoff. The circuit has therefore changed back to its stable state. It will remain in this state until the next trigger pulse arrives at the input.

Bistable Multivibrator

A bistable multivibrator has two stable states of operation. A trigger pulse applied to the input will cause the circuit to assume one stable state. A second pulse will cause it to switch to the alternate stable state. This type of multivibrator will only change states when a trigger pulse is applied. It is often called a *flip-flop*. It flips to one state when triggered and flops back to the other state when triggered. The circuit becomes stable in either state. It will not change states or *toggle* until commanded to do so by a trigger pulse. Figure 15-20 shows a schematic of a bistable multivibrator using bipolar transistors.

When electrical power is first applied to a multivibrator it will assume one of its stable states.

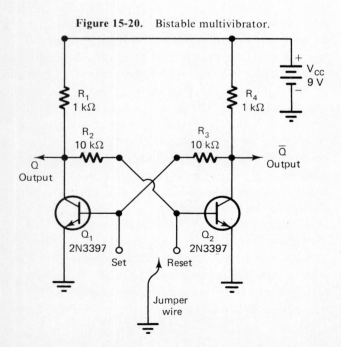

Figure 15-20.   Bistable multivibrator.

One transistor will go into conduction faster than the other. In the circuit of Fig. 15-20 let us assume that Q_1 goes into conduction faster than Q_2. The collector voltage of Q_1 will therefore begin to drop very quickly. Direct coupling between the collector and base will cause a corresponding drop in the base voltage of Q_2. A reduction in Q_2 voltage causes a decrease in I_B and I_C.

The V_C of Q_2 will rise to the value of $+V_{CC}$. This positive-going voltage is coupled back to the base of Q_1 by R_3. This increases the conduction of Q_1, which in turn decreases conduction Q_2. The process continues until Q_1 is saturated and Q_2 is cut off. The circuit remains in this stable state.

To initiate a state change, a trigger pulse must be applied. A negative pulse applied to the base of Q_1 will cause it to go into the cutoff region. A positive pulse applied to the base of Q_2 will cause it to go into conduction. This polarity applies to NPN transistors. The pulse polarity would be reversed for PNP transistors.

In our circuit assume that a negative pulse is applied to the base of Q_1. When this occurs, the I_B and I_C of Q_1 are reduced immediately. The V_C rises toward the value of $+V_{CC}$. This positive-going voltage is coupled back to the base of Q_2. The I_B and I_C of Q_2 rise very quickly. This causes a corresponding drop in the V_C of Q_2. Direct coupling of V_C through R_3 causes a decrease in the I_B and I_C of Q_1. The process continues until Q_1 is cut off and Q_2 reaches saturation. This represents the second stabilized state of operation. The circuit will remain in this state until commanded to change or the power is removed.

IC Waveform Generators

The NE/SE 555 integrated circuit is a multifunction device that is widely used today. It can be modified to respond as an astable multivibrator. This particular circuit can be achieved with a minimum of components and a power source. Circuit design is easy to accomplish and operation is very reliable. This specific chip is available through a number of manufacturers. As a rule, the number 555 usually appears in the manufacturer's part identification number. SN72555, MC14555, SE555, LM555, XR555, and CA555 are some of the common part numbers for this chip.

The internal circuitry of a 555 IC is generally viewed in functional blocks. In this regard, the chip has two comparators, a bistable flip-flop, a resistive divider, a discharge transistor, and an output stage. Figure 15-21 shows the functional blocks of a 555 IC.

The voltage divider of the IC consists of three 5-kΩ resistors. The network is connected internally across the $+V_{CC}$ and ground supply source. Voltage developed by the lower resistor is one-third of V_{CC}. The middle divider point is two-thirds of the value of V_{CC}. This connection is terminated at pin 5. Pin number 5 is designated as the control voltage.

The two *comparators* of the 555 respond as an amplifying switch circuit. A reference voltage is applied to one input of each comparator. A voltage value applied to the other input will initiate a change in output when it differs with the reference value. Comparator I is referenced at two-third of V_{CC} at its negative input. This is where pin 5 is connected to the

middle divider resistor. The other input is terminated at pin 6. This pin is called the *threshold* terminal. When the voltage at pin 6 rises above two-thirds of V_{CC}, the output of the comparator will swing positive. This is then applied to the reset input of the flip-flop.

Comparator 2 is referenced to one-third of V_{CC}. The positive input of comparator 2 is connected to the lower divider network resistor. External pin connection 2 is applied to the negative input of comparator 2. This is called the trigger input. If the voltage of the trigger drops below one-third of V_{CC} the comparator output will swing positive. This is applied to the set input of the flip-flop.

The flip-flop of the 555 IC is a bistable multivibrator. It has reset and set inputs and one output. When the reset input is positive, the output goes positive. A positive voltage to the set input causes

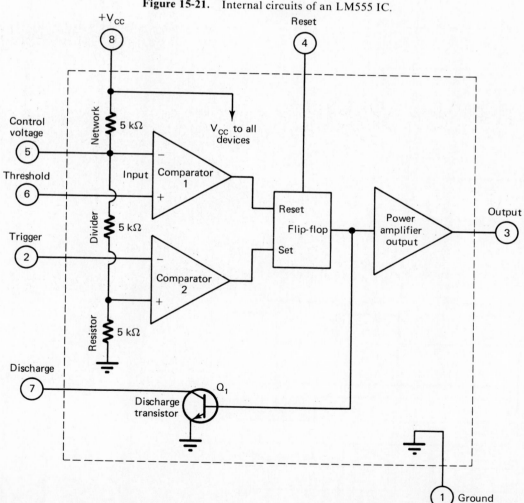

Figure 15-21. Internal circuits of an LM555 IC.

the output to go negative. The output of the flip-flop is dependent on the status of the two comparator inputs.

The output of the flip-flop is applied to both the output stage and the discharge transistor. The output stage is terminated at pin 3. The discharge transistor is connected to terminal 7. The output stage is a power amplifier and a signal inverter. A load device connected to terminal 3 will see either $+V_{CC}$ or ground depending on the state of the input signal. The output terminal switches between these two values. Load current values of up to 200 mA can be controlled by the output terminal. A load device connected to $+V_{CC}$ would be energized when pin 3 goes to ground. When the output goes to $+V_{CC}$ the output would be off. A load device connected to ground would turn on when the output goes to $+V_{CC}$. It would be off when the output goes to ground. The output switches back and forth between these two states.

Transistor Q_1 is called a discharge transistor. The output of the flip-flop is applied to the base of Q_1. When the flip-flop is reset (positive), it forward biases Q_1. Pin 7 connects to ground through Q_1. This causes pin 7 to be grounded. When the flip-flop is set (negative), it reverse biases Q_1. This causes pin 7 to be infinite or open with respect to ground. Pin 7 therefore has two states, shorted to ground or open.

We will now see how the internal circuitry of the 555 responds as a multivibrator.

IC Astable Multivibrator

When used as an astable multivibrator, the 555 is an *RC* oscillator. The shape of the waveform and its frequency are determined primarily by an *RC* network. The astable multivibrator circuit is self-starting and operates continuously for long periods of time. Figure 15-22(a) shows the LM 555 connected as an astable multivibrator. A common

Figure 15-22. Astable multivibrator: (a) circuit; (b) waveforms.

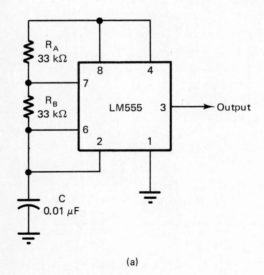

(a)

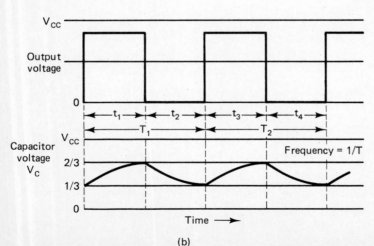

(b)

application of this circuit is the time base generator for clock circuits and computers.

Connection of the 555 IC as an astable multi-vibrator requires two resistors, a capacitor, and a power source. The output of the circuit is connected to pin 3. Pin 8 is $+V_{CC}$ and pin 1 is gound. The supply voltage can be from 5 to 15 V dc. Resistor R_A is connected between $+V_{CC}$ and the discharge terminal (pin 7). Resistor R_B is connected between pin 7 and the threshold terminal (pin 6). A capacitor is connected between the threshold and ground. The trigger (pin 2) and threshold (pin 6) are connected together.

When power is first applied, the capacitor will charge through R_A and R_B. When the voltage at pin 6 (threshold) rises slightly above two-thirds of V_{CC} it will change the state of comparator 1. This will reset the flip-flop and cause its output to go positive. The output (pin 3) will go to ground and the base of Q_1 will be forward biased. Q_1 will discharge C through R_B to ground.

When the charge voltage of C drops slightly below one-third of V_{CC} it energizes comparator 2. The trigger (pin 2) and pin 6 are still connected together. Comparator 2 causes a positive voltage to go to the set input of the flip-flop. This sets the flip-flop, which causes its output to go negative. The output (pin 3) swings to $+V_{CC}$. The base of Q_1 is reverse biased. This opens the discharge (pin 7). C begins to charge again to V_{CC} through R_A and R_B. The process is repeated from this point. The charge value of C varies between one-third and two-thirds of V_{CC}. See the resulting waveforms of Fig. 15-22(b).

The output frequency of the astable multi-vibrator is represented as $f = 1/T$. This represents the total time needed to charge and discharge C. The charge time is represented by spaces t_1 and t_3. In seconds t_1 is $0.693(R_A + R_B)C$. The discharge time is shown as t_2 and t_4. In seconds t_2 is $0.693R_BC$. The combined time for one operational cycle is therefore $T = t_1 + t_2$ or $t_3 + t_4$. Expressed as frequency, this would be

$$\text{frequency} = \frac{1}{\text{time}} \quad \text{or} \quad f = \frac{1}{T}$$

Combining t_1 and t_2 or t_3 and t_4 makes the frequency formula

$$f = \frac{1}{T} \quad \text{or} \quad \frac{1.44}{(R_A + 2R_B)C}$$

The resistance ratio of R_A and R_B is quite critical in the operation of an astable multivibrator. If R_B is more than half the value of R_A, the circuit will not oscillate. Essentially, this would prevent the trigger from dropping in value from two-thirds of V_{CC} to one-third of V_{CC}. This means that the IC would not be capable of retriggering itself. It would therefore be unprepared for the next operational cycle. Most IC manufacturers provide data charts that would assist the user in selecting the correct R_A and R_B values with respect to C.

Blocking Oscillators

Blocking oscillators are an example of the relaxation principle. The active device, which may be a tube or a transistor, is cut off during most of the operational cycle. It turns on for a short operational period of time to discharge an RC network. In appearance this circuit closely resembles the Armstrong oscillator. Feedback from the output to the input is needed to achieve the blocking function. The output of the oscillator is used to change the shape of a waveform.

Figure 15-23 shows a vacuum-tube blocking oscillator. This circuit was commonly used as the vertical oscillator of a television receiver. The *vertical blocking oscillator* (VBO) transformer provides regenerative feedback from the output to the input. An iron core transformer is used in this circuit because the generated frequency is 60 Hz.

When power is first applied to the circuit the grid of V_1 is at zero. With the plate positive, electrons flow from the cathode to the plate. Current flows through the tube and L_1 of the transformer. This current flow causes an expanding field around L_1. By transformer action, voltage is induced into L_2. The polarity of the winding causes the control grid to become positive. This causes the tube to be more conductive. Grid current causes C_1 to charge to its indicated polarity. The process continues until V_1 reaches saturation. At this point there is no further change in plate current (I_p). The induced grid voltage stops immediately. The charge on C_1 immediately drives the grid well below cutoff. No I_p flows in the tube at this time. It is in its relaxation state.

With V_1 at cutoff, C_1 begins to discharge through R_1. The negative grid voltage gradually decreases in value. After a short period of time, V_1 become conductive again. The time (T) is based on

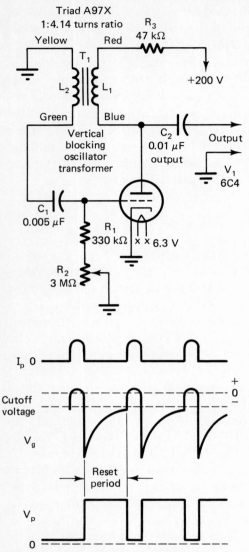

Figure 15-23. Vacuum-tube vertical blocking oscillator: (a) circuit; (b) waveforms.

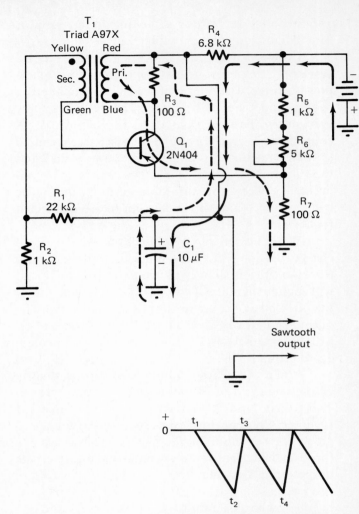

Figure 15-24. Transistor vertical blocking oscillator. Solid arrows denote C_1 charge path; dashed arrows denote C_1 discharge path.

the value of R and C. When V_1 begins to conduct, I_p flows through the tube and L_1. Induced positive grid voltage eventually drives the tube into saturation. The process then repeats itself by driving the grid to cutoff. The design of the circuit is such that V_1 conducts for only a short period of time and is cut off for a long period. Note the I_p, V_g, and V_p waveforms of the circuit.

A transistor version of the blocking oscillator is shown in Fig. 15-24. This circuit is somewhat similar to the vacuum-tube blocking oscillator. A sawtooth-forming capacitor (C_1) charges when the source voltage is initially applied. It discharges through the

transistor when it becomes conductive. The output across the capacitor is a sawtooth waveform.

When the supply voltage is initially applied, the charge of C_1 is zero. This is represented by T_1 on the output waveform. The transistor is cut off by the reverse bias voltage developed by emitter resistor R_7. C_1 begins to charge to $-V_{CC}$ as indicated by the solid arrows. As C_1 charges, voltage is applied to the base of Q_1 through R_1 and the secondary of T_1. This voltage forward biases Q_1. The time of the wave from T_1 to T_2 represents the charging action of C_1. At T_2, the base voltage overcomes the reverse bias emitter voltage. Q_1 goes into conduction and current flows through the primary winding of T_1. Feedback from primary to secondary drives the base more negative. The process continues until Q_1 saturates.

When the transistor is conductive, it provides a low-resistance discharge path for C_1. See the dashed-line arrows of the discharge path. C_1 discharges very quickly because of the low-impedance path. This is represented by the space between T_2 and T_3 of the waveform. When C_1 is fully discharged Q_1 is cut off again by R_7. The process then repreats itself with C_1 charging again.

The frequency of the sawtooth output is dependent on the reverse bias voltage of Q_1. An increase in negative voltage from R_6 will increase the cutoff time for Q_1. This increases the charge time that C_1 needs to develop voltage to bring Q_1 into conduction. Value changes in R_6 alter the sawtooth frequency of the oscillator.

Review

1. Explain what is meant by the term "feedback oscillator."
2. What is a relaxation oscillator?
3. What are the fundamental parts of an oscillator?
4. Explain how the LC circuit responds when dc is applied to it.
5. What is the frequency of an LC circuit using a 0.5-μF capacitor and 100-mH inductor?
6. What causes the wave of an LC circuit to be damped?
7. Explain how the Armstrong oscillator operates.
8. What distinguishes a Hartley oscillator from other oscillator circuits?
9. What function does the capacitor divider of a Colpitts oscillator perform?
10. Explain how a crystal oscillator achieves stabilized oscillation.
11. How much charge voltage does a capacitor develop in one time constant? Two time constants? Five time constants?
12. Why does the $E-B_1$ junction of a UJT oscillator appear as a sawtooth wave?
13. Explain how an astable multivibrator changes states.
14. How does a trigger pulse cause a state change in a monstable multivibrator?
15. What is the polarity of the trigger pulse needed to cause a state change in a bistable multivibrator?
16. What are the internal functions of a 555 timer IC?
17. What is the frequency of an LM 555 IC using 200-kΩ R_A and 200-kΩ R_B resistors and a 1-μF capacitor?
18. When does a blocking oscillator have its relaxing time?

Activities

Armstrong Oscillator
1. Construct the Armstrong oscillator of Fig. 15-7.

2. Q_1 should be a transistor with a gain of 100. A 2N3397 or a 2N2405 transistor are typical. T_1 can be a 12-V step-down transformer.

3. Apply power to the circuit.

4. Connect an oscilloscope to the output to observe the generated waveform.

5. Adjust R_3 to produce the best sine wave.

6. Measure and record the dc voltage at the emitter, base, and collector.

7. If a frequency counter is available, determine the generated output. Some oscilloscopes may be used to determine frequency.

Hartley Oscillator

1. Construct the Hartley oscillator of Fig. 15-8.

2. Turn on the power source. If the circuit is operating properly, test point B and ground should indicate a negative dc voltage.

3. Prepare an oscilloscope for operation. Connect the vertical probe to point B and the ground probe to circuit ground. Make a sketch of the observed waveform.

4. Observe and record the waveforms observed at test points C, D, and A.

5. Turn off the circuit power. Remove C_1 and replace it with a 365-pF variable capacitor.

6. Turn on the circuit power. Adjust the variable capacitor through its range while observing the waveform at test point A. Which position of the capacitor produces the highest and lowest frequency?

7. If an AM radio is available, turn it on and place it near the oscillator circuit. Set the receiver near the center of the dial that is not occupied by a station.

8. Adjust the variable capacitor of the oscillator through its range while listening for a whistling sound on the receiver. What frequency does the receiver indicate?

Colpitts Oscillator

1. Construct the Colpitts oscillator of Fig. 15-9.

2. Test the circuit for operation.

3. Observe the waveforms with an oscilloscope at test points A, B, and C with respect to ground. Make a sketch of the observed waves.

4. With a VOM measure and record the dc voltage at the transistor's emitter, base, and collector.

Crystal Oscillator

1. Construct the Pierce crystal oscillator of Fig. 15-12.

2. Test the circuit for operation.

3. Observe the waveforms with an oscilloscope at test points A and B with respect to ground.

4. With a VOM measure and record the dc voltage at B and C of the transistor.

5. Determine the frequency of the oscillator with an oscilloscope or frequency counter.

UJT Oscillator

1. Construct the UJT oscillator of Fig. 15-16.
2. Test the circuit for operation.
3. Observe the waveform with an oscilloscope at test point E with respect to ground.
4. Exchange R_1 (2.7 K-Ω) for a 10-kΩ resistor. How does this alter the sound output?
5. Remove the speaker and replace it with a 120-Ω resistor.
6. With an oscilloscope observe the waveforms at test points E and B_1 with respect to ground. Make a sketch of the observed waveforms.

Astable Multivibrator

1. Construct the astable multivibrator of Fig. 15-17.
2. Turn on the circuit.
3. Observe the waveforms with an oscilloscope at points Q, A, B, and $\overline{Q}$ with respect to ground. Make a sketch of the waveforms.
4. Turn off the circuit. Exchange C_1 and C_2 for two large capacitors of 10 to 100 μF.
5. Turn on the circuit and measure the voltage at Q, A, B, and $\overline{Q}$ for an operational cycle. Explain your findings.

Monostable Multivibrator

1. Construct the monostable multivibrator of Fig. 15-18.
2. Turn on the circuit.
3. Test the dc voltages at Q, A, B, and $\overline{Q}$.
4. Connect a voltmeter to point $\overline{Q}$. Momentarily touch the base (B) of Q_2 to ground. How does the voltage at $\overline{Q}$ respond?
5. Feed a square-wave generator trigger pulse into the trigger input. (*Note*: The astable multivibrator of Fig. 15-17 or 15-22 may be used for the trigger input.)
6. With an oscilloscope observe the waveforms at points Q, A, B, and $\overline{Q}$.

Bistable Multivibrator

1. Construct the bistable multivibrator of Fig. 15-20.
2. Apply power to the circuit.
3. Measure the dc voltage at Q and $\overline{Q}$. What is the initial status of Q_1 and Q_2? If Q_1 is set, Q will be the value of V_{CC} and $\overline{Q}$ will be 0.6 V. If Q_2 is set, $\overline{Q}$ will be the value of V_{CC} and Q will be 0.6 V.
4. Touch a jumper ground wire to the input of the transistor, which has a low output. This will turn the transistor off. Its output will go to V_{CC}.
5. Alternately test the set and resets of Q_1 and Q_2 while measuring the respective output voltage values.
6. Touch the ground lead to the set input. Measure and record the voltage at Q and $\overline{Q}$.
7. Touch the ground lead to the reset input. Measure and record the voltage at Q and $\overline{Q}$.

8. Explain how the circuit responds when set and reset.

555 Astable Multivibrator

1. Construct the IC multivibrator of Fig. 15-22.
2. Turn on the circuit power.
3. With an oscilloscope, observe the waveform at the input (pin 3) with respect to ground.
4. Calculate the frequency of the output.
5. Measure the frequency with an oscilloscope or frequency counter.
6. Observe the waveform across C.
7. Turn off the power.
8. Make the following changes: $R_A = 200$ kΩ; $C = 50$ μF. Turn on the power and test the circuit.
9. Measure and record the dc voltage at the output through an operational cycle. How does it respond?
10. Measure the voltage at pins 2 and 7. How does it respond?

Blocking Oscillators

1. Construct the blocking oscillator circuit of Fig. 15-23 or Fig. 15-24. The indicated transformer may be used for either circuit.
2. Turn on the power source.
3. Measure dc voltages at the base and collector or plate and control grid.
4. With an oscilloscope observe the base and collector waveforms with respect to ground. In the vacuum-tube circuit observe the grid and plate waveforms.
5. Adjust R_B or R_2 and observe the change in the waveforms. Make a sketch of the waveform change that occurs.

16

Communication Systems

In a previous chapter we were able to make sound louder by using electronic amplifiers. High-level sound amplification made it possible to communicate over a rather substantial distance. A public address amplifier system, for example, permits an announcer to communicate with a large number of people in a stadium or an arena. Even the most sophisticated sound system, however, has some limitations. Sound waves moving away from the source have a tendency to become somewhat weaker the farther they travel. An increase in signal strength does not, therefore, necessarily solve this problem. People near large speakers usually become very uncomfortable when the sound is increased to a high level. It is possible, however, to communicate with people over long distances without increasing sound levels. *Electromagnetic waves* make this type of communication possible. Radio, television, and long-distance telephone communication are achieved by this process.

Electromagnetic wave communication systems play a very important role in our daily lives. We listen to radio receivers, watch television, talk to friends over long distances on the telephone and even communicate with astronauts by electromagnetic waves. This type of communication is widely used today. It is important that we have some basic understanding of this type of communication.

In this chapter we investigate how sound is transmitted through the air by electromagnetic waves. This type of communication uses high-frequency alternating current or radio-frequency energy for its operation. These signals travel through the air at 186,000 miles per second. Electromagnetic communication permits sound to travel long distances instantaneously.

IMPORTANT TERMS

In the study of communication systems one frequently encounters a number of new, and somewhat unusual terms in common usage. These terms play an important role in the presentation of this mate-

409

rial. As a rule, it is very helpful to review these terms before proceeding with the chapter.

Beat frequency.   A resulting frequency that develops when two frequencies are combined in a nonlinear device.

Beat frequency oscillator.   An oscillator of a CW receiver. Its output beats with the incoming CW signal to produce an audio signal.

Blanking pulse.   A part of the TV signal where the electron beam is turned off during the retrace period. There is both vertical and horizontal blanking.

Buffer amplifier.   An RF amplifier that follows the oscillator of a transmitter. It isolates the oscillator from the load.

Carrier wave.   An RF wave to which modulation is applied.

Center frequency.   The carrier wave of an FM system without modulation applied.

Chroma.   Short for "chrominance." Refers to color in general.

Compatible.   A TV system characteristic in which broadcasts in color may be received in black and white on sets not adapted for color.

Deflection.   Electron beam movement of a TV system that scans the camera tube or picture tube.

Deflection yoke.   A coil fixture that moves an electron beam vertically and horizontally.

Diplexer.   A special TV transmitter coupling device that isolates the audio carrier and the picture carrier signals from each other.

Field, even lined.   The even-numbered scanning lines of one TV picture or frame.

Field, odd lined.   The odd-numbered scanning lines of one TV picture or frame.

Frame.   A complete electronically produced TV picture of 525 horizontally scanned lines.

Ganged.   Two or more components connected together by a common shaft—a three-ganged variable capacitor.

Heterodyning.   The process of combining signals of independent frequencies to obtain a different frequency.

High-level modulation.   Where the modulating component is added to an RF carrier in the final power output of the transmitter.

Hue.   A color, such as red, green, or blue.

Interlace scanning.   An electronic picture production process in which the odd lines are all scanned, then the even lines, to make a complete 525-line picture.

Intermediate frequency (IF).   A single frequency that is developed by heterodyning two input signals together in a superheterodyne receiver.

I-signal (I).   A color signal of a TV system that is in phase with the 3.58-MHz color subcarrier.

Line-of-sight transmission.   An RF signal transmission that radiates out in straight lines because of its short wavelength.

Modulating component.   A specific signal or energy used to change the characteristic of an RF carrier. Audio and picture modulation are very common.

Modulation.   The process of changing some characteristic of an RF carrier so that intelligence can be transmitted.

Monochrome.   Black-and-white television.

Negative picture phase.   A video signal characteristic where the darkest part of a picture causes the greatest change in signal amplitude.

Phasor.   A line used to denote value by its length and phase by its position in a vector diagram.

Plumbicon.   A television camera tube that operates on the photoconduction principle.

Q signal.   A color signal of a TV system that is out of phase with the 3.58-MHz color subcarrier.

Retrace.   The process of returning the scanning beam of a camera or picture tube to its starting position.

Saturation.   The strength or intensity of a color used in a TV system.

Scanning.   In a TV system, the process of moving an electron beam vertically and horizontally.

Selectivity.   A receiver function of picking out a desired radio frequency signal. Tuning achieves selectivity.

Sidebands.   The frequencies above and below the carrier frequency that are developed because of modulation.

Skip. An RF transmission signal pattern that is the result of signals being reflected from the ionosphere or the earth.

Sky-wave. An RF signal radiated from an antenna into the ionosphere.

Sync. An abbreviation for synchronization.

Synchronization. A control process that keeps electronic signals in step. The sweep of a TV receiver is synchronized by the transmitted picture signal.

Telegraphy. The process of conveying messages by coded telegraph signals.

Trace time. A period of the scanning process where picture information is reproduced or developed.

Vestigial sideband. A transmission procedure where part of one sideband is removed to reduce bandwith.

Videcon. A TV camera tube that operates on the photoconductive principle.

Wavelength. The distance between two corresponding points that represents one complete wave.

Y signal. The brightness or luminance signal of a TV system.

Zero beating. The resulting difference in frequency that occurs when two signals of the same frequency are heterodyned.

SYSTEMS

A communication system is very similar to any other electrical system. It has an energy source, transmission path, control, a load device, and one or more indicators. These individual parts are all essential to the operation of the system. The physical layout of the communication system is one of its most distinguishing features.

Figure 16-1 shows a block diagram of a radio-frequency communication system. The signal source of the system is a radio-frequency (RF) transmitter. The transmitter is the center or focal point of the system. The RF signal is sent to the remaining parts of the system through space. Air is the transmission path of the system. RF finds air to be an excellent signal path. The control function of the system is directly related to the signal path. The distance that the RF signal must travel has a great deal to do with its strength. The load of the system is an infinite number of radio receivers. Each receiver picks the

Figure 16-1. Radio-frequency communication system.

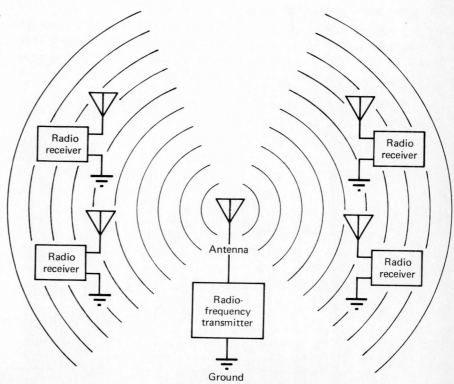

signal out of the air and uses it to do work. Any number of receivers can be used without directly influencing the output of the transmitter. System indicators may be found at a number of locations. Meters, indicating lamps, and waveform monitoring oscilloscopes are typical indicators. These basic functions apply in general to all RF communication systems.

A number of different RF communication systems are in operation today. As a general rule, these systems are classified according to the method by which signal information or intelligence is applied to the transmitted signal. Three very common communication systems are continuous wave (CW), amplitude modulation (AM), and frequency modulation (FM). Each system has a number of unique features that distinguishes it from the others. The transmitter and receiver of the system are quite different. They are all similar to the extent that they use RF signal energy source as a source of operation.

A more realistic way to look at an RF communication system is to divide the transmitter and receiver into separate systems. The transmitter then has all of its system parts in one discrete unit. Its primary function is to generate an RF signal that contains intelligence, and to radiate this signal into space. The receiver can also be viewed is an independent system. It has a complete set of system parts. Its primary function is to pick up the RF transmitted signal, extract the intelligence, and develop it into a usable output. A one-way communication system has one transmitter and an infinite number of receivers. Commercial AM, FM, and TV communication is achieved by this method. Two-way mobile communication systems have a transmitter and a receiver at each location. This type of system permits direct communication between each location. Citizen's band (CB) radio is considered to be two-way communication.

Figure 16-2 shows a simplification of the transmitter and receiver as independent systems. The transmitter has an RF oscillator, amplifiers, and a power source. The antenna-ground at the output circuit serves as the load device for the system. Intelligence is applied according to the design of the system. The system may be CW, AM, or FM.

The RF receiver function is represented as an independent system. It will not to be operational unless a signal is sent out by the transmitter. The receiver has tuned RF amplification, detection, reproduction, and a power supply. The detection function picks out the intelligence from the received signal. The reproduction unit, which is usually a speaker, serves as the load device. The receiver will only respond to the RF signal sent out by the transmitter.

ELECTROMAGNETIC WAVES

Electromagnetic communication systems rely on the radiation of high-frequency energy from an antenna-ground network. This particular principle was discovered by Heinrich Hertz around 1885. He found

Figure 16-2. Transmitter-receiver systems.

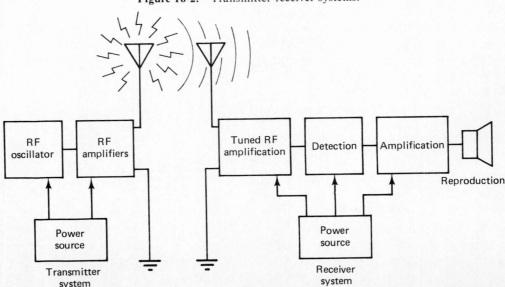

that high-frequency waves produced by an electrical spark caused electrical energy to be induced in a coil of wire some distance away. He also discovered that current passing through a coil of wire produces a strong electromagnetic field. The fundamental unit of frequency is expressed in hertz. This is in recognition of his electromagnetic discoveries.

When direct current is applied to a coil of wire it causes the field to remain stationary as long as current is flowing. When dc is first turned on the field expands. When it is turned off, the field collapses and cuts across the coil. As a general rule, dc electromagnetic field development is of no significant value in radio communication.

When low-frequency ac is applied to an inductor it produces an electromagnetic field. Since ac is in a constant state of change, the resulting field is also in a state of change. It changes polarity twice during each operational cycle.

Figure 16-3 shows how the electromagnetic field of a coil changes when ac is applied. In step (a), the field is expanding during the first half of the positive alternation. At the peak of the alternation, coil current begins to decrease in value. Step (b) shows the field collapsing. The current value decreases to the zero baseline at this time. This completes the positive alternation or first 180° of the sine wave. Step (c) shows the field expanding for the negative alternation. The polarity of this field is now reversed. At the peak of the negative alternation the field begins to collapse. Step (d) shows the field change for the last part of the alternation. The cycle is complete at this point. The sequence is repeated for each new sine wave.

When high-frequency ac is applied to an inductor it also produces an electromagnetic field. The directional change of high frequency ac, however, occurs very quickly. The change is so rapid that there is not enough time for the collapsing field to return to the coil. As a result, a portion of the field becomes detached and is forced away from the coil. In a sense, the field *radiates out* from the coil. The word *radio* was developed from these two terms.

The electromagnetic wave of an RF communication system radiates out from the antenna of the transmitter. This wave is the end result of an interaction between electric and magnetic fields. The resulting wave is invisible, cannot be heard, and travels at the speed of light. These waves become weaker after leaving the antenna. Stronger electromagnetic waves can be effectively used in a radio

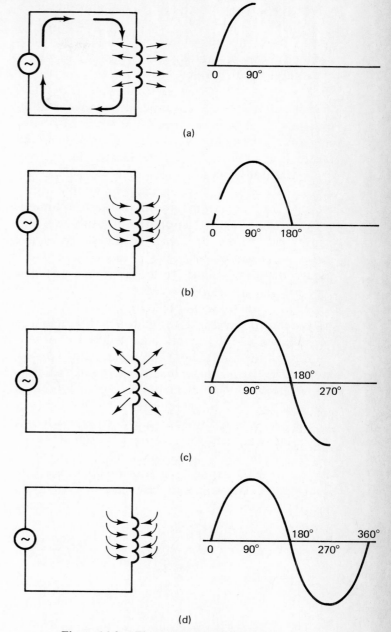

Figure 16-3. Electromagnetic field changes.

communication system. These waves do not effect human beings like high-powered sound waves. RF waves are also easier to work with in the receiver circuit.

Electromagnetic wave radiation is directly related to its frequency. Low-frequency (LF) waves of 30 to 300 kHz and medium-frequency (MF) waves of 300 to 3000 kHz tend to follow the curvature of the earth. These waves are commonly called ground waves. They are usable during the day or night for distances up to 100 miles. Daytime reception of commercial AM radio is largely through ground waves.

A different part of the electromagnetic wave radiated from an antenna is called a *sky wave*. Depending on its frequency, sky waves are reflected back to the earth by a layer of ionized particles called the ionosphere. These waves permit signal reception well beyond the range of the ground wave. Sky-wave signal patterns change according to ion density and the position of the ionized layer. During the daylight hours the ionosphere is very dense and near the surface of the earth. Signal reflection or *skip* of this type is not very suitable for long-distance communication. In the evening hours the ionosphere is less dense and moves to higher altitudes. Sky-wave reflection patterns have larger angles and travel greater distances. Figure 16-4 shows some sky-wave patterns and the ground wave.

Very high frequency (VHF) signals of 30 to 300 MHz tend to move in a straight line. The height of the antenna and its radiation angle has a great deal to do with the direction of the signal path. *Line-of-sight transmission* describes this type of radiation pattern. FM, TV, and satellite communication are achieved by VHF signal radiation. The physical size of the wave or its *wavelength* decreases with an increase in frequency. High-frequency signals have a short wavelength. These signals pass very readily through the ionosphere without being reflected or distorted. Communication between an FM or TV transmitter and a receiver on earth is limited to a few

hundred miles. Communication between an earth station and a satellite can involve thousands of miles.

CONTINUOUS-WAVE COMMUNICATION

Continuous-wave communication is the simplest of all communication systems. It consists of a transmitter and a receiver. The transmitter employs an oscillator for the generation of an RF signal. The oscillator generates a continuous sine wave. In most systems the CW signal is amplified to a desired power level by an RF power amplifier. The output of the final power amplifier is then connected to the antenna-ground network. The antenna radiates the signal into space. Information or intelligence is applied to the CW signal by turning it on and off. This causes the signal to be broken into a series of pulses or short bursts of RF energy. The pulses conform to an intelligible code. The international Morse Code is in common usage today. Figure 16-5 shows a simplification of the Morse Code. A short burst of RF represents a dot. A burst three times longer represents a dash. Signals of this type are called keyed or coded continuous waves. The Federal Communications Commission (FCC) classifies keyed CW signals as type A1. This is considered to be radio *telegraphy* with on–off keying. Figure 16-6 shows a comparison of a continuous-wave signal and a keyed CW signal.

Figure 16-4.   Sky-wave and ground-wave patterns.

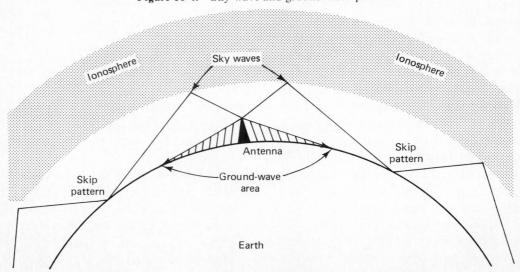

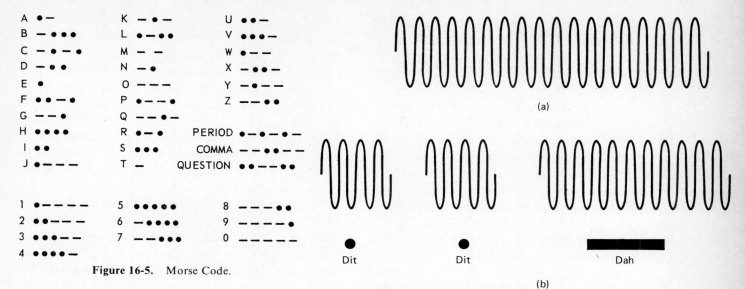

A	● —	K	— ● —	U	● ● —
B	— ● ● ●	L	● — ● ●	V	● ● ● —
C	— ● — ●	M	— —	W	● — —
D	— ● ●	N	— ●	X	— ● ● —
E	●	O	— — —	Y	— ● — —
F	● ● — ●	P	● — — ●	Z	— — ● ●
G	— — ●	Q	— — ● —		
H	● ● ● ●	R	● — ●	PERIOD	● — ● — ● —
I	● ●	S	● ● ●	COMMA	— — ● ● — —
J	● — — —	T	—	QUESTION	● ● — — ● ●

1	● — — — —	5	● ● ● ● ●	8	— — — ● ●
2	● ● — — —	6	— ● ● ● ●	9	— — — — ●
3	● ● ● — —	7	— — ● ● ●	0	— — — — —
4	● ● ● ● —				

Figure 16-5. Morse Code.

Dit   Dit   Dah

(b)

Figure 16-6. Comparison of CW signals: (a) continuous-wave signal; (b) coded or keyed continuous wave, letter D.

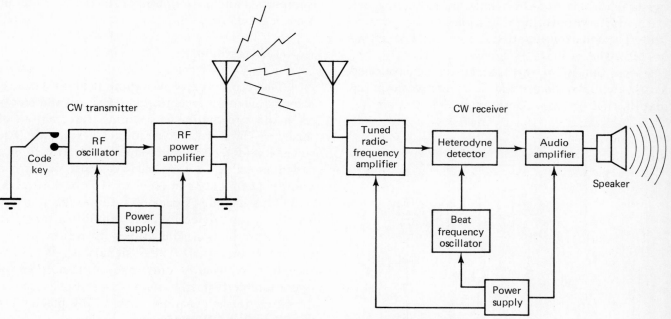

Figure 16-7. Block diagram of a CW communication system.

The receiver function of a CW communication system is somewhat more complex than the transmitter. It has an antenna-ground network, a tuning circuit, an RF amplifier, a heterodyne detector, a beat frequency oscillator, an audio amplifier, and a speaker. The receiver is designed to pick up a CW signal and convert it into a sound signal that drives the speaker. Figure 16-7 shows a block diagram of the CW communication system.

CW Transmitter

A CW transmitter in its simplest form has an oscillator, a power supply, an antenna-ground network, and a keying circuit. The power output developed by the circuit is usually quite small. In this type of system, the active device serves jointly as an oscillator and a power amplifier. This is done to reduce the number of circuit stages in front of the antenna. An oscillator used in this manner generally has reduced frequency stability. As a rule, frequency stability is a very important operational consideration. Single-stage transmitters are rarely used today, because of their instability.

A single-stage CW transmitter is shown in Fig. 16-8. This particular circuit is a tuned base Hartley oscillator. The output frequency is adjustable up to 3.5 MHz. The power output is approximately 0.5 W. When the output of the oscillator is connected to a load, it usually causes a shift in frequency. The load should therefore be of a constant impedance value in order for this type of transmitter to be effective.

Before a CW transmitter can be placed into operation it is necessary to close the code key. This operation supplies forward bias voltage to the base through the divider network of R_1 and R_2. With the transistor properly biased, a feedback signal is supplied to the T_1–C_1 tank circuit. This charges C_1 and causes the circuit to oscillate. Opening the key causes the oscillator to stop functioning. Intelligence can be injected into the CW signal in the form of code. The code key simply responds as a fast-acting on–off switch.

The output of the transmitter is developed across a coil wound around T_1. Inductive coupling of this type is very common in RF transmitter circuits. A low-impedance output is needed to match the antenna-ground impedance. When this is accomplished there is a maximum transfer of energy from the oscillator to the antenna.

A single-stage CW transmitter has a number of shortcomings. The power output is usually held to a low level. The frequency stability is rather poor. These two problems can be overcome to some extent by adding a power amplifier after the oscillator. Systems of this type are called master oscillator power amplifier (MOPA) transmitters.

Figure 16-9 shows the circuitry of a low-power MOPA transmitter system. Q_1 is a modified Pierce oscillator. The crystal (Y_1) provides feedback from the collector to the base. Transformer T_1 is used to match the collector impedance of Q_1 to the base of transistor Q_2. Q_2 is the power amplifier. It operates as a class C amplifier. The oscillator is therefore isolated from the antenna through Q_2. This provides a fixed load for the oscillator. Operational stability is improved and power output is increased with this modification.

CW Receiver

The receiver of a CW communication system is responsible for intercepting an RF signal and recovering the information it contains. Information of this type is in the form of a radiotelegraph code. The signal is an RF carrier wave that is interrupted by a coded message. A CW receiver must perform a number of functions in order to achieve this operation. This includes signal reception, selection, RF amplification, detection, AF amplification, and sound reproduction. These functions are also used in the reception of other RF signals. In fact, CW reception is usually only one of a number of operations performed by a communications receiver. This function is generally achieved by placing the receiver in its CW mode of operation. A switch is usually needed to perform this operation.

Figure 16-10 shows the functions of a CW communication receiver in a block diagram. Note that electrical power is needed to energize the active components of the system. Not all blocks of the diagram are supplied operating power. This means that some of the functions can be achieved without a transistor or a vacuum tube. The antenna, tuning circuit, and detector do not require power supply energy for operation. These functions are achieved by signal energy. A radio receiver circuit has a signal

Figure 16-8.　　Single-stage CW transmitter.

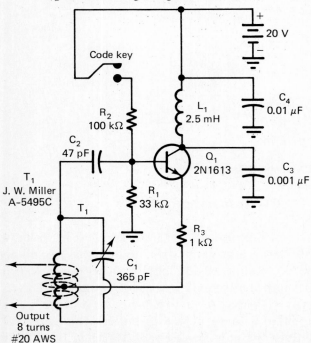

Code key

R_2 100 kΩ

C_2 47 pF

T_1 J. W. Miller A-5495C

T_1

L_1 2.5 mH

Q_1 2N1613

R_1 33 kΩ

R_3 1 kΩ

C_1 365 pF

20 V

C_4 0.01 μF

C_3 0.001 μF

Output 8 turns #20 AWS wire

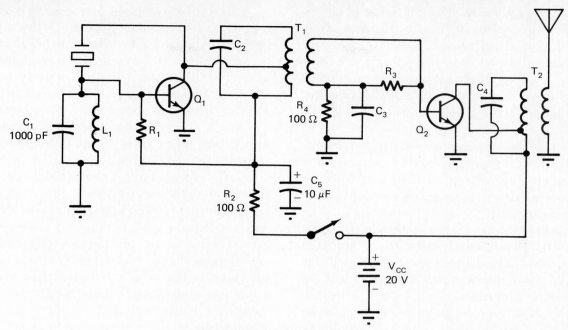

Figure 16-9. MOPA transmitter.

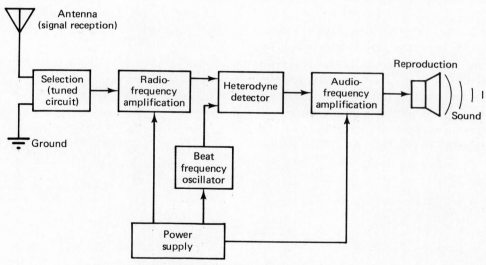

Figure 16-10. CW receiver functions.

energy source and operational energy source. The power supply provides operational energy directly to the circuit. The signal source for RF energy is the transmitter. This signal must be intercepted from the air and processed by the receiver during its operation.

ANTENNAS. The antenna of a CW receiver is responsible for the reception function. It intercepts a small portion of the RF signal that is sent out by the transmitter. Ordinarily, the receiving antenna only

develops a few microwatts of power. In a strict sense, the antenna is a transducer. It is designed to convert electromagnetic wave energy into RF signal voltage. The signal voltage then causes a corresponding current flow in the antenna-ground network. RF antenna power is the product of signal voltage and current.

Receiving antennas come in a variety of styles and types. In two-way communication systems, the transmitter and receiving antenna are of the same unit. The antenna is switched back and forth between the transmitter and receiver. In one-way communication systems antenna construction is not particularly critical. A long piece of wire can respond as a receiving antenna. In weak signal areas, antennas may be tuned to resonate at the particular frequency being received. Portable radio receivers use small antenna coils that are attached directly to the circuit.

When the electromagnetic wave of a transmitter passes over the receiving antenna, ac voltage is induced into it. This voltage causes current to flow as shown in Fig. 16-11. Starting at point A of the waveform, the electrons are at a standstill. The induced signal then causes the voltage to rise to its positive peak. The resulting current flow is from the antenna to the ground. It rises to a peak, then drops to zero at point C on the curve. The polarity of the induced voltage changes at this point. Between points C and E it rises to the peak of the negative alternation and returns again to zero. The resulting current flow is from the ground into the antenna. This induced signal is repeated for each succeeding alternation.

With radio frequency induced into the antenna, a corresponding RF is brought into the receiver by the antenna coil. The antenna coil serves as an impedance-matching transformer. It matches the impedance of the antenna to the impedance of the receiver input circuit. When outside antennas are used, the primary winding is an extension of the antenna. The loop antenna coil of a portable receiver is attached directly to the input circuit.

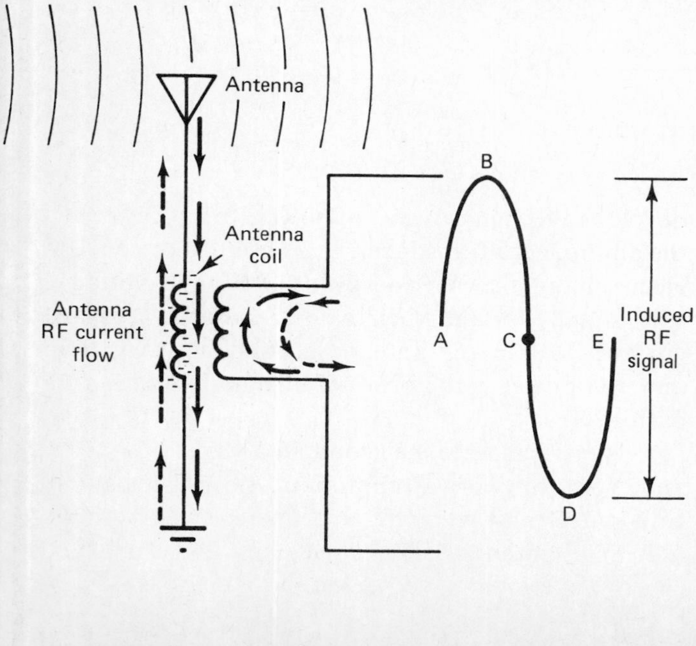

SIGNAL SELECTION. The signal selection function of a radio receiver refers to its ability to pick out a desired RF signal. In receiver circuits, the antenna is generally designed to intercept a band or range of different frequencies. The receiver must then select the desired signal from all those intercepted by the antenna. Signal selection is primarily achieved by an *LC* resonant circuit. This circuit provides a low-impedance path for its resonant frequency. Non-resonant frequencies see a very high impedance. In effect, the resonant frequency signal is permitted to pass through the tuner without opposition.

Figure 16-12 shows the input tuner of a CW receiver. In this circuit L_2 is the secondary winding of the antenna transformer. C_1 is a variable capacitor. By changing the value of C_1, the resonant frequency of L_2–C_1 is tuned to the desired frequency. The selected frequency is then permitted to pass into the remainder of the receiver. Signal selection for AM, FM, and TV receivers all respond to a similar tuning circuit.

Figure 16-12. (a) Input tuning circuit; (b) tuner frequency response curve.

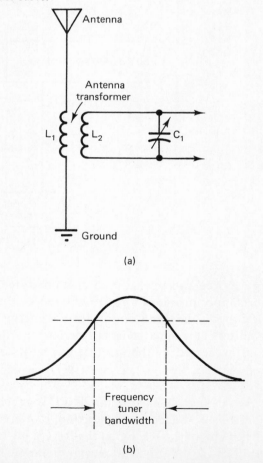

(a)

(b)

Figure 16-11. Antenna signal voltage and current.

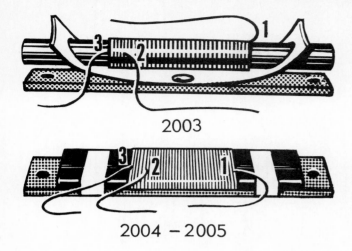

2003

2004 – 2005

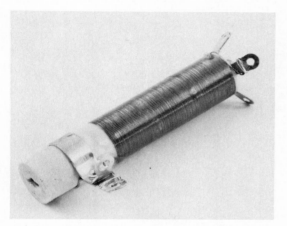

Figure 16-13. Antenna coils. (Courtesy of J. W. Miller Div./ Bell Industries.)

In communication receiving circuits a number of tuning stages are needed for good signal selection. Each stage of tuning permits the receiver to be more selective of the desired frequency. The tuning response curve of one tuned stage is shown in Fig. 16-12. In a crowded radio-frequency band, several stations operating near the selected frequency might be received at the same time. Additional tuning improves the selection function by narrowing the response curve. Figure 16-13 shows some representative loop antenna coils.

RF AMPLIFICATION. Most CW receivers employ at least one stage of radio-frequency amplification. This specific function is designed to amplify the weak RF signal intercepted by the antenna. Some degree of amplification is generally needed to boost the signal to a level where its intelligence can be recovered. RF amplification can be achieved by a variety of active devices. Vacuum tubes have been used for a number of years. Solid-state circuits can employ bipolar transistors and MOSFETs in their design. Several manufacturers have now developed IC RF amplifiers. Any of these active devices can be used to achieve RF signal amplification.

Two representative RF amplifier circuits are shown in Fig. 16-14. The primary function of each circuit is basically the same. The device employed usually determines the level of amplification developed. Note that the amplifier has both input and output tuning circuits. This circuit is called a tuned radio-frequency (TRF) amplifier. The broken line between the two variable capacitors indicates that they are *ganged* together. When one capacitor is tuned, the second capacitor is adjusted to a corresponding value.

HETERODYNE DETECTION. Heterodyne detection is an essential function of the CW receiver. *Heterodyning* is simply a process of mixing two ac signals in a nonlinear device. In the CW receiver, the nonlinear device is a diode. The applied signals are called the fundamental frequencies. The resulting output of this circuit has four frequencies. Two of these are the original fundamental frequencies. The other two are *beat frequencies*. Beat frequencies are the addition and the difference in the fundamental frequencies. Heterodyne detectors used in other receiver circuits are commonly called mixers.

Figure 16-15 shows the signal frequencies applied to a heterodyne detector. Fundamental frequency F_1 is representative of the incoming RF signal. This signal has been keyed on and off with a telegraphic code. Fundamental frequency F_2 comes from the beat frequency oscillator (BFO) of the receiver. It is a CW signal of 1.001 MHz. Both RF signals are applied to the receiver's diode.

The resulting output of a heterodyne detector is F_1, F_2, $F_1 + F_2$, and $F_2 - F_1$. F_1 is 1.0 MHz and F_2 is 1.001 MHz. Beat frequency B_1 is 2.001 MHz and beat frequency B_2 is 1000 Hz. The output of the detector contains all four frequencies. The difference beat frequency of 1000 Hz is in the range of human hearing. F_1, F_2, and F_3 are RF signals.

When two RF signals are applied to a diode the resulting output will be the sum and difference signals. If the two signals are identical, the output frequency will be the same as the input. This is generally called *zero beating*. There is no developed beat frequency when this occurs.

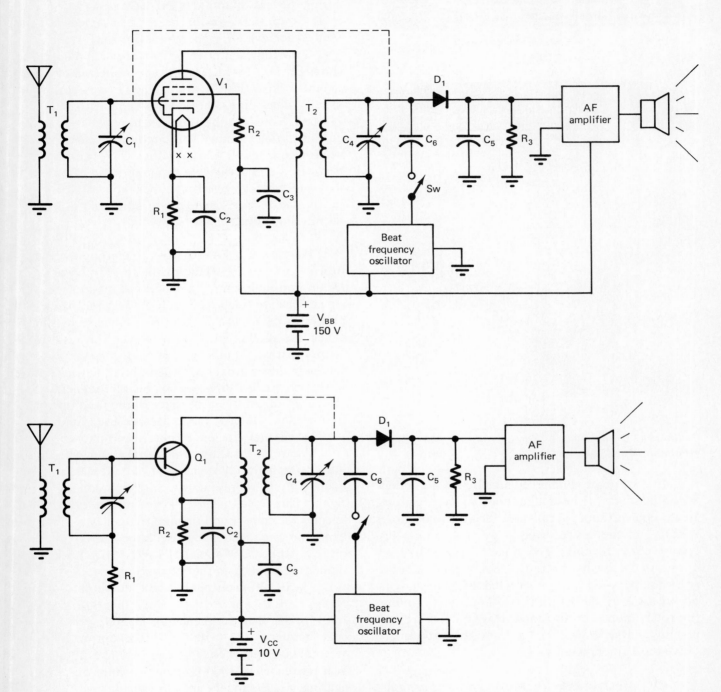

Figure 16-14. Tuned radio frequency receivers.

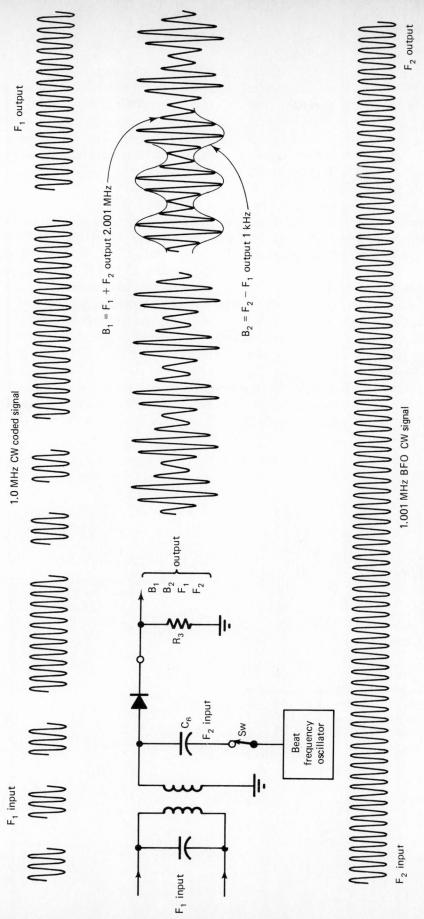

Figure 16-15. Heterodyne detector frequencies.

When the two RF input signals are slightly different, beat frequencies will occur. At one instant the signals will move in the same direction. The resulting output will be the sum of the two frequencies. When one signal is rising and one falling at a different rate, there will be times when the signals cancel each other. The addition and difference of the two signals are therefore frequency dependent. B_1 and B_2 are combined in a single RF wave. The amplitude of this wave varies according to frequency difference in the two signals. In a sense, the mixing process causes the amplitude of the RF wave to vary at an audio rate.

The diode of a heterodyne detector is also responsible for rectification of the applied signal frequencies. The rectified output of the added beat frequency is of particular importance. It contains both RF and AF. Figure 16-16 shows how the diode responds. Initially, the RF signals are rectified. The output of the diode would be a series of RF pulses varying in amplitude at an AF rate. C_5 added to the output of the diode responds as a C input filter. Each

RF pulse will charge C_5 to its peak value. During the off period C_5 will discharge through R_3. The output will be a low-frequency AF signal of 1000 Hz. This signal will occur only when the incoming CW signal has been keyed. The AF signal is representative of the keyed information imposed on the RF signal at the transmitter.

BEAT FREQUENCY OSCILLATOR. Nearly any basic oscillator circuit could be used as the BFO of a CW receiver. Figure 16-17 shows a Hartley oscillator. Feedback is provided by a tapped coil in the base circuit. C_1 and T_1 are the frequency-determining components. Note that this particular oscillator has variable-frequency capabilities. C_1 is usually connected to the front panel of the receiver. Adjustment of C_1 is used to alter the tone of the beat frequency signal. When the BFO signal is equal to the coded CW signal no sound output will occur. This is where zero beating takes place. When the frequency of the BFO is slightly above or below the incoming signal frequency a low AF tone is produced. Increasing the

Figure 16-16. Heterodyne diode output.

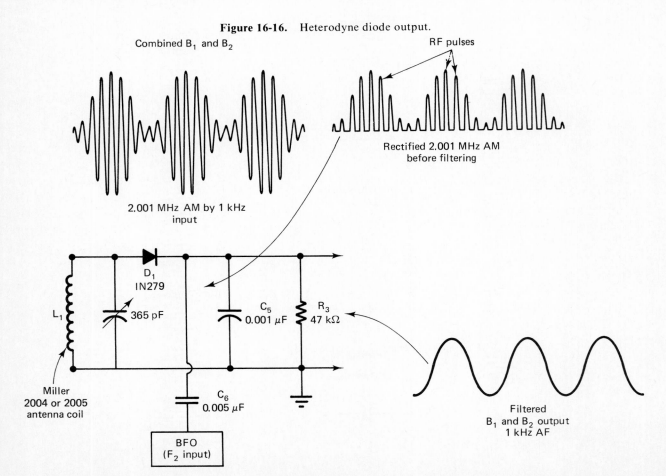

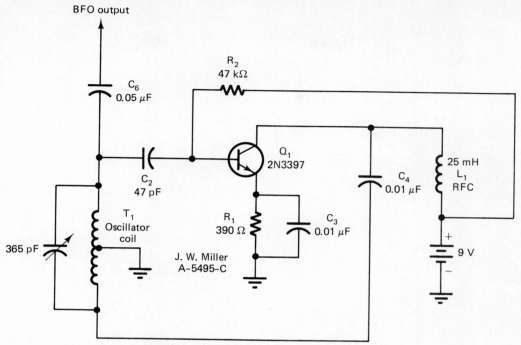

BFO output

C_6 0.05 µF

R_2 47 kΩ

C_2 47 pF

Q_1 2N3397

T_1 Oscillator coil

365 pF

R_1 390 Ω

C_3 0.01 µF

J. W. Miller A-5495-C

C_4 0.01 µF

25 mH L_1 RFC

9 V

Figure 16-17. Hartley oscillator used as a BFO.

BFO frequency causes the pitch of the AF tone to increase. Adjustment of the AF output is a matter of personal preference. In practice, an AF tone of 400 to 1 kHz is common.

Output of the BFO is coupled to the diode through capacitor C_6. In a communications receiver, BFO output is controlled by a switch. With the switch on, the receiver will produce an output for CW signals. With the switch off, the receiver will respond to AM and possibly FM signals. The BFO is only needed to receive coded CW signals.

AUDIO-FREQUENCY AMPLIFICATION. The AF amplifier of a CW receiver is responsible for increasing the level of the developed sound signal. The type and amount of signal amplification varies a great deal among different receivers. Typically, a small-signal amplifier and a power amplifier are used. The small-signal amplifier responds as a voltage amplifier. This amplifier is designed to increase the signal voltage to a level that will drive the power amplifier. Power amplification is needed to drive the speaker. The power output of a communications receiver rarely ever exceeds 5 W. A number of the AF amplifier circuits in Chapter 14 could be used in a CW receiver.

AMPLITUDE MODULATION COMMUNICATION

Amplitude modulation or AM is an extremely important form of communication. It is achieved by changing the physical size or amplitude of the RF wave by the intelligence signal. Voice, music, data, and picture intelligence can be transmitted by this method. The intelligence signal must first be changed into electrical energy. Transducers such as microphones, phonograph cartridges, tape heads, and photoelectric devices are designed to achieve this function. The developed signal is called the *modulating component*. In an AM communication system, the RF transmitted signal is much higher in frequency than the modulating component. The RF component is an uninterrupted CW wave. In practice, this part of the radiated signal is called the *carrier wave*.

An example of the signal components of an AM system are shown in Fig. 16-18. The unmodulated RF carrier is a CW signal. This signal is generated by an oscillator. In this example, the carrier is 1000 kHz or 1.0 MHz. A signal of this frequency would be in the standard AM broadcast band of 535 to 1620 kHz. When listening to this

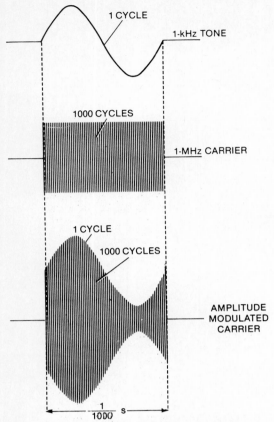

Figure 16-18. AM signal components.

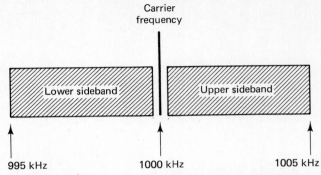

Figure 16-19. Sideband frequencies of an AM signal.

station, a receiver would be tuned to the carrier frequency. Assume now that the RF signal is modulated by the indicated 1000-Hz tone. The RF component will change 1000 cycles for each AF sine wave. The amplitude of the RF signal will vary according to the frequency of the modulating signal. The resulting wave is called an amplitude-modulated RF carrier.

In order to achieve amplitude modulation the RF and AF components are applied to a nonlinear device. A vacuum-tube or solid-state device operating in its nonlinear region can be used to produce modulation. In a sense, the two signals are mixed or heterodyned together. This operation causes beat frequencies to be developed. For the signals of Fig. 16-18, there will be two beat frequencies. Beat frequency 1 is the sum of the two signal frequencies. B_1 is 1,000,000 Hz plus 1000 Hz or 1,001,000 Hz. B_2 is 1,000,000 Hz minus 1000 Hz or 999,000 Hz. The resulting AM signal therefore contains three RF signals. These are the 1.0-MHz carrier, 0.999 MHz, and 1.001 MHz.

When the modulating component of an AM signal is music, the resulting beat frequencies become quite complex. As a rule, this involves a range or a band of frequencies. In AM systems these are called *sidebands*. B_1 is the upper sideband and B_2 is the lower sideband.

The space that an AM signal occupies with its frequency is called a channel. The bandwidth of an AM channel is twice the highest modulating frequency. For our 1-kHz modulating component, a 2-kHz bandwidth is needed. This would be 1 kHz above and below the carrier frequency of 1 MHz. In commercial AM broadcasting a station is assigned a 10-kHz channel. This limits the AM modulation component to a frequency of 5 kHz. Figure 16-19 shows the sidebands produced by a standard AM station.

It is interesting to note that the carrier wave of an AM signal contains no modulation. All the modulation appears in the sidebands. If the modulating component is removed, the sidebands will disappear. Only the carrier will be transmitted. The sidebands are directly related to the carrier and the modulating component. The carrier has a constant frequency and amplitude. The sidebands vary in frequency and amplitude according to the modulation component. In AM radio, the receiver is tuned to the carrier wave.

Percentage of Modulation

In AM radio it is not permitted by law to exceed 100% modulation. Essentially, this means that the modulating component cannot cause the RF component to vary over 100% of its unmodulated value. Figure 16-20 shows an AM signal with three different levels of modulation.

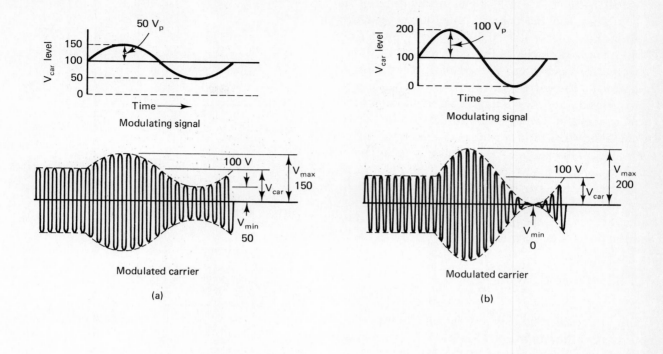

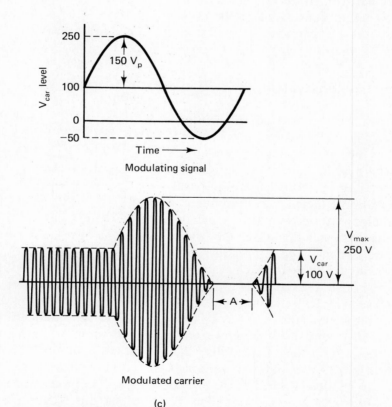

Figure 16-20. AM modulation levels: (a) undermodulation; (b) 100% modulation; (c) overmodulation.

When the peak amplitude of modulating signal is less than the peak amplitude of the carrier, modulation is less than 100%. If the modulation component and carrier amplitudes are equal, 100% modulation is achieved. A modulating component greater than the carrier will cause overmodulation. An overmodulated wave has an interrupted spot in the carrier wave. Overmodulation causes increased signal bandwidth and additional sidebands to be generated. This causes interference with adjacent channels.

Modulation percentage can be calculated or observed on an indicator. When operating voltage values are known, the percentage of modulation can be calculated. The formula is

$$\% \text{ modulation} = \frac{V_{\max} - V_{\min}}{2 \text{ V carrier}}$$

Using the values of Fig. 16-20(a), compute the percentage of modulation. Compute the percentage of modulation for part (b). If the $V_{\min}$ value of the overmodulated signal is considered to be 0 V, compute the modulation percentage.

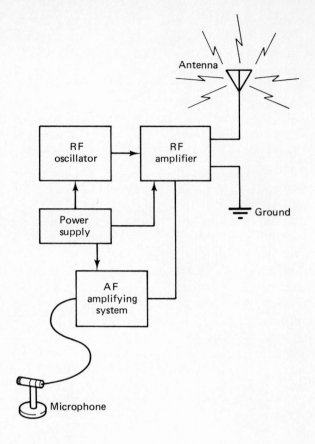

AM Communication System

A block diagram of an AM communication system is shown in Fig. 16-21. The transmitter and receiver respond as independent systems. In a one-way communication system there is one transmitter and an infinite number of receivers. Commercial AM radio is an example of one-way communication. Two-way communication systems have a transmitter and receiver at each location. CB radio systems are of the two-way type. The operating principles are basically the same for each system.

The transmitter of an AM system is responsible for signal generation. The RF section of the transmitter is primarily the same as that of a CW system. The modulating signal component is, however, a unique part of the transmitter. Essentially, this function is achieved by an AF amplifier. A variety of amplifier circuits can be used. Typically, one or two small-signal amplifiers and a power amplifier are suitable for low-power transmitters. The developed modulation component power must equal the power level of the RF output. Modulation of the two

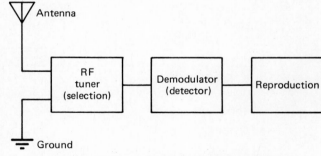

Figure 16-21.   AM communication system.

signals can be achieved at a number of places. High-level modulation occurs in the RF power amplifier. Low-level modulation is achieved after the RF oscillator. All amplification following the point of modulation must be linear. Only high-level modulation will be discussed in this chapter.

The receiver of an AM system is responsible for signal interception, selection, demodulation, and reproduction. Most AM receivers employ the heterodyne principle in their operation. This type of receiver is known as a superheterodyne circuit. It is somewhat different from the heterodyne detector of the CW receiver. A large part of the circuit is the same as the CW receiver. We will only discuss the new functions of the AM receiver.

AM TRANSMITTERS. A wide range of AM transmitters are available today. Toy "walkie-talkie" units with an output of less than 100 mW are very popular. Citizen's band (CB) transmitters are designed to operate at 27 MHz with a 5-W output. AM amateur radio transmitters are available with a power output of up to 1 kW of power. Commercial AM transmitters are assigned power output levels from 250 W to 50 kW. These stations operate between 535 and 1620 kHz. In addition to this, there are 2-meter mobile communication systems, military transmitters, and public service radio systems that all use AM. As a general rule, frequency allocations and power levels are assigned by the Federal Communications Commission.

An AM transmitter has a number of fundamental parts regardless of its operational frequency or power rating. It must have an oscillator, an audio signal component, an antenna, and a power supply. The oscillator is responsible for the RF carrier signal. The audio component is responsible for the intelligence being transmitted. The modulation function can be achieved in a variety of ways. The signal is radiated from the antenna.

A simplified AM transmitter with a minimum of components is shown in Fig. 16-22. Transistor Q_1 is an AF signal amplifier. The sound signal being amplified is developed by a crystal microphone. It is then applied to transistor Q_2. This transistor is a Colpitts oscillator. L_1 and C_4 determine the RF frequency of the oscillator. The amplitude of the oscillator varies according to the AF component. Resistor R_3 is used to adjust the amplitude level of the AF signal. The developed AM output signal is applied to the transmitting antenna. With proper design of the transmitter, it can be tuned to the standard AM broadcast band. The frequency of the system is adjusted by capacitor C_4. The operating range of this unit is several hundred feet.

A schematic of an improved low-power AM transmitter is shown in Fig. 16-23. Transistors Q_1,

Figure 16-22. Simplified AM transmitter.

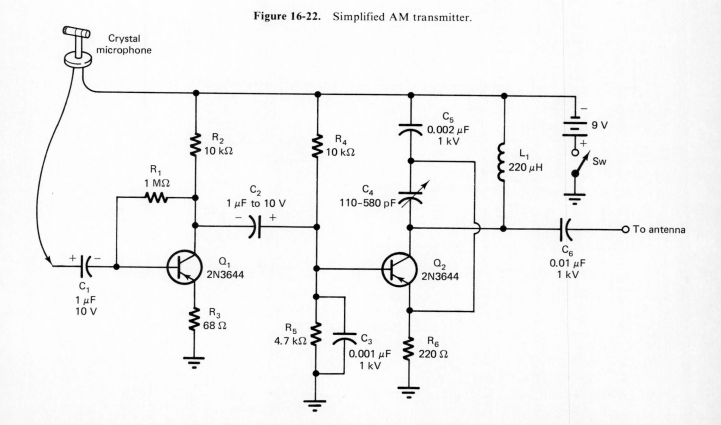

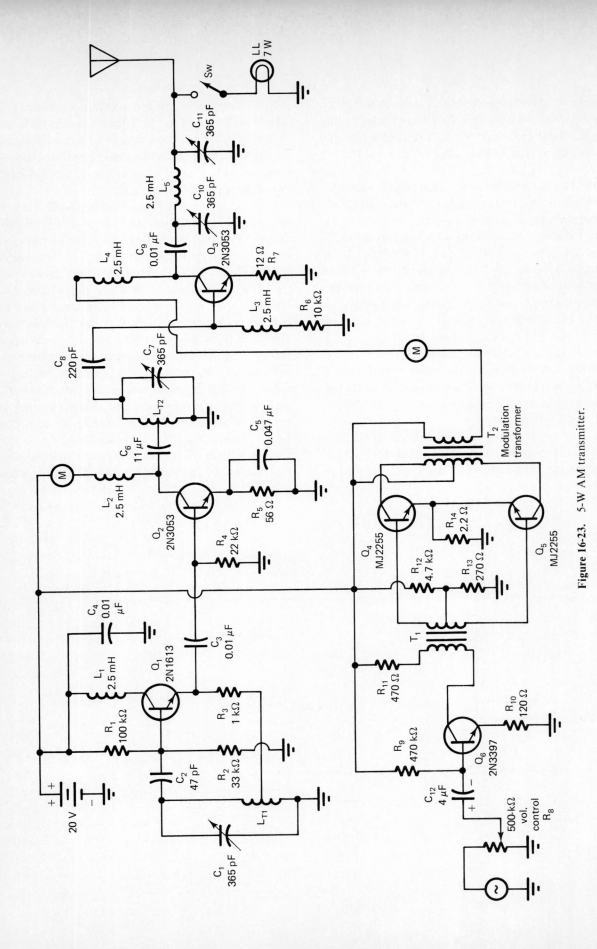

Figure 16-23. 5-W AM transmitter.

Q_2, and Q_3 of this circuit are responsible for the RF signal component. The AF component is developed by Q_4, Q_5, and Q_6. High-level modulation is achieved by this transmitter. The AF signal modulates the RF power amplifier Q_3.

Transistor Q_1 is the active device of a Hartley oscillator. This particular oscillator has a variable-frequency output of 1 to 3.5 MHz. The frequency-determining components are C_1 and L_{T1}. The output of the oscillator is coupled to the base of Q_2 by capacitor C_3. The emitter-follower output of the oscillator has a very low impedance.

Transistor Q_2 is an RF signal amplifier. This transistor is primarily responsible for increasing the signal level of the oscillator. It is also used to isolate the RF load from the oscillator. This is needed to improve oscillator frequency stability. When Q_2 is used in this regard, it is called a *buffer amplifier*.

The signal output of Q_2 must be capable of driving the power amplifier. In high-power transmitters several RF signal amplifiers may be found between the oscillator and the power amplifier. Each stage is responsible for increasing the signal level to a suitable level. RF amplifiers of this type are often called drivers. Q_3 is an RF power amplifier. It is designed to increase the power level of the RF signal applied to its input. The output is used to drive the transmitting antenna. The load of Q_3 is a tuned circuit composed of C_{10}, L_5, and C_{11}. This is a pi-section filter. A filter of this type is used to remove signals other than those of the resonant frequency. C_{11} is the output capacitor of the filter. It is adjusted to match the impedance of the antenna. C_{10} is the input capacitor. It is used to resonate the filter to the applied carrier frequency. Resonance of the tuning circuit occurs when the collector current meter dips to its lowest value. In the broadcasting field an adjustment of this type is called "dipping the final." Q_3 is operated as a class C power amplifier.

The modulating component of the transmitter is developed by an AF amplifier. Q_4 and Q_5 are push-pull AF power amplifiers. The developed AF signal is applied to the modulation transformer T_2. This signal causes the collector voltage of Q_3 to vary at an AF rate instead of being dc. This causes the RF output to vary in amplitude according to the AF component. The output signal has a carrier and two sidebands. The power output is approximately 5 W.

Do not connect the output of this transmitter to an outside antenna unless you hold a valid radio-telephone operator's license. A load lamp is used for operational testing. The intensity of the load lamp is a good indication of the RF power developed by the transmitter.

A SIMPLE AM RECEIVER. An AM receiver has four primary functions that must be achieved in order for it to be operational. No matter how complex or involved the receiver is, it must accomplish these functions. They are: signal interception, selection, detection, and reproduction. Figure 16-24 shows a diagram of an AM receiver that accomplishes these functions. Note that the circuit does not employ an amplifying device. No electrical power source is needed to make this receiver operational. The signal source is intercepted by the antenna-ground network. The receiver is energized by the intercepted RF signal energy. A receiver of this type is generally called a crystal radio. The detector is a crystal diode.

The functional operation of our crystal diode radio receiver is very similar to the CW receiver. This particular circuit does not employ an RF amplifier, a beat frequency oscillator, or an AF amplifier. A strong AM signal is needed to make this receiver operational. Signal interception and selection are achieved in the same way as in the CW receiver.

The detection or demodulation function of an AM recever is responsible for removing the AF component from the RF signal. The detector is essentially a half-wave rectifier for the RF signal. Germanium diodes are commonly used as detectors. They are more sensitive to RF signals than a silicon diode.

The waveforms of Fig. 16-24 show how the crystal diode responds. The selected AM signal is applied to the diode. Detection is accomplished by rectification and filtering. The detected wave is a half-wave rectification version of the input. C_2 responds as an RF filter. It charges to the peak value of each RF pulse. It then discharges through the resistance of the earphone when the diode is reverse biased. The average value of the RF component appears across C_2. This is representative of the AF signal component. It energizes a small coil in the earphone. A thin metal disk in the earphone fluctu-

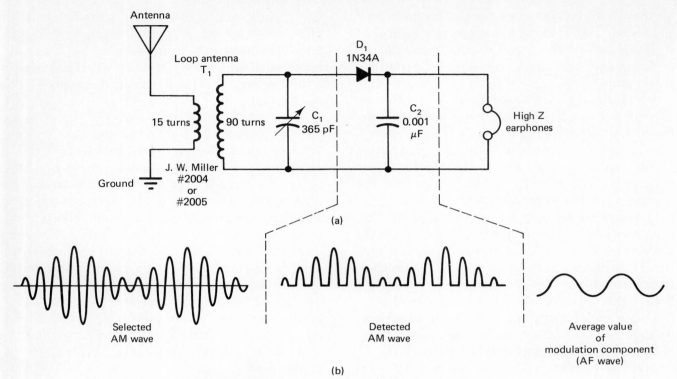

Figure 16-24.   Crystal radio receiver: (a) circuit; (b) waveforms.

ates according to coil energy. The earphone therefore changes electrical energy into sound waves. AF signal reproduction is achieved by the earphone.

SUPERHETERODYNE RECEIVERS.   Practically all AM radio receivers in operation today are of the superheterodyne type. This type of system accomplishes all of the basic receiver functions. It has a number of circuit modifications which provide improved reception capabilities. It has excellent selectivity and is very sensitive to long-distance signal reception. Figure 16-25(a) shows a block diagram of an AM superheterodyne receiver. This particular receiver is designed for signal reception between 550 and 1605 kHz.

Superhetrodyne operation is somewhat unusual compared with other methods of radio reception. Special RF amplifiers that are tuned to a fixed frequency are used in this circuit. These amplifiers are called *intermediate-frequency* (IF) amplifiers. Each incoming station frequency is changed into the IF. The IF amplifier responds in the same manner to all incoming signals. Each receiver has a tunable CW oscillator. This local oscillator generates an unmodulated RF signal of constant amplitude. This signal and the selected station are then mixed together.

Mixing or heterodyning these two signals produces the IF signal. The IF contains all the modulation characteristics of the incoming RF signal. The IF signal is the difference beat frequency. After suitable amplification, the IF signal is applied to the detector. The resulting AF output signal is then amplified and applied to the speaker for sound reproduction.

Figure 16-26 shows a schematic diagram of a representative AM superheterodyne receiver. Transistor Q_1 is a tuned RF amplifier. Adjustment of capacitor C_1 selects the desired RF signal frequency. Q_2 is the mixer stage. The selected RF signal and the local oscillator are mixed together in Q_2. Q_3 is the local oscillator. The oscillator signal is fed to the base Q_2 through C_8. Transistor Q_4 is the IF amplifier. The output of Q_2 is coupled to Q_4 by transformer T_4. This transformer is tuned to pass only an IF of 455 kHz. T_5 is the output IF transformer. The IF signal is coupled to the diode detector D_1 through this transformer. The detector rectifies the IF signal and develops the AF modulating component. C_{15} is the RF bypass filter capacitor and R_{12} is the volume control. Q_5 is the first AF signal amplifier. Push-pull AF power amplification is achieved by transistors Q_6 and Q_7. The AF output signal is changed into sound by the speaker. The entire circuit is energized

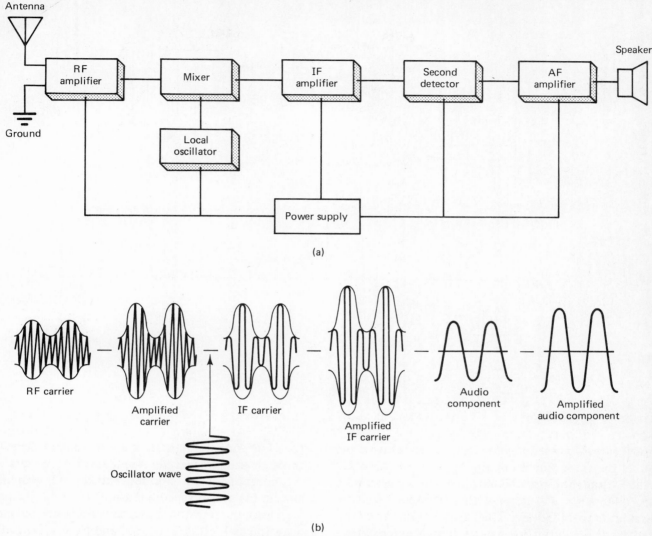

Figure 16-25. Superheterodyne receiver: (a) block diagram; (b) waveforms.

by a 9-V battery. Switch SW-1 is connected to the volume control and turns the circuit on and off.

Practically all of the superheterodyne receiver circuits have been discussed in conjunction with other system functions. The local oscillator, for example, is a variable-frequency Hartley oscillator. Mixing of the oscillator signal and selected RF signal has been described as heterodyning. Its output contains two input frequencies and two beat frequencies. The detector is essentially a crystal diode. An AF amplifying system is used to develop the audio signal component. The speaker was previously described as a transducer. All of these operations remain the same when used in the superheterodyne receiver. Operational frequencies and circuit performance are generally somewhat different than

previously described. We will therefore direct our attention to specific circuit performance.

Using Figs. 16-25 and 16-26, we will now see how the superheterodyne responds when tuned to a specific frequency. Assume now that a number of AM signals are intercepted by the antenna-ground network. The desired station to be received occupies a frequency of 1000 kHz. The tuner dial is adjusted 1000 kHz. The AF signal component then produces sound from the speaker. The volume control is adjusted to a desired signal level. The receiver is operational and performing its intended function.

In order for our receiver to develop a sound signal, a number of operational steps must be performed. Adjustment of the tuning dial changes the *LC* circuit of the RF amplifier. Capacitors C_1, C_2

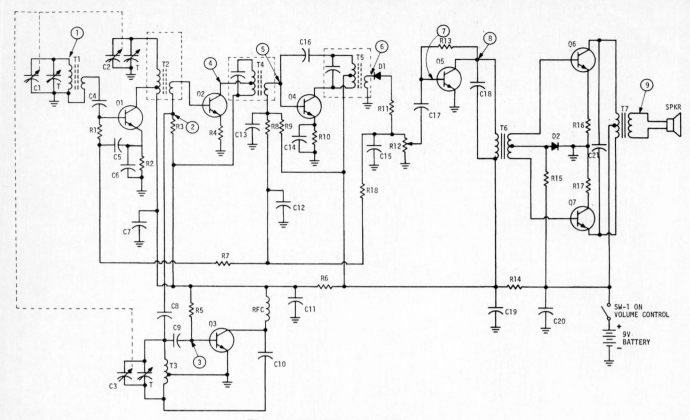

Figure 16-26. Transistor AM receiver.

and C_3 are all ganged together. One tuner adjustment alters the three tuned circuits at the same time. C_1 and C_2 tune the input and output of the RF amplifier to 1000 kHz. C_2 adjusts the frequency of the oscillator to 1455 kHz. The LC components of the oscillator are designed to generate a frequency that will be 455 kHz higher than the selected RF signal.

The 1000-kHz signal now passes through the input tuner and is applied to the base of Q_1. This common-emitter amplifier increases the voltage level of the applied signal. C_2 and T_2 are also tuned to pass the signal to the base of Q_2. The oscillator signal is coupled to the base of Q_2 through capacitor C_8. The incoming signal is AM and the oscillator signal is CW. By heterodyning action the collector of Q_2 has four signals. The fundamental frequencies are 1000 kHz AM and 1455 kHz CW. The beat frequencies are 2455 kHz AM and 455 kHz AM.

The IF amplifier has fixed tuning in its input and output circuits. T_4 is the input IF transformer. It is tuned to resonant at 455 kHz. This frequency will pass into the base of Q_4 with a minimum of opposition. The other three signals will encounter a very high impedance. 455 kHz will be amplified by Q_4. The signal is then coupled to the detector through transformer T_5. The detector recovers the AF component and applies it to the AF amplifier system for sound reproduction.

Assume now that the receiver is tuned to select a station at 1340 kHz. C_1, C_2, and C_3 all changed to the new signal frequency. The oscillator now develops a CW signal of 1795 kHz. The mixer has 1340 kHz and 1795 kHz applied to its input. The collector of Q_2 has these two fundamental frequencies plus 3135 kHz and 455 kHz. The IF amplifier section will process the 455-kHz signal and apply it to the detector. The AF component is then recovered and processed for reproduction.

Suppose now that the receiver is tuned to a station located at 600 kHz. This frequency would be amplified and applied to the mixer. The oscillator would now send a CW signal of 1055 kHz to the mixer. The collector of Q_2 will have signals of 600, 1055, 1655, and 455 kHz. The IF amplifier will again only pass the 455-kHz signal for reproduction.

The basic operation of an AM superheterodyne receiver should be obvious by this time. The oscillator is designed so that it will always be higher in

Figure 16-27. AM receiver IF transformers. (Courtesy of J. W. Miller Div./Bell Industries.)

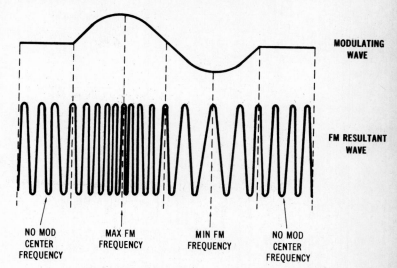

Figure 16-28. Frequency modulation waves.

frequency than the incoming station frequency. In effect, its frequency is the incoming station frequency plus the IF. The standard IF for AM receivers is 455 kHz. Through this type of circuit design the IF can be adjusted to a specific frequency. Figure 16-27 shows two IF transformers that are used in transistor AM receivers. The metal can surrounding the transformer shields it from stray electromagnetic fields. This type of transformer is designed for printed-circuit-board installations.

FREQUENCY MODULATION COMMUNICATION

Frequency modulation (FM) has a number of unique advantages over other communication systems. It uses very low audio signal levels for modulation, has excellent frequency reproduction capabilities, and is nearly immune to noise and interference. FM is commonly used in the VHF range of 30 to 300 MHz and extends into the UHF range of 300 to 3000 MHz. The commercial FM band is 88 to 108 MHz. FM is the dominant form of communication for private two-way mobile communication services.

In an FM communication system, intelligence is superimposed on the carrier by variations in frequency. An FM signal does not effectively change in amplitude. The modulating component does, however, cause the carrier to shift above and below its *center frequency*. Each FM station has an assigned center frequency. Receivers are tuned to this frequency for signal reception. When the carrier is unmodulated it rests at the center frequency.

Frequency allocations for FM are made by the FCC.

Examine the FM signal of Fig. 16-28. Note that the modulating component is low-frequency AF and the carrier is RF. In commercial FM, the modulating component could be an audio signal of 20 to 15 kHz. The carrier would be of some RF value between 88 and 108 MHz. As the modulating component changes from 0 to 90° it causes an increase in the carrier frequency. The carrier rises above the center frequency during this time. Between 90 and 180° of the audio component, the carrier decreases in frequency. At 180° the carrier returns to the center frequency. As the audio signal changes from 180° to 270° it causes a decrease in carrier frequency. The carrier drops below center frequency during this period. Between 270 and 360° the carrier rises again to the center frequency. This shows us that without modulation applied the carrier rests at the center frequency. With modulation, the carrier will shift above and below the center frequency.

We have seen that the modulating component of an FM system causes the carrier to shift above and below its center frequency. The amplitude of the modulating component is also extremely important in this operation. The larger the modulating signal, the greater the frequency shift. For example, if the center frequency of an FM signal is 100 MHz, a weak audio signal may cause a 1-kHz change in frequency. The carrier would change from 100.001 to 99.999 MHz. A strong audio signal could cause the carrier to change as much as 75 kHz. This action would cause a shift of 100.075 to 9.925 MHz. It is important to note that the amplitude of the carrier

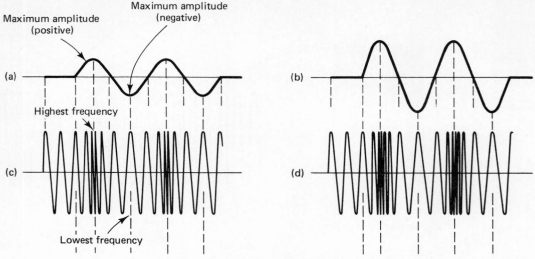

Figure 16-29.   Influence of amplitude on FM.

does not change. Figure 16-29 shows how the amplitude of the audio component changes the frequency of the carrier.

The frequency of the modulation component is another important consideration in FM. Modulating frequency determines the rate at which the frequency change takes place. With a 1-kHz modulating component the carrier will swing back and forth 1000 times per second. A modulating component of 10 kHz will cause the carrier to shift 10,000 times per second.

Figure 16-30 shows how the FM carrier responds to two different frequencies. Here t_1 has two cycles of modulation in the same time that t_2 has three cycles; t_2 is of slightly higher frequency than t_1.

Note how the carrier changes during the same period of time for each modulating component.

FM Communication System

A block diagram of an FM communication system is shown in Fig. 16-31. The transmitter and receiver respond as independent systems. In commercial FM each station has a transmitter. There can be an infinite number of FM receivers. In two-way mobile communication systems there is a transmitter and receiver at each location. FM mobile communications systems are classified as narrowband FM. This type of system only transmits voice signals. The carrier of a narrowband FM system

Figure 16-30.   FM due to frequency changes in modulating component.

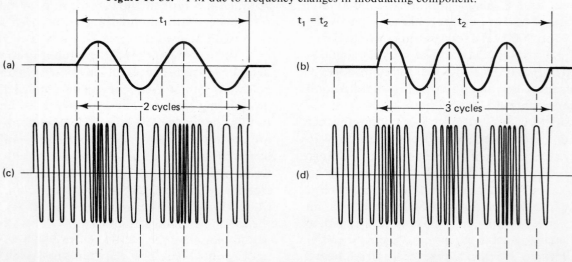

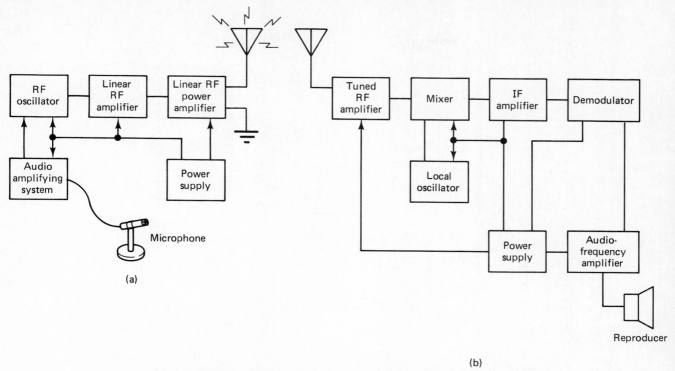

Figure 16-31. FM communication system: (a) direct FM transmitter; (b) superheterodyne FM receiver.

only deviates ±5 kHz. Commercial FM is considered to be wideband FM. The carrier of a commercial FM system deviates ±75 kHz at 100% modulation. The operational principles of narrowband and wideband FM are primarily the same for each system.

The transmitter of an FM system is primarily responsible for signal generation [see Fig. 16-31(a)]. Note that it employs an oscillator, an RF signal amplifier, and a power amplifier. The modulating component of an FM system is applied directly to the oscillator. AF changes in the applied signal cause the oscillator to deviate. Amplifiers following the oscillator are classified as linear devices.

The receiver of an FM system is responsible for signal interception, selection, demodulation, and reproduction [see Fig. 16-31(b)]. Most FM receivers are of the superheterodyne type. These receivers are similar in operation to the AM superheterodyne circuit. The selection frequency is much higher (88 to 108 MHz) for commercial FM. A standard IF of 10.7 MHz is used in an FM receiver today. The demodulator of an FM receiver is uniquely different from an AM detector. The FM demodulator is designed to respond to changes in the carrier frequency. The modulating component is recovered

from these frequency changes. The remainder of the circuit is the same as the AM superheterodyne.

FM Transmitters. There is a rather extensive list of different FM transmitters available today. Wireless microphones have an output of approximately 100 mW. Maritime mobile FM communication systems operate with a few watts of power. Two-way land mobile communication employs low-power transmitters. Educational FM transmitters can operate with a power of several hundred thousand watts. Commercial FM transmitters have a similar power rating. In addition to these types are public service radio and military FM. The frequency allocation and power capabilities of an FM transmitter are allocated and regulated by the FCC.

An FM transmitter regardless of its power rating or operational frequency has a number of fundamental parts. The oscillator is responsible for the development of the RF carrier. A large number of FM transmitters employ direct oscillator modulation. This means that the frequency of the oscillator is shifted by direct application of the modulation component. The RF signal is then processed by a number of amplifiers. The power amplifier ulti-

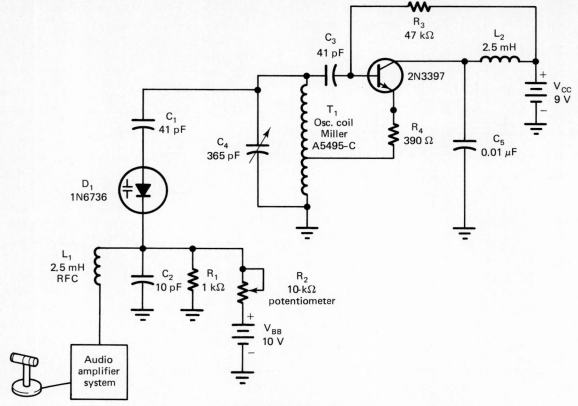

Figure 16-32. Direct FM transmitter.

mately feeds the FM carrier into the antenna for radiation.

A simplified direct FM transmitter is shown in Fig. 16-32. A basic Hartley oscillator is used as the RF signal source. With power applied, the oscillator will generate the RF carrier. The frequency of the oscillator is determined by the value of C_4 and T_1. Without modulation, the oscillator would generate a constant frequency. This is representative of the unmodulated carrier wave.

Modulation of the oscillator is achieved by voltage changes across a *varicap diode*. The capacitance of this diode changes with the value of the reverse bias voltage. Bias voltage is used to establish a dc operating level for this diode. A change in audio signal voltage causes the capacitance of the diode to change at an audio rate. Note that the series network of C_1, D_1, and C_2 is connected in parallel with capacitor C_4 of the oscillator tank circuit. A change in audio signal level will cause a corresponding change in tank circuit capacitance. This in turn causes the oscillator frequency to change according

to the modulating component. The modulating component is applied directly to the oscillator's frequency determining components.

Direct FM can be achieved in a number of ways. The varicap diode method is commonly found in new transmitter equipment. This application is only one of a large number of circuit variations of the varicap diode. Circuits of this type are particularily well suited for voice communication in narrowband FM.

The power output of an FM transmitter is dependent primarily on its application. In some systems the oscillator output can be applied directly to the antenna for radiation. In other systems the power level must be raised to a higher level. The output of the oscillator would be applied to linear amplifiers for processing. The amplification function would be primarily the same as that achieved by CW and AM systems. We will not repeat the discussion of RF power amplification for FM transmitters. The primary difference in FM power amplification is the resonant frequency of the tuned circuits. FM usually operates in the VHF band. This generally calls for smaller capacitance and inductance values in the resonant circuits.

FM Receivers. FM signal reception is achieved in basically the same way as AM and CW. An FM superheterodyne circuit is designed to respond to frequencies in the VHF band. Commercial FM signal reception is in the range 88 to 108 MHz. Higher-frequency operation generally necessitates some change in the design of antenna, RF amplifier, and mixer circuits. These differences are due primarily to the increased frequency rather than the FM signal. The RF and IF sections of an FM receiver are somewhat different. They must be capable of passing a 200-kHz bandwidth signal instead of the 10-kHz AM signal. The most significant difference in FM reception is in the demodulator. This part of the receiver must pick out the modulating component from a signal that changes in frequency. In general, this circuit is more complicated than the AM detector. The AF amplifier section of an FM receiver is generally better than an AM receiver. It must be capable of amplifying frequencies of 30 to 15 kHz. Figure 16-33 shows a block diagram of an FM superheterodyne receiver. Some differences in FM reception are indicated by each functional block.

FM Demodulation. The demodulator of an FM receiver is the primary difference between AM and FM superheterodyne receivers. A number of demodulators have been used to achieve this receiver function. Today the ratio detector seems to be the most widely used. This circuit is an outgrowth of ealier FM discriminators. Solid-state diodes and transformer design have made the ratio detector a very practical circuit.

A basic ratio detector is shown in Fig. 16-34. The operation of this detector is based on the phase relationship of an FM signal that is applied to the input. This signal comes from the IF amplifier section of the receiver (see Fig. 16-33). It is 10.7 MHz and deviates in frequency. Design of the circuit causes two IF signal components to appear in the secondary winding of the transformer. One part of this component is coupled by capacitor C_1 to the center connection of the transformer. This resulting voltage is considered to be V_1. Note the location of V_1 in the circuit and voltage line diagrams.

The second IF signal component is developed inductively by transformer action. Design of the secondary winding causes two voltage values to be developed. These voltages are labeled V_2 and V_3 in the circuit and voltage line diagrams. They are of equal amplitude and 180° out of phase with each other. The center-tap connection of the secondary winding is used as the common reference point for these voltage values.

The resulting secondary voltage is based on the combined component voltage values of V_1, V_2, and V_3. V_{D1} and V_{D2} are the resulting secondary voltage. These two voltage values will appear across diodes D_1 and D_2. The developed voltage for each diode is

Figure 16-33. Block diagram of an FM receiver.

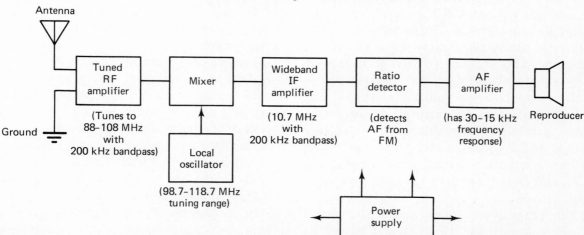

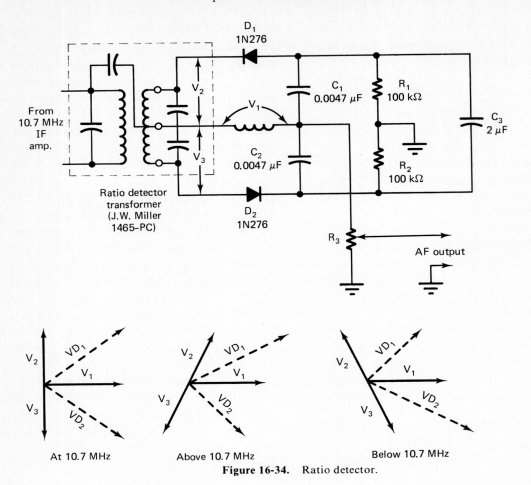

Figure 16-34.   Ratio detector.

based on the phase relationship of the two IF components. Note the location of V_{D1} and V_{D2} in the circuit and the voltage line diagram.

Operation of the ratio detector is based on voltage developed by the transformer for diodes D_1 and D_2. With no modulation applied to the FM carrier, the transformer will have a 10.7-MHz IF applied to the transformer. The developed voltage values are shown by the center-frequency voltage line diagram. This method of voltage display is generally called a *phasor* diagram. A phasor shows the relationship of voltage values (line length) and their phase relationship (direction). At the center frequency note that the resulting diode voltage values (V_{D1} and V_{D2}) are of the same length. This means that each diode will receive the same voltage value. D_1 and D_2 will conduct an equal amount of current.

When the carrier swings above the center frequency it causes the IF to swing above its resonant frequency. Note the resulting phasor for this change in frequency. V_{D1} is longer than V_{D2}. This means that D_1 will conduct more current than D_2. In the same regard, note how the phasor changes when the frequency swings below resonance. This condition will cause D_1 to be less conductive and D_2 to be more conductive. Essentially, this means that the IF signal is translated into different diode voltage values.

Let us now see how the input RF voltage is used to develop an AF signal. The two diodes of the ratio detector are connected in series with the secondary winding and capacitors C_1 and C_2. For one alternation of the input signal the two diodes are reverse biased. No conduction occurs during this alternation [see Fig. 16-35(a)]. For the next alternation both diodes are forward biased. The input signal voltage is then rectified [see Fig. 16-35(b)]. Essentially, this means that the incoming signal is changed into a pulsating waveform for one alternation.

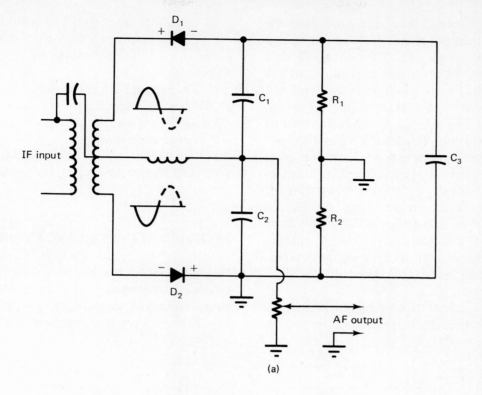

(a)

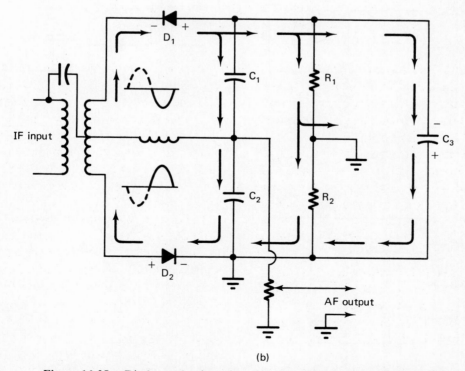

(b)

Figure 16-35. Diode conduction of a ratio detector: (a) nonconduction; (b) conduction.

Figure 16-36 shows how the ratio detector responds when the input signal deviates above and below the center frequency. Keep in mind that conduction occurs only for one alternation of the input. Part (a) shows how the circuit will respond when the input is at its 10.7-MHz center frequency. Each diode has the same input voltage value for this condition of operation. Capacitors C_1 and C_2 will charge to equal voltages as indicated. With respect to ground, the output voltage will be -2 V. Note this point on the output voltage waveform in part (d).

Assume now that the input IF signal swings to 10.8 MHz. This condition causes D_1 to receive more voltage than D_2. C_1 charges to -3 V while C_2 charges to -1 V. With respect to ground, the output voltage rises to -1 V. Note this point on the output voltage waveform.

The input IF signal then swings to 10.6 MHz. This condition causes D_2 to receive more voltage

than D_1. C_2 charges to -3 V while C_1 charges to -1 V. With respect to ground, the output voltage drops to -3 V. See this point on the output voltage waveform.

The output of the ratio detector is an AF signal of 2 V p-p. This signal corresponds to the frequency changes placed on the carrier at the transmitter. In effect, we have recovered the AF component from the FM carrier signal. This signal can then be amplified by the AF section for reproduction.

TELEVISION COMMUNICATION

Nearly everyone has had an opportunity to view a television communication system in operation. This communication process plays a very important role in our lives. Very few people spend a day without

Figure 16-36. Ratio detector response to frequency.

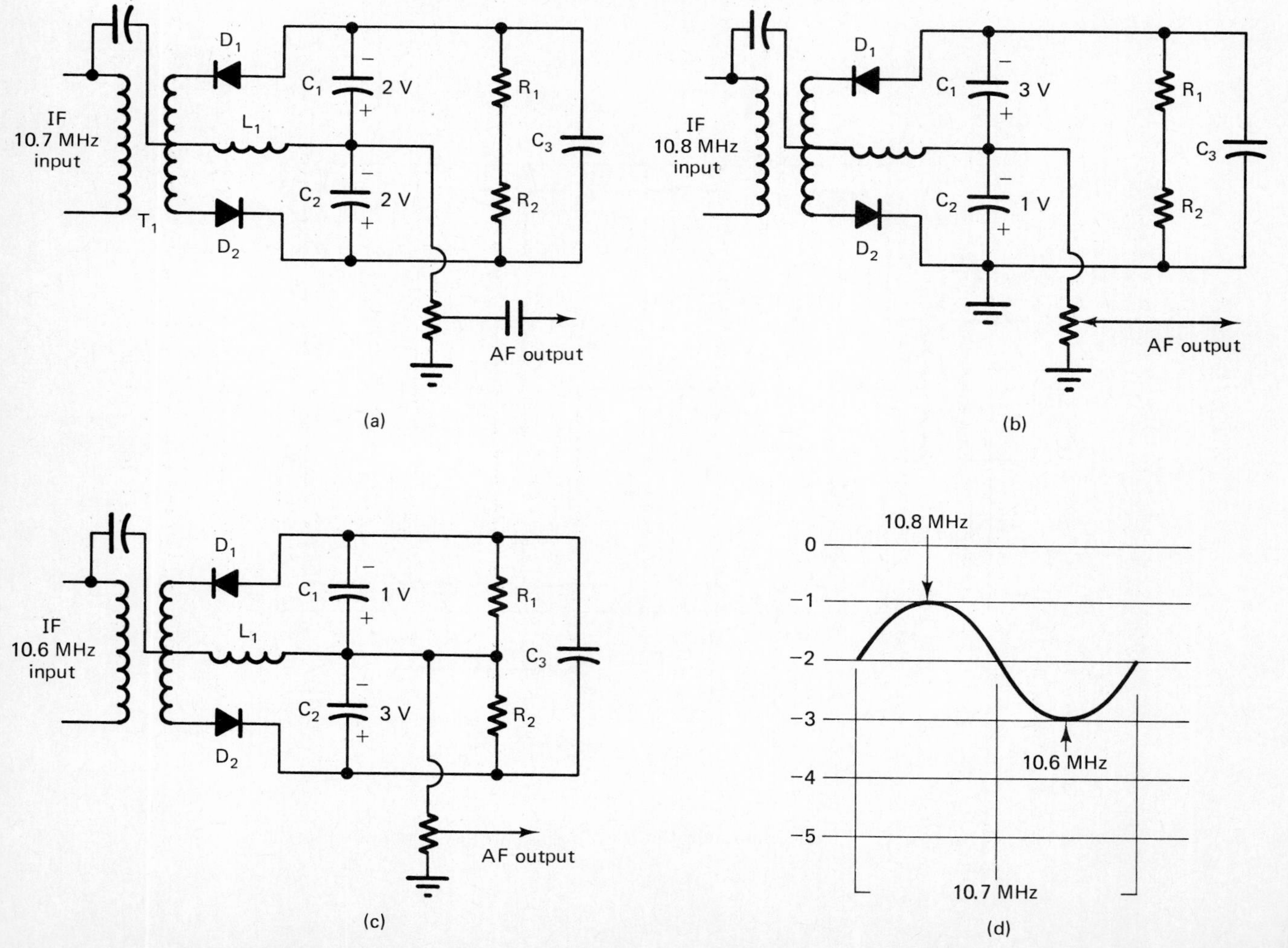

(a)

(b)

(c)

(d)

watching television. It is probably the most significant application of electronics today.

The signal of a television system is quite complex. It is made up of a number of unique parts or components. Basically, the transmitted signal has a picture carrier and a sound carrier. These two signals are transmitted at the same time from a single antenna. The picture carrier is amplitude modulated. The sound carrier is frequency modulated. A television receiver intercepts these two signals from the air. They are tuned, amplified, demodulated, and ultimately reproduced. Sound is reproduced by a speaker. Color and picture information are reproduced by a picture tube. Television is primarily a one-way communication system. There is a central transmitting station and an infinite number of receivers. A simplification of the television communication system is shown in Fig. 16-37.

The television camera of our system is basically a transducer. It changes the light energy of a televised scene into electrical signal energy. Light energy falling on a highly sensitive surface varies the conduction of current through a resistive material. The resulting current flow is proportional to the brightness of the scene. An electron beam scans horizontally and vertically across the light-sensitive surface. Vidicon tubes and plumbicons are employed in TV cameras today.

Figure 16-38 show a simplification of the television camera tube. Note that a complete circuit exists between the cathode and power supply. Electrons are emitted from the heated cathode. These are formed into a very thin beam and directed toward the back side of the photoconductive layer. Conduction through the layer is based on the intensity of light from the scene being televised. A bright or intense light area becomes low resistant. Dark areas have a higher resistance. As the electron beam scans across the back of the photoconductive layer it sees different resistance values. Conduction through the layer is based on this resistance. A discrete area with low resistance causes a large current through the layer. Dark areas cause less current flow. Current flow is directly related to the light intensity of the televised scene. Output current flow appears across the load resistor (R_L). Voltage developed across R_L is amplified and ultimately used to modulate the picture carrier. In practice, the developed camera tube voltage is called a video signal. "Vide" means something to see. A camera tube sees things electronically.

The scene being televised by a camera tube must be broken into very small parts called picture elements. For this to take place, it is necessary to scan the light-sensitive surface of a camera tube with a stream of electrons. This process is very similar to

Figure 16-37. TV communication system: (a) transmitter; (b) receiver.

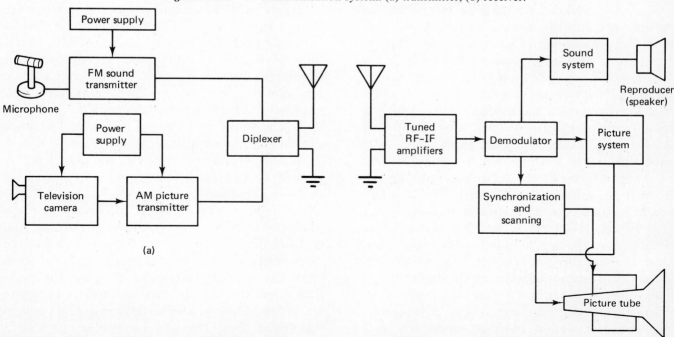

(a)

(b)

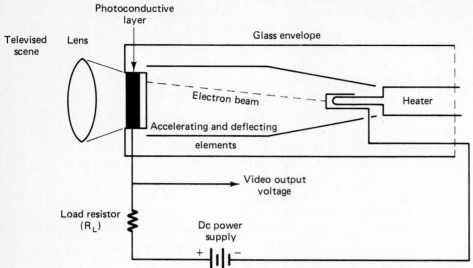

Figure 16-38. Simplification of a TV camera tube.

reading a printed page. Letters, words, and sentences are placed on the page by printing. We do not determine what is on a printed page at one instant. Our eyes must scan the page starting at the upper left-hand corner one line at a time. They move left to right, drop down one line and quickly return to the left, then scan right again for the next line. The process continues until all lines are scanned.

In a similar way the electron beam of a camera tube scans the back surface of the photoconductive layer. The electron beam is deflected horizontally and vertically by an electromagnetic field. A coil fixture known as the *deflection yoke* is placed around the neck of the camera tube. This coil deflects the electron beam. Current flow in the deflection yoke is varied so that the field rises to a peak value, then drops to its starting value. Figure 16-39 shows the deflection yoke current and the resulting electron beam scanning action. Notice that each line has a *trace* and a *retrace time*. During the trace period the line is scanned from left to right. This takes a rather large portion of the complete sawtooth wave. Retrace occurs when the beam returns from right to left. Notice that this takes only a small portion of the total waveform. The same condition applies to the vertical sweep waveform.

Figure 16-39 shows another rather unique difference in the scanning lines. During the trace time, the scanning line is solid. This indicates that the electron beam is conducting during this time. It also shows a broken line during the retrace period. This indicates that the electron beam is nonconductive during this period. In effect, conduction occurs

during the trace time and no conduction occurs during retrace. This same condition applies to the vertical trace and retrace time.

The scanning operation of a camera tube requires two complete sets of deflection coils. One set is for horizontal deflection and one for vertical. The current needed to produce deflection comes from two sawtooth oscillators. A vertical blocking oscillator and a horizontal multivibrator could be used for this operation. In U.S. television, the horizontal sweep frequency is 15750 Hz and the vertical frequency is 60 Hz.

To produce a moving television scene there must be at least 30 complete pictures produced per second. In a TV system a complete picture is called a frame. The U.S. television system has 525 horizontal lines in a *frame*. These lines are not scanned progressively, such as 1, 2, 3, 4, 5, and so on. They are, however, divided into two fields. One field contains 262.5 odd-numbered lines. Lines 1, 3, 5, 7, 9, 11, and so on, make up the odd field. The other field has 262.5 even-numbered lines. Lines 2, 4, 6, 8, 10, and so on, are included in the even field. A complete picture or frame has one odd-lined field and one even-lined field.

Picture production in a TV system employs *interlace scanning*. To produce one frame, the odd-line field is scanned first. After scanning all the odd-numbered lines, the electron beam is deflected from the bottom position to the top. The even-numbered-line field is then scanned. A complete frame has 262.5 odd-numbered lines and 262.5 even-numbered lines. The odd-line field starts with a

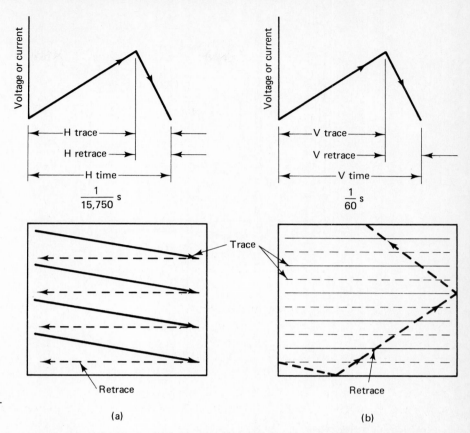

Figure 16-39. Scanning and sweep signals: (a) horizontal; (b) vertical.

(a)   (b)

complete line and ends with a half line. The even field starts with a half line and ends with a complete line.

The picture repetition rate of U.S. television is 30 frames per second. Since two fields are needed to produce one frame, the vertical frequency is 2×30, or 60 Hz. This particular frequency was chosen to coincide with the ac power line frequency. In some foreign countries the vertical frequency is 50 Hz.

In television signal production, the vertical and horizontal sweep circuits must be properly synchronized in order to produce a picture. The signal sent out by the transmitter must contain *synchronization* or *sync* information. This signal is used to keep the oscillator of the receiver in step with the correct signal frequency. Separate generators are used to develop the sync signal. This is added to the video signal developed by the camera.

Picture Signal

The picture signal of a TV transmitter contains a number of important parts. Each part of the signal plays a specific roll in the operation of the system. The video signal, for example, is developed by the camera tube. It represents instantaneous variations in scene brightness. Figure 16-40(a) shows the video signal for one horizontal line.

The video signal of a television system has *negative picture phase*. This means that the highest-amplitude part of the signal corresponds to the darkest picture area. Bright picture areas have the lowest amplitude level. Signal levels that are 75% of the total amplitude range are considered to be in the black region. All light disappears in this region. Signal-level amplitude percentages are shown in Fig. 16-40(a).

In addition to the video information, the picture signal must also provide some way of cutting off the electron beam at certain times. When scanning occurs the electron beam is driven to cut off during the retrace period. Horizontal retrace occurs at the end of one line and vertical retrace occurs at the end of each field. A rectangular pulse of sufficient amplitude is needed to reach cutoff. This condition permits the electron beam to retrace without producing unwanted lines. This part of the signal is called *blanking*. A composite signal has both horizontal and vertical blanking pulses. Figure 16-40(a) shows the location of a horizontal blanking

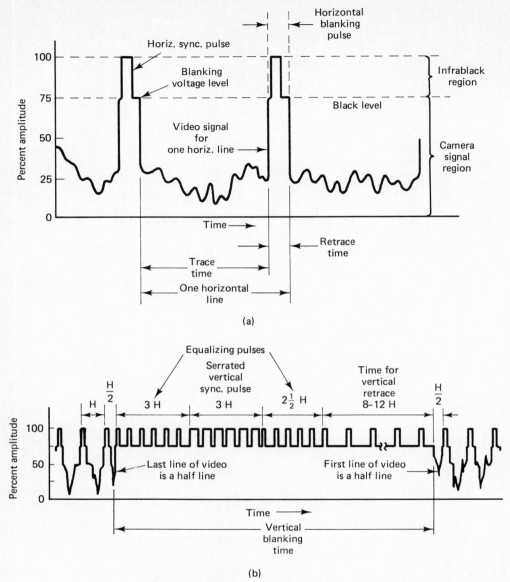

Figure 16-40.   Composite TV picture signal.

pulse. Vertical blanking occurs after 262.5 horizontal lines. Figure 16-40(b) shows the vertical blanking time.

A composite TV picture signal also has vertical and horizontal synchronization pulses. These pulses ride on the top of the blanking pulses. Horizontal sync pulses are shown in both parts of Fig. 16-40. The serrated pulses of part (b) provide continuous horizontal sync during the vertical retrace time. The vertical sync pulse is made up of six rectangular pulses near the center of the vertical blanking time. The width of a vertical sync pulse is much greater than that of the horizontal sync pulse. All of these pulses, plus blanking and the video signal, are described as a composite picture signal.

Television Transmitter

A television transmitter is divided into two separate sections or divisions that feed outputs into a common antenna. The video section is responsible for the picture part of the signal. A crystal oscillator is used for carrier wave generation. As a general rule, the frequency is multiplied to bring it up to an allocated channel in the VHF or UHF band. An intermediate power amplifier and a final power amplifier follow the last multiplier. The modulating component is a composite picture signal. It contains video, blanking pulses, sync, and equalizing pulses. The composite signal is amplified and ultimately applied to the final power amplifier. The final is amplitude modulated by the composite picture signal. This section of the transmitter is essentially the same as that of a commercial AM station. There is an obvious difference in frequency and power output. Figure 16-41 shows a block diagram of a black-and-white TV transmitter.

A unique difference in TV and commercial AM transmitter circuitry appears after the final power amplifier. The TV output signal is applied to a *vestigial sideband* filter. This filter is designed to remove all lower sideband frequencies 1.25 MHz below the carrier frequency. The entire upper sideband of the signal is transmitted. A large portion of the lower sideband is suppressed. This is purposely done to reduce the frequency occupied by a channel. With the lower sideband suppressed, the bandwidth of a TV channel is 6 MHz.

The vestigial sideband signal is then applied to a diplexer. The *diplexer* is a filter circuit that isolates the picture and sound carriers. Essentially, the sound carrier will not pass into the picture section and the picture carrier will not pass into the sound section. This prevents undesirable interaction between the two carrier signals.

The sound section of a TV system is primarily an FM transmitter. It is very similar to a commercial FM transmitter. The center frequency of the FM carrier is always 4.5 MHz above the picture carrier. Carrier deviation is ±25 kHz in a TV system.

Figure 16-41. Monochrome TV transmitter.

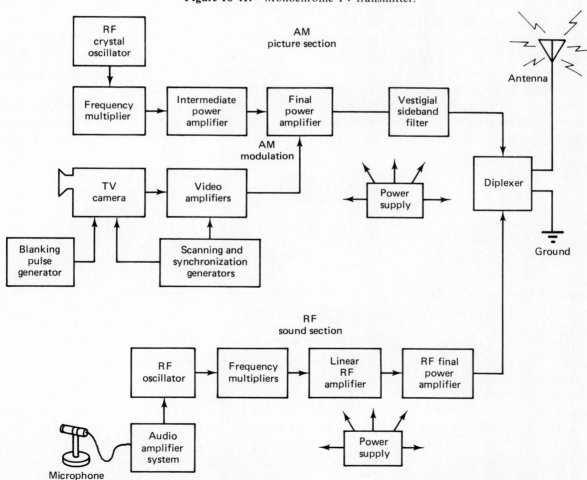

Modulation is normally applied to the oscillator of the FM sound system. The remainder of the sound section is similar to that of a commercial FM transmitter. See the FM sound section of the transmitter in Fig. 16-41.

Television Receiver

A television receiver is designed to intercept the electromagnetic waves sent out by the transmitter and use them to develop sound and a picture. The received signals are in the VHF or UHF band. The FCC has allocated a 6-MHz bandwidth for each TV channel. Channels 2 through 13 are the VHF band. These frquencies are from 54 to 216 MHz. Channels 14 to 83 are in the UHF band. This ranges from 470 to 890 MHz. All channels in the immediate area induce a signal into the antenna. A desired station is selected by altering a tuning circuit. This *LC* circuit passes only the selected channel and rejects the others.

A functional block diagram of a black-and-white television receiver is shown in Fig. 16-42. Practically all of this circuitry has been used in other communication systems. The front end of a TV receiver for example, is a superheterodyne circuit. It has a tuned RF amplifier, a mixer, and an oscillator. This section of the receiver is called the tuner. It is housed in a shielded metal container to reduce

interference. The output of the tuner is an IF signal. A standard IF for TV is 41.25 MHz for the sound and 45.75 MHz for the picture. The IF must pass both sound and picture carriers. This necessitates a 6-MHz bandpass for the IF amplifiers.

The demodulation function of a TV receiver is achieved by a diode. This circuit is primarily an AM detector. The output of the detector has all the picture information placed on the carrier at the transmitter. After demodulation the picture carrier is discarded. Three different kinds of signal information appear at the output of the detector. It recovers the video signal and sync signals for immediate use. The sound carrier passes into the sound IF section.

The video signal recovered by the detector is processed by the video amplifier. The output of this section is then used to control the brightness of the electron beam of the picture tube. A dark spot detected by the TV camera will cause a corresponding reduction in picture tube brightness. A bright spot or an intense picture element will cause the picture tube to conduct very heavily. This will cause a corresponding bright spot to appear on the picture tube face. The video signal developed by the camera of the transmitter will be faithfully reproduced on the picture tube screen.

The sync signal of the video detector is used to synchronize the vertical and horizontal sweep oscillators. These oscillators develop the sawtooth waves

Figure 16-42.   Monochrome TV receiver.

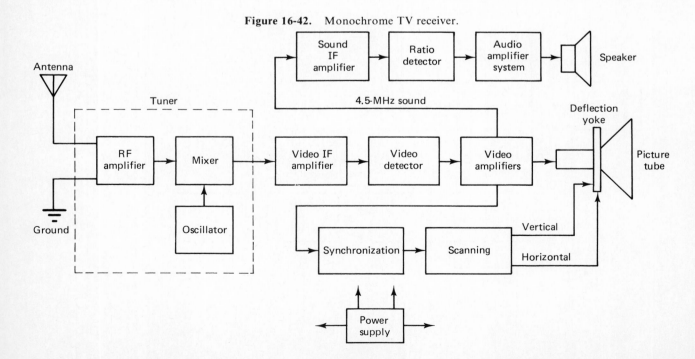

that are used to deflect the electron beam. The electron beam must be in step with the transmitted signal in order for the picture to be usable. The transmitted sync signal is recovered and used to trigger the two receiver sweep oscillators.

An AM video detector does not effectively respond to frequency changes in the IF sound carrier. It does, however, heterodyne the two signals together. The difference between the 45.75- and 41.25-MHz IF signals is 4.5 MHz. This signal takes on the FM modulation characteristic of the sound carrier. 4.5 MHz is called the sound IF. The sound IF deviates ±25 kHz above and below 4.5 MHz.

A ratio detector can be used to *demodulate* the sound IF signal. The recovered audio component is then processed by an AF amplifier. It is ultimately used to drive a loudspeaker for sound reproduction. The FM sound section of a TV receiver is very similar to that of a standard FM broadcast receiver.

The picture tube of a TV receiver is responsible for changing an electrical signal into light energy. A picture tube is also called a cathode-ray tube (CRT). The intensity of an electron beam changes as it scans across the face of the tube. The inside face of the tube is coated with phosphor. When the electron beam strikes tiny grains of phosphor, it produces light. A combination of different light and dark phosphor grains causes a picture to appear on the inside of the

face area. The resulting picture can be observed by viewing the front of the face area.

Figure 16-43 shows a simplification of the CRT of a TV receiver. The tube is divided into three parts. The gun area is responsible for electron beam production. The coil fixture attached to the neck of the tube deflects the electron beam. The viewing area changes electrical energy into light energy.

A CRT is a thermionic emission device. Heat applied to the cathode causes it to emit electrons. These electrons form into a fine beam and move toward the face of the tube. A high positive charge is placed on a conductive coating on the inside of the glass housing. This coating serves as the anode. The electron beam is attracted to this area of the tube because of its positive charge. Voltage is supplied to the coating by an external anode connection terminal. The anode voltage of a black and white CRT is normally 15 kV.

After leaving the cathode, electrons pass through the control grid. Voltage applied to the grid controls the flow of electrons. A strong negative voltage will cause the electron beam to be cut off. Varying values of grid voltage will cause the screen to be illuminated. The video signal developed by the receiver can be used to control the intensity level of the electron beam. Focusing of the electron beam is achieved by the next two electrodes. They shape the

Figure 16-43. Cathode-ray tube.

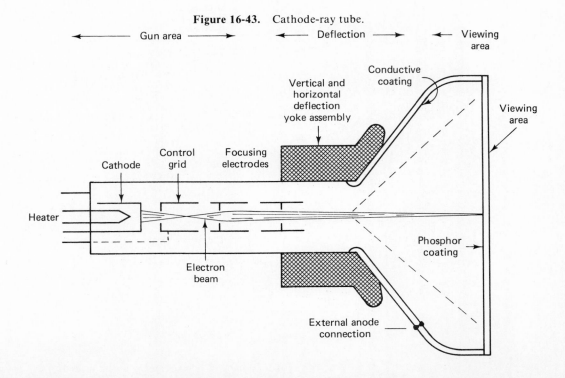

electron beam into a very fine trace. These electrodes are generally called focusing anodes.

Electromagnetic deflection of the electron beam is achieved by the yoke assembly. This coil fixture produces an electromagnetic field and the electron beam causes deflection of the beam. Both vertical and horizontal deflection are needed to produce scanning.

Color Television

The transmission of a color picture by television complicates the system to some extent. The principles involved in transmission are primarily the same as those of *monochrome* or black-and-white television. One of the problems of color TV is that it must be *compatible*. This means that programs designed for color reception must also be received on a monochrome receiver. The transmitted signal must therefore contain color information as well as the monochrome signal. All of this must fit into the 6-MHz channel allocation.

The picture portion of the color signal is basically the same as the monochrome signal. The video signal is, however, made up of three separate color signals. Each color signal is produced by an independent camera tube. One camera tube is used to develop a signal voltage that corresponds to the red content of the scene being televised. Blue and green camera tubes are used in a similar arrangement to produce the other two color signals. Red, green, and blue are considered to be the primary colors of the video signal. White is a mixture of all three colors. Black is the absence of all three primary colors.

COLOR CAMERAS. Figure 16-44 shows a simplified color television camera. The scene being televised is focused by a lens onto a special mirror. This mirror reflects one-third of the light and passes two-thirds of the light. The reflected image goes through a filter and is applied to the blue camera tube. Two-thirds of the image passing through the mirror is applied to a divider mirror. One-third of it

Figure 16-44.   Simplified color TV camera.

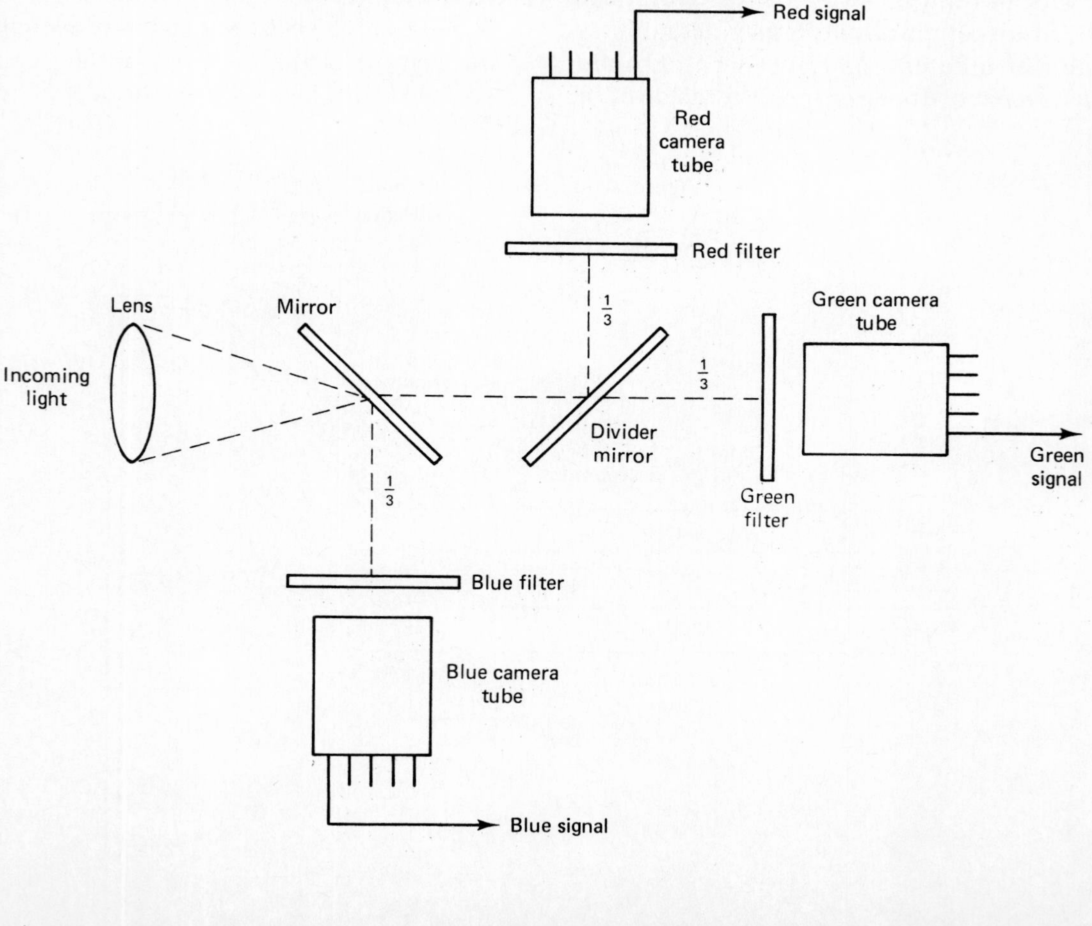

is applied to the red camera and one-third to the green camera. Each camera tube provides an output signal that is proportional to the light level of the primary color. The brightness or *luminance* signal is a mixture of the three primary colors. These proportions of the color signal are 59% green, 30% red, and 11% blue. The luminance signal is generally called a *Y signal*. A monochrome receiver responds only to the Y signal and produces a standard black-and-white picture.

The human eye needs two stimuli to perceive color. One of these is called *hue*. Hue refers to a specific color. Green leaves have a green hue and a red cap has a red hue. The other consideration is called *saturation*. This refers to the amount or level of color present. A vivid color is highly saturated. Light or weak colors are diluted with white light. A scene being televised by a color camera has hue, saturation, and brightness.

COLOR TRANSMITTERS.   Figure 16-45 shows a block diagram of a color transmitter. The color section adds significantly to the transmitter. The remainder of the diagram has been reduced to a few blocks. This part of the transmitter is essentially the same for either a color or a black-and-white system.

In a color transmitter three separate color signals are developed by the TV camera. These colors are applied to a matrix and a mixer circuit. The mixer develops the luminance signal. Green, red, and blue are mixed in correct proportions. Luminance is the equivalent of a black-and-white signal. The output of the mixer (Y signal) goes directly to the carrier modulator. This signal is the same as the modulating component of a black-and-white transmitter.

The matrix is a rather specialized mixing circuit. It combines the three color signals into an I and a Q signal. These two signals can be used to represent any color developed by the system. The *Q signal* corresponds to green or purple information. The *I signal* refers to orange or cyan signals. These two signals are used to amplitude modulate a 3.58-MHz subcarrier signal. The carrier part of this signal is suppressed to prevent interference. The I signal is kept in phase with the 3.58-MHz subcarrier. The Q signal is one quadrant or 90° out of phase with the *subcarrier*. The I and Q signals are sideband frequencies of the subcarrier. Only this sideband information is used to modulate the transmitter picture carrier.

A synchronization signal is also generated by

Figure 16-45.   Color TV transmitter block diagram.

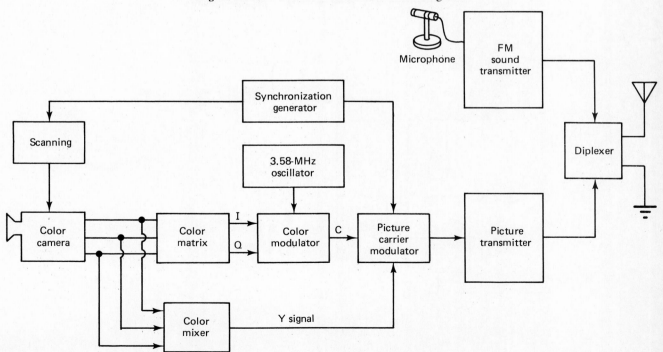

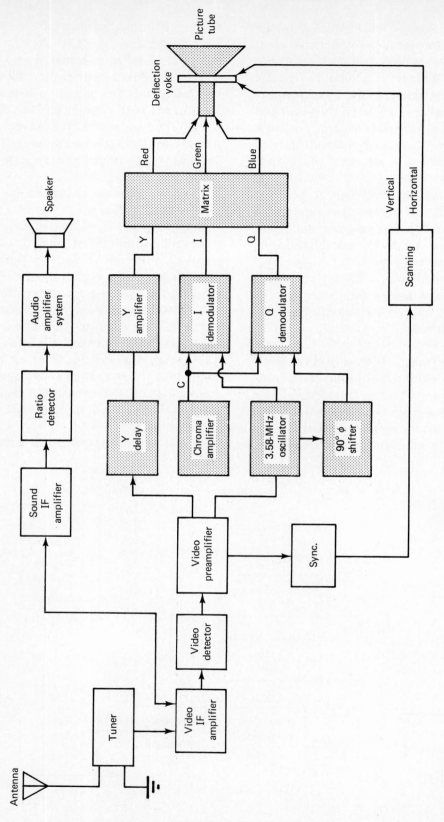

Figure 16-46. Color TV receiver block diagram.

the color transmitter. This signal is applied to both the modulator and the camera scanning circuit. Scanning of the three camera tubes must be synchronized with the modulated carrier output. The sync signal is an essential part of the picture carrier modulation component. The modulating component contains I, Q, Y, and sync signals.

COLOR RECEIVERS. A color television receiver picks up the transmitted signal and uses it to develop sound and a picture. It has all the basic parts of a monochrome receiver plus those needed to recover the original three color signals. Figure 16-46 shows a block diagram of the color TV receiver. The shaded blocks denote the color section of the receiver. This is where the primary difference in color and monochrome TV receivers occurs.

Operation of the color receiver is primarily the same as that of a black-and-white receiver up to the video demodulator. After demodulation, the composite video signal is then divided into chroma (C) and Y signals. The Y or luminance signal is coupled to a delay circuit. This slows down the Y signal. It will now get to the matrix at the same time as the chroma signal. The C signal is applied to the chroma amplifier. This signal must pass through a great deal more circuitry before reaching the matrix.

Remember that the chroma signal contains I and Q color and a 3.58-MHz suppressed carrier. To demodulate this signal, the 3.58-MHz signal must be reinserted. Notice that the 3.58-MHz oscillator is connected to the I and Q demodulators. This is where the carrier reinsertion function takes place.

The I demodulator receives a 3.58-MHz signal directly from the oscillator. This is where the in-phase signal is derived. The 3.58-MHz signal fed into the Q demodulator is shifted 90°. This is where the quadrature signal is derived. I and Q color signals appear at the output of the demodulator.

The matrix of the receiver has three signals applied to its input. The Y signal contains the luminance information. The I and Q signals are the demodulated color signals. The matrix combines these three signals in proper proportions. Its output comprises the original red, green, and blue color signals. These signals are applied to the R, G, and B guns of the picture tube. The electron beam of each gun strikes closely spaced red, green, and blue phosphor spots on the face of the tube. These spots glow in differing amounts according to the signal level. Any color can be reproduced on the face of the tube by the three primary colors.

Review

1. Why does RF radiate away from an antenna better than AF?
2. How is information placed on the carrier wave of a CW system?
3. What are the system parts of a CW transmitter?
4. Explain how the receiver of a CW system recovers the transmitted information.
5. How is the RF signal selection function accomplished?
6. Explain the heterodyning function.
7. Explain how voice signals are sent out by an AM transmitter.
8. What is meant by the following AM terms?
    a. Carrier wave
    b. Modulating component
    c. Sidebands
    d. Percentage of modulation
9. What are the primary functions of an AM receiver?
10. Describe how AM reception is achieved by a superheterodyne receiver.

11. What causes the IF of a superheterodyne receiver to be the same for all incoming signals?

12. Describe some of the primary differences between AM and FM communication systems.

13. How is the carrier modulated in an FM system?

14. Explain how an FM signal is demodulated.

15. What is vestigial-sideband transmission?

16. How is interlaced scanning achieved?

17. Describe the relationship of fields, frames, and the transmitted picture signal.

18. What is the primary difference between monochrome and color cameras?

19. Explain how the picture component of a TV transmitter is developed.

20. Explain what occurs during the trace and retrace time of one horizontal line.

21. Describe the negative picture phase concept.

22. Why is horizontal and vertical blanking important in picture production?

23. What is the luminance signal?

24. How does a color transmitter differ from a monochrome transmitter?

25. Explain how a color signal modulates the RF carrier.

26. Describe how a color signal develops a picture in the receiver.

Student Activities

CW Transmitter

1. Construct the CW transmitter of Fig. 16-8.

2. Wrap a few turns of wire around T_1. Connect an oscilloscope to the coil.

3. Close the code key and observe the waveform. Open and close the key while observing the waveform.

4. Place an AM radio receiver near the antenna coil. Turn it on and tune the receiver for an audio tone at the speaker.

5. Key the transmitter to see if it is producing the tone.

6. If the signal is close to a local station, adjust C_1 to a different frequency.

7. Send several messages in Morse Code.

8. Test the transmitting range of the system.

9. Retain the CW transmitter circuit for addition activities.

Beat Frequency Oscillator

1. Construct the BFO of Fig. 16-17.

2. Apply power to the circuit. Measure and record the dc voltage at the emitter, base, and collector of Q_1.

3. With an oscilloscope, observe the output waveform.

4. Make a sketch of the observed waveforms.

5. Retain the BFO oscillator circuit.

Heterodyne Detection

1. Construct the Heterodyne detector circuit of Fig. 16-16.

2. Use the BFO of Fig. 16-17 for the F_2 input.

3. Use the CW transmitter of Fig. 16-8 for the CW input. Only one CW transmitter is needed for the entire class.

4. Place the antenna coil near the CW transmitter. Key the transmitter.

5. Tune the detector input and BFO for an AF output signal. An oscilloscope may be used to observe the signal.

6. It may be possible to hear the detected signal with a set of high-impedance earphones. Remove R_3 and connect the earphone in its place.

7. Adjust the BFO for a zero beat and for frequencies above and below the beat frequency.

AM Transmitter

1. Construct the AM transmitter of Fig. 16-22.

2. Turn on the switch and test the circuit for operation.

3. Connect 4 to 5 ft. of wire to the antenna connection.

4. Tune an AM receiver to the transmitted signal.

5. The transmitter can be tuned to a different frequency by adjusting C_4.

6. Record the operating frequency.

7. Retain an operating AM transmitter for the next activity.

AM Crystal Radio Receiver

1. Construct the crystal radio receiver of Fig. 16-24. Note: T_1 may be constructed by winding 15 turns and 90 turns of No. 28 wire on a 1-in.-diameter paper tube.

2. Place the receiving antenna close to the AM transmitting antenna of Fig. 16-22.

3. Ask another person to speak into the transmitter microphone.

4. Tune the receiver to the transmitted signal.

5. If an outside antenna is available, connect the receiver to the outside antenna and a ground terminal.

6. Tune in a local AM station.

AM Receiver Circuitry

1. If a superheterodyne AM receiver is available, you may want to investigate its circuit.

2. Locate such things as the antenna coil, tuning capacitor, IF transformers, and speaker interstage coupling transformers.

3. If a schematic is available, compare the circuit and the location of its components.

4. Describe some of the obvious features of the receiver.

AM Receiver Circuit Construction

1. A number of commercial companies have AM radio receiver kits available for construction. These kits are fairly inexpensive and rather easy to assemble.

2. Investigate the availability of these kits as a take-home class project.

Direct FM Transmitter

1. Construct the direct FM transmitter of Fig. 16-32.

2. Turn on the dc power and test the circuit.

3. Connect an oscilloscope across T_1.

4. Speak into the microphone while observing the RF waveform on the oscilloscope.

5. R_2 is adjusted to set the sweep range of D_1.

6. Describe how modulation changes the oscilloscope waveform.

FM Wireless Microphone

1. Construct the FM wireless microphone of Fig. 16-47.

2. Tune in an FM receiver to the transmitted signal.

3. Record the transmitting frequency.

4. Change the frequency by adjusting C_6.

Figure 16-47.   Wireless FM microphone.

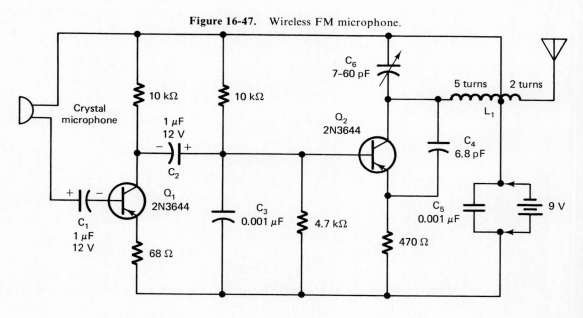

All resistors are $\frac{1}{2}$ W 10%

C_1 and C_2: 1-μF 12-V Mallory 12 X 1

C_3 and C_5: 0.001-μF 1-kV Spragus 56A–D10

C_4 : 6.8-pF 1-kV Sprague 5gA–V68

C_6 : 7-60 pF ARCO 404

L_1 : 7 turns of #19 AWG wire in a $\frac{5}{16}$ -in. coil, tapped at 2 turns

(a)

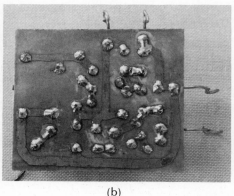
(b)

(c)

Figure 16-48. Wireless FM microphone: (a) component layout; (b) PC board; (c) completed project.

FM Wireless Microphone Project

1. Develop a printed circuit board for the wireless microphone of Fig. 16-47.
2. Figure 16-48 shows the component layout (a), sample PC board (b), and the completed project in a box (c). The board size is 2 in. $\times$ $3\frac{1}{2}$ in.
3. The placement of L_1 and C_6 is quite important. Keep them as close together as possible.
4. Solder in all components and assemble the unit.
5. Test its operating range.

Ratio Detector Circuit

1. Construct the ratio detector of Fig. 16-34.
2. Apply a 10.7-MHz FM input to the circuit.
3. Without modulation, measure the voltage at the AF output.
4. Modulate the input and measure the AF output voltage.
5. Measure the voltage across C_1 and C_2 with and without modulation.

Monochrome Television

1. Remove the back cover of a monochrome TV receiver. If possible, remove the chassis of the receiver from the cabinet. Use safety goggles if the CRT is removed. Avoid striking the CRT. Replace the knobs to each control. Be very careful to avoid contact with the chassis and the electrical parts of the receiver. In some cases it may be necessary to use an isolation transformer to reduce ground problems.
2. Locate the picture tube or CRT. Note the front viewing area, back side of the tube, neck, and socket. Look for the electron gun in the neck of the CRT.

3. Locate the deflection yoke. Describe the structure of the yoke and its location.

4. Locate the tuner. There should be two separate chassis units. One is for VHF and one for UHF. What channels are controlled by each tuner? How is a channel selected?

5. Make a list of the receiver operational controls. Briefly explain what each control achieves. Turn on the receiver and carefully tune in a local station. Adjust each operator control. Describe its function.

6. Locate the service controls of the receiver. What function does each control achieve?

7. Examine the speaker. What is its approximate size?

8. How is ac power supplied to the receiver chassis?

9. Explain the type of receiver examined. Is it solid state, vacuum tube, or a combination?

10. Briefly describe the general circuit construction. Is it a wired chassis, does it have printed circuit boards, or has it a combination of both?

11. Turn off the power and return the chassis to the cabinet.

Color TV Receivers

1. Using a color TV receiver, note the location of the operator controls. Make a list of these controls.

2. What are the convenient service controls at the rear of the chassis? Make a list of these controls.

3. Turn on the receiver and adjust each operator control. Describe the function of each control.

4. Adjust the service controls. Describe the function of each control.

5. If time permits, remove the back cover. With the power off, note the location of the CRT, chassis, speaker, and tuner.

6. Describe the type of chassis employed. Is it solid state, vacuum tube, or a combination of both?

7. Note the location of the deflection yoke. How does it compare with the monochrome receiver?

8. How does the tuner compare with that of a monochrome receiver?

9. Return the cover to the receiver. Turn it on and adjust it to produce a good picture.

17

Digital Electronic Systems

Digital electronics is undoubtedly the fastest-growing area in the field of electronics today. Computers, calculators , watches, clocks, video games, and test instruments are only a few of the applications. Most of these things were unheard of a few years ago. Digital electronics now plays an important role in our daily lives.

Our discussion of electronics up to this point has centered primarily around analog applications. An *analog* is a method of determining one quantity in terms of another. Temperature, for example, is often determined by the position of a column of mercury. Voltage, current, and resistance can be determined by a coil of wire that interacts with a magnetic field. Analog devices are usually concerned with continuously changing values. An analog value could be 1, 3, 5, 8, 10.3, 117, or any one of an infinite number of values.

Digital electronics is considered to be a counting operation. A digital watch tells time by counting generated pulses. The resulting count is then displayed as numbers representing hours, minutes, and seconds. A computer also has an electronic clock that generates pulses. These pulses are counted and in many cases manipulated to perform a control function. Digital circuits can store signal data, retrieve it when needed, and make operational decisions. Signal values are generally represented by two-state data. A pulse is either present or it is not. Data are either of a high value or a low value. There is nothing between these two values. Digital electronics is dependent primarily on the manipulation of two-state data.

IMPORTANT TERMS

In this chapter we are going to investigate some of the basic principles of digital electronics. You will have a chance to become familiar with a number of basic digital system functions. New words such as flip-flop, gates, counters, and decoding will begin to have some meaning for you. A few of these terms are singled out for study. As a rule, the chapter will be more meaningful if these terms are reviewed before proceeding with the text.

Alphanumeric. A numbering system containing numbers and letters.

Analog. Being continuous or having a continuous range of values.

Assembly language. A computer programming language that uses abbreviations in mnemonic form.

Base. The number of symbols in a number system. A decimal system has a base of 10.

Binary. A two-state or two-digit numbering system.

Binary-coded decimal. A code of 10 binary numbers for the decimal values from 0 to 9.

Central processing unit (CPU). A functional part of a computer that achieves control, arithmetic logic, and signal manipulation.

Counter. A digital circuit capable of responding to state changes through a series of binary changes.

Decade counter. A counter that achieves 10 states or discrete values.

Decoding. A digital circuit that changes coded data from one form to a different form.

Digital. A value or quantity related to digits or discrete values.

Encoding. A converter that changes an input signal into binary data.

Fetch. A computer instruction that retrieves data from memory.

Firmware. Permanently installed instructions in computer hardware that control operation.

Flip-flop. A circuit that changes back and forth between two states.

Gate. A circuit that performs special logic operations such as AND, OR, NOT, NAND, AND NOR.

Hardware. Electronic circuits, parts, and physical parts of a computer.

High-level language. A form of computer programming that uses words and statements.

Interface. A process or piece of equipment in a computer that brings two things together.

Keyboard. A device with keys or switches that are used in digital equipment.

Logic. A decision-making capability of computer circuitry.

Machine language. A form of computer programming that involves direct coding of operations and memory addresses.

Memory. The storage capability of a device or circuit.

Mnemonic. An abbreviation that resembles a word or reminds one of a word.

Negative logic. A system where low or "0" has voltage and high or "1" has little or no voltage.

Nonvolatile memory. A computer storage function that retains data without the use of electrical power. Disk and tape memory are nonvolatile.

Positive logic. A system where low or "0" has no voltage and high or "1" has voltage.

Programmable read-only memory (PROM). A memory device that can be programmed by the user.

Radix. The base of a numbering system.

Reading memory. Sensing binary information that has been stored on a magnetic tape, disk, or IC chip.

Read-only memory. A form of stored data that can be sensed or read. It is not altered by the sensing process.

Software. Instructions, such as a program that controls the operation of a computer.

Toggle. A switching mode that changes between two states, such as on and off.

Truth table. A graph or table that displays the operation of a logic circuit with respect to its input and output data.

Volalite memory. Stored binary data that are lost or destroyed when the electrical power is removed.

Writing memory. Placing binary data into the storage cells of a memory device.

DIGITAL SYSTEMS

A digital system is somewhat unique compared with other electrical systems. As in all electrical systems, an energy source is needed to make the digital system operational. Dc electricity is primarily used as an operational energy source. This may be obtained from the ac power line and changed into a usable form of dc. Rectifier power supplies are commonly used in most systems. Portable systems are energized by batteries. Small digital systems can operate for a rather long period of time from a single battery. The dc electrical source is primarily used to energize the active components of the system. Today digital systems are largely of the IC type.

A digital system usually has internal signal generation. The digital signal is primarily a number of electrical pulses that occur in a given unit of time. Pulse generators and electronic clock circuits are used for this function. This part of the system is generated by an oscillator. As a rule, this type of generator produces square waves or spiked pulses. This signal is then processed through the system to achieve its counting function. In a sense, a digital system has an operational energy source and a digital signal source. Both sources are developed and applied directly to the system when it is made operational.

The other functions of a digital system are somewhat unusual. The *path* of electrical energy and the digital signal is primarily achieved by the metal foil of a printed circuit board. There is very little hand wired electrical circuitry in a digital system today. *Control* is primarily achieved by logic gates. Gates contain transistors, MOSFETs, and diodes. As a rule, most gates are achieved by ICs. The *load* of a digital system is quite unusual. The load on the operational source is all the components that are energized by the supply. The load of the digital signal is determined by the number of gates it supplies. The load in both cases does work. *Indicators* are included in nearly every system. Digital displays and cathode-ray tubes are widely used as indicators. This function is the most obvious part of the system. Operation is generally based on the response of the display.

Decimal Numbering Systems

Digital information has been used by human beings almost all of their history. Parts of the body were first used as a means of counting. Fingers and toes were often used to represent numbers. In fact, the word "digitus" in Latin means finger or toe. This term is the basis of the word *digital*.

Most counting that we do today is based on groups of 10. This is probably an outgrowth of our dependence on fingers and toes as a counting tool. Counting with 10 as a base is called the decimal system. Ten unique symbols or numbers are included in this system. The numbers 0, 1, 2, 3, 4, 5, 6, 7, 8, and 9 are the base values of the decimal system. In general, the number of discrete values or symbols in a counting system is called the *base* or *radix*. A decimal system has a base of 10.

Nearly all numbering systems have place value. This refers to the value that a number has with respect to its location in the counting sequence. The largest number value that can be represented at a specific location is determined by the base of the system. In the decimal system, the first position to the left of the decimal point is called the units place. Any number from 0 to 9 can be used in this place. Number values greater than 9 are expressed by using two places. The next location to the left of the units place is the 10's position. Any number value from 0 to 99 can be expressed by two places. Each succeeding place added to the left extends the numbering capability by the power of 10. Three places would permit a count of $10 \times 10 \times 10$, or 1000. Four places would be $10 \times 10 \times 10 \times 10$, or 10,000. The system continues for 100,000, 1,000,000, 10,000,000, and so on.

A specific number value of any base can be expressed by the addition of each weighted place value. The decimal number 2319 would be expressed as $(1000 \times 2) + (100 \times 3) + (10 \times 1) + (1 \times 9)$. Note that the weight of each number increases by the power of 10 for each place to the left of the decimal point. In a numbering system, place values can also be expressed as a power of the base. For the decimal system, this would be 10^3, 10^2, 10^1, and 10^0. Each succeeding place would be an expression of the next power value of the base.

The base 10 or decimal numbering system is extremely important and widely used today. Electronically, however, the decimal system is rather difficult to use. Each number would require a specific value to distinguish it from the others. Number detection would also require some unique method of distinguishing each value from the others. The electronic circuitry of a decimal system would be rather complex. In general, it is difficult to achieve and awkward to maintain.

Binary Numbering Systems

Nearly all digital electronic systems are of the *binary* type. This type of system uses 2 as its base or radix. The largest number that can be expressed at a place location is the number 1. Only the numbers 0 and 1 are used in a binary system. Electronically, only two voltage values are needed to express binary numbers. This can be a voltage value and no voltage. The number one is usually referenced as some voltage value greater than zero. Binary systems that use voltage as 1 and no voltage as 0 are described as having *positive logic*. *Negative logic* has voltage for 0 and no voltage for 1. Positive logic is more readily used today. Only positive logic will be expressed in this discussion.

The two operational states or conditions of an electronic circuit can be expressed as on and off. An off circuit usually has no voltage applied. This would be representative of the "0" or "off" state. An on circuit has voltage applied. This could be used as the "1" or "on" state. With the use of electronic devices it is possible to change states in a microsecond or less. Millions of "1" and "0" can be manipulated by a digital system in a second.

The operational basis of a binary system is very similar to that of the decimal system. The base of the binary system is 2. This means that only the numbers 0 and 1 are used to denote a specific value. The first place to the left of the binary point is the units or 1's location. Place values to the left are expressed as

powers of 2. Representative values are $2^0 = 1$, $2^1 = 2, 2^2 = 4, 2^3 = 8, 2^4 = 16, 2^5 = 32, 2^6 = 64$, and so on.

A binary number such as 101 is the equivalent of the decimal number 5. Starting at the binary point, the place values would be $1 \times 2^0 + 0 \times 2^1 + 1 \times 2^2$ or $1 + 0 + 4 = 5$. The conversion of a binary number to a decimal value is shown in Fig. 17-1.

A shortcut version of the binary-to-decimal conversion process is shown in Fig. 17-2. In this conversion method, first write the binary number. In this example, 1001101 is used. Starting at the binary point, indicate the decimal value for each binary place location with a 1. Do not indicate a value for the zero places. Add the indicated place value assignments. Record the decimal equivalent of the binary number. Practice this procedure on several different binary numbers. With a little practice the conversion process is very easy to achieve.

Changing a decimal number to a binary number is achieved by repetitive steps of division by the number 2. When the quotient is even with no remainder a "0" is recorded. A remainder is also recorded. In this case, it will always be the number "1." The steps needed to convert a decimal number into a binary number are shown in Fig. 17-3.

The conversion process is achieved by first recording the decimal value. The decimal number 30 is used in this example. Divide the recorded number by 2: 30 ÷ 2 equals 15. The remainder is "0." Record the "0" as the first binary place value. Transfer 15 to

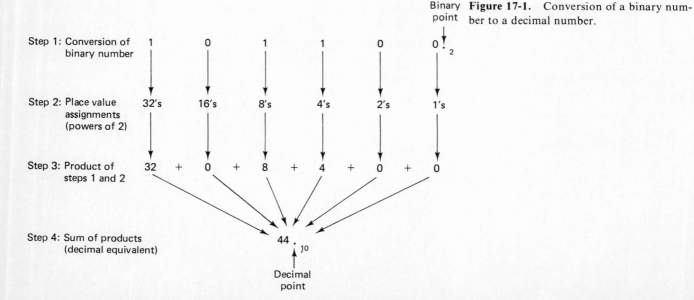

Figure 17-1. Conversion of a binary number to a decimal number.

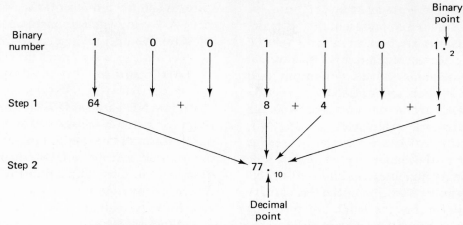

Figure 17-2. Binary-to-decimal conversion shortcut.

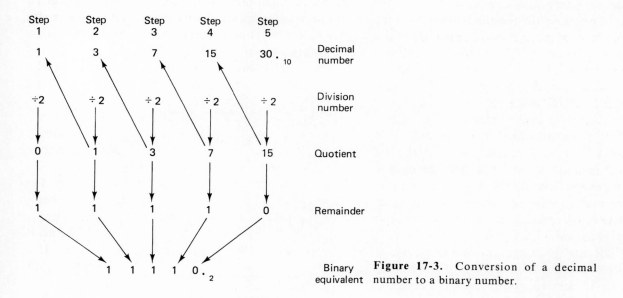

Binary equivalent **Figure 17-3.** Conversion of a decimal number to a binary number.

position 2. Divide this value by 2. 15 ÷ 2 equals 7 with a remainder of 1. Record the remainder. Transfer 7 to position 3. 7 ÷ 2 equals 3 with a remainder of 1. Record the remainder and transfer 3 to position 4. 3 ÷ 2 equals 1 with a remainder of 1. Record the remainder and transfer 1 to position 5. 1 ÷ 2 equals "0" with a remainder of 1. The conversion process is complete when the quotient equals a value of "0." The binary equivalent of a decimal is indicated by the recorded remainders. For this example the binary equivalent of 30 is 11110.

Practice the decimal-to-binary conversion process with several different number values. With a little practice the process becomes relatively easy to accomplish.

SYSTEM OPERATIONAL STATES

Digital electronic systems must employ very precise operational states or values in order to be useful. As a rule, binary signals are commonly used today. This type of signal can be processed very quickly through the system. It has two operational states. These can be defined as on–off, 1 or 0, voltage or no voltage, up or down, right or left, or any other type of two-state designation. There is no in-between condition of operation for a binary designation. The two states must be uniquely different and easy to recognize.

The operational states of a binary system are commonly indicated by letters, numbers, and symbols. In positive logic, a voltage, an on condition, or a letter such as A are used to represent the "1" state. No voltage, off, and $\overline{A}$ are used to indicate the "0" state. A discrete device can be set to either operational state. It usually remains in this condition until something causes it to change states.

A variety of electronic devices have two-state operational capabilities. Transistors, diodes, ICs, relays, lamps, and switches are commonly used. A bistable device has the capability of storing one binary digit. A binary digit is essentially a "1" or "0." The word *bit* is a shorted version of the term *binary digit*. A bistable device has the capability of storing one bit of data. By employing a number of these devices, it is possible to develop a digital instrument that will make decisions. The output of this type of device is directly related to the data applied to its input. Bistable devices are commonly used in binary logic circuits.

BINARY LOGIC FUNCTIONS

Any bistable device can be used to achieve a binary logic function. This term refers to its ability to respond to a two-state input signal and develop a corresponding output signal. This type of circuit has the ability to make logical decisions. Three basic circuits are used in digital electronics to make simple logic decisions. These are generally called gates. AND, OR, and NOT gates are defined as basic logic functions. A gate is made up of a bistable device and all its circuit components. The logic decision made by each gate is uniquely different from the others.

The logic function of a gate is usually described by some type of an "if–then" statement. For example, *if* the input to an AND gate has two 1's, *then* its output will be 1. *If* a one is applied to either input of an OR gate, the output will *then* be a 1. If a 1 is applied to the NOT gate, the output will *then* be 0.

Digital electronics is directly related to the operation of basic logic gates. Anyone working with this type of equipment must be familiar with basic gate functions. In this case we are concerned primarily with input–output characteristics. A single-pole, single-throw toggle switch is used as a two-state input device. A lamp serves as the output of the gate. An "on" condition is considered to be a "1"

state. When the lamp is off the output is in the "0" state. A closed switch is considered to be on or in the "1" state. An open switch is off or in the "0" state.

AND Gates

An AND gate has two or more inputs and one output. The logic decision of an AND gate is based on the status of its input. If both or all inputs are on, the output will be on. Mathematically, this is described as a multiplication function. One times one equals one. Zero times one, one times zero, and zero times zero all produce zero outputs.

Figure 17-4(a) shows a simple switch-lamp AND logic circuit. The switches achieve the input function. The lamp responds as the output. Each input has two states. A two-input circuit potentially has 2×2 or four possible input combinations. A resulting output signal will be developed for each input combination. An operational table for the

Figure 17-4. AND gates: (a) logic circuit; (b) truth table; (c) symbol.

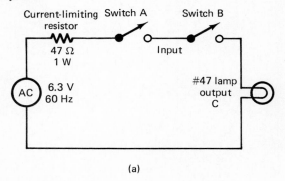

(a)

Input		Output
Switch A	Switch B	Lamp C
0	0	0
0	1	0
1	0	0
1	1	1

(b)

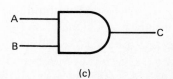

(c)

AND gate is shown in Fig. 17-4(b) and a logic symbol is shown in part (c). Operational tables are commonly called truth tables.

The input combinations of our lamp-switch AND gate are 00, 01, 10, and 11. This corresponds to the binary count 0, 1, 2, and 3 in order. Note that there is only an output at C when the inputs are both 1. This is expressed as $A \cdot B = C$. This gate will only develop an output when inputs A AND B are both 1.

OR Gates

An OR gate is designed to have two or more inputs and one output. With two input control devices, there will be 2×2, or four, possible output combinations. There is a corresponding output for each input combination. Figure 17-5 shows a simple lamp-switch OR gate in part (a), its logic symbol in part (b), and a truth table for the circuit in part (c).

Figure 17-5. OR gates: (a) logic circuit; (b) truth table; (c) symbol.

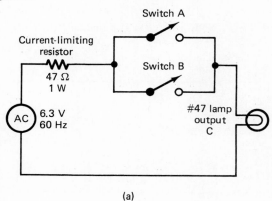

(a)

Input		Output
Switch A	Switch B	Lamp C
0	0	0
0	1	1
1	0	1
1	1	1

(b)

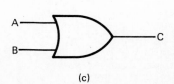

(c)

Functionally, an OR gate will produce a "1" output when either A or B or both inputs are 1. Mathematically, this is expressed as $A + B = C$. This is called OR addition. OR gates are primarily used to decide whether or not a 1 appears at the input. The overhead lamp or interior lights of an automobile are controlled by an OR circuit. Individual door switches and the dash panel switch are all used to achieve control. Each switch serves as an input device with the lamp responding as the output.

NOT Gates

The NOT function is somewhat unusual compared with other gate functions. A NOT gate has one input and one output. The input and output both have two states or conditions of operation. The output of a NOT gate is the reverse of its input. A "1" input, for example, produces a "0" output. A "0" input will generate a "1" output. In effect, this gate has a polarity changing or inverting function.

Figure 17-6(a) shows a simple switch-lamp NOT gate. When the switch is off, the lamp is on.

Figure 17-6. NOT gates: (a) logic circuit; (b) truth table; (c) symbols.

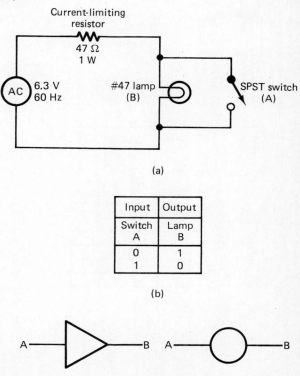

(a)

Input	Output
Switch A	Lamp B
0	1
1	0

(b)

(c)

Closing the switch shorts out the lamp. NOT gates are commonly used for logic signal inversion. The truth table and logic symbol of a NOT gate are shown in parts (b) and (c).

Mathematically, a NOT gate inverts or complements a number. This is expressed in words as an A input causes the output to NOT be A. This is expressed in equation form as $A = \overline{A}$. Applications of the NOT function are rather apparent when used in logic gate combination.

Combination Logic Gates

When two or more of the three basic logic gates are connected, we have combinational logic gates. An AND gate followed by a NOT gate is called a NOT-AND or NAND gate. Essentially, the output is an inversion of the AND function.

A lamp-switch NAND gate is shown in Fig. 17-7(a). Control of this circuit is achieved by switches A and B. The output is represented by the lamp.

Note that the lamp will be on for three of the input combinations. When switches A and B are both "on" the lamp will be off. Mathematically, this is expressed by the equation $A \cdot B = \overline{C}$. The bar over a letter indicates the inversion of the function. NAND gates are frequently used to achieve other combinational logic functions. The truth table and logic symbol of a NAND are shown in parts (b) and (c). Compare this truth table with the AND gate of Fig. 17-4.

When an OR gate is followed by a NOT gate the NOT-OR or NOR function is achieved. This function is primarily an inversion of the OR output. Remember that an OR gate has a "1" output whenever a "1" appears in the input. A NOR gate only has a "1" output when the two inputs are both "0."

A lamp-switch version of the NOR gate is shown in Fig. 17-8(a). Note that the input switches are in parallel with lamp C. Whenever a switch is in the "1" state it shorts out the lamp. This means that the lamp will be on or in the "1" state only when both

Figure 17-7. NAND gates: (a) logic circuit; (b) truth table; (c) symbol.

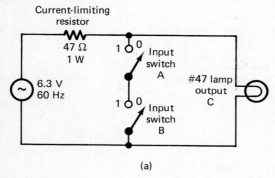

Input		Output
Switch A	Switch B	Lamp C
0	0	1
0	1	1
1	0	1
1	1	0

(b)

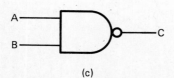

(c)

Figure 17-8. NOR gates: (a) logic circuit; (b) truth table; (c) symbol.

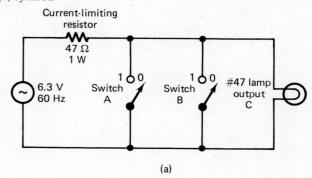

Input		Output
Switch A	Switch B	Lamp C
0	0	1
0	1	0
1	0	0
1	1	0

(b)

(c)

switches are off or "0." A NOR truth table and symbol are also shown in parts (b) and (c). Comparing the truth table of the NOR gate with the OR gate shows the inversion function.

GATE CIRCUITS

Gate circuits can be achieved today in a variety of ways. The switch-lamp version of the gate circuit was used here only to demonstrate logic functions. Actual logic gates are usually quite different. Relays, transistors, diodes, and ICs can all be used in the construction of a logic gate. The switching function can be achieved by one or more discrete components. Conduction is primarily based on forward and reverse biasing of a device. The switching rate of a solid-state gate can be achieved very quickly.

Logic circuits were first built with discrete components. Integrated-circuit technology has, however, caused a rather decided change in gate circuit construction. Complete gate circuits, and in many cases several individual gates, are now built in one IC chip. As a general rule, these ICs are now mass produced at a very nominal price. Complex logic functions can be readily accomplished by connecting several ICs. An understanding of different logic functions has become extremely important today. When selecting an IC for a specific application, one simply looks at the pin connections and truth table. The logic symbols and truth tables used in this presentation are representative of those found in a manufacturer's data manual.

TIMING AND STORAGE ELEMENTS

Digital electronics involves a number of items which are not classified as gates. Circuits or devices of this type have a unique role to play in the operation of a system. Included in this are such things as timing devices, storage elements, counters, decoders, memory, and registers. Truth tables, symbols, operational characteristics, and applications of these items will be presented here. Today, these circuits or devices are primarily built on an IC chip. The internal construction of the chip cannot be effectively altered. Operation is controlled by the application of an

external signal to the input. As a rule, very little can be done to control operation other than altering the input signal.

FLIP-FLOPS

In Chapter 15 we discussed bistable multivibrators. This type of device was used to generate a square wave. It could also be triggered to change states when an input signal is applied. A bistable multivibrator, in a strict sense, is a flip-flop. When it is turned on, it assumes a particular operational state. It does not change states until the input is altered. A flip-flop has two outputs. These are generally labeled Q and $\overline{Q}$. They are always of an opposite polarity. Two inputs are usually needed to alter the state of a flip-flop. A variety of names are used for the inputs. These vary a great deal between different flip-flops.

Figure 17-9 shows the truth table and logic symbol of an R-S flip-flop. R stands for the reset input and S represents the set input. Notice that the truth table is somewhat more complex than that of a gate. It shows, for example, the applied input, previous output, and resulting output. Two of the input conditions cause an unpredictable output status.

To understand the operation of an R-S flip-flop, we must first look at the previous outputs. This is the status of the output before a change is applied to the input. The first four items of the previous

Figure 17-9. R-S flip-flop: (a) symbol; (b) truth table.

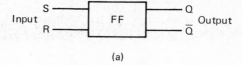

(a)

Applied inputs		Previous outputs		Resulting outputs	
S	R	Q	$\overline{Q}$	Q	$\overline{Q}$
0	0	1	0	1/0	0/1
0	1	1	0	1	0
1	0	1	0	0	1
1	1	1	0	1	0
0	0	0	1	0/1	1/0
0	1	0	1	1	0
1	0	0	1	0	1
1	1	0	1	0	1

(b)

outputs are Q1 and $\overline{Q}0$. The second four states have Q0 and $\overline{Q}1$.

In the second line of the truth table note the previous outputs of Q and $\overline{Q}$. When the S is "0" and R is "1" the resulting output does not change. Refer now to the third row. Again note the previous output status of Q and $\overline{Q}$. When S is "1" and R is "0" the resulting outputs change status. The resulting output has three predictable states and one unpredictable state for each of the four previous input combinations.

A variety of different flip-flops are used in digital electronic systems today. In general, each flip-flop type has some unique characteristic to distinguish it from the others. An R-S-T flip-flop, for example, is a triggered R-S flip-flop. It will not change states when the R and S inputs assume a value until a trigger pulse is applied. This would permit a large number of flip-flops to change states all at the same time. Figure 17-10 shows the logic symbol and truth table of the RST flip-flop.

Another very important flip-flop has J-T-K inputs. A J-K flip-flop of this type does not have an unpredictable output state. The J and K inputs have set and clear input capabilities. These inputs must be present for a short time before the clock or trigger input pulse arrives at T. In addition to this, J-K flip-flops may employ preset and preclear functions. This is used to establish sequential timing operations. Figure 17-11 shows the logic symbol and truth table of a J-K flip-flop.

Figure 17-10.   R-S-T flip-flop: (a) symbol; (b) truth table.

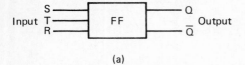

(a)

Applied inputs			Previous outputs		Resulting outputs	
S	T	R	Q	$\overline{Q}$	Q	$\overline{Q}$
0	0	0	1	0	1/0	0/1	←Unpredictable
0	1	1	1	0	1	0
1	0	0	1	0	0	1
1	1	1	1	0	1	0
0	0	0	0	1	0/1	1/0	←Unpredictable
0	1	1	0	1	1	0
1	0	0	0	1	0	1
1	1	1	0	1	0	1

(b)

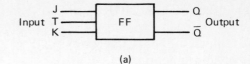

(a)

Applied inputs			Previous outputs		Resulting outputs	
J	T	K	Q	$\overline{Q}$	Q	$\overline{Q}$
0	0	0	0	1	0	1
0	1	1	0	1	0	1
1	0	0	0	1	1	0
1	1	1	0	1	1	0	←Toggle state
0	0	0	1	0	1	0
0	1	1	1	0	0	1
1	0	0	1	0	1	0
1	1	1	1	0	1	0

(b)

Figure 17-11.   J-K flip flop: (a) symbol; (b) truth table.

A flip-flop has a memory, it can be used in switching operations, and it can count pulses. A series of interconnected flip-flops is generally called a *register*. Each register can store one binary digit or bit of data. Several flip-flops connected form a counter. Counting is a fundamental digital electronic function.

For an electronic circuit to count, a number of things must be achieved. Basically, the circuit must be supplied with some form of data or information that is suitable for processing. Typically, electrical pulses that turn on and off are applied to the input of a counter. These pulses must initiate a state change in the circuit when they are received. The circuit must also be able to recognize where it is in the counting sequence at any particular time. This requires some form of memory. The counter must also be able to respond to the next number in the sequence. In digital electronic systems flip-flops are primarily used to achieve counting. This type of device is capable of changing states when a pulse is applied, has memory, and will generate an output pulse.

There are several types of counters used in digital circuitry today. Probably the most common of these is the binary counter. This particular counter is designed to process two-state or binary information. J-K flip-flops are commonly used in binary counters.

Refer now to the single J-K flip-flop of Fig. 17-11. In its *toggle* state this flip-flop is capable of achieving counting. First, assume that the flip-flop is in its reset state. This would cause Q to be 0 and $\overline{Q}$ to be 1. Normally, we are only concerned with Q output in counting operations. The flip-flop is now connected for operation in the toggle mode. J and K must both be made high or in the 1 state. When a pulse is applied to the "T" or clock input, Q changes to "1." This means that with one pulse applied, a "1" is generated in the output. The flip-flop has, therefore, counted one time. When the next pulse arrives, Q resets or changes to "0." Essentially, this means that two input pulses produce only one output pulse. This is a divide-by-two function. For binary numbers, counting is achieved by a number of divide-by-two flip-flops.

In order to count more than one pulse, additional flip-flops must be employed. For each flip-flop added to the counter, its capacity is increased by the power of 2. With one flip-flop the maximum count was 2^0, or 1. For two flip-flops it would count two places, such as 2^0 and 2^1. This would reach a count of 3 or a binary number of 11. The count would be 00, 01, 10, and 11. The counter would then clear and return to 00. In effect, this counts four state changes. Three flip-flops would count three places, or 2^0, 2^1, and 2^2. This would permit a total count of eight state changes. The binary values are 000, 001, 010, 011, 100, 101, 110, and 111. The maximum count is seven, or 111. Four flip-flops would count four places, or 2^0, 2^1, 2^2, and 2^3. The total count would make 16 state changes. The maximum count would be 15, or the binary number 1111. Each additional flip-flop would cause this to increase one binary place.

Figure 17-12 shows a four-bit binary counter. Four J-K flip-flops are included in this circuit. Sixteen counts can be made with this circuit. The counts are listed under the diagram. A light-emitting diode is used as an output display for each flip-flop. When a particular LED is "on" it indicates the "1" state. An off LED indicates the "0" state.

Operation of the counter is based on the number of clock pulses applied to its input. The flip-flops will only trigger or change states on the negative-going part of the clock pulse. Note this on the input waveform. The output of flip-flop one (FF1) will change back and forth between "1" and "0" for each pulse. One pulse applied to the input of

Figure 17-12. Four-bit binary counter.

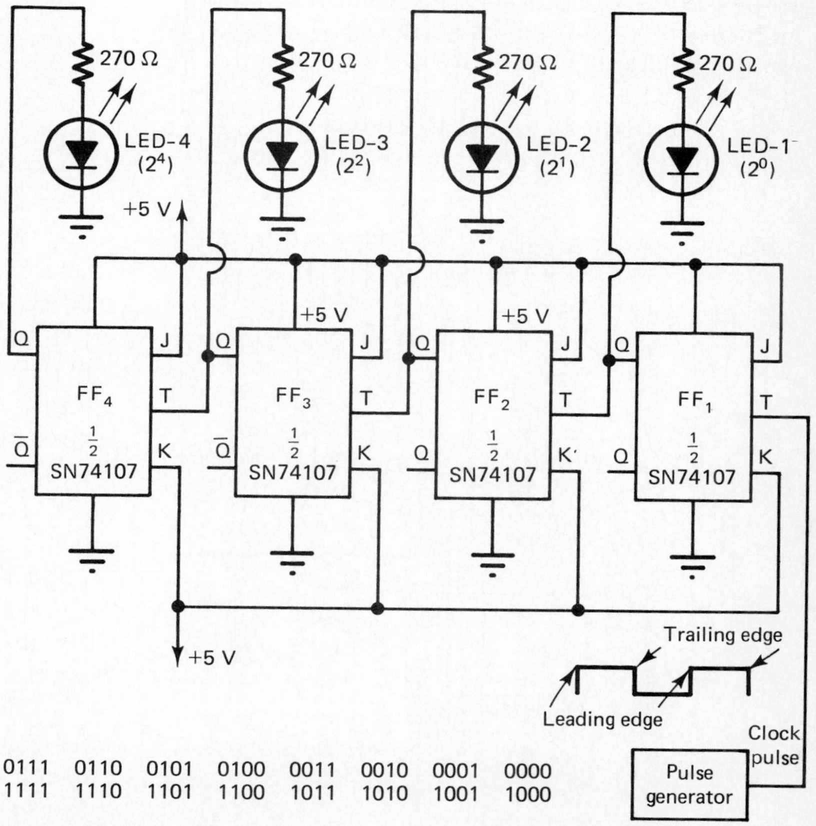

FF1 will cause its Q output to be "1." The 2^0 LED will turn on, indicating a "1" count. The display will be 0001. The second pulse will turn off FF1. The negative-going output of Q will turn on FF2. The Q output of FF2 will then become a "1." This will light LED-2 or the 2^1 display. The entire display will now be 0010, or the binary 2 count. The next pulse will trigger FF1. LED-1 will light again because the Q output of FF1 is "1" again. The display will show 0011, or the binary 3 count. The fourth pulse will reset FF1 and FF2 will set FF3. The Q output of FF3 will go positive and turn on LED-3. The display will show 0100, or the binary 4 count. The operation is then repeated, the counts being 0101, 0110, and 0111. The next pulse will clear FF1, FF2, and FF3. The state change of FF3 will set FF4. The count will now be 1000, or the binary 8 count. The operation is again repeated. The count display will be 1001, 1010, 1011, 1100, 1101, 1110, and 1111. This reaches the maximum count that can be achieved with four flip-flops. The next pulse will clear all four flip-flops. The LED display will then be 0000. The entire process will repeat itself starting with the first pulse.

A four-bit IC binary counter is shown in Fig. 17-13. This particular circuit achieves the same thing as the counter of Fig. 17-12. Four J-K flip-flops are built into this chip. The Q output of each flip-flop

Figure 17-13.   Four-bit IC binary counter.

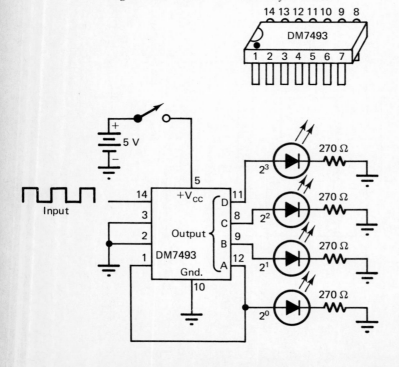

appears at terminals A, B, C, and D. These serve as the 2^0, 2^1, 2^2, and 2^3 or 1, 2, 4, and 8 outputs. LEDs connected to each output will respond as the display. This chip essentially has one input and four outputs. This chip employs interconnected transistors in its operations. It belongs to the transistor-transistor logic family of ICs.

DECADE COUNTERS

Decade counters are widely used in digital electronic systems. This type of counter has ten counting states in its output. The output is in binary numbers. In practice, this is more commonly called a *binary-coded decimal* (BCD) counter. Its output is 000, 0001, 0010, 0011, 0100, 0101, 0110, 0111, 1000, and 1001. It is used to change binary numbers into decimal form. One BCD counter is needed for each place of a decimal number. One counter of this type can only count to 9. Two would be required to count to 99. Three are needed for 999. The place value increases with each additional counter.

BCD counters are primarily binary counters that clear all flip-flops after the 9 count, or 1001. The first seven counts appear in the normal binary sequence. Figure 17-14 shows how the J-K flip-flops of the BCD counter are connected. Notice that FFA, FFB, and FFC are connected the same as those of Fig. 17-12. The resulting binary count is shown under the diagram.

At count 7, or binary 0111, a "1" is applied to the AND gate from the Q outputs of FFC and FFB. Two "1's" applied to an AND gate causes the J of FFD to go "1." The next pulse clears FFA, FFB, and FFC and sets FFD. The counter now reads 8, or a binary 1000. Note that the $\overline{Q}$ output of FFD is returned to FFB. With FFD set to a "1," $\overline{Q}$ is "0." This prevents FFB from further triggering until FFD is cleared.

Arrival of the next clock pulse causes FFA to be set. The count is now 9, or binary 1001. The next pulse clears FFA and FFD at the same time. The count is now 0, or binary 0000. With FFB and FFC already clear, the counter is reset to zero and is ready for the next count.

Figure 17-15 shows a BCD counter constructed from an IC chip. The ABCD output will be in binary code for numbers 0 through 9. The input is applied to pin 14. As a general rule, the output of a BCD

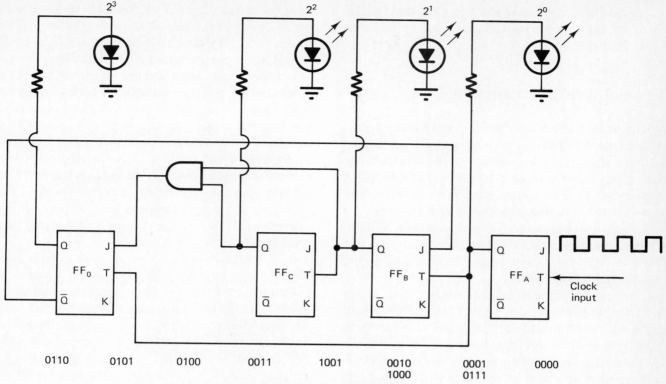

Figure 17-14. BCD flip-flop circuit.

Figure 17-15. BCD counter.

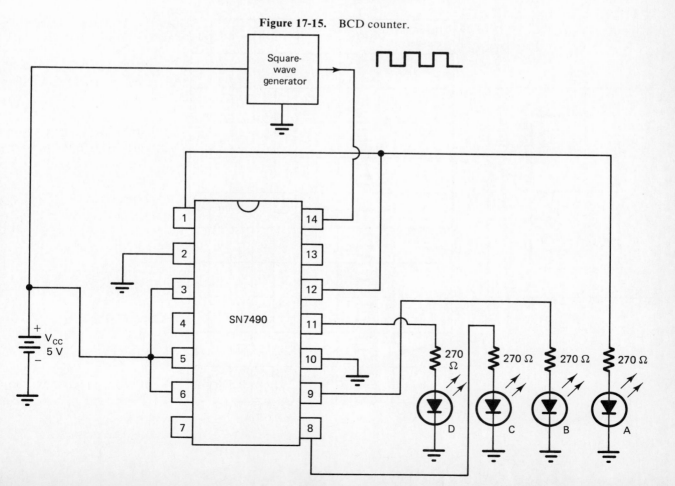

counter is usually applied to a decoding device. The decoder changes binary data into an output that drives a display device.

DIGITAL SYSTEM DISPLAYS

Many people who need to read the output of a digital system may not be familiar with the BCD method of number display. The system must therefore change its output to a more practical method of display. A variety of display devices have been developed to perform this function. One that is primarily used for numbers is called a seven-segment display. Calculators, watches, and test instruments employ this type of display. It is widely used in nearly all digital electronic systems today.

A seven-segment number display divides the decimal numbers 0 through 9 into lines. Only four vertical lines and three horizontal lines are needed to display a specific number. Each line or segment is designed to be illuminated separately. Figure 17-16(a) shows a typical seven-segment display. Each

segment may contain several LEDs in a common bar. The size of the display generally determines the number of LEDs in a specific segment. All the LEDs in a segment are energized at the same time.

Individual lines or bars of the display are labeled a, b, c, d, e, f, and g. A seven-bit binary number is applied to the display. Each segment is fed by a specific part of the binary number. In a common-anode type of display, when the input goes to ground or "0" it illuminates the segments. In a common-cathode display, the input must be "1" or +5 V to illuminate a segment. The truth table of Fig. 17-16(b) is for a common-anode display. For a common-cathode display the polarity of each segment would be inverted.

A common-anode seven-segment display test circuit is shown in Fig. 17-17. Pins 14, 9, and 3 are the anode connections. These are connected to the positive side of the power source. Pins 1, 2, 6, 7, 8, 10, 11, and 13 are the segment connections. Note that each segment is connected to a resistor. This limits the current to the LEDs in each segment to a reasonable value. Connecting one resistor to the

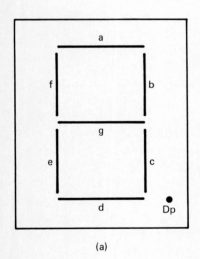

(a)

Decimal number	Display	a	b	c	d	e	f	g
0		0	0	0	0	0	0	1
1		1	0	0	1	1	1	1
2		0	0	1	0	0	1	0
3		0	0	0	0	1	1	0
4		1	0	0	1	1	0	0
5		0	1	0	0	1	0	0
6		1	1	0	0	0	0	0
7		0	0	0	1	1	1	1
8		0	0	0	0	0	0	0
9		0	0	0	1	1	0	0

(b)

Figure 17-16. Seven-segment display data: (a) display; (b) truth table.

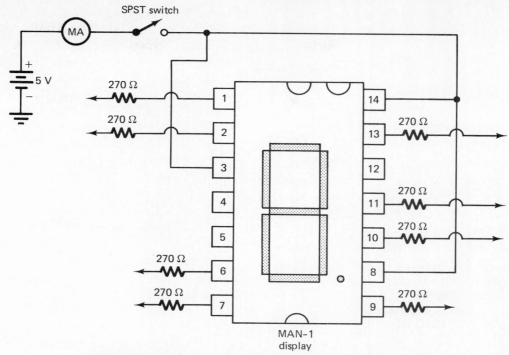

Figure 17-17. Common-anode seven-segment display test circuit.

DECODING

ground will energize a specific segment. Connecting them all to ground will produce an 8. display. Seven segments and a decimal point are included in this display.

DECODING

For a display to be operational, it must receive an appropriate input signal. This signal is normally in binary form. The input of a seven-segment display requires seven bits of data. The number 2, for example, is 0010010. See the data for the decimal number 2 in Fig. 17-16. This does not correspond directly to the output of a counter. A counter usually contains binary data in a specific sequence. A BCD output for the number 2 would be 0010. These data therefore need to be changed into the seven-segment code to cause a 2 to appear on the display. The process of changing data of one form into that of another form is called decoding.

A logic diagram of a BCD-to-seven segment decoder-driver is shown in Fig. 17-18. The BCD input is applied to terminals A, B, C, and D. The seven-segment output appears at a, b, c, d, e, f, and g.

The circuit also has a lamp test input, blanking input, and ripple blanking input. A low or "0" applied to the lamp test will cause all segments of the display to be illuminated. A low or "0" applied to the blanking input turns off all segments. This control is also used when several displays are grouped together for a multidigit number. It blanks out all leading zeros automatically. If this were not used, a multi-digit display would show a number of unnecessary zeros. For example, the number 52 on an eight-digit display would be 00000052. With automatic leading zero blanking, the number displayed would be 52.

For driving a common-anode display the outputs of the decoder must go low or to ground to light a segment. Common-cathode displays must go high or to +VCC to produce segment illumination. The selected decoder/driver chip would depend on the type of display employed. Note the truth table of Fig. 17-18. "0" indicates the low or grounded output. The "1" output condition represents an open circuit. The decoder shown here is used to drive a common-anode type of display. The number zero, for example, would illuminate all segments except g. A representative pin connection diagram of a decoder-driver is also shown in the diagram.

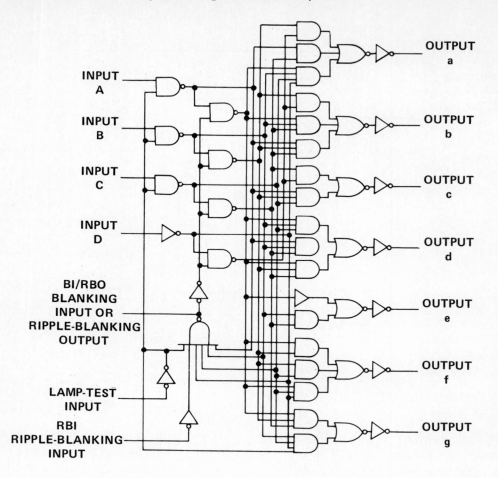

INPUT A

INPUT B

INPUT C

INPUT D

BI/RBO BLANKING INPUT OR RIPPLE-BLANKING OUTPUT

LAMP-TEST INPUT

RBI RIPPLE-BLANKING INPUT

OUTPUT a

OUTPUT b

OUTPUT c

OUTPUT d

OUTPUT e

OUTPUT f

OUTPUT g

BCD INPUTS				7-SEGMENT OUTPUTS						
A	B	D	D	a	b	c	d	e	f	g
0	0	0	0	0	0	0	0	0	0	1
0	0	0	1	1	0	0	1	1	1	1
0	0	1	0	0	0	1	0	0	1	0
0	0	1	1	0	0	0	0	1	1	0
0	1	0	0	1	0	0	1	1	0	0
0	1	0	1	0	1	0	0	1	0	0
0	1	1	0	0	1	0	0	0	0	0
0	1	1	1	0	0	0	1	1	1	1
1	0	0	0	0	0	0	0	0	0	0
1	0	0	1	0	0	0	0	1	0	0

"0" OUTPUT = GROUNDED ELEMENT
"1" OUTPUT = OPEN CIRCUIT

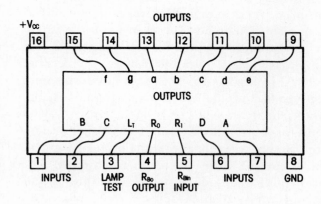

Figure 17-18. BCD-to-seven segment decoder-driver.

A counting timer circuit with a digital display is shown in Fig. 17-19. This circuit demonstrates the operation of a decoder-driver and the seven-segment display. The 555 square-wave generator can be adjusted to produce one pulse per second or minute. The display will count 0 to 9 as an indication of the pulse rate. R1 is adjusted for time in seconds or minutes. Operational time is based on the formula

$$time = 0.693(RA + 2RB)C$$

For counting in seconds, C should be a value of $2\mu F$. One-minute counting is achieved with C in the range 100 μF.

The 0 to 9 counter can be extended to a 0 to 99 counter by adding an additional BCD counter, decoder, and display. Pin 11 of the low-order or units display is connected to the input (pin 14) of the tens display. The common or ground of each display unit must also be connected together. Each additional counter would be connected in the same manner.

The count would be 0, 1, 2, 3, 4, 5, 6, 7, 8, 9, 10, 11 and continue up to 99. The next count would clear the counter to 00.

DIGITAL COUNTING SYSTEMS

Digital counting systems are very common today. This application of the counting process is used to determine a number of pulses occurring in a given unit of time. Each pulse simply triggers the input when it occurs. The input source generates a two-state data signal. In its simplest form, a light beam can be broken as an indication of a signal change. Figure 17-20 shows a production line counter circuit that employs this principle in its operation.

The fundamental parts of a counting system are the basis of nearly all digital display devices. The primary difference in the system is the input device. The display or readout device is essentially the same

Figure 17-19. Counting timer circuit.

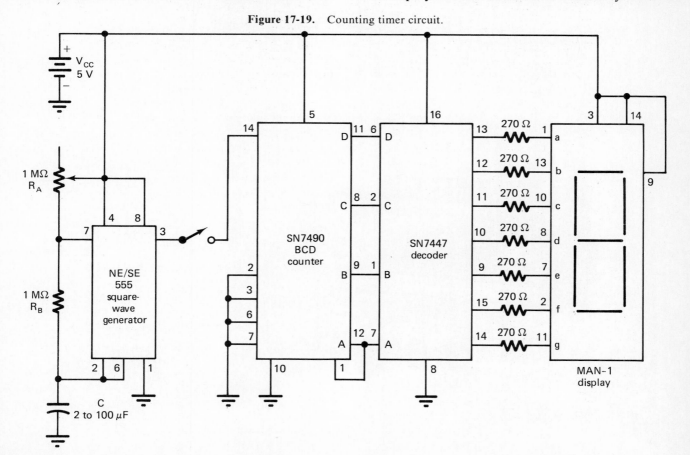

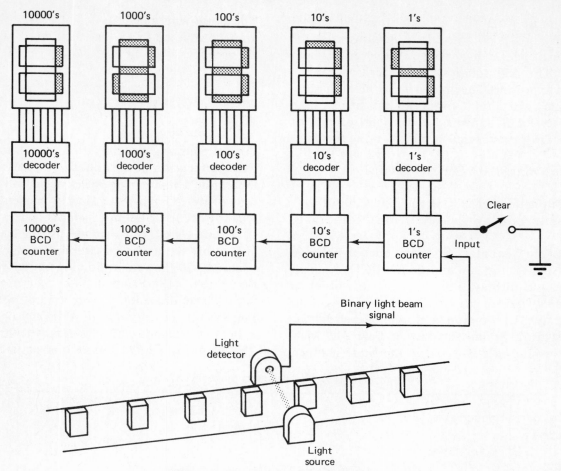

Figure 17-20.   Production line counter.

Figure 17-21.   Frequency counter. (Courtesy of Sencore.)

for all digital instruments. A BCD counter, decoder, and a display are used for each digit. In operation, a two-state or (on and off) signal is simply applied to the input of the counter assembly. Each on pulse or "1" causes the counter to advance one count. The output of the counter is then applied to a decoder-driver. The display is driven by the decoder. One input pulse change will cause a 1 to appear on the display. The 1's or unit counter can only progress from 0 to 9 counts. A count of 10 or more would activate both the 1's and 10's display units. Counts in excess of 99 would require three or more displays. Figure 17-21 shows a frequency meter that counts to 1 gigahertz (GHz) per second. A gigahertz is 1000 MHz or 10^9 Hz.

COMPUTERS

Computers are primarily responsible for the rapid expansion of the digital electronics field that we are experiencing today. This technology has had a rather decided impact on our society. Computers provide a wide range of services. Complex calculations can be performed, data can be manipulated, deductions can be made, and a variety of operational processes can be performed. Commercial inventory, accounting, math calculations, manufacturing operations, and video games have caused us to become very dependent on the computer. Computer applications in the future will be much more significant than they are today.

There is a great deal more to the operation of a computer than the average person realizes. When we look at a digital computer we see primarily its hardware. Hardware includes such things as logic gates, counters, data porocessing circuits, signal generation, and displays. As a general rule, computer hardware by itself is rather useless. It cannot function or operate unless it is told what to do. It has to be instructed to do something for it to be of any value. *Firmware* and *software* are needed to communicate with the computer. Essentially, these two things tell the hardware what to do. Firmware is a permanently installed set of instructions in the computer's hardware. It tells the hardware how to function and in what sequence to perform. The manufacturer of the system develops the firmware. Software is used to

instruct the computer how to do a specific job. After an operational sequence has been completed the computer may be instructed to perform another task. The software function of a computer is called a *program*. One computer can employ an unlimited number of programs.

Hardware

The hardware of a computer can take on a variety of shapes and sizes depending on its design. Regardless of its complexity, a computer has a number of basic functions. These are the basis of its operation. The functions are arranged in a specific pattern. Most computers conform to the same basic operational pattern. The end result of each computer is essentially the same. A block diagram of basic computer functions is shown in Fig. 17-22. Notice that it has an input/output device, an arithmetic/logic unit, control, memory and a power source.

The input of a computer is often considered as an interface between the real world and the machine. *Interface* refers to the process of bringing two or more things together. *Keyboards* are frequently used

Figure 17-22. Digital computer block diagram.

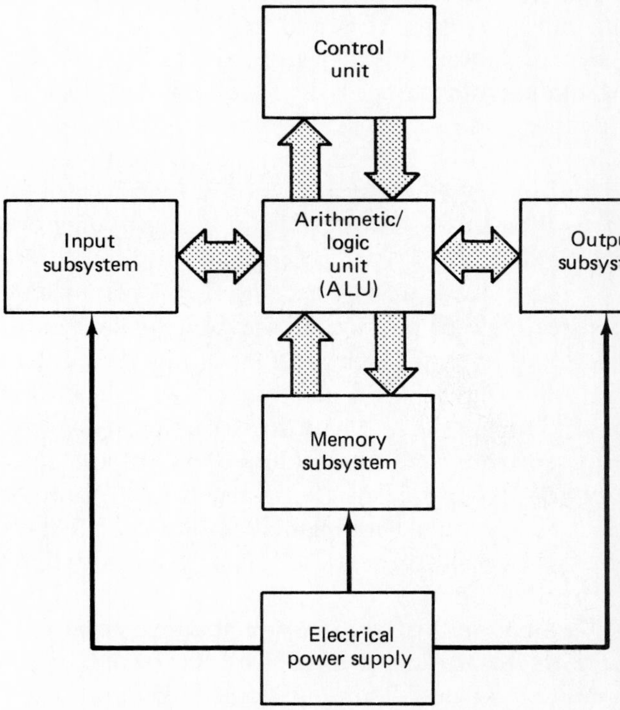

as an interface device. This device accepts real-world input and converts it into something that the computer understands. Basically, the input of a computer is responsible for *encoding*. This refers to the process of changing decimal numbers and/or alphabetic characters into a coded signal. The resulting signal is generally some type of a binary data. Typewriter keyboards and calculator keyboards generally serve as the input for most digital electronic systems. In addition to this, data may also be supplied by punched cards, punched tape, magnetic tape, and disks.

Coded data developed by the computer input are applied to the arithmetic/logic unit or ALU. The ALU manipulates these data according to instructions. Normally, the incoming data are placed in a *data register*. This serves as a short-term storage circuit for binary data. Registers are usually capable of storing one data *word*. A word may be 8 or 16 bits of data in a group. Registers are generally made of a number of flip-flops. Each flip-flop will store one bit of data. These data can be applied to all flip-flop at one time or may be shifted into the register one bit at a time.

To demonstrate the operation of the ALU, assume that the computer needs to add two numbers. Initially, the first number is applied to the input. It is transported to the ALU and applied to register A. The second number is then entered into register B. A command from the control section tells the ALU to add register A to B. The sum is then placed in register A, replacing the first number. In performing this operation the ALU responds only to binary data words. The ALU is responsible for all math functions and logic decisions.

The control unit of a computer is responsible for directing the operation of the ALU. In a sense, all functional operations are influenced by the control unit. Control is basically achieved by two operational cycles. During the *fetch cycle* instructional data are retrieved from the memory unit. The *execute* cycle tells how to carry out the instruction. Computer operation is based on repeated fetch and execution cycles. The operational time of each cycle is directed by the control unit.

After the ALU has completed an operational cycle its accumulated data are transferred to the computer output. This information or data must then be translated back into something that the real world understands. Normally, this operation is called *decoding*. Essentially, decoding is responsible for changing data into alphabetical or digital information. This information is then used to actuate the output device. Printers, typewriters, cathode-ray tube terminals, and digital displays are typical computer output devices.

Data Information

To communicate with a computer, we need numbers, letters, and other symbols. This information must be changed into some type of a standardized code. Codes that represent this type of information are called *alphanumeric*. They must represent 10 decimal digits, 26 letters of the alphabet, and several punctuation marks. One very important code is the American Standard Code for Information Interchange (ASCII). This is a seven-bit code. It has a three-bit group and a four-bit group. The three-bit group is a column select group. It is the most significant or first three bits on the left. The four-bit group is the row select binary number. Figure 17-23 shows the ASCII code. Note that the column numbers are 000 through 111 at the top of the chart. The row numbers are 000 through 111. The letter A, for example, would be 100 0001. This shows its location as column 100 and row 0001. The number 4 would be 011 0100. When an ASCII keyboard is used, a seven-bit code will occur for each letter or code button depressed. Note also that there are several abbreviations included in the ASCII code. These are defined in the chart.

Memory

A very important function of a computer system is its ability to store and recall data. In general terms this is called memory. An individual element or cell in a memory unit has the capability of storing a single bit of binary data. The total capacity of a memory unit is dependent on the number of cells in the structure. Normally, data are placed into memory in binary words.

Storing or placing binary data in memory is called *writing*. Sensing the information stored in memory is called *reading*. In most systems it is essential to have both reading and writing memory capabilities. Memory of this type is called read/write

American Standard Code for Information Interchange.

	000	001	010	011	100	101	110	111
0000	NULL	DC_0	b	0	@	P		
0001	SOM	DC_1	!	1	A	Q		
0010	EOA	DC_2	"	2	B	R		
0011	EOM	DC_3	#	3	C	S		
0100	EOT	DC_4	$	4	D	T		
0101	WRU	ERR	%	5	E	U		
0110	RU	SYNC	&	6	F	V		
0111	BELL	LEM	,	7	G	W		
1000	FE_0	S_0	(	8	H	X		
1001	HT / SK	S_1	)	9	I	Y		
1010	LF	S_2	*	:	J	Z		
1011	V_{tab}	S_3	+	;	K	[		
1100	FF	S_4	,	<	L	\		ACK
1101	CR	S_5	–	=	M	]		②
1110	SO	S_6	⋆	>	N	↑		ESC
1111	SI	S_7	/	?	O	←		DEL

Definitions of control abbreviations:

ACK	Acknowledge	LEM	Logical end of media
BELL	Audible signal	LF	Line feed
CR	Carriage return	RU	"Are you...?"
DC_0–DC_4	Device control	SK	Skip
DEL	Delete idle	SI	Shift in
EOA	End of address	SO	Shift out
EOM	End of message	S_0–S_7	Separator (space)
EOT	End of transmission	SOM	Start of message
ERR	Error	V_{tab}	Vertical tabulation
ESC	Escape	WRU	"Who are you?"
FE	Format effector	②	Unassigned control
FF	Form feed	SYNC	Synchronous idle
HT	Horizontal tabulation		

Example of code format:

B_7 B_1

1000100 is the code for D

three-bit group   four-bit group

Figure 17-23.   ASCII code.

random access memory (RAM). In addition to this, a computer also has read-only memory (ROM). Firmware instructions are placed in this memory.

Memory can be achieved in a variety of ways in a computer system. It is generally classified as being either *volatile memory* or *nonvolatile memory*. Volatile memory requires electrical power to retain stored data. Integrated-circuit chips are commonly used for this function. Random access memory is of this type. Data in the form of "1's" and "0's" can be written or placed in memory. We can then read or retrieve these data in the same form. RAMs are used to store data program data. Memory of this type is easy to achieve, requires very little space, and consumes a very small amount of electrical power. Electrical energy must be supplied continuously to the memory device in order to retain its contents.

Nonvolatile memory stores data without the need of electrical power. Disk memories store data as magnetized spots on a rotating disk. Tape memories place data on a thin film of magnetic tape. Data of this type can be stored for long periods of time without being altered.

The ROM of an IC chip is also of the nonvolatile type. To retain this type of memory specific elements are built into the chip. Once the element has been "burned in" it cannot be altered. Chips of this type are generally formed by the manufacturer. Math tables and firmware data can be placed on a chip in this manner.

Programmable read-only memory (PROM) chips are also available today. These devices are considered to be field programmable. They ultimately respond as a ROM. Programming is achieved by the user according to specific circuit needs. PROMs have special emitter resistors in each memory cell. When a particular cell is addressed, a pulse of current burns out the emitter resistor. Each cell can be programmed only one time. The programmed pattern is permanent. The chip then responds as a ROM. PROMs are essentially non-volatile memory devices.

A simplificiation of the memory process is shown in Fig. 17-24. This particular IC chip is described as a 64-bit random access read/write memory. As noted here, individual cells of the chip are organized in a pattern of rows and columns. This particular pattern has 16 rows of four-bit data words.

To write these data into the chip a particular cell row must be selected. The row address decoder is used for this function. Note that row 12 has been selected. The number 1100 would be entered into the row address decoder to select this location. Binary data are then applied to the four data inputs. To write this data word into row 12, the write enable and memory enable inputs must both go to low or "0" logic level. The input data would then appear at row 12. To read data, a specific address must be selected. Then the write enable and memory enable switches are placed in the logical "1" level. The output will have the same binary data as that written into memory.

Software

For a computer to function, it must be supplied with a set of operating instructions. Firmware is instructional data that are built into the system. These data tell the computer how to operate. Software is instructional data that tells the computer what to do during a specific operation. Software can be altered to conform with a wide range of different procedures. Programs are developed to make the computer conform to a variety of different operations.

Software for a computer can be achieved at three different levels. *Machine language* deals with direct coding of binary data at the input and output of the system. Programming with a machine language requires a good understanding of the system's memory capacity and its internal connections. This method of programming is rather efficient. It does not require a great deal of computer memory to achieve. A second level of instruction uses *assembly language* programs. This type of data is supplied in *mnemonic* (nē-mon´-ik) form. A mnemonic is an abbreviation that brings to mind a word. CLR, for example, is the mnemonic for clear. Assembly language programming is rather easy to achieve. It generally requires less knowledge of the computer. More memory space is needed to convert mnemonic inputs into binary data. Many computers are designed for a *high-level language*. Real-world statements are used in this programming level. In general, the firmware of a high-level machine is more complex than the others. It requires very little knowledge of the computer. More space is required for programmed memory. Programming of this type is easier to achieve and usually requires less time to learn.

Computer Operation

A computer is designed to follow a fixed sequence of operations. When power is first applied, it is designed to retrieve information from a predetermined memory location. The first information is a firmware instruction. The manufacturer of the system has these data built into a read-only memory unit. Retrieving something from memory is generally called a *fetch* instruction. After data have been obtained from memory, the computer must then *execute* the instruction. The arithmetic/logic unit

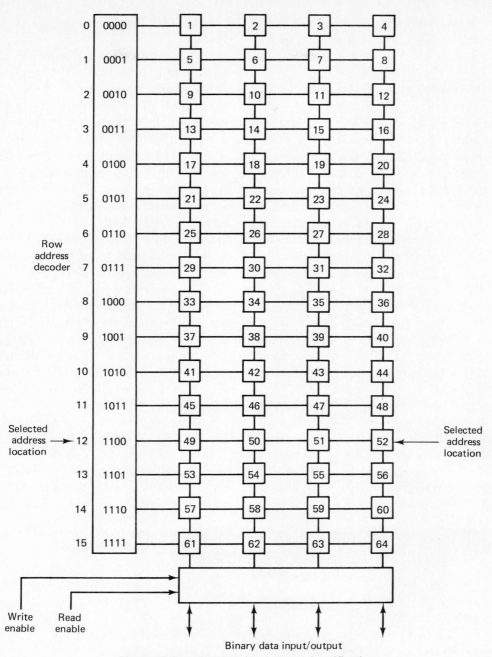

Figure 17-24. Read–write memory simplification.

then examines the data. Gates are actuated by the data. Signals directed to the control unit are routed to other parts of the computer to perform given operations.

Because of the high rate of operational performance, the next fetch instruction is usually being prepared. Unless the system is told to do something different, the next instruction is processed. Nor-mally, the second instruction is related in some way to the first instruction. Some instructions may require data from a previous operation in order to continue. To proceed from this point the computer must be supplied with input data. A program is normally used to direct operation of the system from this point. Operation continues on a repeated steps of fetch, execute, memory manipulation, and input output data transfer. The operational process continues until the computer is instructed to stop its operational sequence.

Microcomputer Systems

Microcomputer systems are probably the most significant development that has taken place since the beginning of the computer. A microcomputer is simply a small version of a regular computer. It has a rapid operational cycle, is small and compact, has increased reliability with reduced cost, and has a high level of serviceability. Advances in IC technology have made the microcomputer possible. The capabilities of this computer have not yet been fully realized. Figure 17-25 shows an operational microcomputer system.

A microcomputer system appears to be quite simple when viewed as a block diagram. It has a microprocessor, memory, an interface adapter, distribution buses, input/output, and a power source. A block diagram of the microcomputer system is shown in Fig. 17-26.

The microprocessor unit (MPU) is considered to be the heart of the microcomputer. It has a central processing unit built into the chip. The CPU consists of all the circuitry needed to locate memory addresses and to interpret operating instructions. The execution of an instruction is also accomplished within the chip. An arithmetic/logic unit, control

section, and data registers are all included in the MPU.

The input/output of a microcomputer is shown as a combinational unit. The input or output can be called up for operation by the MPU. The input is usually called up first in order to make the system operational. Keyboards teletypewriters, analog-to-digital converters, and cassette tape decks are commonly used as input devices. Normally, input data are first applied to an interface adaptor. This device transforms the input signal into compatible digital form for the microprocessor. This signal is then applied to the MPU through the data bus. One specific input device is selected by the MPU at a time. The desired input device is a selected binary word applied to the address bus.

The memory unit of a microcomputer is used primarily to store binary data. It also tells the MPU what to do during the execution cycle. MPU requests memory by sending out an address code on the address bus. Data are either written into this address or read. The function being performed by the MPU determines the read/write operation. Data flow is only through the data bus.

The output unit converts binary data into some useful form of output energy. Output devices can be printers, tape punches, cathode-ray-tube displays, and digital-to-analog converters. Output data come from the MPU over the data bus. The interface adaptor accepts these data and changes them to an

Figure 17-25. Microcomputer system. (Courtesy of Radio Shack, a division of Tandy Corp.)

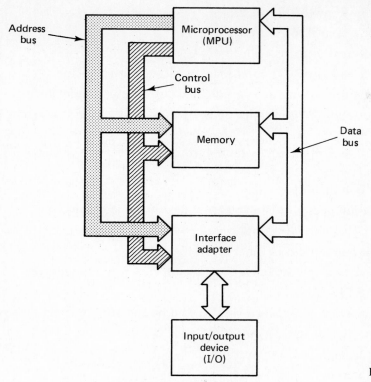

Figure 17-26. Microcomputer system block diagram.

appropriate output signal. A specific output device is selected by the MPU through the input/output address bus. During operation the input/output devices are alternately switched into operation by the system.

The operation and control of a microcomputer is much more complex than outlined here. As a rule, a computer operator does not need to know all the details of its operation. Most units have an instructional set that introduces the user to the system. With some practice, a person can become very proficient in the operation of a microcomputer system. A number of personal computers are now available at a fraction of the cost of a full-size computer. Microcomputers applications are expanding at an extremely rapid pace today.

Review

1. Explain the difference between a digital value and an analog value.

2. What are the fundamental parts of a digital system?

3. What is the radix of the decimal system and the binary system?

4. What is meant by the place value of a numbering system?

5. Express the binary number equivalents of 14, 23, 61, 72, 128, and 264.

6. Change the following binary numbers to decimal equivalents 1011, 10001, 11110, 1000000, and 1000011.

7. Develop a truth table that shows the equivalent relationship of AND, OR, NAND, and NOR gates.

8. What is the mathematical equivalent of AND, OR, NOT, NAND, and NOR gates?

9. How does a flip-flop store data?
10. Explain how one flip-flop counts.
11. Explain how four flip-flops count to 1111.
12. Explain how decade or BCD counting is achieved with flip-flops.
13. Explain how a number, such as 5, is indicated on a seven-segment display.
14. What are the letter designations of individual segments in a seven-segment display?
15. Why is decoding needed for a binary counter with a seven-segment display?
16. What are the fundamental parts of a computer?
17. Explain the differences between software, hardware, and firmware.
18. What are encoding and decoding?
19. What are a bit and a word in binary terms?
20. What is achieved during the fetch and execute cycles of a computer?
21. What is the ASCII code?
22. Describe the differences between volatile and nonvolatile memory.
23. What is meant by read-only memory?
24. Explain what a random access memory is.
25. What is the function of an ALU?
26. How does a microcomputer differ from a regular computer?

Student Activities

Switch-Lamp Gates

1. Connect the AND gate of Fig. 17-4.
2. Prepare a truth table showing the relationship of the two inputs and the output.
3. Connect the OR gate of Fig. 17-5. Test the circuit and prepare a truth table.
4. Connect the NOT gate of Fig. 17-6. Prepare a truth table of its operation.
5. Connect the NAND gate of Fig. 17-7. Test the circuit and prepare a truth table.
6. Connect the NOR gate of Fig. 17-8. Test the circuit and prepare a truth table.

IC Logic Gate Test Circuit

1. In evaluating an IC logic gate, we will use the following test circuit. It can be used to evaluate any of the logic gates. Switch A and LED A are for one input. Switch B and LED B are for the other input. LED C is used to observe the state of the output.
2. Connect the test circuit of Fig. 17-27.

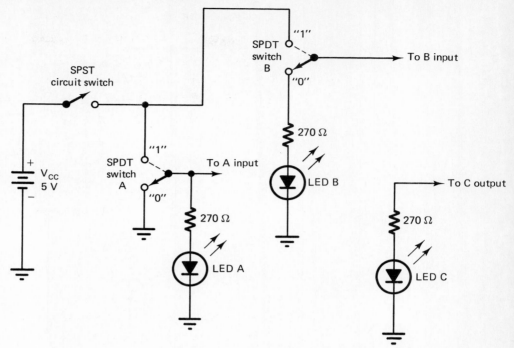

Figure 17-27. IC logic gate test circuit.

3. To test the operation of a logic gate, look up the connection diagram in a manufacturer's data manual.

4. Connect an input A or B to the gate input terminals. Connect the C output to the gate output.

5. Apply 5 V dc to the test circuit and the IC.

6. When switch A is "1" LED A should light. Switch B and LED B should respond in the same manner. In the "0" or off state the respective LEDs will be off.

7. When the output of the gate is "1," the C LED will light; "0" causes LED C to be off.

Logic Gate Testing

1. Look up the pin connections for an SN7404, an SN7408, a DM7400, an SN7402, and an SN7432.

2. Make a drawing of the pin connections for each chip being used in the test procedure.

3. Connect the A and B inputs and the C output to one gate. By a switching sequence determine if the gate is good or faulty. Four gates are included on each chip except the SN7404. It has six gates.

4. Test each gate on the chip. Note the status of each gate.

R-S Flip-Flop

1. Using the logic gate test circuit of Fig. 17-27, connect the R-S flip-flop circuit of Fig. 17-28(a). Note that one additional LED is needed. The inputs are labeled R and S and the outputs are Q and $\overline{Q}$.

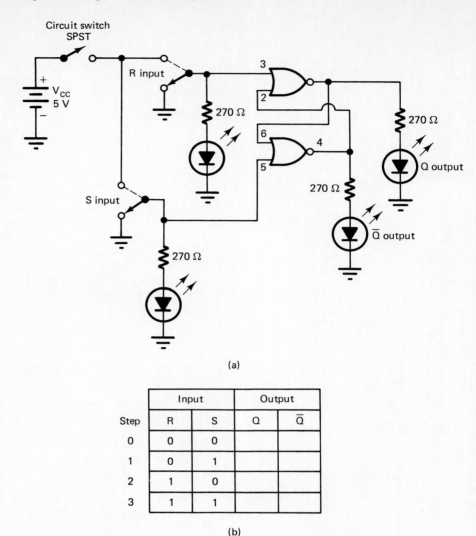

(a)

	Input		Output	
Step	R	S	Q	$\overline{Q}$
0	0	0		
1	0	1		
2	1	0		
3	1	1		

(b)

Figure 17-28.   (a) R-S flip-flop circuit; (b) truth table.

2.   Any two of the four gates on the SN7402 chip can be used.

3.   Turn on the power supply. An "on" LED indicates the "1" state and off is "0." Be certain to apply power to the +VCC and ground pins of the IC.

4.   What is the initial state of Q and $\overline{Q}$?

5.   Alter the R-S switches so that both inputs are "0."

6.   Record the Q and $\overline{Q}$ output in a truth table.

7.   Alter the R-S switches to conform with steps 1, 2, and 3 of the truth table [Fig. 17-28(b)]. Record the Q and $\overline{Q}$ outputs for each step.

J-K Flip-Flop

1.   Using the modified logic gate tester of Fig. 17-27, test one of the J-K flip-flops in SN74107. Look up the chip in a data manual to find the pin connections.

2.   Initially, connect the clear input (CLR) at +5 V.

3.   Place the J-K inputs both at "0." Note the state of Q and $\overline{Q}$.

4.   Then connect the T input to +5 V. Does the resulting output change?

5. Set the J-K inputs as in Fig. 17-11(a) and check the previous and resulting outputs of Q and $\overline{Q}$. Then apply the designated T input and note the outputs.

6. Make a truth table of your data.

Four-Bit Binary Counter

1. Construct the four-bit binary counter of Fig. 17-13. FF-1, FF-2, FF-3, and FF-4 are all built on the DM7493 chip.

2. The counting pulse may be achieved with a SPDT switch. If the count is irregular, it is probably due to contact bouncing of the switch.

3. A square-wave generator or a pulse generator may be used in place of the switch. The NE555 square-wave generator of Fig. 17-18 may also be constructed for this purpose.

4. Cycle the counter through its range following the binary count. An on LED indicates "1" and an off LED represents the "0" state.

BCD Counter

1. Construct the BCD counter of Fig. 17-15. The SN7490 has an AND gate and four flip-flops built on a chip.

2. The counting pulse may be achieved with a SPDT switch. If the count is irregular, it is probably due to contact bouncing of the switch.

3. A square-wave generator or a pulse generator should be used in place of the switch. The NE555 square-wave generator of Fig. 17-19 can be used for this purpose.

4. Cycle the counter through its range, noting counting sequence.

Seven-Segment Display

1. Construct the seven-segment display test circuit of Fig. 17-17.

2. Turn on the power supply and adjust it to 5 V dc.

3. Turn on the circuit switch.

4. Alternately connect each of the 270-Ω resistors to the ground side of the power supply.

5. Make a sketch of the segment illuminated and note the pin number that is energized.

6. Record the current needed to produce illumination.

7. Using the appropriate segment combinations, form the numbers 0, 1, 2, 3, 4, 5, 6, 7, 8, and 9. Note the segment combination used for each number.

Counting Timer

1. Construct the counting timer of Fig. 17-19.

2. Turn on the power supply and input switch.

3. The speed of the count is based on the setting of RA and the capacitor selection of C. Adjust RA to a rather fast timing rate.

4. Observe the counting cycle of the timer.

5. Calibrate RA for a 1-s or a 1-min count.

6. Test the operation of the circuit for its count duration.

18

Electronic Power Control

Control is one of the basic functions of all electronic systems. It is primarily responsible for altering the flow of electrical energy between the source and load. Anytime that a system is placed into operation some form of control is involved. The process of turning a system on and off is a prime example of control. This action is usually called full or complete control. All systems have some form of full control. In addition to this, control may involve some type of an adjustment. Partial control describes this function. Electronically, partial control can be achieved in a number of ways. Adjusting the volume of a stereo amplifier, the speed of a motor, or the brightness of a lamp is an example of partial control.

Partial control can be achieved in a variety of ways. Essentially, the last half of this book has been dealing in some way with partial control. In this regard, we have been concerned with control that alters the shape, amplitude, and frequency of an electronic signal. Rectification is used to alter the shape of a waveform. Amplification changed the amplitude of a signal. Frequency control is achieved by a tuning circuit or an oscillator. These control processes are all extremely important electronic functions.

Electronic power control deals with devices that alter the conduction time of a waveform. This function is usually quite unique when compared with other methods of control. It may be used, for example, to change only one *alternation* of a sine wave. The conduction may be varied anywhere between 0 and 100% of the alternation. Both alternations of a sine wave may also be controlled. They may be changed between 0 and 100% of their conduction time. Control of this type is accomplished by a number of rather specialized electronic devices.

IMPORTANT TERMS

Electronic power control has a number of rather new and, in some cases, unusual terms. These terms play an important role in the presentation of this material. It is usually helpful to review these terms before proceeding with the main text of the chapter.

Alternation. Half of an ac sine wave. There is a positive alternation and a negative alternation for each ac cycle.

Bidirection triggering. The triggering or conduction that can be achieved in either direction of an applied ac wave. Diacs and triacs are of this classification.

Breakover voltage. The voltage across the anode–cathode of an SCR or terminal 1 and terminal 2 of a triac that causes it to turn on.

Circuit switching. An on–off control function. Conduction occurs during the on time with no conduction during the off time.

Commutation. The changing from on to off or from one state to the opposite state.

Conduction time. The time when a solid-state device is turned on or in its conductive state.

Contact bounce. When a mechanical switch is turned on, the contacts are forced together. This causes the contacts to open and close several times before making firm contact.

Delay time. The time that an alternation is held in an off state before it turns on.

Diac. A solid-state diode device that will trigger in either direction when its breakover voltage is exceeded. A bidirectional triggering device.

Forward breakover voltage. The voltage where an SCR or triac goes into conduction in quadrant I of its $I–V$ characteristics.

Gate. The control electrode of a triac, SCR, or FET.

Gate current (I_G). The resulting current that flows in the gate when it is forward biased.

Holding current (I_H). A current level that must be achieved when an SCR or triac latches or holds in the conductive state.

Internal resistance. The resistance between the anode–cathode of an SCR or $T_1–T_2$ of a triac. Also called the dynamic resistance.

Latch. A holding state where an SCR or triac turns on and stays in conduction when the gate current is removed.

Main 1 or 2. The connection points on a triac where the load current passes. Also called terminal 1 and terminal 2.

Off-state resistance. The resistance of an SCR or triac when it is not conducting.

Phase. A position or part of an ac wave that is compared with another wave. A wave can be in phase or out of phase with respect to another.

Phase shifter. A circuit that changes the phase of an ac waveform.

Quadrant. One-fourth or a quarter part of a circle or a graph.

Reverse breakdown. The conduction that occurs in quadrant III of the $I–V$ characteristic of a triac.

Rheostat. The two-terminal variable resistor.

Snubber. A resistor–capacitor network placed across a triac to reduce the effect of an inductive voltage.

Switching device. An electronic device that can be turned on and off very rapidly. This type of device is capable of achieving control by switching.

Terminal (T_1 *or* T_2). The connecting points where load current pass through a triac. These two connection points are also called main 1 or 2.

Triggering. The process of causing an NPNP device to switch states from off to on.

Turn-on time. The time of an ac waveform when an alternation occurs.

ELECTRONIC POWER CONTROL SYSTEM

Electric power is commonly used as an energy source for such things as welding, electric heat, and a variety of other operations. These items, in general, refer to the load of a power system. A certain amount of work is achieved by the load device when it is energized. Controlling the load of an electrical power system has been a rather significant problem through the years. Control is usually not achieved very efficiently.

Figure 18-1 shows how a *rheostat* is used to control an electrical power circuit. The rheostat in this case is used to change the brightness of an incandescent lamp. An increase in resistance will

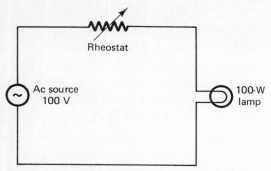

Figure 18-1.   Rheostat-controlled lamp circuit.

normally cause the lamp to be dim. A decrease in resistance will cause it to become brighter. With the rheostat set to zero resistance the circuit will have 1 A of current flow. This is determined by the formula

$$\text{current } (I) = \frac{\text{power } (P)}{\text{voltage } (V)}$$

or

$$I = \frac{P}{V} = \frac{100 \text{ W}}{100 \text{ V}}$$
$$= 1 \text{ A}$$

This shows when the lamp would be operating at its maximum output level. The resistance of the circuit in this condition is

$$\text{resistance } (R) = \frac{\text{voltage } (V)}{\text{current } (I)}$$

or

$$R = \frac{V}{I} = \frac{100 \text{ V}}{1 \text{ A}}$$
$$= 100 \text{ }\Omega$$

We know that the current flow of the lamp circuit is directly related to resistance. An increase in resistance will reduce the total circuit current. Assume now that the rheostat is adjusted to produce a resistance of 100 Ω. The total resistance of the circuit will now change to 200 Ω. This would be the original resistance of 100 Ω plus the rheostat resistance of 100 Ω. The resulting circuit current flow will decrease in

value. This is shown to be

$$\text{current } (I) = \frac{\text{voltage } (V)}{\text{total resistance } (R_t)}$$

or

$$I = \frac{V}{R_t} = \frac{100 \text{ V}}{200 \text{ }\Omega}$$
$$= 0.5 \text{ A}$$

Note that the circuit current has decreased to half of its original value.

A rather interesting change in the lamp circuit occurs when the total resistance increases. Some of the source voltage will appear across the rheostat. This will be deducted from the lamp voltage. With the lamp and the rheostat both being 100 Ω, there will now be 50 V across each. The lamp will obviously be operating with reduced power. This is determined to be

$$\text{power } (P) = \text{current } (I) \times \text{voltage } (V)$$

or

$$P = I \cdot V = 0.5 \text{ A} \times 50 \text{ V}$$
$$= 25 \text{ W}$$

The lamp will therefore receive only 25 W, or 25% of its original power. The brightness will be only one-fourth as great as it was initially. Resistance can therefore be used effectively to control power.

It should be pointed out that an equal amount of power appears across the rheostat of our lamp control circuit. In this case the rheostat power is

$$\text{power } (P) = \text{current } (I) \times \text{voltage } (V)$$

or

$$P = I \cdot V = 0.5 \text{ A} \times 50 \text{ V}$$
$$= 25 \text{ W}$$

This means that the rheostat must consume 25 W of power to reduce the lamp to one-fourth of its normal brightness. The rheostat in this case is consuming as much power as the load device. Power consumed by

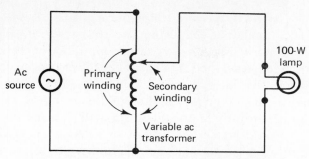

Figure 18-2. Voltage-controlled lamp circuit.

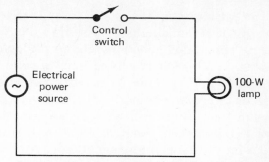

Figure 18-3. Switch-controlled lamp circuit.

the rheostat appears as heat. Since the purpose of our circuit is to control light, heat is not a desirable feature. In effect, any heat dissipated by *R* is considered to be wasted energy. We must pay for this power in order to control lamp brightness. Control of a power system by changes in current is therefore not very efficient.

As an alternative, electrical power can be controlled by variations in circuit voltage. Figure 18-2 shows a voltage-controlled lamp dimmer circuit. In this circuit, lamp brightness is controlled by a variable ac transformer. A reduction in voltage will produce a corresponding change brightness. This type of control is more efficient than the rheostat circuit. In effect, the lamp is the primary resistive device. Power consumed by the circuit will appear only at the lamp. Some power is consumed by the primary winding of the transformer. This power is usually rather nominal compared to the rheostat circuit.

Power control by a variable transformer is more efficient than the rheostat method. A variable transformer is, however, a rather expensive item. When large amounts of power are being controlled this device also becomes quite large. Control of this type is more efficient but has a number of limitations.

One of the most efficient methods of electrical power control that we have today is through *circuit switching*. When a switch is turned on, power is consumed by the load. When the switch is turned off, no power is consumed. The switching method of power control is shown in Fig. 18-3. If the switch is low resistant, it consumes very little power. When the switch is open, it does not consume any power. By switching the circuit on and off very rapidly the average current flow can be reduced. The brightness

of the lamp will also be reduced. In effect, lamp brightness is controlled by the switching speed. Power in this method of control is not consumed by the control device. This type of control is efficient and very effective.

The switching method of electric power control cannot be achieved very effectively with a manual switch. The mechanical action of the switch will not permit it to be turned on and off very quickly. A switch would soon wear out if used in this manner. Electronic switching can be used to achieve the same result. Switching action of this type can be accomplished in a lamp circuit without a noticable flicker. Electronic power control deals with the switching method of controlling power.

Silicon-Controlled Rectifiers

A silicon-controlled rectifier or SCR is probably the most popular electronic power control device today. The SCR is used primarily as a switching device. Power control is achieved by switching the SCR on and off during one alternation of the ac source voltage. For 60 Hz ac the SCR would be switched on and off 60 times per second. Control of electrical power is achieved by altering or delaying the *turn-on time* of an alternation.

An SCR, as the name implies, is a solid-state rectifier device. It conducts current in only one direction. It is similar in size to a comparable silicon power diode. SCRs are usually small, rather inexpensive, waste very little power, and require practically no maintenance. The SCR is available today in a full range of types and sizes to meet nearly any power control application. Presently, they are available in current ratings from less than 1 A to over 1400 A. Voltage values range from 15 to 2600 V.

SCR CONSTRUCTION. An SCR is a solid-state device made of four alternate layers of P- and N-type silicon. Three P-N junctions are formed by the structure. Each SCR has three leads or terminals. The anode and cathode terminals are similar to those of a regular silicon diode. The third lead is called the *gate*. This lead determines when the device switches from its off to on state. An SCR will usually not go into conduction by simply forward biasing the anode–cathode. The gate must be forward biased at the same time. When these conditions occur, the SCR becomes conductive. The internal resistance of a conductive SCR is less than 1 Ω. Its reverse or *off-state resistance* is generally in excess of 1 MΩ.

A schematic symbol and the crystal structure of an SCR are shown in Fig. 18-4. Note that the device has a PNPN structure from anode to cathode. Three distinct P-N junctions are formed. When the anode is made positive and the cathode negative, junctions 1 and 3 are forward biased. J_2 is reverse biased. Reversing the polarity of the source alters this condition. J_1 and J_3 would be reverse biased and J_2 would be forward biased. This would not permit conduction. Conduction will occur only when the anode, cathode, and gate are all forward biased at the same time.

Some representative SCRs are shown in Fig. 18-5. Only a few of the more popular packages are shown here. As a general rule, the anode is connected to the largest electrode if there is a difference in their physical size. The gate is usually smaller than the other electrodes. Only a small gate current is needed to achieve control. In some packages, the SCR symbol is used for lead identification.

Figure 18-4. SCR crystal: (a) symbol; (b) structure.

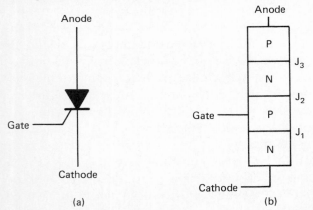

SCR OPERATION. Operation of an SCR is not easily explained by using the four-layer PNPN structure of Fig. 18-4. A rather simplified method has been developed that describes the crystal structure as two interconnected transistors. Figure 18-6 shows this imaginary division of the crystal. Notice that the top three crystals form a PNP transistor. The lower three crystals form an NPN device. Two parts of each transistor are interconnected. The two-transistor circuit diagram is an equivalent of the crystal structure division.

Figure 18-7 shows how the two-transistor equivalent circuit and the PNPN structure of an SCR will respond when voltage is applied. Note that the anode is made positive and the cathode negative. This condition forward biases the emitter of each transistor and reverse biases the collector. Without base voltage applied to the NPN transistor the equivalent circuit is nonconductive. No current will flow through the load resistor. Reversing the polarity of the voltage source will not alter the conduction of the two circuits. The emitter of each transistor will be reverse biased by this condition. This means that no conduction will take place when the anode and cathode are reversed biased. With only anode–cathode voltage applied, no conduction will occur in either direction.

Assume now that the anode and cathode of an SCR are forward biased, as indicated in Fig. 18-7. If the gate is momentarily made positive with respect to the cathode, the emitter–base junction of the NPN transistor becomes forward biased. This action will immediately cause collector current to flow in the NPN transistor. As a result of this, base current will also flow into the PNP transistor. This in turn causes a corresponding collector current to flow. PNP collector current now causes base current to flow into the NPN transistor. The two transistors therefore *latch* or hold when in the conductive state. This action continues even when the gate is disconnected. In effect, when conduction occurs, the device will latch on its "on" state. Current will continue to flow through the SCR as long as the anode–cathode voltage is of the correct polarity.

In order to turn off a conductive SCR, it is necessary to momentarily remove or reduce the anode–cathode voltage. This action will turn off the two transistors. The device will then remain in this state until the anode, cathode, and gate are all

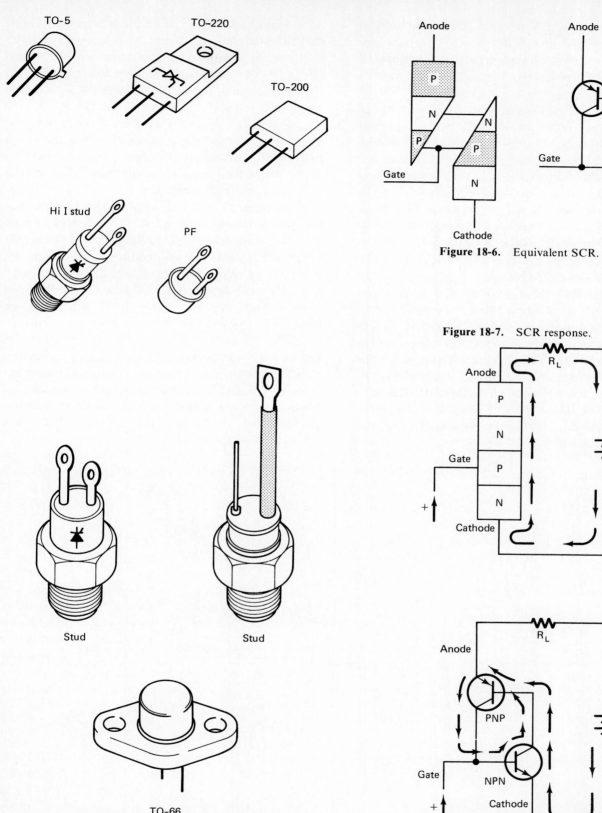

Figure 18-6. Equivalent SCR.

Figure 18-7. SCR response.

TO-66

Figure 18-5. Representative SCR packages.

forward biased again. With ac applied to an SCR, it will automatically turn off during one alternation of the input. Control is achieved by altering the turn-on time during the conductive or "on" alternation.

SCR *I–V* Characteristics.

The current–voltage characteristics of an SCR tell a great deal about its operation. The *I–V* characteristics of Fig. 18-8 shows that an SCR has two conduction states. *Quadrant* I shows conduction in the forward direction. This shows how conduction occurs when the *forward breakover voltage* (V_{BO}) is exceeded. Note that the curve returns to approximately zero after the V_{BO} has been exceeded. When conduction occurs, the internal resistance of the SCR drops to an extremely small value. This value is similar to that of a forward-biased silicon diode. The conduction current (I_{AK}) must be limited by an external resistor. This current, however, must be great enough to maintain conduction when it starts. The *holding current* or I_H level must be exceeded in order for this to take place. Note that the I_H level is just above the knee of the I_{AK} curve after it returns to the center.

Quadrant III of the *I–V* characteristic curve shows the *reverse breakdown* condition of opera-tion. This characteristic of an SCR is very similar to that of a silicon diode. Conduction occurs when the peak reverse voltage (PRV) value is reached. Normally, an SCR would be permanently damaged if the PRV is exceeded. Today, SCRs have PRV ratings of 25 to 2000 V.

For an SCR to be used as a power control device, the forward V_{BO} must be altered. Changes in gate current will cause a decrease in the V_{BO}. This occurs when the gate is forward biased. An increase in I_G will cause a large reduction in the forward V_{BO}. An enlargement of quadrant I of the *I–V* characteristic is shown in Fig. 18-9. This shows how different values of I_G change the V_{BO}. With zero I_G it takes a V_{BO} of 400 V to produce conduction. An increase in I_G reduces this quite significantly. With 7 mA of I_G the SCR conducts as a forward-biased silicon diode. Lesser values of I_G will cause an increase in the V_{BO} needed to produce conduction.

The gate current characteristic of an SCR shows a very important electrical operating condition. For any value of I_G there is a specific V_{BO} that must be reached before conduction can occur. This means that an SCR can be turned on when a proper combination of I_G and V_{BO} are achieved. This

Figure 18-8.   *I–V* characteristics of an SCR.

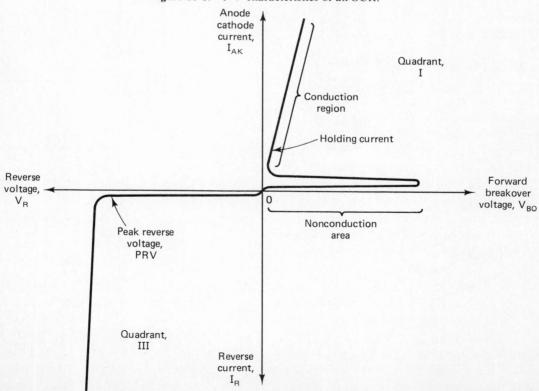

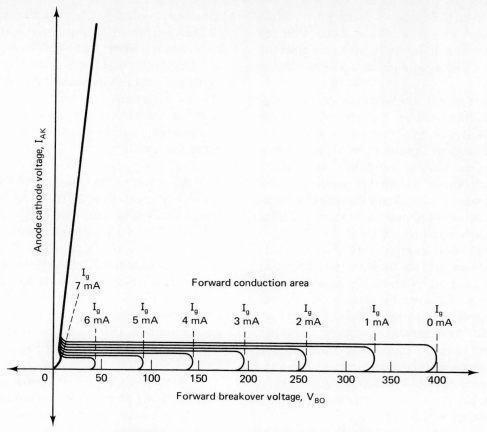

Figure 18-9. Gate current characteristics of an SCR.

characteristic is used to control the conduction when the SCR is used as a power control device.

DC POWER CONTROL WITH SCRs. When an SCR is used as a power control device it responds primarily as a switch. When the applied source voltage is below the forward breakdown voltage, control is achieved by increasing the gate current. Gate current is usually made large enough to ensure that the SCR will turn on at the proper time. Gate current is generally applied for only a short period of time. In many applications this may be in the form of a short-duration pulse. Continuous I_G is not needed to trigger an SCR into conduction. After conduction occurs, the SCR will not turn off until the I_{AK} drops to zero.

Figure 18-10 shows an SCR used as a dc power control switch. In this type of circuit, a rather high load current is controlled by a small gate current. Note that the electrical power source (V_s) is controlled by the SCR. The polarity of V_s must forward bias the SCR. This is achieved by making the anode positive and the cathode negative.

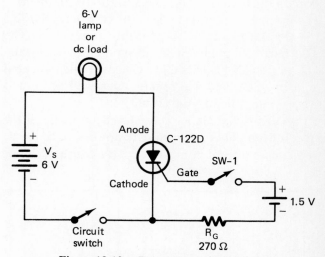

Figure 18-10. Dc power control switch.

When the circuit switch is turned on initially, the load is not energized. In this situation the V_{BO} is an excess of V_s voltage. Power control is achieved by turning on SW-1. This forward biases the gate. If a suitable value of I_G occurs, it will lower the V_{BO} and turn on the SCR. The I_G can be removed and the

SCR will remain in conduction. To turn the circuit off, momentarily open the circuit switch. With the circuit switch on again, the SCR will remain in the off state. It will go into conduction again by closing SW-1.

Dc power control applications of the SCR require two switches to achieve control. This application of the SCR is not very practical. The circuit switch would need to be capable of handling the load current. The gate switch could be rated at an extremely small value. If several switches were needed to control the load from different locations, this circuit would be more practical. More practical dc power circuits can be achieved by adding a number of additional components. Figure 18-11 shows a dc power control circuit with one SCR being controlled by a second SCR. SCR_1 would control the dc load current. SCR_2 controls the conduction of SCR_1. In this circuit, switching of a high current load is achieved with two small low-current switches. SCR_1 would be rated to handle the load current. SCR_2 could have a rather small current-handling capacity. Control of this type could probably be achieved for less than a circuit employing a large *electrical contactor switch*.

Operation of the dc control circuit of Fig. 18-11 is based on the conduction of SCR_1 and SCR_2. To turn on the load, the "on" pushbutton is momentarily closed. This forward biases the gate of SCR_1. The V_{BO} is reduced and SCR_1 goes into conduction. The load current latches SCR_1 in its conduction state. This action also causes C_1 to charge to the indicated polarity. The load will remain energized as long as power is supplied to the circuit.

Turn-off of SCR_1 is achieved by pushing the stop button. This momentarily applies I_G to SCR_2 and causes it to be conductive. This causes C_1 to be connected across SCR_1. The charge on C_1 is momentarily applied to the anode–cathode of SCR_1. This reduces the V_{AK} of SCR_1 and causes it to turn off. The circuit will remain in the off state until it is energized by the start button. An SCR power circuit of this type can be controlled with two small pushbuttons. As a rule, control of this type would be more reliable and less expensive than a dc electrical contactor circuit.

AC POWER CONTROL WITH SCRs. Ac electrical power control applications of an SCR are very common. As a general rule, control is very easy to achieve. The SCR automatically turns off during one alternation of the ac input. This eliminates the turn-off problem with the dc circuit. The load of an ac circuit will see current only for one alternation of the input cycle. In effect, an SCR power control circuit has halfwave output. The conduction time of an alternation can be varied with an SCR circuit. We can have variable output through this method of control.

A simple SCR power control switch is shown in Fig. 18-12. Connected in this manner, conduction of ac will only occur when the anode is positive and the cathode negative. Conduction will not occur until SW-1 is closed. When this takes place, there is gate current. The value of I_G lowers the V_{BO} to where the SCR becomes conductive. R_g of the gate circuit limits the peak value of I_G. Diode (D_1) prevents reverse voltage from being applied between the gate and cathode of the SCR. With SW-1 closed, the gate will be forward biased for only one alternation. This is the same alternation that forward biases the anode–cathode. With a suitable value of I_G and correct anode–cathode voltage (V_{AK}), the SCR will become conductive.

Figure 18-11. Dc power control circuit.

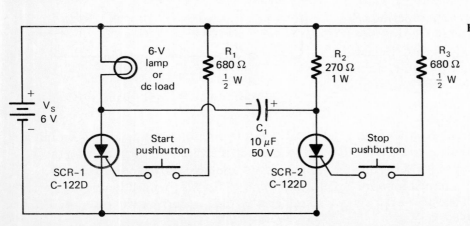

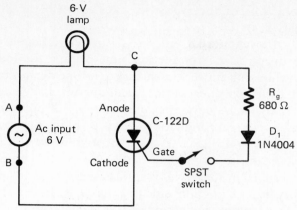

Figure 18-12. SCR power control switch.

The ac power control switch of Fig. 18-12 is designed primarily to take the place of a mechanical switch. With a circuit of this type, it is possible to control a rather large amount of electrical power with a rather small switch. Control of this type is very reliable. The switch does not have contacts that spark and arc when changes in load current occur. Control of this type, however, is only an on–off function.

SCRs are widely used to control the amount of electrical power supplied to a load device. Circuits of this type respond very well to 60 Hz ac. This type of circuit is called a half-wave SCR phase shifter. Figure 18-13 shows a simplified version of the variable power control circuit.

Figure 18-13. Phase control SCR circuit and waveforms.

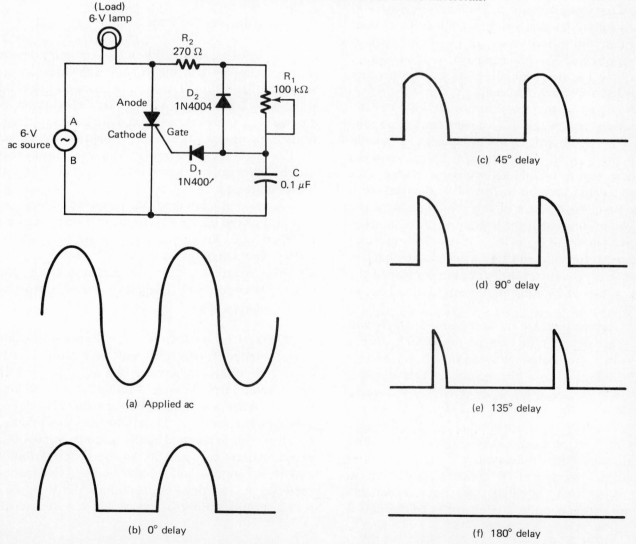

(a) Applied ac

(b) 0° delay

(c) 45° delay

(d) 90° delay

(e) 135° delay

(f) 180° delay

Operation of the variable power control circuit is based on the charge and discharge of capacitor C_1. Assume now that the positive alternation occurs. Point A will be positive and point B negative. This will forward bias the SCR. It will not turn on immediately, however, because there is no gate current. Capacitor C_1 begins to charge through R_1. In a specific time, the gate voltage builds up to where the gate current is great enough to turn on the SCR. Diode D_1 is forward biased by the capacitor voltage. I_G then flows through D_1. Once the SCR goes into conduction it continues for the remainder of the alternation. Current flows through the load. In effect, turn-on of the SCR is delayed by the value of R-C.

When the negative alternation occurs, point A becomes negative and point B goes positive. This reverse biases the SCR. No current flows through the load for this alternation. Diode D_2 is, however, forward biased by this voltage. Capacitor C_1 then discharges very quickly through D_2. This bypasses the resistance of R_1. The capacitor is reset or in a ready state for the next alternation. The process repeats itself for each succeeding cycle.

A change in the resistance of R_1 will alter the conduction time of the SCR power control circuit. When R_1 is increased, C_1 cannot charge as quickly during the positive alternation. It therefore takes more time for C_1 to build up its charge voltage. This means that the turn-on of I_G will be delayed for a longer time. As a result of this, conduction of the SCR will be delayed at the beginning of the alternation. See wavefore (c) of Fig. 18-13. The resulting load current for this condition will be somewhat less than that of waveform (b). Variable control of the load is achieved by altering the conduction time of the SCR.

A further increase in the resistance of R_1 will cause more delay in the turn-on of the SCR during the positive alternation. Waveforms (d), (e), and (f) of Fig. 18-13 shows the result of this change. Waveform (d) has a 90° delay of SCR conduction time. Waveform (e) shows a 135° delay in the conduction time. Waveform (e) shows a 135° delay in the conduction time. If the value of R_1 is great enough, the SCR will delay its turn-on for the full alternation. Waveform (f) shows a 180° turn-on delay. Note that the load does not receive any current for this condition. Control of this circuit is from full conduction as in waveform (b) to 180° delay as in (f).

Power control with an SCR is very efficient. If no delay of the conduction time occurs, the load develops full power. In this case, full power occurs only for one alternation. If conduction is delayed 90°, the load develops only half of its potential power. If the conduction delay is 135°, the load develops only 25% of its potential power. It is important to note that nearly all of the power controlled by the circuit is applied to the load device. Very little of it is consumed by the SCR. In variable power control applications it is extremely important that the load be supplied to its full value of power. With an SCR control circuit, no power is consumed by other components. By delaying the turn-on time of conduction, power can be controlled with the highest level of efficiency.

Triac Power Control

Ac power control can be achieved with a device that switches on and off during each alternation. Control of this type is accomplished with a special solid-state device known as a triac. This device is described as a three-terminal ac switch. Gate current is used to control the conduction time of either alternation of the ac waveform. In a sense, the triac is the equivalent of two reverse-connected SCRs feeding a common load device.

A triac is classified as a gate-controlled ac switch. For the positive alternation it responds as a PNPN device. An alternate crystal structure of the NPNP type is used for the negative alternation. Each crystal structure is triggered into conduction by the same gate connection. The gate has a dual-polarity triggering capability.

TRIAC CONSTRUCTION.   A triac is a solid-state device made of two different four-layer crystal structures connected between two terminals. We do not generally use the terms "anode" and "cathode" to describe these terminals. For one alternation they would be the anode–cathode. For the other alternation they would respond as the cathode–anode. It is common practice, therefore, to use the terms *main 1* and *main 2* or *terminal 1* and *terminal 2* to describe these leads. The third connection is the gate. This lead determines when the device switches from its off

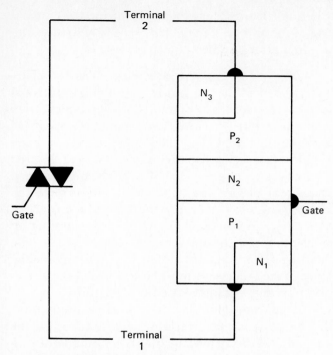

Figure 18-14. Triac symbol and crystal structure.

to its on state. The gate will normally go into conduction when it is forward biased. This is usually based on the polarity of terminal 1. If T_1 is negative, G must be positive. When T_1 is positive, the gate must be negative. This means that ac voltage must be applied to the gate to cause conduction during each alternation of the T_1–T_2 voltage.

A schematic symbol and the crystal structure of a triac are shown in Fig. 18-14. Notice the junction of the crystal structure simplification. Looking from T_1 to T_2 the structure involves crystals N_1, P_1, N_2, and P_2. The gate is used to bias P_1. This is primarily the same as an SCR with T_1 serving as the cathode and T_2 the anode.

Looking at the crystal structure from T_2 to T_1 it is N_3, P_2, N_2, and P_1. The gate is used to bias N_2 for control in this direction. This is similar to the structure of SCR in this direction. Notice that T_1, T_2, and G_1 are all connected to two pieces of crystal. Conduction will take place only through the crystal polarity that is forward biased. When T_1 is negative, for example, N_1 is forward biased and P_1 is reverse biased. Terminal selection by bias polarity is the same for all three terminals.

The schematic symbol of the triac is representative of reverse-connected diodes. The gate is con-

nected to the same end as T_1. This is an important consideration when connecting the triac into a circuit. The gate is normally forward biased with respect to T_1.

TRIAC OPERATION. When ac is applied to a triac, conduction can occur for each alternation. For this to take place T_1 and T_2 must be properly biased with respect to the gate. Forward conduction occurs when T_1 is negative, the gate (G) is positive, and T_2 is positive. Reverse conduction occurs when T_1 is positive, G is negative, and T_2 is negative. Conduction in either direction is similar to that of the SCR. The two-transistor operation used with the SCR applies to the triac in both directions. Figure 18-15 shows an example of the operational conduction polarities of a triac. Note that dc is used as a power source. This is used here only to denote operational polarities. In practice, dc would not be used as a power source for a triac. It is an ac power control device.

The operation of a triac is primarily the same as that of the SCR. When T_1 and T_2 are of the correct polarity, gate current must occur in order to reduce the breakover voltage. Small values of I_G will cause the breakover voltage to be quite large. An increase in I_G decreases the breakover voltage. This applies equally to both forward and reverse conduction. After conduction occurs, the gate loses its effectiveness.

When a triac goes into conduction its *internal resistance* drops to an extremely low value. Typical resistance values are less than 1 Ω. The device remains in this state for the remainder of the alternation. It is essential that current flowing through the triac be great enough to maintain conduction. This is generally called the holding current or I_H. The resistance of the load device and the value of the source voltage determine conduction current. This current must be slightly greater than I_H to maintain conduction.

Operation of a triac has one rather unique problem that does not occur in the SCR. This refers to its conduction *commutation*. Commutation is used to describe the turn-off of conduction. In triac operation commutation time can be very critical. With a resistive load, turn-off starts when the conduction current drops below the I_H level. It usually continues into the next alternation until the

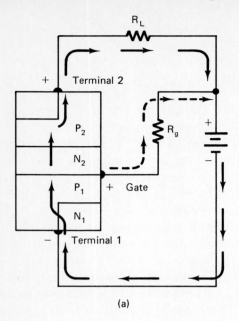

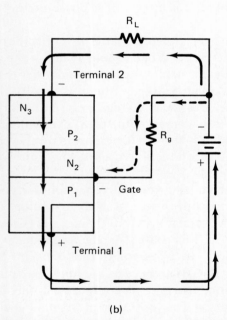

Figure 18-15. Triac condition polarities: (a) forward conduction, T_1 −, T_2+, G+; (b) reverse condition, T_2^-, T_1^+, G^-.

breakover voltage is reached. In most applications the triac has sufficient time to reach turn-off. With an inductive load, conduction time is extended somewhat. This is due to the inductive voltage caused by the collapsing magnetic field. A special *RC* circuit is placed across the triac to reduce this problem. This circuit is called a *snubber*. A 0.1-mF capacitor with a 100-Ω series resistor is placed across

the triac. The capacitor charges during the collapsing of the inductive field. This generally reduces the commutation problem.

TRIAC *I–V* CHARACTERISTICS. The *I–V* characteristic of a triac shows how it responds to forward and reverse voltages. Figure 18-16 is a typical triac *I–V* characteristic. Note that conduction occurs in quadrants I and III. The conduction in each quadrant is primarily the same. With 0 I_G, the breakover voltage is usually quite high. When breakover occurs, the curve quickly returns to the center. This shows a drop in the internal resistance of the device when conduction occurs. Conduction current must be limited by an external resistor. The holding current or I_H of a triac occurs just above the knee of the I_T curve. I_H must be attained or the device will not latch during a specific alternation.

Quadrant III is normally the same as quadrant I. This assures that operation will be the same for each alternation. Since the triac is conductive during quadrant III, it does not have a peak reverse voltage rating. It does, however, have a maximum reverse conduction current value. This is, of course, the same as the maximum forward conduction value. The conduction characteristics of quadrant III are mirror images of quadrant I.

TRIAC APPLICATIONS. Triacs are used primarily to achieve ac power control. In this application the triac responds primarily as a switch. Through normal switching action, it is possible to control the ac energy source for a portion of each alternation. If conduction occurs for both alternations of a complete sine wave, 100% of the power is delivered to the load device. Conduction for half of each alternation permits 50% control. If conduction is for one-fourth of each alternation, the load receives only 25% of its normal power. It is possible through this device to control conduction for the entire sine wave. This means that a triac is capable of controlling from 0 to 100% of the electrical power supplied to a load device. Control of this type is very efficient. Practically no power is consumed by the triac while performing its control function.

Static Switching. The use of a triac as a static switch is primarily an on–off function. Control of this type has a number of advantages over mechanical load switching. A high current energy source can be controlled with a very small switch. There is no

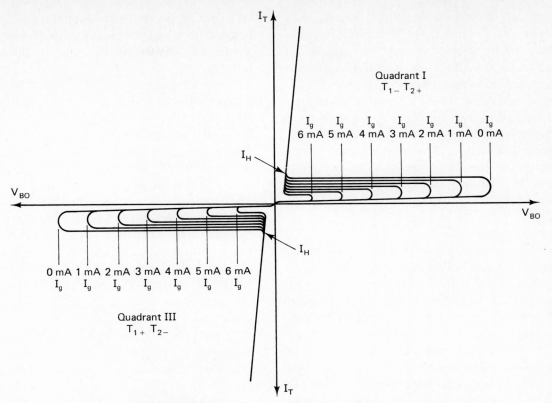

Figure 18-16. *I–V* characteristics of a triac.

contact bounce with solid-state switching. This generally reduces arcing and switch contact destruction. Control of this type is rather easy to achieve. Only a small number of parts are needed for a triac switch.

Two rather simple triac switching applications are shown in Fig. 18-17. Circuit (a) shows the load being controlled by a SPST switch. When the switch is closed, ac is applied to the gate. Resistor R_1 limits the gate current to a reasonable operating value.

Figure 18-17. Triac switching circuits: (a) static triac switch; (b) three-position static switch.

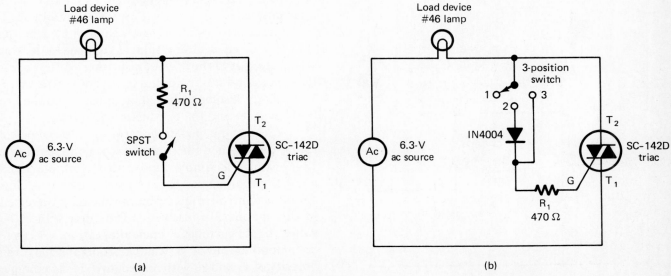

With ac applied to the gate, conduction occurs for the entire sine wave. The gate of this circuit requires only a few milliamperes of current to turn on the triac. Practically any small switch could be used to control a rather large load current.

Circuit (b) of Fig. 18-17 is considered to be a three-position switch. In position 1, the gate is open and the power is off. In position 2, gate current flows for only one alternation. The load receives power during one alternation. This is the half-power operating position. In position 3, gate current flows for both alternations. The load receives full ac power in this position.

Start–Stop Triac Control. Some electrical power circuits are controlled by two pushbuttons or start–stop switches. Control of this type begins by momentarily pushing the start button. Operation then continues after releasing the depressed button. To turn off the circuit, a stop button is momentarily pushed. The circuit then resets itself in preparation for the next starting operation. Control of this type is widely used in motor control applications and for lighting circuits. A triac can be adapted for this type of power control.

A start–stop triac control circuit is shown in Fig. 18-18. When electrical power is first applied to this circuit, the triac is in its nonconductive state. The load does not receive any power for operation.

All the supply voltage appears across the triac because of its high resistance. No voltage appears across the RC circuit, the gate, or the load device initially.

To energize the load device, the start pushbutton is momentarily pressed. C_1 charges immediately through R_1 and R_2. This, in turn, causes I_G to flow into the gate. The V_{BO} of the triac is lowered and it goes into conduction. Voltage now appears across the load, R_2–C_1, and the gate. The charging current of R_2–C_1 and the gate continue and are at peak value when the source voltage alternation changes. The gate then retriggers the triac for the next alternation. C_1 is recharged through the gate and R_2. The next alternation change causes I_G to again flow for retriggering of the triac. The load receives full power from the source. The process will continue into conduction as long as power is supplied by the source.

To turn off the circuit, the stop button is momentarily depressed. This action immediately bypasses the gate current around the triac. With no gate current, the triac will not latch during the alternation change. As a result, C_1 cannot be recharged. The triac will then remain off for each succeeding alternation change. Conduction can be restored only by pressing the start button. This circuit is a triac equivalent of the ac motor electrical contactor.

Triac Variable Power Control. The triac is widely used as a variable ac power control device. Control of this type is generally called full-wave control. "Full-wave" refers to the fact that both alternations of a sine wave are being controlled. Variable control of this type is achieved by delaying the start of each alternation. This process is very similar to that of the SCR. The primary difference is that triac conduction applies to the entire sine wave. For this to be accomplished, ac must be applied to both the gate and the conduction terminals.

Variable ac power control can be achieved rather easily when the source is low voltage. Figure 18-19 shows a simple low-voltage variable lamp control circuit. Note that the gate current of this circuit is controlled by a potentiometer. Connected in this manner, adjustment of R_1 determines the value of gate current for each alternation.

Conduction of a triac is controlled by the polarity of T_1 and T_2 with respect to the gate voltage.

Figure 18-18.   Start–stop triac control.

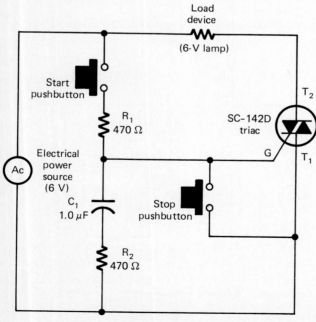

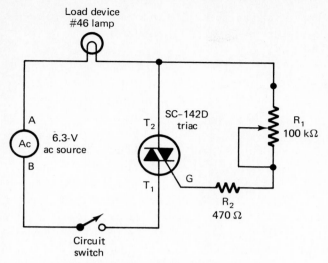

Figure 18-19.   Variable lamp control.

For the positive alternation, assume that point *A* is positive and *B* is negative. This causes a $+T_2$, a $-T_1$, and a $+I_G$. The value of the circuit gate current is determined by the resistance of R_1 and R_2. For a high-resistance setting of R_1, the triac may not go into conduction at all. For a smaller resistance value, conduction can be delayed in varying amounts. Generally, conduction delay will occur only during the first 90° of the alternation. If conduction does not occur by this time, it will be off for the last 90° of the alternation.

Variable control of the same type also occurs during the negative alternation. For this alternation, point *A* is negative and point *B* is positive. This causes a $-T_2$, a $+T_1$, and a $-I_G$. Gate current will flow and cause conduction during this alternation. The resistance setting of R_1 influences I_G in the same manner as it did for the positive alternation. Both alternations will therefore be controlled equally. Variable control of this type applies to only 50% of the source voltage. If conduction does not occur in the first 90° of an alternation, no control will be achieved.

Triac Phase Control.   With ac phase control of a triac it is possible to achieve nearly 100% control of the conduction time. For this to be achieved, two distinct ac voltage values are applied to the triac. The ac source voltage is applied to terminals T_1 and T_2. This voltage must not be changed. A second ac signal is applied to the gate. Gate current is developed by this voltage. Control of the triac is achieved by shifting the phase of the gate voltage.

Changes from 0 to 90° or 0 to 180° are possible depending on circuit design. The gate voltage must lag behind the source voltage in order to initiate control.

Figure 18-20 shows five different operating states of a triac phase control circuit. In part (a), the T_1–T_2 source voltage and the gate voltage are in phase. This causes 0° delay and full conduction of the triac. The load device receives 100% of the source voltage and full power. Part (b) shows where the gate voltage is shifted 45° behind the source voltage. Power to the load is reduced to approximately 75% of its full value. A 90° delay is shown in part (c). The gate voltage has shifted behind the source voltage by 90°. This causes the load to receive only 50% of its full-power capabilities. Parts (d) and (e) show the remaining two conditions. In (d), the load receives only 25% of its full power. Part (e) shows 180° delay and no power to the load. With the phase shifter, it is possible to achieve full control of power to the load device.

Only four components are used in the triac phase control circuit of Fig. 18-21. Resistor R_1 and capacitor C_1 serve as a single-element *phase shifter*. The phase of the gate voltage is shifted by R_1. The *diac* is a voltage-controlled trigger device. It will conduct in either direction when the applied voltage reaches a specific value. Diacs were purposely designed to trigger the gate of a triac. This particular diac will conduct when gate voltage reaches 28 V in either direction.

Operation of the single-element triac phase-shifter circuit is based on the time constant of R_1 and C_1. For the positive alternation, C_1 charges through R_1. When C_1 reaches the breakover voltage of the diac, it turns on. C_1 then discharges through the diac and the gate of the triac. Gate current lowers the breakover voltage of the triac, and conduction occurs. Conduction continues for the remainder of the alternation.

During the negative alternation C_1 charges in the reverse direction through R_1. When the reverse breakover voltage of the diac is reached, it goes into conduction. C_1 discharges through the diac and the gate of the triac. This triggers the triac into conduction. Conduction continues for the remainder of the negative alternation. The process then repeats itself for each succeeding alternation.

Delay of conduction is based on the time constant of R_1 and C_1. Control with a single-leg

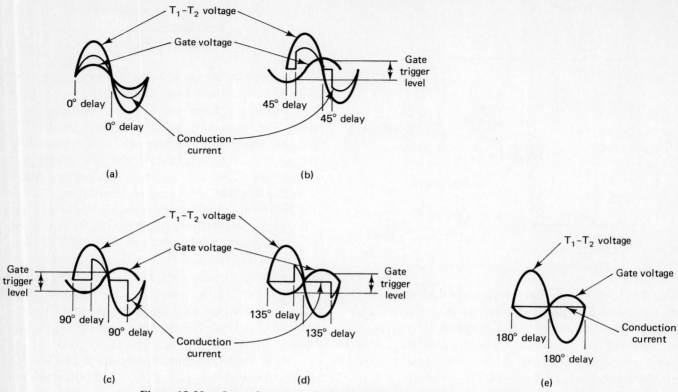

Figure 18-20. Operating states of a triac: (a) 100% power; (b) 75% power; (c) 50% power; (d) 25% power; (e) 0% power.

Figure 18-21. Basic triac phase shifter.

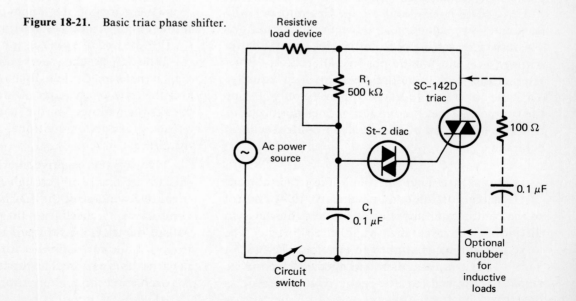

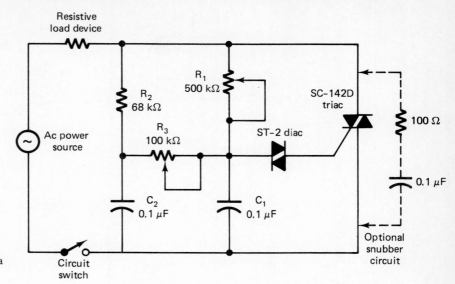

Figure 18-22. Two-stage phase shifter for a triac.

phase shifter is good only for the first 90° of an alternation. If conduction does not occur in the first 90°, it will remain off for the remainder of the alternation. Applications of this circuit are limited to lamp control, heater operation, and fan speed control.

An improved triac power control circuit is shown in Fig. 18-22. This circuit employs a double-stage RC phase shifter. Control of this circuit is in the approximate range 10 to 170°. Power control is from 16 to 95% of full power.

Operation of the double-stage RC phase shifter is based on alternate charge and discharge of two RC circuits. When power is first applied, C_2 charges very quickly through R_2. A short time later C_1 begins to charge from the developed voltage across C_2. When the charge on C_1 reaches the breakover voltage of the diac, it goes into conduction. C_1 discharges through the diac and the gate of the triac. This triggers the triac into conduction. Conduction then continues for the remainder of the alternation. The charge time of C_1 is based on adjustment of R_1 and R_3.

For the next alternation, C_2 recharges through R_2. C_1 begins to charge to the C_2 voltage a short time later through R_3. When the diac is triggered, C_1 discharges through the triac. This increases the gate current and causes the triac to be conductive. Conduction continues for the remainder of the alternation. The process then repeats itself for each succeeding alternation.

The double-stage RC phase shifter has an increased control range. When R_1 is large, C_1 charges to the developed voltage of C_2. R_3 should be adjusted so that the circuit drops out of conduction when R_1 is at its maximum value. R_3 is considered as a trimmer adjustment for the circuit.

Diac Power Control

A diac is a special diode that can be triggered into conduction by voltage. This device is classified as a *bidirectional* trigger diode. This means that it can be triggered into conduction in either direction. The word "diac" is derived from the terms *di*ode for *ac*. This device is used primarily to control the gate current of a triac. It will go into conduction during either the positive or the negative alternation. Conduction is achieved by simply exceeding the breakover voltage.

Figure 18-23 shows the crystal structure, schematic symbol, and I-V characteristics of a diac. Note that the crystal is similar to that of a transistor without a base. The N_1 and N_2 crystals are primarily the same in all respects. A diac will therefore go into conduction at precisely the same negative or positive voltage value. Conduction occurs only when input voltage exceeds the breakover voltage. A rather limited number of diacs are available today. The one shown here has a minimum V_{BO} of 28 V. This

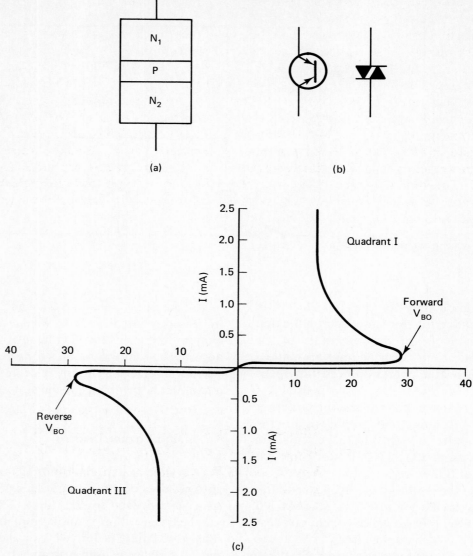

Figure 18-23.   Diac: (a) crystal structure; (b) symbols; (c) *I–V* characteristics.

particular device is a standard trigger for triac control. Note that the voltage across the diac decreases in value after it has been triggered.

ELECTRONIC CONTROL CONSIDERATIONS

Electronic power control with an SCR or a triac is very efficient when used properly. These devices are, however, attached directly to the ac power line. Severe damage to a load device and a potential electrical hazard may occur if this method of control

is improperly connected. As a general rule, SCR and traic control should not be attempted for ac-only equipment. This includes such things as fluorescent lamps, radios, TV receivers, induction motors, and other transformer-operated devices.

SCR and triac control can be used effectively to control resistive-type loads. This includes such things as incandescent lamps, soldering irons or pencils, heating pads, and electric blankets. It is also safe to use this type of control for universal motors. This type of motor is commonly used in portable power tools and some small appliances. Such things as portable drills, saber saws, sanders, electric knives, and mixers are included in this list. When in doubt about a particular device, check the manufacturer's instruction manual. It is also important that the wattage rating of the load not exceed the wattage of the control device. Wattage ratings are nearly always stamped on the device or indicated on a label. Check this rating before attempting to control the device with an SCR or a triac.

It is extremely important that all control circuits be placed in an insulated housing or container. The ac power line is connected directly to the control circuit. Touching any part of the control circuit may cause a severe electrical shock. A common practice is to place the control circuit in an insulated box which has no outside metal parts or knobs.

The procedure for servicing or analyzing the operation of an ac power control circuit is extremely important. Avoid touching all metal parts when in operation. Work with the circuit only on an insulated tabletop or bench. Do not attempt to connect electrically energized test instruments to the circuit. If at all possible, the control device should be energized by an isolation transformer. This will reduce the ground problems and some electrical hazards. Circuit testing should be done when the power control device is disconnected from the power line.

Review

1. Explain why control by switching is more efficient than current or voltage changing.
2. Why is SCR control more efficient than a rheostat?
3. Explain how an SCR conducts when the proper voltage values and polarities are applied.
4. How can two transistors be used to explain SCR operation?
5. What is meant by the term "forward breakover voltage"?
6. How is the forward breakover voltage of an SCR reduced?
7. Why does an SCR operate only in quadrant I of the $I-V$ characteristics?
8. What does the peak reverse voltage tell about an SCR?
9. What is meant by the term "holding current"?
10. When dc power is controlled by an SCR, what is one of the major control problems?
11. Explain why ac is easier to control with an SCR than is dc.
12. Why is the output of an SCR only half-wave or pulsating dc?
13. What causes an SCR to latch into conduction after being triggered into conduction?
14. What are the primary physical differences between an SCR and a triac?
15. What are the electrode polarities needed to cause triac conduction in quadrants I and III of the $I-V$ characteristics?
16. Explain how variable control of a triac is achieved.
17. How can an RC circuit be used to alter the turn-on or conduction of a triac?
18. What is meant by phase control of a triac?
19. Explain why a double-stage RC phase shifter has a better range of triac control than does a single-leg phase shifter.
20. How is conduction achieved in a diac?

Student Activities

SCR Testing with a VOM

1. Prepare a VOM or VTVM to measure resistance in the $R \times 1$ range. Digital meters do not usually respond well to this test because of the low-current capabilities of the ohmmeter.

2. Select any two SCR leads at random. Test the forward and reverse resistance. If a resistance occurs in either direction, the two leads are the cathode and gate. The third lead will be the anode. If the resistance is infinite in both directions, switch one of the two leads. Test again for a resistance indication in one direction. Keep in mind the position of the anode lead identified.

3. A good SCR should indicate lowest resistence when the gate is made positive and the cathode negative. Determine which of the two leads are the gate and cathode.

4. We will now verify the correctness of the identified leads. Connect the negative ohmmeter lead to the cathode and the positive lead to the anode. Then momentarily touch the gate to the positive anode lead. This should cause a low-resistance A–K indication on the ohmmeter. Disconnect the gate from the anode lead. The ohmmeter should continue to indicate low resistance. This shows the latching function of an SCR. If the test responds in this manner, identification of A, K, and G is correct. If it does not respond, the SCR is faulty or the wrong leads have been identified.

5. Test several SCRs to become familiar with the identification procedure.

6. Make a sketch of lead placement for the SCRs that will be used in future activities.

DC Power Control Switch

1. Construct the static dc switch of Fig. 18-10.

2. Turn off both switches.

3. Turn on the circuit switch. What happens?

4. Turn on the gate switch SW-1. What happens?

5. Turn off the gate switch (SW-1). How does the lamp respond?

6. Turn off the circuit switch. Then turn on the circuit switch again. How does the lamp respond?

7. Turn off the circuit switch.

8. Repeat the procedure for steps 2 to 5.

9. The circuit can be used with a greater V_S voltage if desired. The lamp should be changed when using a larger V_S. For 12 V_S use a 12-V lamp.

DC Start–Stop Power Control

1. Construct the start–stop dc power control circuit of Fig. 18-11.

2. Turn on the circuit by momentarily pushing the start button. How does the lamp respond?

3. Push the stop button. How does the lamp respond?

4. Test the operation of the circuit several times.

5. Turn off the power source and replace R_2 with a 6-V lamp.

6. Repeat steps 2 to 4. Describe how the two lamps respond in the circuit.

Static AC Power Control with an SCR

1. Construct the ac power control switch of Fig. 18-12.

2. Close the gate switch. How does the load respond?

3. Open the gate switch. How does the load respond? Why does it respond in this manner?

4. A different source voltage can be used. The lamp must be capable of handling the voltage.

AC Phase Control of an SCR

1. Construct the SCR phase control circuit of Fig. 18-13.

2. Turn on the ac source. Adjust R_1 through its range. Describe the response of the lamp.

3. If the ac source is from a low-voltage transformer, connect an oscilloscope across the lamp.

4. Adjust R_1 through its range while observing the waveform. Describe the waveform.

Triac Testing

1. Prepare a VOM or VTVM to measure resistance on the $R \times 1$ range. We will assume that the leads have straight polarity. A digital VOM or FET VOM does not work well for this test procedure. The current capability is usually quite low.

2. Select any two leads at random. Measure the forward and reverse resistance. T_1 and T_2 will show an infinite or extremely high resistance in either direction. If this does not occur, select the alternate lead. Check the resistance ratio until T_1 and T_2 are found.

3. After finding T_1 and T_2, the third lead is obviously the gate.

4. Using one of the T leads and the gate, measure the forward and reverse resistance. T_1 and G should have low resistance in both the forward and reverse directions. T_2 and G will show an infinite resistance in both directions. This should permit identification of T_1, G, and T_2.

5. Make a sketch of the identified leads.

6. To varify the correctness of the identified leads, we will test the latching condition.

7. Connect the positive ohmmeter lead to T_2 and the negative to T_1. Momentarily touch the gate to T_2. This should cause a low resistance reading and latch T_1 and T_2. Disconnect the gate. Conduction between T_1 and T_2 should continue.

8. Connect the positive ohmmeter lead to T_1 and the negative lead to T_2. Momentarily touch the gate to T_2. This should cause a low resistance reading and latch T_1 and T_2. Disconnect the gate. The low resistance should remain between T_1 and T_2.

9. Repeat the lead identification and latching procedure on several different triacs.

Start–Stop Triac Control

1. Construct the start–stop triac control circuit of Fig. 18-18.
2. Use 6 V or 12 V ac and a corresponding lamp for the circuit.
3. Momentarily push the start pushbutton. How does the load respond?
4. Depress the stop pushbuttom. How does the load respond?
5. Try the circuit several times to verify its operation.

AC Variable Lamp Control

1. Connect the low-voltage variable lamp control circuit of Fig. 18-19.
2. Turn on the circuit switch.
3. Adjust R_1 through its operating range while observing the brightness of the lamp.
4. Describe how the circuit responds.

Basic Phase-Shifter Triac Control

1. Construct the phase-shifter triac control circuit of Fig. 18-21. The ac power source will be 120 V. The load could be a 6-W lamp, a 12-W lamp, or a 100-W 120-V lamp.
2. The control R_1 should be adjusted with an insulated knob.
3. Turn on the circuit switch. Adjust R_1 through its range. How does the lamp respond?
4. Turn off the circuit switch and disconnect the low-voltage power source.
5. If desired, construct the circuit using 120 V ac. Use an isolated ac source if at all possible. Use a 120-V lamp of any wattage up to 250 W.
6. Have the instructor check the 120-V circuit.
7. Turn on the circuit switch. Adjust R_1 through its range while observing the lamp. Notice how the control is adjusted to product the lowest light level.
8. Turn off the circuit switch and remove the 120-V ac power source.

Motor Speed Control

1. Construct the basic triac phase shifter of Fig. 18-21. The power source in this will be 120 V ac. A small universal motor of less than $\frac{1}{4}$ hp will be used for the load device.
2. If at all possible, use an isolated ac power source or an isolation transformer for this activity.
3. Connect the optional snubber circuit across the triac.
4. Adjust R_1 to produce the lowest load current. This will be the same as step 8 of the preceding activity.
5. Ask the instructor to approve your circuit.
6. Turn on the circuit switch.

7. Adjust R_1 slowly. If the circuit is responding properly, the rotational speed of the motor can be changed by R_1.

8. Turn off the circuit switch and disconnect the ac power source.

Two-Stage Phase-Shifter Triac Control Circuit

1. Construct the improved phase-shifter circuit of Fig. 18-22. Use the same procedures outlined in the basic phase-shifter activity. Follow the same precautions.

2. Does the circuit have any better control capabilities than the basic phase-shifter circuit?

3. Turn off the circuit switch and remove the 120-V ac power source.

APPENDIX

1

The Elements

PERIODIC TABLE OF THE ELEMENTS

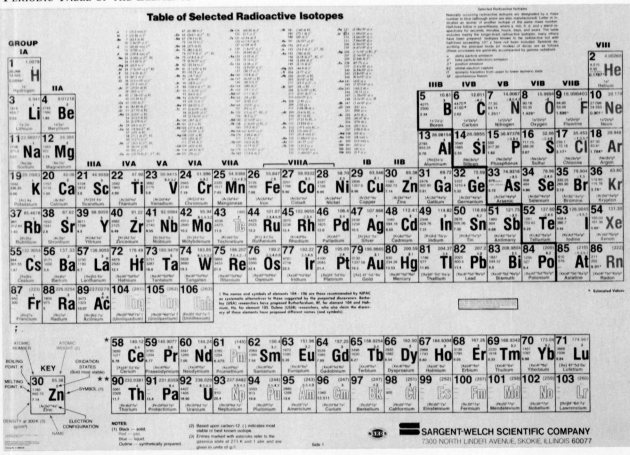

(Courtesy of Sargent-Welch Scientific Co.)

510

AN ALPHABETICAL LIST OF THE ELEMENTS

Element	Symbol	Atomic no.	Atomic weight	Element	Symbol	Atomic no.	Atomic weight	Element	Symbol	Atomic no.	Atomic weight
Actinium	Ac	89	227*	Hafnium	Hf	72	178.6	Praseodymium	Pr	59	140.92
Aluminum	Al	13	26.97	Hahmium	Ha	105	262*	Promethium	Pm	61	145*
Americium	Am	95	243*	Helium	He	2	4.003	Protactinium	Pa	91	231*
Antimony	Sb	51	121.76	Holmium	Ho	67	164.94	Redium	Ra	88	226.05
Argon	Ar	18	39.944	Hydrogen	H	1	1.0080	Rodon	Rn	86	222
Arsenic	As	33	74.91	Indium	In	49	114.76	Rhenium	Re	75	186.31
Astatine	At	85	210*	Iodine	I	53	126.91	Rhodium	Rh	45	102.91
Barium	Ba	56	137.36	Iridium	Ir	77	192.2	Rubidium	Rb	37	85.48
Berkelium	Bk	97	247*	Iron	Fe	26	55.85	Rutherium	Ru	44	101.1
Beryllium	Be	4	9.013	Krypton	Kr	36	83.8	Rutherfordium	Rf	104	260*
Bismuth	Bi	83	209.00	Lanthanum	La	57	138.92	or Kurchatonium	or Ku		
Boron	B	5	10.82	Lawrencium	Lw	103	257*	Samarium	Sm	62	150.43
Bromine	Br	35	79.916	Lead	Pb	82	207.21	Scandium	Sc	21	44.96
Cadmium	Cd	48	112.41	Lithium	Li	3	6.940	Selenium	Se	34	78.96
Calcium	Ca	20	40.08	Lutetium	Lu	71	174.99	Silicon	Si	14	28.09
Californium	Cf	98	251*	Magnesium	Mg	12	24.32	Silver	Ag	47	107.880
Carbon	C	6	12.01	Manganese	Mn	25	54.94	Sodium	Na	11	22.997
Cerium	Ce	58	140.13	Mendelevium	Mv	101	256*	Strontium	Sr	38	87.63
Cesium	Cs	55	132.91	Mercury	Hg	80	200.61	Sulfur	S	16	32.066
Chlorine	Cl	17	35.457	Molybdenum	Mo	42	95.95	Tantalum	Ta	73	180.95
Chromium	Cr	24	52.01	Neodymium	Nd	60	144.27	Technetium	Tc	43	97*
Cobalt	Co	27	58.94	Neon	Ne	10	20.183	Tellurium	Te	52	127.61
Copper	Cu	29	63.54	Neptunium	Np	93	237*	Terbium	Tb	65	158.93
Curium	Cm	96	247	Nickel	Ni	28	58.69	Thallium	Tl	81	204.39
Dysprosium	Dy	66	162.46	Niobium	Nb	41	92.91	Thorium	Th	90	232.12
Einsteinium	E	99	254*	Nitrogen	N	7	14.008	Thulium	Tm	69	168.94
Erbium	Er	68	167.2	Nobelium	No	102	253	Tin	Sn	50	118.70
Europium	Eu	63	152.0	Osmium	Os	76	190.2	Titanium	Ti	22	47.90
Fermium	Fm	100	255*	Oxygen	O	8	16.000	Tungsten	W	74	183.92
Fluorine	F	9	19.00	Palladium	Pd	46	106.7	Uranium	U	92	238.07
Francium	Fr	87	233*	Phosphorus	P	15	30.975	Vanadium	V	23	50.95
Gadolinium	Gd	64	156.9	Platinum	Pt	78	195.23	Xenon	Xe	54	131.3
Gallium	Ga	31	69.72	Plutonium	Pu	94	244	Ytterbium	Yb	70	173.04
Germanium	Ge	32	72.60	Polonium	Po	84	210	Yttrium	Y	39	88.92
Gold	Au	79	197.0	Potassium	K	19	39.100	Zinc	Zn	30	65.38
								Zirconium	Zr	40	91.22

*Mass number of the longest-lived of the known available forms of the element.

Soldering

Soldering is an important skill for electrical technicians. Good soldering is important for the proper operation of equipment.

Solder is an alloy of tin and lead. The solder that is most used is 60/40 solder. This means that it is made from 60% tin and 40% lead. Solder melts at a temperature of about 400° F.

For solder to adhere to a joint, the parts must be hot enough to melt the solder. The parts must be kept clean to allow the solder to flow evenly. Rosin flux is contained inside the solder. It is called rosin-core solder.

A good mechanical joint must be made when soldering. Heat is then applied until the materials are hot. When they are hot, solder is applied to the joint. The heat of the metal parts (not the soldering tool) is used to melt the solder. Only a small amount of heat should be used. Solder should be used sparingly. The joint should appear smooth and shiny. If it does not, it could be a "cold" solder joint. Be careful not to move the parts when the joint is cooling. This could cause a "cold" joint.

When parts that could be damaged by heat are soldered, be very careful not to overheat them. Semiconductor components, such as diodes and transistors, are very heat sensitive. One way to prevent heat damage is to use a *heat sink*, such as a pair of pliers. A heat sink is clamped to a wire between the joint and the device being soldered. A heat sink absorbs heat and protects delicate devices. Printed circuit boards, such as the one shown in Fig. A2-1, are also very sensitive to heat. Care should be taken not to damage PC boards when soldering parts onto them.

Several types of soldering irons and soldering guns are shown in Figs. A2-2 to A2-7. Small, low-wattage irons should be used with PC boards and semiconductor devices.

Below are some rules for good soldering:

1. Be sure that the tip of the soldering iron is clean and tinned.

2. Be sure that all the parts to be soldered are heated. Place the tip of the soldering iron so

that the wires and the soldering terminal are heated evenly.

3. Be sure not to overheat the parts.
4. Do not melt the solder onto the joint. Let the solder flow onto the joint.
5. Use the right kind and size of solder and soldering tools.
6. Use the right amount of solder to do the job but not enough to leave a "blob."
7. Be sure not to allow parts to move before the solder joint cools.

Figure A2-1. Printed circuit (PC) board. (Courtesy of TRW/ UTC Transformers.)

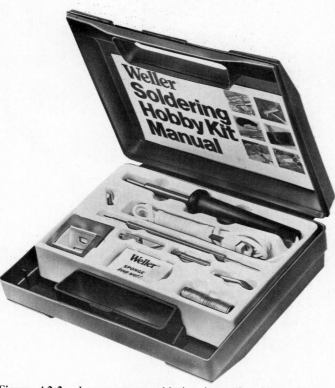

Figure A2-2. Low-wattage soldering iron kit with several soldering tips. (Courtesy of The Cooper Group.)

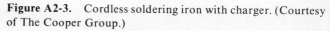

Figure A2-3. Cordless soldering iron with charger. (Courtesy of The Cooper Group.)

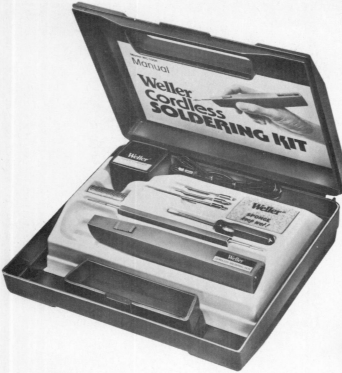

Figure A2-4. Cordless soldering iron kit. (Courtesy of The Cooper Group.)

Figure A2-5. High-wattage soldering gun kit with 60/40 solder and spare tips. (Courtesy of The Cooper Group.)

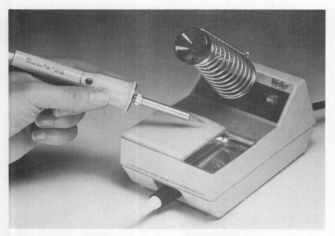

Figure A2-6. Soldering iron with temperature control station and cleaning pad. (Courtesy of The Cooper Group.)

Figure A2-7. (a) Desoldering tool; (b) desoldering bulb. (Courtesy of The Cooper Group.)

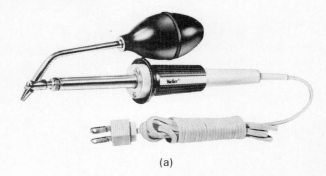

(a)

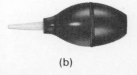

(b)

APPENDIX

Electrical Tools

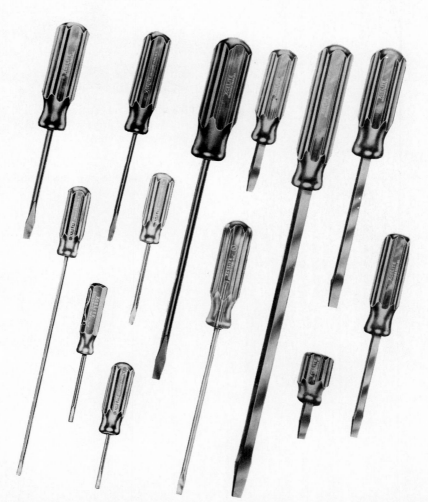

Figure A3-1. Various sizes of standard screwdrivers. (Courtesy of The Cooper Group.)

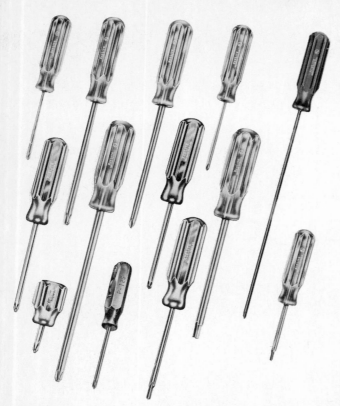

Figure A3-2. Various sizes of Phillips screwdrivers. (Courtesy of The Cooper Group.)

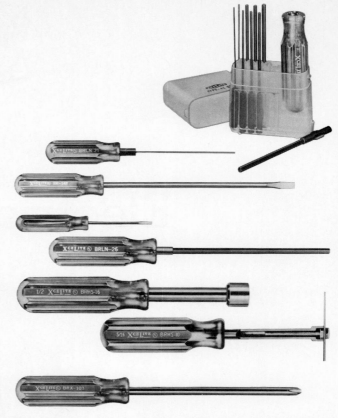

Figure A3-3. (a) Allen setscrew drivers; (b) standard screwdrivers; (c) nut drivers; (d) Phillips screwdriver. (Courtesy of The Cooper Group.)

Figure A3-4. Driver combination kits. (Courtesy of The Cooper Group.)

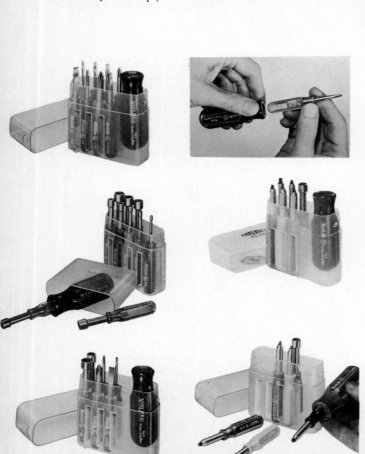

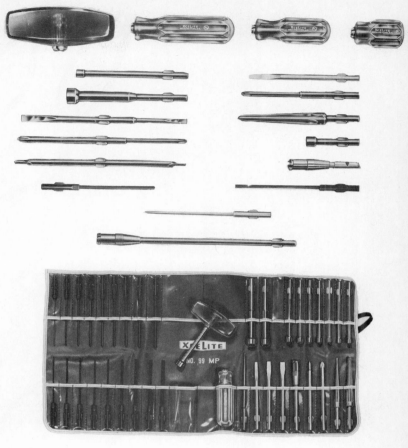

Figure A3-5. Screwdriver kits. (Courtesy of The Cooper Group.)

Figure A3-6. Nut-driver kits. (Courtesy of The Cooper Group.)

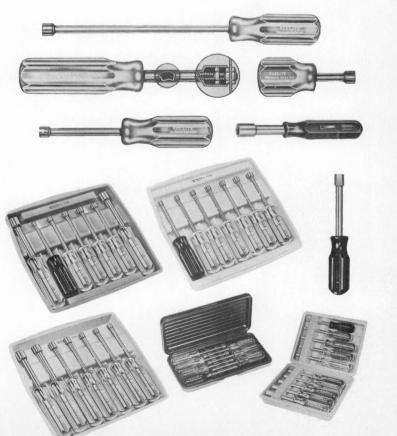

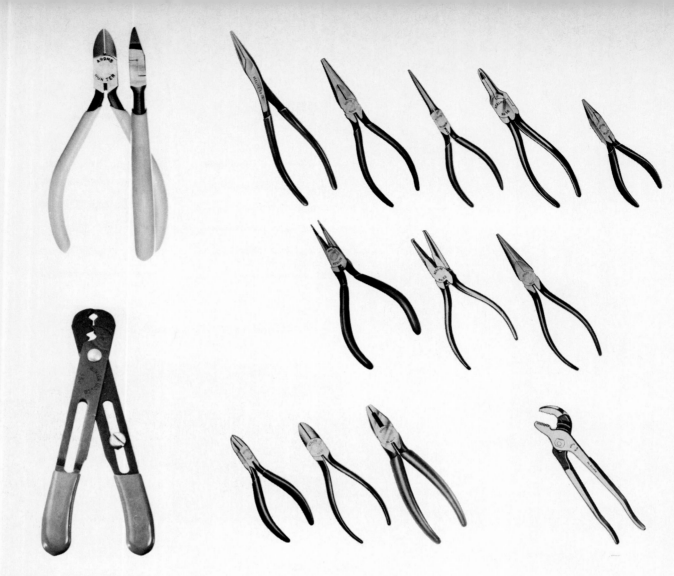

Figure A3-7. (a) Round-nose diagonal wire cutters; (b) wire strippers. (Courtesy of Hunter Tools, Division of K-D Manufacturing Co.)

Figure A3-8. (a) Long-nose or needle-nose pliers; (b) diagonal cutting pliers; (c) channel lock pliers. (Courtesy of The Cooper Group.)

Figure A3-9. Wire crimping tool with stripping and bolt cutting capability. (Courtesy of Vaco Products Co.)

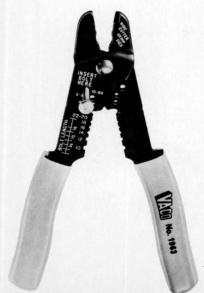

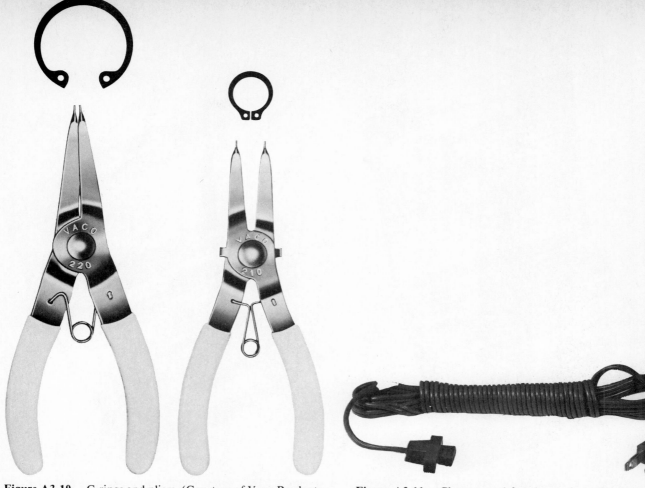

Figure A3-10. C-rings and pliers. (Courtesy of Vaco Products Co.)

Figure A3-11. Cheater cord for electronic equipment repair. (Courtesy of GC Electronics.)

Figure A3-12. Roll of tools for servicing. (Courtesy of The Cooper Group.)

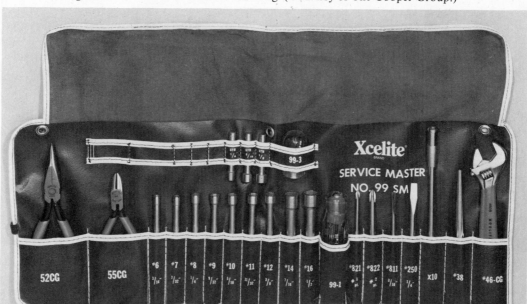

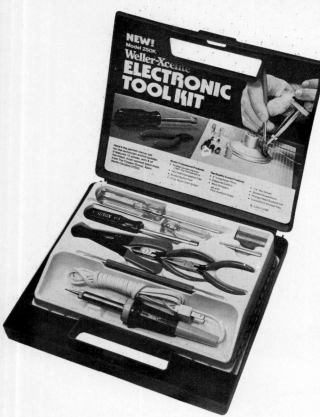

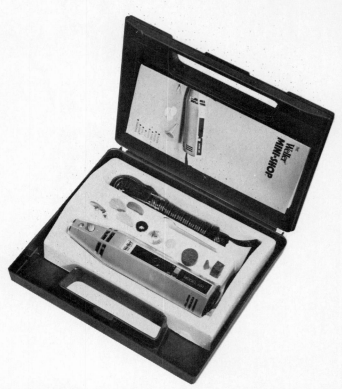

Figure A3-13. Electronic tool kit. (Courtesy of The Cooper Group.)

Figure A3-14. Portable multipurpose tool. (Courtesy of The Cooper Group.)

Symbols for Electricity and Electronics

SYMBOLS FOR ELECTRICITY

Fixed resistor		Ground		Fuse	
Tapped resistor		Contacts (normally closed)		Circuit breaker (single pole)	
Variable resistor (potentiometer)		Contacts (normally open)		Circuit breaker (three pole)	
Thermistor		Switch (single-pole, single-throw)		Coil (air core)	
Fixed capacitor		Switch (single-pole, double-throw)		Coil (iron core)	
Variable capacitor		Switch (double-pole, single-throw)		Coil (tapped)	
Polarized capacitor (electrolytic)		Switch (double-pole, double-throw)		Coil (adjustable)	
Battery				Transformer (air core)	
Alternating current source		Multiposition selector switch (any number of positions may be shown.)		Transformer (iron core)	

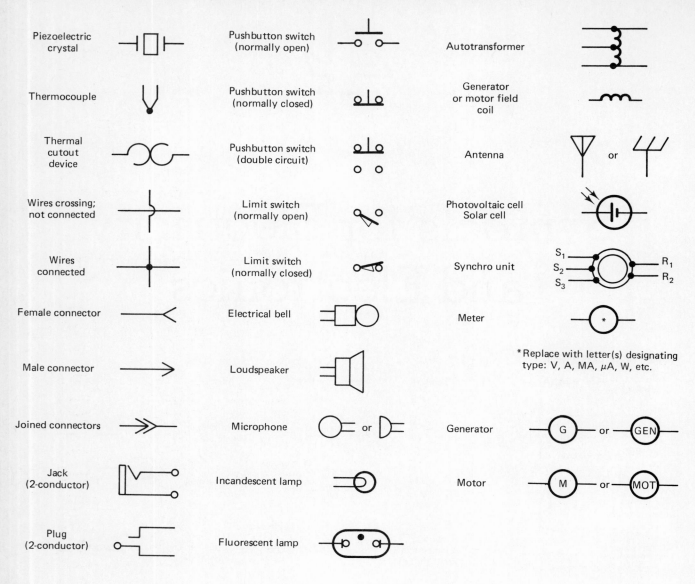

Piezoelectric crystal	Pushbutton switch (normally open)	Autotransformer
Thermocouple	Pushbutton switch (normally closed)	Generator or motor field coil
Thermal cutout device	Pushbutton switch (double circuit)	Antenna
Wires crossing; not connected	Limit switch (normally open)	Photovoltaic cell Solar cell
Wires connected	Limit switch (normally closed)	Synchro unit
Female connector	Electrical bell	Meter
Male connector	Loudspeaker	*Replace with letter(s) designating type: V, A, MA, μA, W, etc.
Joined connectors	Microphone	Generator
Jack (2-conductor)	Incandescent lamp	Motor
Plug (2-conductor)	Fluorescent lamp	

SYMBOLS FOR ELECTRONICS

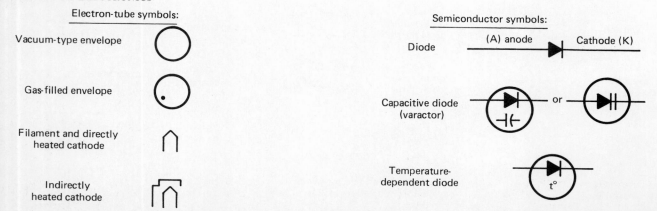

Electron-tube symbols:

Vacuum-type envelope

Gas-filled envelope

Filament and directly heated cathode

Indirectly heated cathode

Semiconductor symbols:

Diode — (A) anode — Cathode (K)

Capacitive diode (varactor)

Temperature-dependent diode

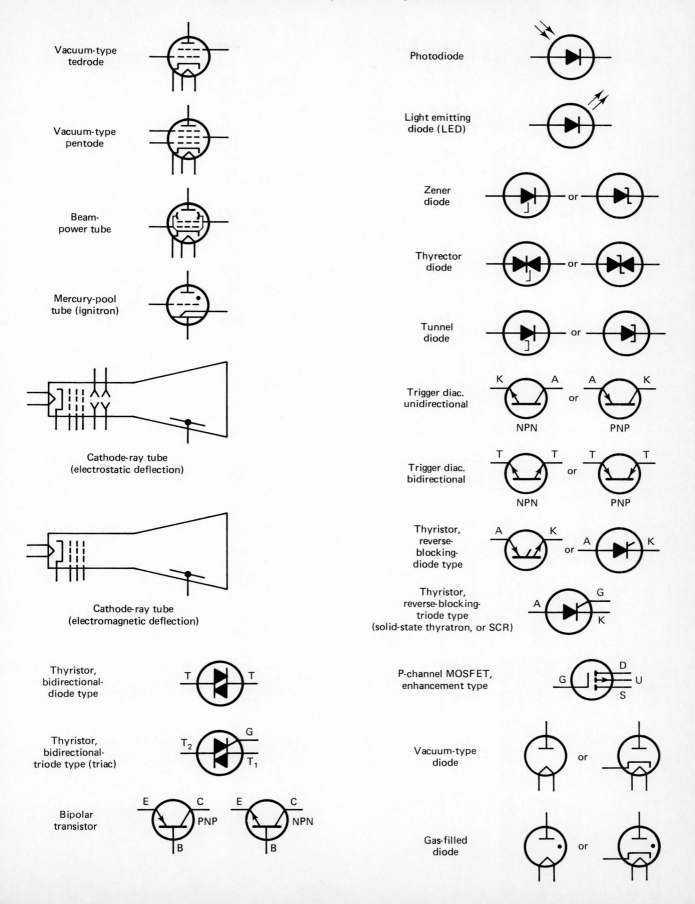

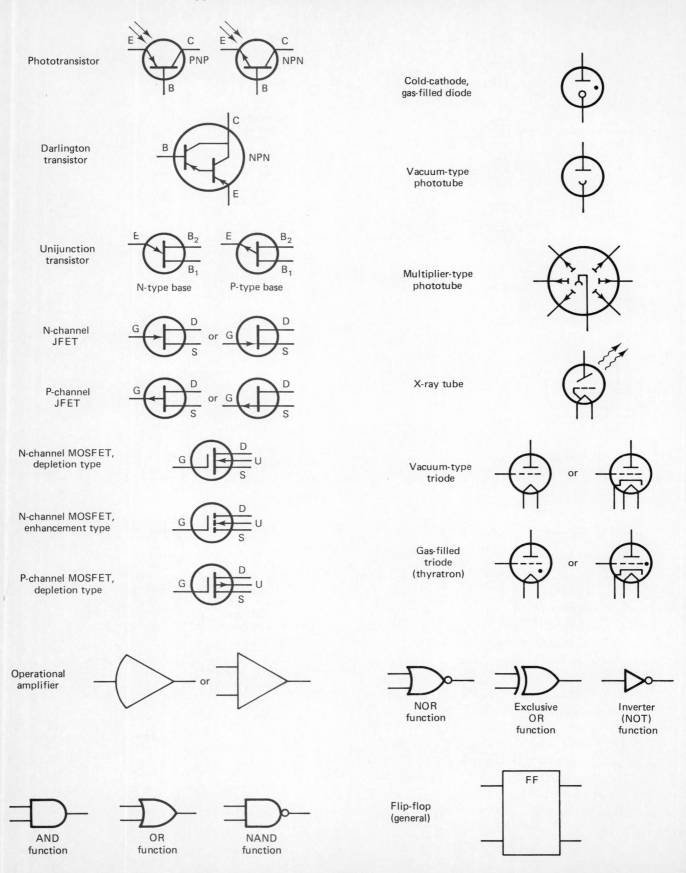

Phototransistor

Darlington transistor

Unijunction transistor

N-type base   P-type base

N-channel JFET

P-channel JFET

N-channel MOSFET, depletion type

N-channel MOSFET, enhancement type

P-channel MOSFET, depletion type

Operational amplifier

AND function

OR function

NAND function

Cold-cathode, gas-filled diode

Vacuum-type phototube

Multiplier-type phototube

X-ray tube

Vacuum-type triode

Gas-filled triode (thyratron)

NOR function

Exclusive OR function

Inverter (NOT) function

Flip-flop (general)

Subscripts

It is common to use subscripts to identify electrical components in circuit diagrams. The circuit shown in Fig. A5-1 has three resistors and a battery. The resistors are labeled R_1, R_2, and R_3. The subscripts identify each of these three resistors. Subscripts also aid in making measurements. The voltage drop across resistor R_1 is called voltage drop E_1. The term "total" is represented by the subscript T, such as E_T. E_T is total voltage applied to a circuit. The current measurement I_2 is the current through resistor R_2 measured at point B. Total current (I_T) is measured at point A. The voltage drop across R_3 is called E_3.

Subscripts are also valuable in troubleshooting and repair of electrical equipment. It would be impossible to isolate problems in equipment without components which are easily identified.

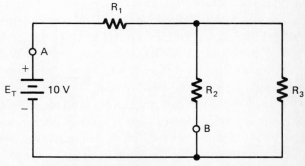

Figure A5-1. Circuit used to show subscripts.

Right Triangles
and Trigonometry

Right triangles and trigonometry are very valuable in the study of electricity and electronics. Trigonometry, or "trig," is a form of mathematics that deals with angles and triangles, particularly the right triangle. A right triangle has one angle of 90°. An example of a right triangle is shown in Fig. A6-1. This example illustrates how resistance, capacitive reactance, and impedance are related in an ac series RC circuit. Resistance (R) and capacitive reactance (X_C) are 90° apart. Their angle of intersection forms a *right angle*. The law of right triangles, known as the Pythagorean theorem, can be used to solve circuit problems. The theorem states that "in a right triangle, the square of the hypotenuse is equal to the sum of the squares of the other two sides." With reference to Fig. A6-1, the Pythagorean theorem is used as $Z^2 = R^2 + X_C^2$ or $Z = \sqrt{R^2 + X_C^2}$.

By using trig relationships, problems dealing with phase angles, power factor, and reactive power in ac circuits can be solved. The three most used trig functions are the sine, the cosine, and the tangent. These functions show the ratios of the sides of a triangle. These functions show the ratios of the sides of a triangle. They determine the size of the angle.

Figure A6-2 shows how these ratios are expressed. Their values are found in a table of trigonometric functions. Table A6-1 lists the sines, cosines, and

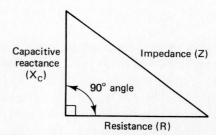

Figure A6-1. Right triangle for an ac series RC circuit.

Figure A6-2. Sine, cosine, and tangent ratios.

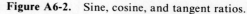
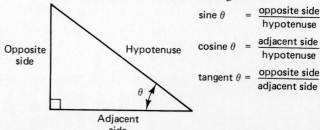

$$\text{sine } \theta = \frac{\text{opposite side}}{\text{hypotenuse}}$$

$$\text{cosine } \theta = \frac{\text{adjacent side}}{\text{hypotenuse}}$$

$$\text{tangent } \theta = \frac{\text{opposite side}}{\text{adjacent side}}$$

Table A6-1
TABLE OF TRIGONOMETRIC FUNCTIONS

Angle	sin	cos	tan	Angle	sin	cos	tan
0°	0.0000	1.000	0.0000	45°	0.7071	0.7071	1.0000
1	.0175	.9998	.0175	46	.7193	.6947	1.0355
2	.0349	.9994	.0349	47	.7314	.6820	1.0724
3	.0523	.9986	.0524	48	.7431	.6691	1.1106
4	.0698	.9976	.0699	49	.7547	.6561	1.1504
5	.0872	.9962	.0875	50	.7660	.6428	1.1918
6	.1045	.9945	.1051	51	.7771	.6293	1.2349
7	.1219	.9925	.1228	52	.7880	.6157	1.2799
8	.1392	.9903	.1405	53	.7986	.6018	1.3270
9	.1564	.9877	.1584	54	.8090	.5878	1.3764
10	.1736	.9848	.1763	55	.8192	.5736	1.4281
11	.1908	.9816	.1944	56	.8290	.5592	1.4826
12	.2079	.9781	.2126	57	.8387	.5446	1.5399
13	.2250	.9744	.2309	58	.8480	.5299	1.6003
14	.2419	.9703	.2493	59	.8572	.5150	1.6643
15	.2588	.9659	.2679	60	.8660	.5000	1.7321
16	.2756	.9613	.2867	61	.8746	.4848	1.8040
17	.2924	.9563	.3057	62	.8829	.4695	1.8807
18	.3090	.9511	.3249	63	.8910	.4540	1.9626
19	.3256	.9455	.3443	64	.8988	.4384	2.0503
20	.3420	.9397	.3640	65	.9063	.4226	2.1445
21	.3584	.9336	.3839	66	.9135	.4067	2.2460
22	.3746	.9272	.4040	67	.9205	.3907	2.3559
23	.3907	.9205	.4245	68	.9272	.3746	2.4751
24	.4067	.9135	.4452	69	.9336	.3584	2.6051
25	.4226	.9063	.4663	70	.9397	.3420	2.7475
26	.4384	.8988	.4877	71	.9455	.3256	2.9042
27	.4540	.8910	.5095	72	.9511	.3090	3.0777
28	.4695	.8829	.5317	73	.9563	.2924	3.2709
29	.4848	.8746	.5543	74	.9613	.2756	3.4874
30	.5000	.8660	.5774	75	.9659	.2588	3.7321
31	.5150	.8572	.6009	76	.9703	.2419	4.0108
32	.5299	.8480	.6249	77	.9744	.2250	4.3315
33	.5446	.8387	.6494	78	.9781	.2079	4.7046
34	.5592	.8290	.6745	79	.9816	.1908	5.1446
35	.5736	.8192	.7002	80	.9848	.1736	5.6713
36	.5878	.8090	.7265	81	.9877	.1564	6.3138
37	.6018	.7986	.7536	82	.9903	.1392	7.1154
38	.6157	.7880	.7813	83	.9925	.1219	8.1443
39	.6293	.7771	.8098	84	.9945	.1045	9.5144
40	.6428	.7660	.8391	85	.9962	.0872	11.43
41	.6561	.7547	.8693	86	.9976	.0698	14.30
42	.6691	.7431	.9004	87	.9986	.0523	19.08
43	.6820	.7314	.9325	88	.9994	.0349	28.64
44	.6947	.7193	.9657	89	.9998	.0175	57.29
				90	1.0000	.0000	

tangents of angles from 0 to 90°. Use the heading at the top of the table and the degree listing from top to bottom in the left-hand column. Read the correct value of sine, cosine, and tangent across from the angle value. This process can be reversed to find the size of an angle when the value of the sides are known. The term "inverse" is used for this process. For example, "inverse sine 0.5 = 30°" means that 30° is the angle whose sine is 0.5. When given "inverse sine 0.866 = ●," look through the listings of sine functions in Table A6-1. Find that the angle whose sine is 0.866 is equal to 60°.

Trig ratios hold true for angles of any size. Angles in the first quadrant of a standard graph are from 0 to 90°. They are used as a reference. To solve for angles greater than 90° (second-, third-, and fourth-quadrant angles), an angle must be converted to a first-quadrant angle (refer to Fig. A6-3). All

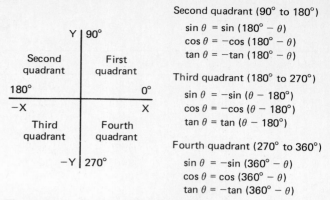

Figure A6-3. Standard graph and the method used to find sine, cosine, and tangent values.

first-quadrant angles have positive values. Angles in the second, third, and fourth quadrants have two negative values and one positive value.

Capacitor Color Code

The most common color-coded capacitor is the *mica capacitor*. This capacitor is usually rectangular in shape. It has a series of colored dots which allows its value to be determined. An example of a mica capacitor is shown in Fig. A7-1.

The EIA (Electronic Industries Association) standard mica capacitor color code table assigns each color dot a letter (A, B, C, etc.). Each letter represents a specific number. Table A7-1 shows the EIA standard for mica capacitor color coding. It is similar to the resistor color code.

Other shapes of color-coded capacitors may also be used. In addition to the EIA capacitor color code standards, the Military Standard is also used. It is almost the same as the EIA code. When a different type of color-coded capacitor is found, use an electronics data manual to find the value.

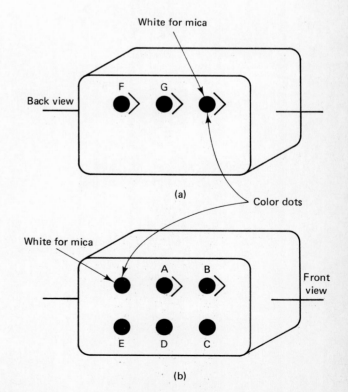

Figure A7-1. Mica color-coded capacitor: (a) back view; (b) front view.

Table A7-1

EIA MICA CAPACITOR COLOR CODE

Color	Capacitance		Multiplier	Tolerance ±%	Characteristics	Working voltage dc	Operating temperature (°C)
	A	B	C	D	E	F	G
Black	0	0	1	20	This includes such		
Brown	1	1	10	1	data as temperature	100	
Red	2	2	100	2	coefficient, draft,		−55 to +85
Orange	3	3	1000	3	and insulation	300	
Yellow	4	4	10000		resistance		−55 to +125
Green	5	5		5		500	
Blue	6	6					
Violet	7	7					
Gray	8	8					
White	9	9					
Gold			0.1			1000	
Silver			0.01	10			

APPENDIX

8

Powers of 10 and Common Logarithms

POWERS OF 10

10^{10}	=	10,000,000,000
10^{9}	=	1,000,000,000
10^{8}	=	100,000,000
10^{7}	=	10,000,000
10^{6}	=	1,000,000
10^{5}	=	100,000
10^{4}	=	10,000
10^{3}	=	1,000
10^{2}	=	100
10^{1}	=	10
10^{0}	=	1
10^{-1}	=	.1
10^{-2}	=	.01
10^{-3}	=	.001
10^{-4}	=	.000 1
10^{-5}	=	.000 01
10^{-6}	=	.000 001
10^{-7}	=	.000 000 1
10^{-8}	=	.000 000 01
10^{-9}	=	.000 000 001
10^{-10}	=	.000 000 000 1

531

COMMON LOGARITHMS

N	0	1	2	3	4	5	6	7	8	9
0	....	0000	3010	4771	6021.	6990	7782	8451	9031	9542
1	0000	0414	0792	1139	1461	1761	2041	2304	2553	2788
2	3010	3222	3424	3617	3802	3979	4150	4314	4472	4624
3	4771	4914	5051	5185	5315	5441	5563	5682	5798	5911
4	6021	6128	6232	6335	6435	6532	6628	6721	6812	6902
5	6990	7076	7160	7243	7324	7404	7482	7559	7634	7709
6	7782	7853	7924	7993	8062	8129	8195	8261	8325	8388
7	8451	8513	8573	8633	8692	8751	8808	8865	8921	8976
8	9031	9085	9138	9191	9243	9294	9345	9395	9445	9494
9	9542	9590	9638	9685	9731	9777	9823	9868	9912	9956
10	0000	0043	0086	0128	0170	0212	0253	0294	0334	0374
11	0414	0453	0492	0531	0569	0607	0645	0682	0719	0755
12	0792	0828	0864	0899	0934	0969	1004	1038	1072	1106
13	1139	1173	1206	1239	1271	1303	1335	1367	1399	1430
14	1461	1492	1523	1553	1584	1614	1644	1673	1703	1732
15	1761	1790	1818	1847	1875	1903	1931	1959	1987	2014
16	2041	2068	2095	2122	2148	2175	2201	2227	2253	2279
17	2304	2330	2355	2380	2405	2430	2455	2480	2504	2529
18	2553	2577	2601	2625	2648	2672	2695	2718	2742	2765
19	2788	2810	2833	2856	2878	2900	2923	2945	2967	2989
20	3010	3032	3054	3075	3096	3118	3139	3160	3181	3201
21	3222	3243	3263	3284	3304	3324	3345	3365	3385	3404
22	3424	3444	3464	3483	3502	3522	3541	3560	3579	3598
23	3617	3636	3655	3674	3692	3711	3729	3747	3766	3784
24	3802	3820	3838	3856	3874	3892	3909	3927	3945	3962
25	3979	3997	4014	4031	4048	4065	4082	4099	4115	4133
26	4150	4166	4183	4200	4216	4232	4249	4265	4281	4298
27	4314	4330	4346	4362	4378	4393	4409	4425	4440	4456
28	4472	4487	4502	4518	4533	4548	4564	4579	4594	4609
29	4624	4639	4654	4669	4683	4698	4713	4728	4742	4757
30	4771	4786	4800	4814	4829	4843	4857	4871	4886	4900
31	4914	4928	4942	4955	4969	4983	4997	5011	5024	5038
32	5051	5065	5079	5092	5105	5119	5132	5145	5159	5172
33	5185	5198	5211	5224	5237	5250	5263	5276	5289	5302
34	5315	5328	5340	5353	5366	5378	5391	5403	5416	5428
35	5441	5453	5465	5478	5490	5502	5514	5527	5539	5551
36	5563	5575	5587	5599	5611	5623	5635	5647	5658	5670
37	5682	5694	5705	5717	5729	5740	5752	5763	5775	5786
38	5798	5809	5821	5832	5843	5855	5866	5877	5888	5899
39	5911	5922	5933	5944	5955	5966	5977	5988	5999	6010
40	6021	6031	6042	6053	6064	6075	6085	6096	6107	6117
41	6128	6138	6149	6160	6170	6180	6191	6201	6212	6222
42	6232	6243	6253	6263	6274	6284	6294	6304	6314	6325
43	6335	6345	6355	6365	6375	6385	6395	6405	6415	6425
44	6435	6444	6454	6464	6474	6484	6493	6503	6513	6522
45	6532	6542	6551	6561	6571	6580	6590	6599	6609	6618
46	6628	6637	6646	6656	6665	6675	6684	6693	6702	6712
47	6721	6730	6739	6749	6758	6767	6776	6785	6794	6803
48	6812	6821	6830	6839	6848	6857	6866	6875	6884	6893
49	6902	6911	6920	6928	6937	6946	6955	6964	6972	6981
50	6990	6998	7007	7016	7024	7033	7042	7050	7059	7067
N	0	1	2	3	4	5	6	7	8	9

COMMON LOGARITHMS (CONTINUED)

N	0	1	2	3	4	5	6	7	8	9
50	6990	6998	7007	7016	7024	7033	7042	7050	7059	7067
51	7076	7084	7093	7101	7110	7118	7126	7135	7143	7152
52	7160	7168	7177	7185	7193	7202	7210	7218	7226	7235
53	7243	7251	7259	7267	7275	7284	7292	7300	7308	7316
54	7324	7332	7340	7348	7356	7364	7372	7380	7388	7396
55	7404	7412	7419	7427	7435	7443	7451	7459	7466	7474
56	7482	7490	7497	7505	7513	7520	7528	7536	7543	7551
57	7559	7566	7574	7582	7589	7597	7604	7612	7619	7627
58	7634	7642	7649	7657	7664	7672	7679	7686	7694	7701
59	7709	7716	7723	7731	7738	7745	7752	7760	7767	7774
60	7782	7789	7796	7803	7810	7818	7825	7832	7839	7846
61	7853	7860	7868	7875	7882	7889	7896	7903	7910	7917
62	7924	7931	7938	7945	7952	7959	7966	7973	7980	7987
63	7993	8000	8007	8014	8021	8028	8035	8041	8048	8055
64	8062	8069	8075	8082	8089	8096	8102	8109	8116	8122
65	8129	8136	8142	8149	8156	8162	8169	8176	8182	8189
66	8195	8202	8209	8215	8222	8228	8235	8241	8248	8254
67	8261	8267	8274	8280	8287	8293	8299	8306	8312	8319
68	8325	8331	8338	8344	8351	8357	8363	8370	8376	8382
69	8388	8395	8401	8407	8414	8420	8426	8432	8439	8445
70	8451	8457	8463	8470	8476	8482	8488	8494	8500	8506
71	8513	8519	8525	8531	8537	8543	8549	8555	8561	8567
72	8573	8579	8585	8591	8597	8603	8609	8615	8621	8627
73	8633	8639	8645	8651	8657	8663	8669	8675	8681	8686
74	8692	8698	8704	8710	8716	8722	8727	8733	8739	8745
75	8751	8756	8762	8768	8774	8779	8785	8791	8797	8802
76	8808	8814	8820	8825	8831	8837	8842	8848	8854	8859
77	8865	8871	8876	8882	8887	8893	8899	8904	8910	8915
78	8921	8927	8932	8938	8943	8949	8954	8960	8965	8971
79	8976	8982	8987	8993	8998	9004	9009	9015	9020	9025
80	9031	9036	9042	9047	9053	9058	9063	9069	9074	9079
81	9085	9090	9096	9101	9106	9112	9117	9122	9128	9133
82	9138	9143	9149	9154	9159	9165	9170	9175	9180	9186
83	9191	9196	9201	9206	9212	9217	9222	9227	9232	9238
84	9243	9248	9253	9258	9263	9269	9274	9279	9284	9289
85	9294	9299	9304	9309	9315	9320	9325	9330	9335	9340
86	9345	9350	9355	9360	9365	9370	9375	9380	9385	9390
87	9395	9400	9405	9410	9415	9420	9425	9430	9435	9440
88	9445	9450	9455	9460	9465	9469	9474	9479	9484	9489
89	9494	9499	9504	9509	9513	9518	9523	9528	9533	9538
90	9542	9547	9552	9557	9562	9566	9571	9576	9581	9586
91	9590	9595	9600	9605	9609	9614	9619	9624	9628	9633
92	9638	9643	9647	9652	9657	9661	9666	9671	9675	9680
93	9685	9689	9694	9699	9703	9708	9713	9717	9722	9727
94	9731	9736	9741	9745	9750	9754	9759	9763	9768	9773
95	9777	9782	9786	9791	9795	9800	9805	9809	9814	9818
96	9823	9827	9832	9836	9841	9845	9850	9854	9859	9863
97	9868	9872	9877	9881	9886	9890	9894	9899	9903	9908
98	9912	9917	9921	9926	9930	9934	9939	9943	9948	9952
99	9956	9961	9965	9969	9974	9978	9983	9987	9991	9996
100	0000	0004	0009	0013	0017	0022	0026	0030	0035	0039
N	0	1	2	3	4	5	6	7	8	9

Index

Index